Automotive Engine Repair and Rebuilding

Second Edition

By Chek-Chart Publications,
a Division of
H. M. Gousha

William J. Turney, *Editor*
Kevin D. Roberts, *Contributing Editor*

HarperCollinsCollegePublishers

Acknowledgments

In producing this series of textbooks for automobile technicians, Chek-Chart has drawn extensively on the technical and editorial knowledge of the nation's carmakers and suppliers. Automotive design is a technical, fast-changing field, and we gratefully acknowledge the help of the following companies and organizations in allowing us to present the most up-to-date information and illustrations possible. These companies and organizations are not responsible for any errors or omissions in the instructions or illustrations, or for changes in procedures or specifications made by the carmakers or suppliers, contained in this book or in any other Chek-Chart product:

ABS Products, South Gate, CA
Audi of America, Inc.
BHJ Products, Inc.
Bayco Division, VSB, Inc., Sacramento, CA
Champion Spark Plug Co.
Chaves Performance Engines, San Jose, CA
Chrysler Motors Corporation
Competition Cams
Clayton Industries
DCM Tech, Inc.
DeAnza College, Cupertino, CA
Evergreen Valley College, San Jose, CA
Federal-Mogul Corporation
Fel-Pro, Inc.
Ford Motor Company
Furtado's Machine Shop, San Jose, CA
General Motors Corporation
    AC-Delco Division
    Buick Motor Division
    Cadillac Motor Division
    Chevrolet Motor Division
    Oldsmobile Division
    Pontiac Motor Division
Griffin Auto Parts, Inc., Santa Clara, CA
Hines Industries, Inc.
Hopper Shop Equipment Sales, Stanton, CA
The Hotsy Corporation, Engelwood, CO
Iskenderian Racing Cams
Honda Motor Co., Inc.
Kent-Moore Tool Group
MAC Tools, Inc.
Mazda Motor Corporation
Melling Automotive Products, Jackson, MI
Mitsubishi Motors Corporation
Mobil Oil Corporation
Nissan Motor Corporation
Norton/TRW Ceramics
Penniman & Richards, San Jose, CA
Porsche Cars North America, Inc.
Rottler Manufacturing
Sim-Test Division of Hawk Instruments, Inc.
Snap-On Tools Corporation
Storm Vulcan
Sunnen Products Company
Total Seal
Toyota Motor Sales, U.S.A., Inc.
TRW, Inc.
Volkswagen of America

The authors have made every effort to ensure that the material in this book is as accurate and up-to-date as possible. However, neither Chek-Chart nor HarperCollins nor any related companies can be held responsible for mistakes or omissions, or for changes in procedures or specifications made by the carmakers or suppliers.

The comments, suggestions, and assistance of the following reviewers were invaluable:

    Ronald M. Davis, Stockton, CA
    Kevin D. Roberts, San Jose, CA

At Chek-Chart, Daniel L. Doornbos and Ramona Torres managed the production of this book. Type was set by Maria Glidden and Diane Maurice. Original art and photographs were produced by John Badenhop, Dave Douglass, Gerald A. McEwan, Richard K. DuPuy and William J. Turney. The project is under the direction of Roger L. Fennema.

AUTOMOTIVE ENGINE REPAIR AND REBUILDING, Second Edition, Classroom Manual and Shop Manual Copyright © 1993 by Chek-Chart, H.M. Gousha Company, a Division of Simon & Schuster Inc.

All rights reserved. Printed in the United States of America. No part of this publication may be reproduced, stored in a retrieval system, or duplicated in any manner without the prior written consent of Chek-Chart, H. M. Gousha Company, a Division of Simon & Schuster, Inc., P.O. Box 49006, San Jose, CA 95161-9006

Library of Congress Cataloging-in-Publication Data is available.
CIP 93-25184

Contents

INTRODUCTION v

HOW TO USE THIS BOOK vi

PART ONE — SAFETY, TOOLS, AND SPECIFICATIONS 1

Chapter 1 — Safety 2
Personal Safety 2
Know Your Shop 3
Fire 3
Solvents and Chemicals 4
Vehicle and Underhood Safety 4
Electrical Safety 4
Engine Repair and Rebuilding Safety 5
Machine Tool Safety 11

Chapter 2 — Hand Tools for Engine Rebuilding 13
General Hand Tools 13
Automobile Rebuilder Tool Set 27

Chapter 3 — Power and Specialized Engine Rebuilding Tools 28
Power and Machine Tools 28
Specialized Engine Tools 30

Chapter 4 — Measurement and Measuring Equipment 57
The Metric System 57
General Measurement Tools 60
Fastener Torque 71

Chapter 5 — Obtaining Specifications, Manuals, Engines, and Kits 75
Manufacturer's Specifications 75
Replacement Parts, Subassemblies, and Engines 81

PART TWO — DIAGNOSIS, IN-CAR REPAIR, AND ENGINE REMOVAL 85

Chapter 6 — Engine Testing and Diagnosis 86
Oil Consumption 86
Engine Oil Pressure 91
Basic Diagnostic Tests 93
Engine Noises 103

On the Cover:
Front — The 3.0-liter DOHC V6 engine used in the Taurus SHO, courtesy of Ford Motor Company.
Rear — The Hines Industries HC500-mc Computer Balancer; and the Rottler FE-24A Programmable Cylinder Boring Machine.

Chapter 7 — Engine Repair in the Vehicle   108
Servicing the Camshaft Drive   108
Valve Train Adjustment and Repair   126
Removing and Replacing the Cylinder Head   137
Servicing the Bottom End   141

Chapter 8 — Engine Removal   145
Preparation   145
Disconnecting the Engine   148
Lifting the Engine   155

Chapter 9 — Disassembly, General Inspection, and Cleaning   159
Inspect As You Go   159
Top End Disassembly and Inspection   160
Bottom End Disassembly and Inspection   163
Engine Cleaning   176

Chapter 10 — Repairing Cracks and Damaged Threads   186
Removing Broken Fasteners   186
Repairing Damaged Threads   193
Internal Thread Replacement   198
Crack Detection   202
Crack Repair   208

PART THREE — ENGINE INSPECTION, MACHINING, AND REBUILDING   215

Chapter 11 — Servicing Blocks   216
Tolerances and Oil Clearances   216
Block Inspection   219
Main Bearing Bore Reconditioning   229
Deck Reconditioning   234
Cylinder Reconditioning   240
Lifter Bore Reconditioning   254

Chapter 12 — Servicing Crankshafts, Flywheels, Pistons and Rods   255
Crankshaft Service   255
Piston and Connecting Rod Service   266
Flywheel and Flexplate Service   292

Chapter 13 — Engine Balancing   297
Balancing Precautions   297
Balancing Reciprocating Weight   299
The Bob Weight   300
Balancing Rotating Weight   302

Chapter 14 — Servicing Cylinder Heads and Manifolds   305
Head Reconditioning   305
Head Resurfacing   329
Valve Testing and Assembly   333

Chapter 15 — Servicing Camshafts, Lifters, Pushrods, and Rocker Arms   337
Camshaft Service   337
Valve Lifter and Cam Follower Service   343
Pushrod Service   350
Rocker Arm and Shaft Service   351

PART FOUR — ENGINE ASSEMBLY, REINSTALLATION, AND BREAK-IN   357

Chapter 16 — Engine Assembly   358
General Assembly Practices   358
Oil Pump Service   358
Assembly Preparation   362
Assembling the Block and Bottom End   363
Final Bottom End Assembly   380
Assembling the Top End   385

Chapter 17 — Engine Installation, Break-in, and Delivery   390
Final Assembly Procedures   390
Engine Preoiling   398
Engine Installation   400
Starting the Engine   409
Cycling the Engine   410
Releasing to the Customer   411

INDEX   412

Introduction to Automotive Engine Repair and Rebuilding

Automotive Engine Repair and Rebuilding is part of the HarperCollins/Chek-Chart Automotive Series. The package for each course has two volumes, a *Classroom Manual* and a *Shop Manual*.

Other titles in this series include:
- Automatic Transmissions and Transaxles
- Automotive Brake Systems
- Automotive Heating, Ventilation, and Air Conditioning
- Engine Performance, Diagnosis, and Tune-Up
- Fuel Systems and Emission Controls
- Automotive Electrical and Electronic Systems
- Automotive Steering, Suspension, and Wheel Alignment.

Each book is written to help the instructor teach students to become competent and knowledgeable professional automotive technicians. The two-manual texts are the core of a learning system that leads a student from basic theories to actual hands-on experience.

The entire series is job-oriented, especially designed for students who intend to work in the car service profession. A student will be able to use the knowledge gained from these books and from the instructor to get and keep a job in automotive repair or maintenance. Learning the material and techniques in these volumes is a giant leap toward a satisfying, rewarding career.

The books are divided into *Classroom Manuals* and *Shop Manuals* for an improved presentation of the descriptive information and study lessons, along with representative testing, repair, and overhaul procedures. The manuals are to be used together: the descriptive material in the *Classroom Manual* corresponds to the application material in the *Shop Manual*.

Each book is divided into several parts, and each book is complete by itself. Instructors will find the chapters to be complete, readable, and well thought-out. Students will benefit from the many learning aids included, as well as from the thoroughness of the presentation.

The series was researched and written by the editorial staff of Chek-Chart, and was produced by HarperCollins*Publishers*. For over 60 years, Chek-Chart has provided car and equipment manufacturers' service specifications to the automotive service field. Chek-Chart's complete up-to-date automotive data bank was used extensively to prepare this textbook series.

Because of the comprehensive material, the hundreds of high-quality illustrations, and the inclusion of the latest automotive technology, instructors and students alike will find that these books form the core of the master technician's professional library.

How To Use This Book

Why Are There Two Manuals?

Unless you are familiar with the other books in this series, *Automotive Engine Repair and Rebuilding* will not be like any other textbook you've ever used before. It is actually two books, the *Classroom Manual* and the *Shop Manual*. They have different purposes, but should be used together.

The *Classroom Manual* teaches you what you need to know about automotive engine operation and why the relationship between engine parts is so important. The *Classroom Manual* will be valuable in class and at home, for study and for reference. You can use the text and illustrations for years to refresh your memory — not only about the basics of automotive engine repair, but also about related topics in automotive history, physics, and technology.

In the *Shop Manual*, you learn test procedures, troubleshooting, and how to overhaul the systems and parts you read about in the *Classroom Manual*. The *Shop Manual* provides the practical hands-on information you need to work on automotive engines. Use the two manuals together to fully understand how engines work and how to fix them when they don't work.

What's In These Manuals?

These key features of the *Classroom Manual* make it easier for you to learn, and to remember what you learn:

- Each chapter is divided into self-contained sections for easier understanding and review. The organization shows you clearly which parts make up which systems, and how various parts or systems that perform the same task differ or are the same.
- Most parts and processes are fully illustrated with drawings or photographs. Important topics appear in several different ways, to make sure you can see other aspects of them.
- Important words in the *Classroom Manual* text are printed in **boldface type** and are defined on the same page and in a glossary at the end of the manual. Use these words to build the vocabulary you need to understand the text.
- Review questions are included for each chapter. Use them to test your knowledge.
- Every chapter has a brief summary at the end to help you to review for exams.
- Every few pages you will find sidebars — short blocks of "nice to know" information — in addition to the main text.

How To Use This Book

There is a sample test at the back of the *Classroom Manual*, similar to those given for Automotive Service Excellence (ASE) certification. Use it to help you study and prepare yourself when you are ready to be certified as an expert in one of several areas of automobile technology.

The *Shop Manual* has detailed instructions on overhaul, repair, and rebuilding procedures for modern automotive engines. These are easy to understand, and often have step-by-step explanations to guide you through the procedures. This is what else you'll find in the *Shop Manual*:

• Helpful information that tells you how to use and maintain shop tools and test equipment
• A thorough coverage of the metric system units needed to work on modern engines
• Safety precautions
• Test procedures and troubleshooting hints that will help you work better and faster
• Tips the professionals use that are presented clearly and accurately

Where Should I Begin?

If you already know something about automotive engines and know how to repair them, you will find that this book is a helpful review. If you are just starting in car repair, then the book will give you a solid foundation on which to develop professional-level skills.

Your instructor will design a course to take advantage of what you already know, and what facilities and equipment are available to work with. You may be asked to read certain chapters of these manuals out of order. That's fine. The important thing is to really understand each subject before you move on to the next.

Study the vocabulary words in boldface type. Use the review questions to help you understand the material. When you read the *Classroom Manual*, be sure to refer to your *Shop Manual* to relate the descriptive text to the service procedures. And when you are working on actual car systems and components, look back to the *Classroom Manual* to keep the basic information fresh in your mind. Working on such a complicated piece of equipment as a modern car isn't always easy. Use the information in the *Classroom Manual*, the procedures in the *Shop Manual*, and the knowledge of your instructor to help you.

The *Shop Manual* is a good book for work, not just a good workbook. Keep it on hand while you're working on equipment. It folds flat on the workbench and under the car, and can stand quite a bit of rough handling.

When you perform test procedures and overhaul equipment, you will need a complete and accurate source of manufacturers' specifications, and the techniques for pulling computer trouble codes. Most auto shops have either the carmaker's annual shop service manuals, which lists these specifications, or an independent guide, such as the Chek-Chart *Car Care Guide*. This unique book, with ten-year coverage, is updated each year to give you service instructions, capacities, lubrication recommendations, and troubleshooting tips that you need to work on specific cars.

PART ONE

Safety, Tools, and Specifications

Chapter One
Safety

Chapter Two
Hand Tools for Engine Rebuilding

Chapter Three
Power and Specialized Engine Rebuilding Tools

Chapter Four
Measurement and Measuring Equipment

Chapter Five
Obtaining Specifications, Manuals, Engines, and Kits

Chapter 1

Safety

Safety is important, both to you and to the people who work around you. No repair job is worth a permanent, crippling injury, or the lost time and income resulting from a preventable accident. Even so, technicians who follow poor work practices take these risks regularly, even when the correct procedures require little or no extra time or effort. By following a few common-sense safety practices, you can minimize your chances of being involved in an accident, or of causing accidental injury to others.

In this chapter, we explain some of the basic safety practices that you should keep in mind in an automotive machine shop. Most of these will already be familiar to you, but take some time to refresh your memory.

PERSONAL SAFETY

Automotive engine repair and rebuilding requires the same attention to personal safety as does working on any other automotive system. Most safety practices are simple common sense:

- Tie back long hair, or contain it with a stocking cap. Tuck in any loose clothing, and remove sweaters. Button loose sleeves, or better yet, wear short sleeved shirts when you work.
- Remove any rings, watches, or jewelry before working on an electrical component, or any moving parts. A ring can weld itself to a battery terminal and become red hot in seconds, and a necklace can pull your face into a rotating engine fan.
- Wear closed leather shoes or boots in the rebuilding shop. Aside from the obvious hazards of dropping sharp and heavy objects on your feet, sandals and canvas shoes can't protect you from sharp metal chips, hot particles from grinders, or spills of caustic or toxic chemicals. Professionals often wear steel-toed shoes or boots whenever they work in the shop.

Eye Protection

Eye protection is critically important in engine rebuilding. Your eyes can be permanently damaged when working with stationary machine tools, power drills or impact wrenches, and even when working under a car that has a dirt and sand-encrusted engine compartment. Don't forget that eye protection is also necessary when using compressed air to blow engine components or passages clean.

Make sure that you wear clean, approved safety glasses or goggles whenever necessary,

Safety

Figure 1-1. Always wear safety glasses when using any power or impact tools.

figure 1-1. When working with stationary or hand held grinders, a full face shield gives better protection against flying abrasive particles, figure 1-2.

Work Area

Your work area is important, too:

- Keep the floor in your work area clean and uncluttered. Coil up any air lines, hoses, and electrical cords when they aren't in use, and replace equipment when you are finished using it.
- Keep your work surfaces clean and uncluttered as well. Piles of metal chips or cuttings can easily find their way into an engine you are assembling, in addition to your own skin.
- Use a commercial sweeping compound, oil absorbent, or Portland cement to clean up floor spills, and use shop rags to keep your bench top clean.

KNOW YOUR SHOP

Take the time to become familiar with the shop in which you work:

- Know the location of the first-aid supplies, the nearest telephone in your shop, and the number to call for emergency assistance.
- Know the location of the fire extinguishers in your shop, and any nearby emergency fire alarm switches.
- Know the location of any specific emergency equipment in your shop, such as eyewashes or emergency showers.

Figure 1-2. A full-face shield works even better than safety glasses to protect you from chips and fragments thrown off grinding wheels or brushes.

- Know the location of the emergency exits.
- Know the location of the fuse boxes and the master cut-off switches.
- If your shop pipes in gas for oxyacetylene welding, know the location of the master cutoff valves.

FIRE

Avoiding fire hazards is critical in the confined space of many repair shops:

- Remember that most cars have a fuel tank containing gasoline and highly explosive gasoline vapors. Make sure that any vehicles you work on are parked so that the fuel tank, lines, and filler necks aren't near any source of spark or flame.
- Do not smoke inside the shop, or while you are working. Flammable materials are always present in a repair shop, and you can easily avoid them by smoking outside or in another area.
- Never use gasoline as a cleaning solvent, unless it is specifically recommended. If gasoline is required, make sure you keep a fire extinguisher nearby.

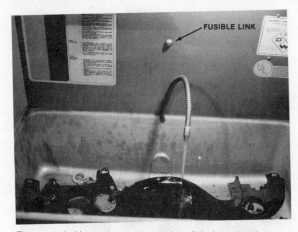

Figure 1-3. Keep the parts washer lid closed unless you are actually using it. This reduces fumes as well as the fire risk.

- Dispose of oil-soaked rags in a metal container. This will contain the flames if the rags catch fire.
- Keep solvent containers, parts washers, and paint or oil containers closed. Parts washers often have a fusible link to close the lid if the solvent inside catches fire, figure 1-3, but avoiding the fire in the first place is much safer.

SOLVENTS AND CHEMICALS

Most shops contain quantities of cleaning solvents, thinners, lubricants, paints, or other chemicals. These are often flammable, and often poisonous.

- Keep any spray cans and solvents away from heat sources. Heat can pressurize the contents until the container explodes.
- Make sure your shop has adequate ventilation. If you have fume hoods or ventilating fans, make sure they work, and use them.
- Always read the instructions before using any new cleaning material.
- Use only the recommended solvents in parts washers. Using the wrong fluids can increase fire risks, expose you to toxic chemicals, and can even destroy the parts you want to clean.
- Do not splash cleaning solvents when putting parts into, or removing them from, a parts washer or cleaning tank. Be sure to wear eye protection.
- Use rubber gloves and eye protection when working with chemicals, such as wire-brushing a solvent-soaked part.
- Use plastic or rubber aprons to protect your skin and clothing. Change any chemical-soaked clothing immediately.
- Clean up any spills right away, and wash any chemicals off your skin with soap and water or hand cleaner. Do not use gasoline, cleaning solvent, or thinner to clean your hands.
- Wash your hands after working with chemicals, especially before eating, drinking, smoking, or using the toilet.
- Dispose of all used chemicals properly. Remember that many cities and states have specific ordinances regulating disposal of hazardous wastes, and that most of the chemicals you use qualify as such. Call your local Fire Department for information on how to dispose of chemicals.

VEHICLE AND UNDERHOOD SAFETY

Working with motor vehicles requires some specific safety practices:

- Do not drive a vehicle into or out of your shop area faster than five miles per hour.
- Do not run the engine inside without ventilation that you know is adequate. Engine exhaust contains toxic carbon monoxide gas that can poison you gradually over a period of hours.
- Before starting an engine, be sure the parking brake is set, the drive wheels are blocked, and the transmission or transaxle is placed in Neutral (manual) or Park (automatic).
- Turn the engine off before working on components under the hood. If you must work with the engine running, be absolutely sure that neither you nor your tools can strike or brush against any moving parts.
- Keep your hands and other body parts away from hot exhaust components. Catalytic converters heat up quickly and retain their heat for a long time after the engine is shut off. If you must work with or around hot engine parts, wear leather gloves — not rubber or plastic. Leather welding gloves can protect your hands from a lot of heat.
- Remember that electric radiator fans can start at any time if the vehicle has been running recently, even if the ignition is off. Always disconnect the fan power lead before working near it, figure 1-4.

ELECTRICAL SAFETY

Many heating, ventilation and air conditioning systems use electric or electronic components or controls, in addition to the battery:

- Make sure your hands, the floor, and your entire work space are dry before touching

Safety

Figure 1-4. Electric cooling fans run on a circuit that is live even when the ignition is off. Be sure to disconnect the power lead before working near them.

any electrical switches or plugs, or using any electrical equipment.
- Use correctly grounded three-prong sockets and extension cords to operate power tools. Some tools use only two-prong plugs. Make sure these are double-insulated.
- Don't use two-prong adapters to plug in a tool requiring a three-hole grounded socket. Use the correct socket instead.
- Don't overload any single circuit with electrical tools or fixtures.
- Keep flames and sparks away from batteries at all times. Batteries give off explosive hydrogen gas, and if ignited, can detonate like a bomb filled with sulfuric acid. Batteries being charged are especially dangerous.
- Always disconnect the battery when working on or near electrical components. Do so by removing the battery ground cable. When installing a battery into a car, always connect the ground cable last.
- Never use the battery top as a tool tray. A wrench touching both battery terminals will melt, and can cause an explosion.

ENGINE REPAIR AND REBUILDING SAFETY

You should be aware of normal tool safety procedures whenever you work, but remember that engine repair and rebuilding requires some specific tools and procedures.

Figure 1-5. Impact sockets are much stronger than those designed for hand-use. Use them for both power and hand impact tools.

Hand Tool Safety

Hand tool safety requires little more than common sense. Keep your tools clean and in good condition. If you damage a tool, either repair it or replace it with one that is safe to use.

Wrenches and Sockets
- Never use a cheater bar or pipe as a wrench extension. If you can't remove a stuck fastener with the standard wrench, then use a stronger, correctly-designed wrench or breaker bar and be prepared for the fastener to break.
- Use the correct-sized wrench on fasteners. The automobile engines you work with might have fasteners in SAE, metric, Whitworth, or British Standard sizes. While some sizes in the different systems are close to each other, the correct tool will usually prevent slipping and rounded fasteners.
- If possible, always pull on the wrench handle, rather than pushing on it.
- Use only impact sockets with impact tools, figure 1-5. Sockets designed for hand-tightening are weaker than impact sockets and will eventually split or shatter if you use them with impact tools. Similarly, use only extensions and flex-joints designed for impact use with an impact tool.
- Open-end wrenches are intended for light-duty, rapid tightening and loosening. When you need to remove a frozen fastener or apply final tightening, use a box-end wrench.
- Use only flare-nut wrenches to turn hose and tubing fittings, figure 1-6. These fitting are often made of soft brass, and an open-end wrench will rapidly destroy them or crimp the tubing.

Hammers, Punches, and Chisels
- Always wear eye protection when using striking tools.

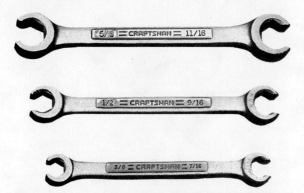

Figure 1-6. Flare nut wrenches grip the fitting on five sides, and give you more leverage without damaging the fitting.

Figure 1-8. Keep the top of the tool squared and chamfered to prevent it from cracking and slipping off-center.

Figure 1-9. Line-up tools are not punches. If you need a punch, buy one you can rely on to work without shattering.

Figure 1-7. Reface chisels and punches regularly on a grinding wheel. Dull punches slip more and a dull chisel cuts slowly and leaves ragged edges.

- Keep chisels and punches sharp by grinding, figure 1-7. Using a mushroomed punch can cause chips of metal to break off, and a chipped or dull chisel edge will cut poorly and leave a ragged surface. Clean up mushroomed tool heads by grinding them to a chamfer, figure 1-8.
- Do not use a line-up tool as a punch. Line-up tools do not have the strength necessary to withstand impact, figure 1-9. Use the correct punch instead.
- Use only the correct hammers for striking metal. Claw hammer faces are hardened for driving soft steel and iron nails, and can shatter if you use them to strike castings, punches, or chisels.
- Keep your hammers in good condition. Handles loosen and hammer faces wear over time. Repair or replace them as necessary.

Hacksaws
- Install the hacksaw blade so that the cutting teeth face forward, away from the handle. American hacksaws are designed to cut on the push stroke, not on the return.
- Use the correct blade for the job. There should always be at least three teeth of the hacksaw blade supported inside the workpiece at all times. Blades that are too coarse cut poorly and snap. If the work is too thin for a hacksaw blade, it is thin enough to cut with tin snips.
- Never force a hacksaw into the work. The blades are fragile and can easily break. Apply slight downward pressure on the forward stroke, and ease up on the return stroke. Let the cutting teeth do the work.

Screwdrivers
Use screwdrivers for what they were intended — driving screws. If you need a chisel, prybar, or gasket scraper, a screwdriver is a poor substitute that can chip, break, and damage mating surfaces.

- Keep your screwdrivers sharp and the tips square by careful grinding. A dull screwdriver can easily slip out of the screw slot.
- Use the correct-sized screwdriver for the job. An undersized slot-type screwdriver can break or gouge the screw slot. The correct size screwdriver to use is the one that is the same width as the screw slot.
- Don't substitute Reed and Prince screwdrivers for Phillips, or vice-versa, figure

Safety

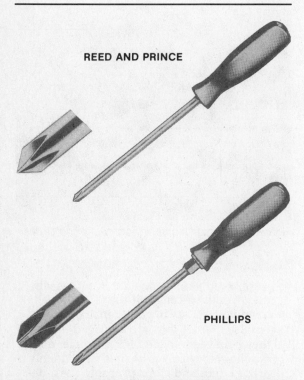

Figure 1-10. Reed and Prince and Phillips screwdrivers are not the same, but are close enough in shape to tempt you. Make sure you also own screw extractors if you choose to interchange the tools.

Figure 1-11. Cherry-picker engine hoists are cheap, easy to use, and are most convenient in small shops.

1-10. Reed and Prince screwdrivers have a sharper point that will bottom out in a Phillips screw before the screwdriver tip fits into the screwhead. Phillips screwdrivers don't fully penetrate Reed and Prince screwheads.
- Use longer screwdrivers whenever you can. Stubby screwdrivers are good for confined spaces, but don't give you as much control or turning power as a longer one.

Figure 1-12. A track-mounted motorized chainfall is even easier to use, because it takes up no floor space. It is also more expensive.

- Don't hold small objects in your hands to use a screwdriver on them. Clamp them in a vise or steady them on the bench instead. A correctly sharpened screwdriver can slash your hands as easily as a sharp knife.

Hoists, Floor Jacks, and Jacking

Hoists and floor jacks are essential tools in removing heavy engines and transmissions. The problems they cause you if they fail are obvious. Be sure to have someone available to help you steady loads if you run into problems, especially if you are removing an engine.

Hoists
The engine hoists you use may be either the common portable cherry-picker design, figure 1-11, or the roof or track-mounted chainfall, figure 1-12. The correct techniques for using them are similar:
- Always begin by centering the lifting hook over the load. If you use a collapsible cherry-picker, make sure the legs are fully extended. This ensures that the hoist is stable, and that an engine won't suddenly swing to one side as it clears the mounts.

Figure 1-13. Floor jacks come in various sizes. For garage work, you will need one at least as large as this.

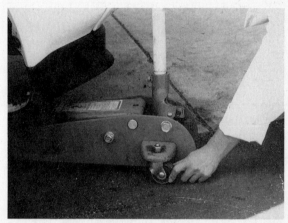

Figure 1-14. Straighten the wheels on the floor jack so it will roll in a straight line as you lift the car.

- Use cherry-pickers on a firm surface, preferably concrete. Sweep the floor before beginning so that pieces of debris can't wedge under the wheels. They should be free to roll slightly as the hoist takes up the load.
- Use a commercial lifting sling or a strong chain attached to the hook. A small-diameter chain simply looped over the hook can slip if the engine is unbalanced under the lifting point. An easy way to keep a chain from slipping is to attach a shackle to it, and attach the hook to the shackle.
- Be sure the lifting points on the engine are sturdy and can take the weight. Most engines have several flanges bolted to the block, cylinder head, or manifolds in the correct position for a balanced lift. If there are none, grade eight bolts threaded *securely* into unused cylinder head holes are usually satisfactory. Slip very large washers over the bolts first, and then screw them into the engine through the chain links. This will keep the chain from pulling over the bolt heads.
- Never overload an engine hoist. All commercial hoists have their capacities marked, and this limit should be respected. Cherry-pickers with extendable booms are especially easy to overload.
- If you use an overhead chainfall, make sure that it is attached to a beam or stand that will support the load.
- Don't lift an engine any higher than absolutely necessary. Cherry-pickers are unstable with a heavy load more than a few feet off the ground.
- Never get underneath any load supported by an engine hoist of any kind. When removing engine or frame mounting hardware, keep your hands out of places where they would be caught if the load slipped.
- Keep your hoists in good repair, and do not use them if they are damaged or if they behave strangely. For example, a hydraulic cherry-picker boom that suddenly drops a foot and stops probably has a deteriorating seal that could fail at any moment.
- Always lower an engine slowly. This will prevent damage to the engine, and will keep the load under your control at all times.

Floor jacks
Floor jacks are often used when removing transmissions with the engine still in the car, or when removing an engine from underneath. Floor jacks vary in size and capacity, but are in general quite similar, figure 1-13. Many of the same guidelines apply to floor jacks that apply to engine hoists. In addition:

- Be sure that the floor jack's lifting pad is placed underneath a part capable of bearing the load. If you need to lift an engine with the lifting pad underneath a sheet metal part such as the oil pan, place several sturdy boards on top of the pad to spread the load.
- Make sure the load is balanced on the lifting pan so that it won't tip off and fall as you lift it.
- Floor jacks roll toward the load as you lift it. Be sure the floor is clean ahead of the wheels. If you twist the wheels so that they are aligned with the jack frame before lifting, then the jack will roll in a straight line without twisting, figure 1-14.
- Use only a properly designed transmission jack with a safety chain to remove and re-

Safety

Figure 1-15. Transmission jacks are the easiest, quickest, and safest way to remove a transmission with the engine still in the car.

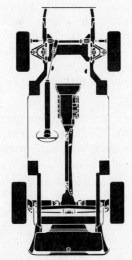

■ LIFT ADAPTER POSITION

Figure 1-16. Manufacturers always specify particular jacking points to avoid damaging or bending the car's frame. This information is also published in aftermarket manuals.

place transmissions, figure 1-15. A floor jack is not stable enough for this work.

Figure 1-17. Use jack stands to support the car if you must work under it or if it must be supported for more than a short time.

Jacking

Jacking a car is not often necessary during engine rebuilding, but sometimes it is necessary to remove an engine or transmission from underneath. Jacking safety is often ignored:

- Make sure that you position the car so that you have enough room to work around it while it is raised.
- Be sure to set the parking brake and place manual transmissions in gear and automatic transmissions in Park.
- Always block both wheels on the side opposite where you will place the jack.
- Be sure to place the jack underneath the correct jacking points. Manufacturers specify jacking points so that the car can be raised safely and without damage, figure 1-16. Unit body cars in particular can be damaged by trying to lift the car at an incorrect point.
- Use the jack only to raise the car high enough to place it on jack stands, figure 1-17. Never get underneath a car supported only by a jack.
- Give the car a firm push after it settles onto the jack stands to make absolutely sure that it won't slip off if someone accidentally bumps the car. If you work in an earthquake-prone region, be extra careful; some of your colleagues have been quite surprised to watch automobiles crash off jack stands during earthquakes. If you feel an earth tremor beginning, get out from under the car.

Figure 1-18. Engine stands keep the engine off the floor and accessible from all sides.

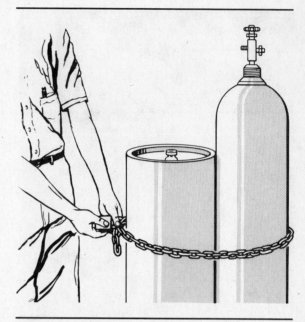

Figure 1-19. Keep the cylinders chained upright at all times.

Hydraulic Lifts

You may occasionally use a hydraulic lift to remove an engine or transmission from underneath a car. Hydraulic lifts are quite safe when properly used:

- Be sure to position the lift pads correctly at the manufacturer's jacking points.
- Lift the vehicle until the wheels are off the ground, and then give it a firm push to make sure it is correctly supported.
- Be sure to set the safety locks if they are manual. If the locks are automatic, watch them to be sure they slip into position as you lift the car.
- Do *not* use the lift if it behaves strangely, if it jerks during raising, slowly settles under load, leaks oil from the exhaust line or cylinder packing, or if it slowly rises after being raised. All these are symptoms of faulty hydraulics. Have the lift repaired.

Engine Stand

Engine stands can be useful for supporting engines during assembly and disassembly, figure 1-18. Some rebuilders prefer to work on the engine on a bench instead. Use the method that works best for you.

- Use the highest quality engine stand you can obtain. Cheap versions of this tool are common.
- Most engine stands are mounted on wheels to let you move the engine around after removing it from the car.
- Remember that they are top-heavy by their nature, and can fall over if given a shove in the wrong direction. Four-point supports are more stable than three-point.
- Lock the rotating bracket in position before using a wrench or lever on an engine mounted in a stand. The engine can unbalance the stand if it rotates suddenly.

Welding Equipment

Both arc and oxyacetylene welding and cutting equipment are used in engine repair and rebuilding. Be sure that you understand the safe procedures for using this equipment. In general:

- Store oxygen and acetylene cylinders upright, either chained to the wall or another immovable object, or in a properly designed cylinder cart, figure 1-19. Remember that acetylene cylinders *must* be stored upright or they can explode.
- Keep the caps screwed onto any cylinders not in use, figure 1-20. The force and power of a compressed gas cylinder with a valve broken off is almost beyond belief. They have been known to hammer their way through brick walls two feet thick, and to rocket through the air for over half a mile.

Safety

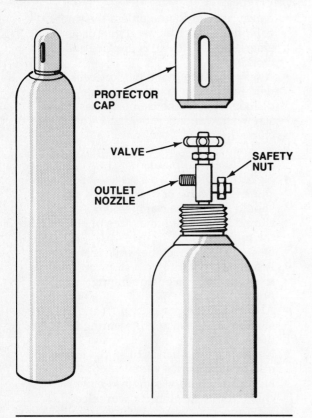

Figure 1-20. The caps protect the valve from damage if the cylinder falls or is struck.

Figure 1-21. Gas welding goggles and arc welding masks are essential when you weld to let you see your work and protect your eyes.

- Keep the correct wrench in position on any gas cylinder being used so that it can be shut off immediately in an emergency.
- Always wear the correct eye protection when using torches or welding equipment, figure 1-21. Goggles designed for gas welding won't protect your eyes from damage caused by the light from an arc welder.
- Never weld or use a torch near fuel or flammable solvents, including gasoline lines.
- Work only in ventilated areas. Welding on some metals gives off poisonous gases.
- Never use a torch on an unprotected concrete floor. The heated concrete can explode. Use firebricks to support your work instead.
- Do not substitute oxygen for compressed air to operate air tools, blow out pipe lines, or remove dust from clothing. A fire or explosion can result.
- Never permit grease or oil to contact cylinder valves, torch fittings, or hose fittings. While oxygen is non-flammable, it greatly aids combustion when it comes in contact with flammable material.

- When opening cylinder valves, always stand to one side of and away from the regulator. A defect in the regulator may allow the gas to blow through and shatter the glass.
- Never move a cylinder by dragging, sliding, or rolling it on its side. To move a cylinder, roll it on its bottom edge.
- Do not strike an arc if someone without eye protection is nearby.
- Avoid welding in damp areas and keep hands and clothing dry. Dampness could cause a severe shock.

Compressed Air

Compressed air is useful for blowing passages clean, for quickly drying freshly-washed castings, and for running air-powered tools such as wrenches and drills.

- Always wear eye protection when using compressed air.
- Don't direct the air jet from blow nozzles at unprotected skin. It can drive metal particles into you, or open and widen small cuts into big ones.
- Don't exceed the recommended pressures when operating air tools.

MACHINE TOOL SAFETY

Working with the machine tools such as boring bars, surface grinders, and drill presses used in automotive engine rebuilding takes special care:

- If you aren't sure how to use any tool, machine, or test equipment, ask your instructor about its safe operation before using it.

Figure 1-22. Always clamp the workpiece in place before using a machine tool on it.

- Never use a machine unless the safety equipment, such as hand guards and emergency switches, is in place and working.
- Always be sure that a workpiece is firmly held in place so that the machine tool can't dislodge it. Use the clamps, vises, and screw attachments provided with the machine, figure 1-22.
- Keep your body clear of any moving parts, and of the workpiece.
- Never handle the workpiece while the machine is running or coasting to a stop. It should already be clamped, and you should have both hands free of both it and the tools.
- Never leave a machine while it is running. A machine coasting to a stop can cause as much damage as one operating under full power.
- Always cut the power and let the machine stop before using any measuring tools on the workpiece.
- Never use your hands to stop a machine or a moving part.
- Never use compressed air to blow metal chips from a machine or workpiece.

Chapter

2

Hand Tools for Engine Rebuilding

In this chapter, you will learn about the tools you will use for automobile engine repair and rebuilding. There are many tools that are essential for mechanical work, but there are many more that are simply convenience items, or that you will use only for a specific job in a specific engine. Professional technicians often have hand tool inventories that are worth as much as five to ten thousand dollars, but they usually purchased them over several years. As a beginning automobile technician, you can start with a basic tool set costing several hundred dollars, but count on spending several thousand dollars more fairly rapidly during your first year or two.

Major tool suppliers often sell their tools through independent salesmen, who operate trucks along a fixed sales route. These dealers specialize in sales to professionals, and will often make a special trip to an automotive shop class if you call them. You can obtain catalogs from these companies by mail or from the dealer.

Some manufacturers of professional-quality hand tools also offer lifetime warranties, which let you trade in the tool for a new one when it breaks or wears out. If you purchase from these manufacturers, then you will never have to buy a particular tool more than once during your career.

Another source for quality tools is your local pawnshop. Tools are expensive, and machinists or rebuilders often pawn their old sets when they upgrade. Be sure that any tools you buy from a pawnshop are in good condition, however, and still have all their accessories. Micrometers, for instance, often become parted from their setting standards.

Before we even get into tools, we should mention that one of your first purchases should be a sturdy, lockable, tool chest. Tools are expensive and easy to lose. A larger tool chest with several drawers lets you keep them organized so that you can immediately see whether anything is missing. The best kind to buy is a roll-away unit that you can wheel right up to the job, and store out of the way when you're finished. Roll-aways usually consist of a cabinet with drawers, mounted on casters, and with a separate tool chest sitting on top, figure 2-1. You can pay several thousand dollars for top quality roll-away tool storage units, but department stores also sell good quality roll-aways for several hundred.

GENERAL HAND TOOLS

When you purchase hand tools, buy the best you can afford. The quality of your work often

Figure 2-1. A roll-away tool storage unit is the most convenient way to keep your tools secure and organized.

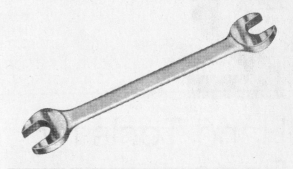

Figure 2-2. Open-end wrenches are quick, handy, and inexpensive.

depends directly on the quality of the tools you use to perform it, and good tools are almost always easier and less fatiguing to use. High-quality hand tools come with precisely-designed and machined working surfaces, and will give excellent service through many years of steady use. Cheap tools, on the other hand, will cause you aggravation for as long as they last, which often isn't very long.

Shop around before you lay out a large amount of money for a set of tools. If you are like many technicians, you will find that you have certain preferences for some tools made by a particular manufacturer, but not for others. For example, you may find that the wrenches from one manufactuerer feel too sharp in your hands, but that you love their screwdrivers and pliers. The socket wrench from another company may have features that you feel are essential, but you prefer to use sockets made by a third.

Wrenches

Wrenches and sockets come in several different standard sizes. Most older American cars use SAE sizes in fractional inches, such as $1/2$, $9/16$, $5/8$, $11/16$, etc. All foreign vehicles, and some domestic ones, use metric-sized fasteners, such as 10 mm, 12 mm, 15 mm, and so on. Both SAE and metric sizes refer to the width of the nut or the bolt across the flats.

Older British machinery may use one of two different size standards: Whitworth, or British Standard. In both systems, bolts are sized according to the diameter of the bolt shank, not the width of the bolt head. Because bolt heads were standardized to the bolt shank, a British mechanic had no more trouble reaching for the correct wrench than did a mechanic used to the SAE system.

Whitworth and British Standard wrenches are interchangeable, but are sized to different types of bolts. Bolts with coarse (Whitworth) threads had one head size, while bolts with fine (British Standard) threads had a bolt head size exactly one size smaller. For instance, $1/4$-inch Whitworth and $5/16$-inch British Standard wrenches are the same size, as are $5/16$-inch Whitworth and a $3/8$-inch British Standard wrenches. A Whitworth $3/16$ wrench is approximately the same size as an SAE $7/16$ wrench.

Open-end wrenches

Open-end wrenches are quick and convenient to use, figure 2-2. They come with two different jaw sizes on each tool, so that you do not need as many tools to handle all the bolt sizes. The head on an open-end wrench sits at an angle, figure 2-3. This allows you to move the wrench until it comes to an obstruction, flip it over, and then move it some more.

The disadvantage of open-end wrenches is that they grip the nut or bolt head only at two widely-separated corners, figure 2-4. This means that if you try to use them to break loose stuck nuts or bolts, you can easily round

Hand Tools for Engine Rebuilding

Figure 2-3. The heads on open-end wrenches are set at an angle so that you can flip the wrench for a new turning position.

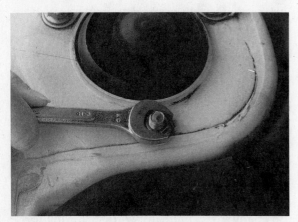

Figure 2-4. Open-end wrenches concentrate the turning force at only two spots on the nut or bolt head, and can round off or distort it.

Figure 2-5. Box-end wrenches hold the hex head at all six corners. Good designs also concentrate the turning force on the flat of the nut, instead of on the corner.

Figure 2-6. Box-end wrench heads mount at an angle from the wrench beam to give you finger clearance while still gripping the nut securely.

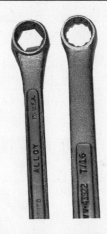

Figure 2-7. Six-point box-end wrenches grip a rounded nut more securely. Twelve-point box-end wrenches offer more turning positions.

off the corners of the fastener. Never try to use an ordinary open-end wrench to remove flare-nut fittings. The wrench will destroy them very quickly, requiring you to cut and reflare the tube to replace the fitting.

Use open-end wrenches for what they were designed for: rapid removal or running down of loose nuts and bolts, or limited use in confined places.

Box-end wrenches

The design of box-end wrenches sidesteps some of the problems that using an open-end wrench presents. Box-end wrench heads surround the nut or bolt, and grip it at all six corners instead of only two, figure 2-5. Like open-end wrenches, box-end wrenches usually come with a different size on each end.

The heads of box-end wrenches mount at an angle of 10 to 15 degrees, figure 2-6, so that you can use them to loosen nuts tightened against a flat surface and still have clearance for your fingers around the wrench. Be sure to allow for this angle when using them, or you can easily twist the wrench off the nut.

Box-end wrenches come in six and twelve-point designs, figure 2-7. The six-point opening spreads the force over more of each flat on

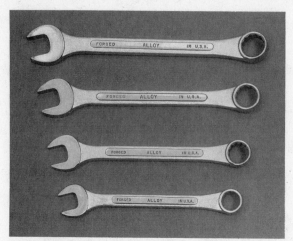

Figure 2-8. Combination wrenches have both open- and box-end heads on the same tool.

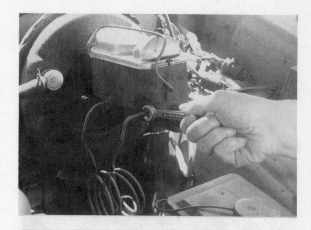

Figure 2-10. Flare nut wrenches grip like a box-end, but the open-end design lets you slip the wrench onto a tubing fitting.

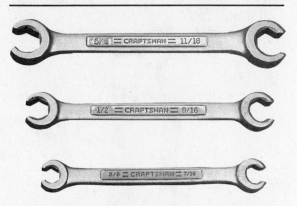

Figure 2-9. Flare nut wrenches can remove flare nuts and line fittings without the damage an open-end wrench will cause.

the fastener, so it is less likely to round the corners of the nut or bolt and will often remove fasteners too rounded for other tools to grip. On the other hand, a twelve-point wrench has twice as many gripping positions as a six-point style, and in confined places this may be more of an advantage.

Use box-end wrenches whenever you need to apply final tightening torque to the fastener, or when the fastener was partly rounded-off by a ham-fisted technician (sometimes this will be you). For delicate fasteners, such as the aluminum or titanium nuts and bolts sometimes used in lightweight gear trains, always use a box-end wrench instead of an open-end.

Both box- and open-end wrench sets often come with multiples of the most common sizes. For instance, a set of six box-end wrenches may have two $1/2$ and $9/16$ sizes, on different wrenches. This lets you loosen a nut while holding the bolt head with another wrench, even if the nut and bolt head are the same size.

Combination wrenches
Combination wrenches are extremely convenient tools that combine an open-end head on one end of the tool with a box-end head of the same size on the other, figure 2-8. These wrenches give you a choice of which style to use at any given moment, without having to reach for a different tool. You can loosen a tight nut or bolt with the box end and then run the fastener off quickly with the open end.

The disadvantage of the combination wrench is that you must own about twice as many wrenches in order to have one of each size. Even so, these are the designs most professionals prefer.

Flare-nut wrenches
Flare nut wrenches, figure 2-9, are designed specifically for removing the soft brass and steel fittings that secure flared fuel and brake lines. These fittings are hollow and extremely delicate. You can't remove them with sockets or box-end wrenches because of the attached tubing, figure 2-10.

Flare-nut wrenches are open-end designs, with either six or twelve-point jaws. The six-point wrenches are the least likely to round the corners of the fittings.

Ratcheting box-end wrenches
Ratcheting box-end wrenches are convenient devices that let you turn a nut or bolt without removing the tool from the fastener, figure 2-11. To switch from tightening to loosening, you simply flip the tool over. These wrenches are also available in six and twelve-point openings. They are very useful when space limitations preclude the use of a socket wrench.

Hand Tools for Engine Rebuilding

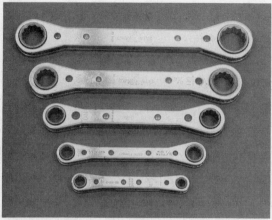

Figure 2-11. Ratcheting box-end wrenches speed disassembly and reassembly, but do not use them for heavy tightening.

Figure 2-12. Socket wrenches come in various drive sizes to handle different jobs.

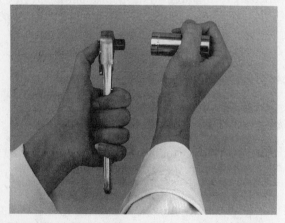

Figure 2-13. Push-button release socket wrenches let you change sockets quickly with one hand.

Socket Wrenches

A socket wrench is usually the most efficient tool for removing nuts and bolts. They consist of a ratcheting handle with a square drive lug that fits into the back of the socket, figure 2-12. You can buy socket wrenches in drive sizes from 1/4-, 3/8-, 1/2-, 3/4-, and one-inch drive,

Figure 2-14. Standard sockets come in all drive sizes.

Figure 2-15. Deep-well sockets let you use a ratchet in situations that would otherwise require slow work with a wrench.

depending on the size of the sockets you need to use. Older engines with cast iron blocks and cylinder heads usually require 1/2-inch drive tools. Modern engines, with aluminum castings and constricted engine compartments, often require 3/8-inch drive tools. You will need both. Metric socket wrenches use the same standard SAE drive sizes, so you can use the same handles for all your sockets.

Socket wrenches come in a variety of styles. Some are available with flex-heads, and most designs come in several lengths. Some designs use a push-button release to let you drop the socket off the ratchet with one hand, figure 2-13, but with most ratchets the sockets simply snap on and off.

Sockets come in a variety of styles, too, but for most purposes a set of standard-depth and deep-well sockets will be sufficient. Standard-depth sockets, figure 2-14, are designed for turning normal nuts and bolts. Deep-well sockets, figure 2-15, are several times as long, and will fit over the end of a protruding bolt to turn a nut that a standard socket can't reach. Like box and combination wrenches, sockets come in six- and twelve-point styles.

Spark plug sockets are special deep-well sockets made specifically for removing and installing spark plugs, figure 2-16. Most spark plug sockets have hexagonal flats at the top so

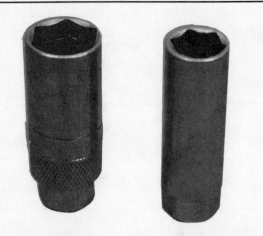

Figure 2-16. A spark plug socket is much more convenient than an ordinary deep well socket for pulling spark plugs.

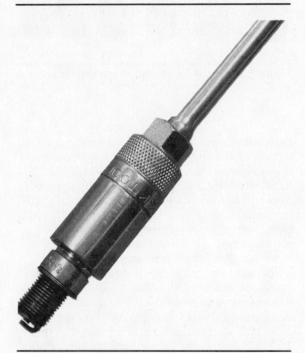

Figure 2-17. The rubber insulator inside grips the plug so that you can remove it from deep plug recesses, and also keeps the plug from burning your fingers.

Figure 2-18. Breaker bars keep you from destroying an expensive ratchet when loosening tight fasteners.

Figure 2-19. Universal joints let you reach nuts and bolts in awkward or confined spaces.

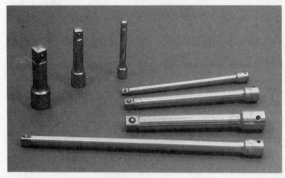

Figure 2-20. Socket wrench extensions come in many lengths for all drive sizes.

that you can turn them with a wrench in restricted areas, and also have a rubber insert inside to grip the ceramic insulator of the plug, figure 2-17.

Accessories

Socket handles and accessories are available in an astonishing variety. The basic tool is the simple reversible ratchet, but you will also need breaker bars, universal joints, and a variety of extensions.

The breaker bar, figure 2-18, is essential. Ratchets aren't designed to take the torque necessary to break a stuck nut or bolt loose, and you can damage them if you try. Use a breaker bar as either a T- or an L-handle to loosen frozen fasteners, then switch to the ratchet to remove them.

Universal joints, figure 2-19, are sturdy flexible joints that let you turn a socket when you can't place the wrench directly above the nut or bolt. You can also buy short extensions with a flexible shaft that allows slight angle changes, but they can't take the amount of torque that a universal joint can.

You can purchase extensions in almost any length from 1 1/2 inch (3 cm) to 3 feet (1 m) for the 1/4, 3/8, and 1/2-inch drive tools that you

Hand Tools for Engine Rebuilding

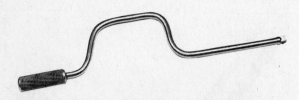

Figure 2-21. Speed handles are great for removing spark plugs, oil pan bolts, and any fasteners that you have already broken loose.

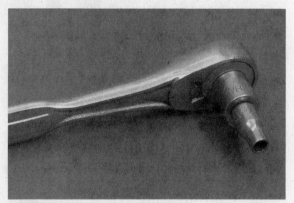

Figure 2-22. You can use adapters to use sockets of one drive size with a socket wrench of another, but the correct tool is a better solution.

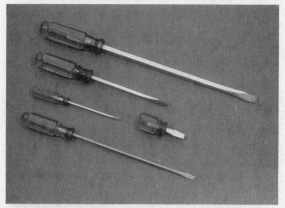

Figure 2-23. Slot-type screwdrivers come in many lengths and blade sizes.

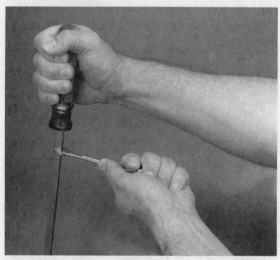

Figure 2-24. You can turn some screwdrivers with a wrench for extra leverage.

will typically use, figure 2-20. Extensions let you reach nuts and bolts in places that you can't reach conveniently with your hands, or let you turn a socket wrench in a location free of obstructions.

Another useful tool in many situations is the speed handle, figure 2-21. Speed handles are essentially cranks with a square socket drive that let you rapidly spin a loose nut or bolt in or out.

You can also purchase adaptors that let you use sockets in one drive size with ratchet wrenches or accesories of another, figure 2-22. Use these adaptors with caution. A large socket wrench can easily over-torque a smaller fastener, and the torque needed to tighten a large fastener might be more than a small socket wrench was designed to deliver.

Screwdrivers

The basic screwdrivers are used to remove and install screws. Many other fasteners can be removed with tools that resemble screwdrivers, however, including small nuts and bolts. You can purchase all types of screwdrivers in a variety of lengths, from stubby sizes only a few inches long, to specialized tools as long as a foot or two. The highest-quality screwdrivers also sometimes come with replaceable blades.

Handles might be round, fluted, square, or triangular — choose the style that fits your hand most comfortably. We discuss the most common screw-head designs below.

Slot-type

Slot-type screwdrivers are sized by their length, and by the width of their shank above the widened portion of the blade. For instance, a slot screwdriver might be described as "$3/16$ by 16 inches." Common sizes range from $1/8$-inch to $1/2$-inch blades, figure 2-23.

Some better-quality slot screwdrivers have square shanks that let you use an open-end wrench to apply extra turning force. Some even have a hexagonal section below the blade so that you can fit a larger box-end wrench onto the shank, figure 2-24.

Phillips and Reed and Prince

These screwdrivers, figure 2-25, are also called cross point or cross slot screwdrivers. The de-

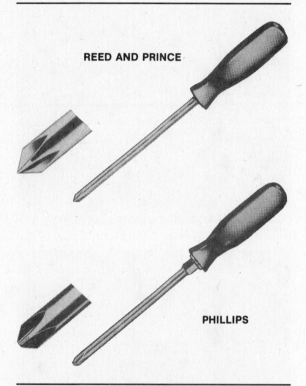

Figure 2-25. Phillips and Reed and Prince screwdrivers work in Phillips and Reed and Prince screws. Make sure you have the correct tool.

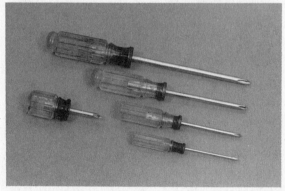

Figure 2-26. Five Phillips screwdrivers will take care of most of your needs.

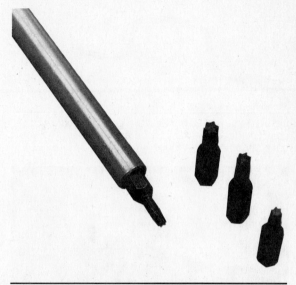

Figure 2-27. Torx screw drivers come either as single-size tools, or as a handle with interchangeable bits.

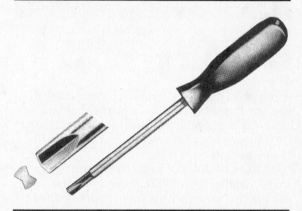

Figure 2-28. Clutch head screws require the correct tool to remove and install them.

signs are very similar, yet are different enough that if you use the wrong screwdriver you will damage the screw head. Cross slot screws used in the United States are generally Phillips.

Phillips screws and screwdrivers come in number sizes, typically from 0 through 4 from the small to large, figure 2-26. Reed and Prince screwdrivers come in fractional inch sizes. When using a cross point screwdriver, look for identification as to type on the handle or shank until you learn to distinguish them by eye.

Torx-head
Manufacturers use Torx-head screws and bolts in applications where they prefer that the vehicle owner leave the fastener alone. Typically, manufacturers use Torx screws to secure seat belt brackets, headlight trim rings, and occasionally interior trim.

A Torx-head screw has a recessed six-cornered hole that is not easily turned by anything except a Torx-head screwdriver, figure 2-27. If you encounter these fasteners, be sure to use the correct tool to remove them. The heads are easily stripped out by slot screwdrivers or Allen wrenches.

Clutch-head
Clutch head screwdrivers are also called figure-eight or butterfly screwdrivers, figure 2-28. Manufacturers use them mainly on body

Hand Tools for Engine Rebuilding

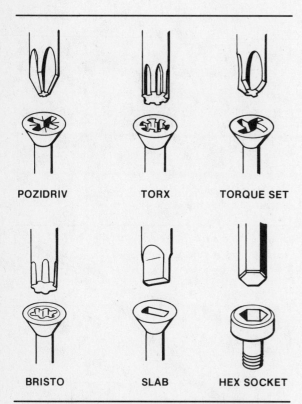

Figure 2-29. Many different types of screw heads have been used over the years in general or specialized applications.

Figure 2-30. Allen-head fasteners are common in engines, especially in gallery plugs or flush-mounted bolts.

Figure 2-31. This Allen tool cannot take as much torque as an L-shaped or socket wrench bit, but it is extremely convenient for most other tasks.

parts and accessories. The deep recess in the screw head is shaped like a butterfly or figure-eight, and prevents the screwdriver from slipping.

Miscellaneous

Besides clutch head, Phillips, and Reed and Prince screwdrivers, there are a number of special recessed-head screws that you may encounter, figure 2-29. Each of these screws require the matching screwdriver made specifically for it. If you try to turn the screw with something else, you will probably damage the screw and make it difficult to remove even with the right screwdriver.

Allen wrenches

Allen-head fasteners, figure 2-30, range in size from small screws to large bolts. They have a recessed hexagonal socket in the top that you turn with a hexagonal wrench. Allen wrenches (or keys) come in many forms, from the simplest L-handles to T-handles, screwdriver handles, socket wrench attachments, and even tools with many wrenches folded inside like a pocket knife, figure 2-31.

Manufacturers use Allen head fasteners in applications with limited space for the screw head. Because of its design, an Allen-head bolt head takes up much less space that would a hexagonal bolt head. They occasionally use them to prevent the technician from over-torquing a casting. An Allen-head bolt will usually strip at a lower torque than would an external hexagonal-head bolt of the same shank diameter.

Allen-head screws are also common as setscrews, lock screws, and small adjuster screws. An advantage of using Allen heads on small screws is that you can hold the screw on the wrench tip and insert it into the hole by holding onto the wrench.

Pliers

Pliers come in many shapes and sizes for almost any job imaginable. Like screwdrivers, they are often abused when technicians substitute them for another gripping tool.

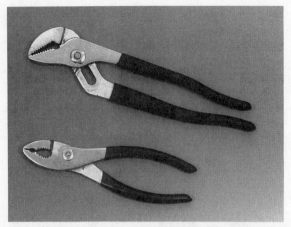

Figure 2-32. Buy the slip-joint and channel lock pliers first. They are by far the most versatile.

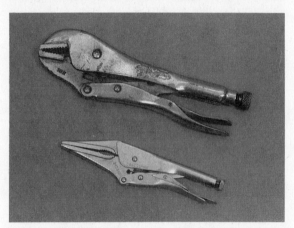

Figure 2-33. Locking pliers come in many different and specialized forms.

Figure 2-34. Locking pliers give you a "third hand" to hold items in place while you do something else.

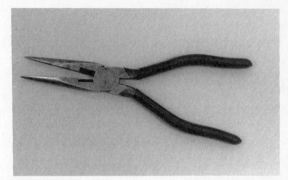

Figure 2-35. Needle-nosed pliers are good for light, delicate gripping work.

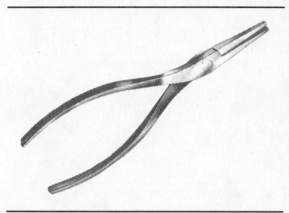

Figure 2-36. The jaws of duckbill pliers meet at a flat, broad, surface.

At the least, you should purchase a pair of slip-joint and channel lock pliers, figure 2-32. Slip joint pliers have machined teeth for gripping irregular or smooth objects, and usually a wire-clipping notch in the base of the jaws. The slip joint permits one tool to hold a wider range of sizes. Channel lock pliers are quite similar, and are useful for gripping larger workpieces.

Locking pliers are also extremely versatile tools, figure 2-33. You can use them anywhere that you need to apply a clamping or gripping force but don't want to keep your hand on the tool. Locking pliers come in a variety of sizes, and with different gripping attachments that let them double as everything from ordinary pliers to C-clamps and vises, figure 2-34. Needle-nose locking pliers are also available.

In addition to these two styles, you should purchase one or more sizes of needle-nosed or snout-nosed pliers, figure 2-35. Needle-nosed pliers permit you to hold very small objects, or to reach into small, confined spaces. Duckbill pliers, figure 2-36, have flat, thin jaws that can get a better grip on wire and other small parts than needle-nosed pliers.

Hose clamp pliers, figure 2-37, look like an ordinary pair of pliers, but have notches cut into the tips of the jaws to enable you to grip the spring-steel wire hose clamps commonly used in OEM heater and coolant hoses. If you have ever struggled with these hose clamps

Hand Tools for Engine Rebuilding

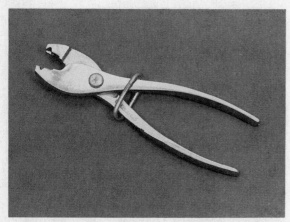

Figure 2-37. Hose clamp pliers are the most convenient tool for removing and installing wire-type hose clamps.

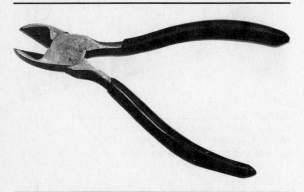

Figure 2-38. Diagonal pliers come with flush or offset cutting jaws. Offset jaws cut with minimum crushing.

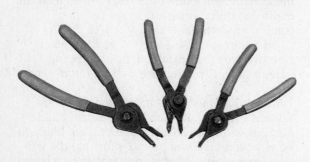

Figure 2-39. Lock-ring pliers come either as a set of different sizes, or with replaceable pins to clamp in the jaws.

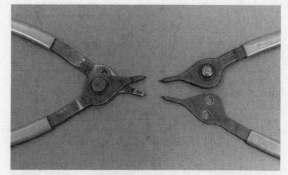

Figure 2-40. Lock rings can be either internal or external, and many lock ring pliers can be adjusted for either.

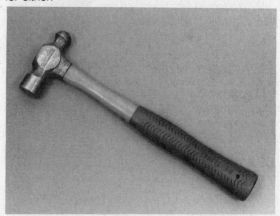

Figure 2-41. Ballpeen hammers come in many weights. You should own several weights.

before, you will appreciate the time savings hose-clamp pliers provide.

Diagonal pliers, also called cutting dykes or sidecutters, figure 2-38, are useful in automotive work for cutting wire. They are also an excellent tool for removing cotter pins, dowel pins, and protruding rivets.

Lockring, or snapring pliers, have pins at the ends of each jaw which you insert into holes in the snapring, figure 2-39. You can then squeeze or expand the ring for removal or insertion. You can adjust some snapring pliers so that you can expand or compress the ring with the same tool, figure 2-40.

There are also pliers for pulling fuses, for removing hose clamps, for pulling spark plug wires, and many other special uses.

Hammers

Many types of hammers are used in auto repair today. The most common types are the ballpeen, plastic and rubber mallets, soft-metal, sledge, and shotfilled hammers. All of these come in a wide variety of sizes and weights, from an eight-ounce ballpeen to a four-pound sledge.

The ballpeen hammer is really two hammers in one, figure 2-41. One side has a flat surface for driving punches, chisels, and similar objects; the other side is rounded and is used to peen rivet heads as well as to form and shape metal.

Figure 2-42. Plastic and rubber hammers let you strike heavy, but cushioned blows to metal castings and stampings.

Figure 2-43. Soft-metal hammers strike sharper and heavier blows than plastic and rubber, but still are easier on work surfaces than a ballpeen.

Figure 2-44. Bricklayer's sledges are compact, short-handled, but heavy hammers useful for heavier duty work in confined places.

Figure 2-45. Dead-blow hammers strike the work without rebounding in your hand. They are easy to control and less tiring to use.

Soft-metal hammers, figure 2-43, are useful for striking soft but heavy blows to large but delicate components. You can purchase soft-metal hammers made of brass, bronze, lead, or malleable iron. Some soft-metal hammers come with detachable tips, just like plastic-faced hammers. Use them for striking precision drifts or seal installers, or for tapping parts together that require a slight interference fit.

Sledge hammers are heavier tools that are seldom used in automobile repair or rebuilding. About the heaviest sledge you are likely to need would be a two-pound short-handle version, sometimes called a rock mason's or bricklayer's hammer, figure 2-44. They are too heavy and inelastic for most work, but when you have to use a large punch or chisel, they are tough to beat.

The shot-filled or dead-blow hammer, figure 2-45, is partially-filled with metal shot so that when the object is struck the hammer doesn't bounce, as a solid rubber or soft metal hammer does. This allows more of the force to go into the object being struck. Most dead blow hammers have faces with plastic or tough urethane covers.

Plastic hammers and rubber mallets, figure 2-42, protect the object being struck. The heads on these hammers are softer than the objects on which they are used. Many plastic hammers have replaceable heads with different hardnesses. Look for a hammer that has one rubber head that you can just dent with a fingernail, and one plastic head that is too hard to indent. Plastic hammers and rubber mallets are useful for tapping valve covers loose, separating engine castings without damaging them, and removing manifolds.

Hand Tools for Engine Rebuilding

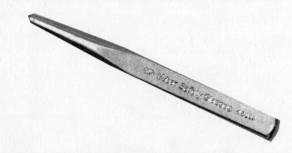

Figure 2-46. Often a single center punch four or five inches long will be all that you need.

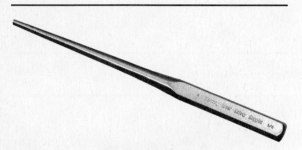

Figure 2-47. Starting punches are stronger than pin punches.

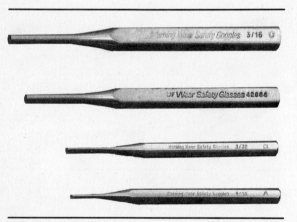

Figure 2-48. Pin punches can fit down into deep holes that a starting punch cannot reach into.

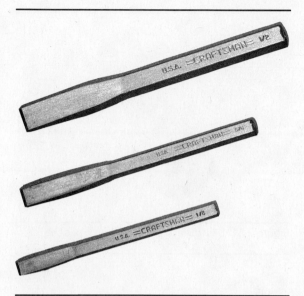

Figure 2-49. Chisels are essential for shaping and moving metal.

Figure 2-50. Chamfer the tops of your chisels and punches to keep then from chipping.

Punches and Chisels

Punches are used for a variety of tasks, such as driving roll pins, staking oil gallery plugs, and center-punching for drilling. There are three main types of punches.

Center punches, figure 2-46, have sharpened conical tips. Use a center punch for upsetting metal surfaces to make non-precision bearings fit tighter, to stake freeze plugs and oil gallery plugs in position, and to punch spots for drilling.

Use a starting punch, figure 2-47, to begin driving rivets and pins from a hole. The tapered tip of a starting punch is strong enough to stand unsupported above the workpiece.

Use a pin punch, figure 2-48, to continue driving a rivet or pin from a hole after the starting punch taper bottoms out in the hole. The straight sides of a pin punch will fit easily down into a deeper hole than a starting punch, but these tools are easy to bend if you try to use them above the workpiece.

Chisels, figure 2-49, have many purposes. You can use them to cut metal stock, to chamfer sharp edges, to remove frozen nuts and bolts, and to knock welding flash off a metal surface.

Keep your punches and chisels sharp for the best results. When the tops begin to mushroom, carefully grind them flat, with a small chamfer, figure 2-50. Grind the working tips of your punches and chisels slowly, cooling them in water frequently to avoid losing the temper. If the metal turns to a blue color, you have

Figure 2-51. Seal installers come in various sizes.

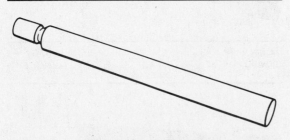

Figure 2-52. Drifts are large-diameter precision punches, which spread the impact force over a larger area.

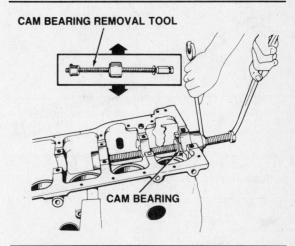

Figure 2-53. Cam bearing drivers have several adjustable heads to fit a variety of engines.

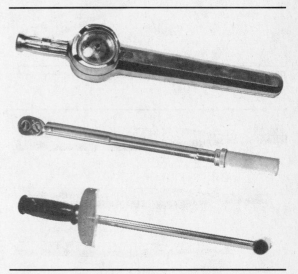

Figure 2-54. You cannot get around owning a torque wrench. It is essential.

overheated it. Chisels work best when you keep the tips ground to a sharp, 45 degree angle.

Seal Installers and Drifts

Seal installers are often necessary for installing oil pump, crankshaft, and other types of oil seals. The installer, figure 2-51, is slightly smaller than the seal. It fits over the seal and helps to drive it in straight.

You use drifts, figure 2-52, to push bearings and other tough metal parts into and out of interference-fit bores, either by tapping with a hammer, or pressing with a hydraulic press. Use a drift when the item is large enough that there is a danger of cocking it at an angle as you drive it. Many technicians use large sockets instead of drifts when they press items on and off.

Other special drifts include drivers for removing and installing cam bearings, figure 2-53.

Torque Wrenches

In engine repair and rebuilding work, you must often tighten a nut or bolt to a specific tension, no more and no less. This is particularly true of fasteners in critical areas, such as head bolts, main bearing bolts, and connecting rod bolts or nuts. Adding a little more torque beyond the manufacturer's specifications ''for good measure'' can stretch the bolt beyond its limit. When this happens, the bolt won't clamp parts together as tightly as it should, it can be severely weakened, or both.

A torque wrench indicates how much force you apply to a nut or bolt. There are three types of torque wrenches in common use, figure 2-54.

- Beam-type
- Dial-type
- Click-type.

Hand Tools for Engine Rebuilding

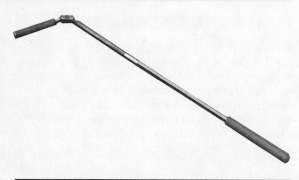

Figure 2-55. Parts retrievers cost only a few dollars, but can save you many minutes or hours of aggravation.

With the beam-type wrench, a pointer indicates the torque on a scale near the handle. With the dial-type wrench, the torque is indicated by a needle and scale on a dial built into the wrench body. With the click-type wrench, the torque is preset by rotating the wrench handle until a pointer aligns with a number marked on the body of the wrench. When the preset torque is reached, the mechanism inside the wrench overcenters and produces a "click" that you can hear as well as feel through the wrench handle.

You can buy torque wrenches graduated in foot-pounds, inch-pounds, Newton-meters or kilogram-meters. Modern torque wrenches usually have both a U.S. Customary and a metric scale on the wrench.

Small Parts Retriever

Parts retrievers, figure 2-55, are so handy and inexpensive that you will wonder how you ever did without them. Basically, all the parts retriever does is let you fish out small items that you have dropped into inconvenient places. There are two kinds. One uses a small magnet fixed to the end of a narrow telescoping handle. These are useful for retrieving loose iron or steel parts that you can't reach with your fingers. The second kind uses a trigger handle that opens and closes a small two- or three-fingered claw to pick up nonmagnetic parts. This kind is often sold with a flexible shaft that lets you reach around corners.

AUTOMOBILE REBUILDER TOOL SET

We have described the types of tools that automotive engine rebuilders and machinists use most frequently. Here is a list of the bare minimal tools that you should have in order to prepare for starting out in automobile engine rebuilding. You will need many other specialized tools that you accumulate over time, but these will get you started:

- Lockable tool box, preferably a roll-away design
- Safety glasses
- Cutting dykes
- Combination pliers
- Locking pliers
- Allen wrench set (standard and metric)
- Punch set
- Chisel set
- Feeler gauges (both blade and spark plug design)
- Screwdrivers: 1/4-, 3/8-, 1/2-inch slot-type
  Phillips: #0, #1, #2, #3
- Plastic or rubber hammer
- Soft-metal hammer: brass or lead
- Ball peen hammer
- Socket set: Standard: 1/2-inch drive:
  3/8- to 1-inch
  Metric 10 mm to 19 mm
  Spark plug sockets
- Socket wrench extensions: 3-, 6-, and 10-inch
- Breaker bar: 1/2-inch drive
- Wrenches: Combination standard 3/8- to 3/4-inch
  Combination metric 10 mm to 19 mm
- Torque wrench: 0-150 foot-pounds (0-200 Nm)
- Gasket scraper
- Parts retriever
- Pocket magnet
- Screw extractor set
- 0 to 1 inch (0 to 25 mm) outside micrometer
- Six-inch pocket scale

Chapter 3

Power and Specialized Engine Rebuilding Tools

POWER AND MACHINE TOOLS

Power tools are hand tools that use an air or electric motor to operate the tool, rather than your own muscles. Machine tools, on the other hand, range from smaller items such as bench grinders to larger stationary pieces of equipment such as drill presses, and milling machines. We discuss specialized machine tools in subsequent chapters on bottom- and top-end repair.

Drill Motor

Electric drills are seldom used in engine repair and rebuilding for drilling holes, but they are very convenient power sources for other purposes. With the correct attachments, you can use drill motors to turn hones, glaze breakers, wire brushes, and many other tools.

You can use drill motors with wire brushes to remove carbon and corrosion from engine parts, figure 3-1. A carbon cleaning brush has a shank that fits the chuck of the drill motor. The bristles of the brush extend almost parallel to the shaft of the motor.

Honing is more often done on precision automatic machinery than it used to be, but hand-held honing is still practiced in smaller shops.

Impact Wrench

Impact wrenches, figure 3-2, use repeated hammerlike blows to turn nuts and bolts. They resemble a drill motor, but instead of a chuck they have a male square drive for turning sockets.

Impact wrenches can be pneumatic or electric. Pneumatic, or compressed air powered, wrenches are usually more powerful than electric wrenches. The torque that these wrenches deliver is often adjustable. Because of their power, use care not to overtighten nuts and bolts — you can twist them right off with a good quality impact wrench.

Sockets used with impact wrenches, figure 3-3, must withstand tremendous pounding, and are specifically designed for this use. They are usually made with 6 points instead of 12 so that the fastener won't be damaged. Impact sockets are usually black or brown instead of being chrome plated. Never use ordinary sockets with an impact wrench. The pounding will damage the socket, and it may shatter. This type of damage is easily detectable, and most tool suppliers won't replace them, even if the sockets are guaranteed.

Power and Specialized Engine Rebuilding Tools

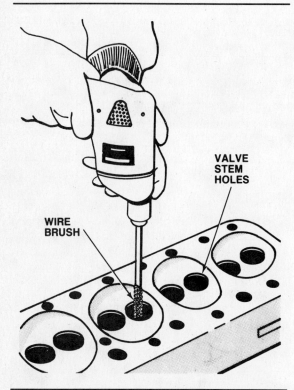

Figure 3-1. Drill motors can power many different engine rebuilding tools, such as brushes, grinding flaps and stones, and even drill bits.

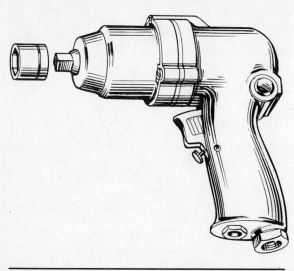

Figure 3-2. Power impact wrenches remove tight fasteners with much less risk of damage than ordinary hand tools.

Impact-resistant extensions and universal joints are also available. Always wear eye protection when using impact tools.

Figure 3-3. Use only impact sockets and accessories with impact tools. Those designed for hand tool use are not strong enough.

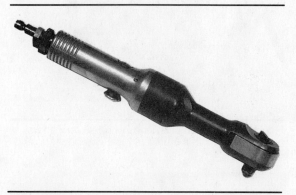

Figure 3-4. Air ratchets make disassembly and assembly very quick.

Air Ratchet

Air ratchets speed up engine disassembly and assembly. They are much smaller than impact wrenches and are sometimes called speed ratchets, figure 3-4. They have no impact feature. The air ratchet often resembles a ratchet wrench and is used in the same manner to loosen or tighten a nut or bolt. You press a lever or button to activate the wrench.

Some air ratchets have power limiters that prevent overtightening. When using an air ratchet that doesn't have a limiter, you can very easily overtighten a small bolt to its breaking point.

You can use the air ratchet to run up nuts and bolts, but you should do the final tightening with a torque wrench.

Bench Grinder

Bench grinders, figure 3-5, usually consist of two grinding wheels on the ends of the shaft of an electric motor. One wheel is usually a coarse grit while the other is fine. The motor is often mounted on a stand or bench. Use grinding wheels to sharpen drills, reshape tools, or rapidly remove metal from parts.

Figure 3-5. You can use bench grinders for general hand-shaping of nuts, bolts, engine components, and castings.

Figure 3-6. High-speed grinders make cylinder head porting or cleanup fast and easy.

Figure 3-7. Air compressors can be mounted outside the shop to reduce noise.

A wire scratch wheel can be mounted on a bench grinder to remove carbon and deposits from valves and other engine parts. Do this only on hard metal parts such as valves, and never on soft metal parts such as pistons.

Rotary Grinder

The hand-held, high-speed, rotary grinder is used to grind inside ports and other areas where the bench grinder can't reach, figure 3-6. You can match ports, smooth combustion chambers, and grind welds or make other repairs with a rotary grinder.

Compressed Air

Almost all shops have a compressed air source easily accessible for pneumatic tools, figure 3-7. Compressed air is also used to pump lubricating oils and greases from remote containers to convenient overhead dispensing tools attached to retractable hoses, figure 3-8.

Quick-release couplers are often used to connect blow guns, air ratchets and other tools to the outlets or hoses. You pull back a sleeve on the coupler, insert the nipple on the tool connector, and release the sleeve to lock the nipple in place. To release the nipple, the sleeve is again pulled back, and the nipple pulled out.

SPECIALIZED ENGINE TOOLS

There are many specialized hand and machine tools that you will use in engine repair and rebuilding. We will group them by the service

Power and Specialized Engine Rebuilding Tools

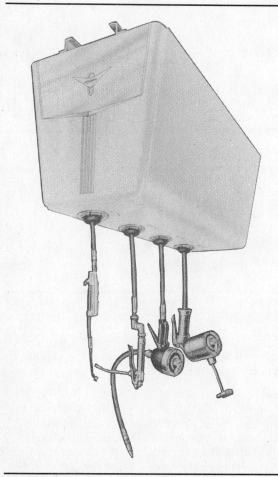

Figure 3-8. Overhead reels keep water and compressed air lines off the floor area.

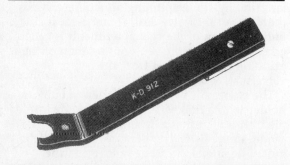

Figure 3-9. You can remove the valve springs with the cylinder head still on the engine with a valve spring compressor like this.

procedures you would follow in a normal engine overhaul or rebuild.

Head and Valve Service Tools

There are many different head and valve tools. You will probably use these more than any others, because many engines come into the

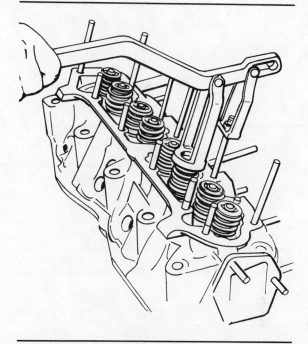

Figure 3-10. You must use this factory tool to remove the valve gear on this OHC engine.

shop for cylinder head work at least once before they are due for a complete rebuild.

Valve spring compressor

There are many tools available for removing and replacing valves. The pry bar in figure 3-9 compresses the spring while compressed air holds the valve closed. Some overhead cam engines require a special tool that bolts to the head and compresses the valve spring so that you can remove the valve keepers, figure 3-10. You can use it with the head on or off the engine. A C-type valve spring compressor is used to compress the valve springs with the heads off the engine, figure 3-11. Other special cylinder head tools include rocker arm stud removers and installers.

Spring tension tester

Two types of spring tension testers are on the market. One type is a scale with a dial reading, figure 3-12. It is similar to an arbor press. The handle moves the ram, which compresses the spring against the top of the scale. The length of the spring is read from a pointer. The force is read directly from the dial.

The other type of spring tension tester uses a torque wrench which has a scale reading, figure 3-13. The micro-adjusting type of torque wrench, which clicks when you reach a set torque, can't be used with this tester.

A wheel on the tester sets spring height. Pressure is then applied with the torque

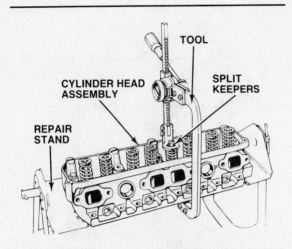

Figure 3-11. The traditional C-clamp type valve spring compressor comes in several different configurations.

Figure 3-13. The torque wrench spring tension tester is fairly inexpensive.

Figure 3-12. The dial reading spring tension tester is the most accurate and precise.

wrench until the tester gives a loud click. At that point, note the reading on the torque wrench and multiply it by two. This will give the spring tension at that length. Springs have both a closed and open specification. The important one is the closed specification. If the spring doesn't measure within ten percent of the closed specification, it may not hold the valve tightly against its seat, and the valves will leak or burn.

Valve seat grinder

Two types of hard-seat grinders are available, the concentric and eccentric types, although probably 98 percent of all modern machine shops use the concentric type. The concentric valve seat grinder uses a large circular grinding stone which grinds the entire circumference of the seat in one motion, figure 3-14. The eccentric grinder employs a smaller circular grinding stone spinning on a different axis than that of the valve seat, figure 3-15. It cuts only a portion of the seat at a time. It is called eccentric because the center of rotation of the grinding stone is offset from the center of the valve seat.

The stones for hard seat grinders mount on a holder with a hollow center. The hollow center fits over a pilot that is inserted into the valve guide, figure 3-16. The pilot, located by the guide, positions the stone holder and the stone. If the guide is worn, the pilot may locate off center or may tip slightly. This will cause a poor fit between the valve and its seat. Before grinding the seats, inspect the guides and repair or replace them as necessary.

The motor for a hard-seat grinder is made to run at high speed. It may look like an ordinary drill motor, but it is quite different. Never try to grind seats with an ordinary drill motor. It doesn't turn fast enough, nor does it have the power to do a good job.

Power and Specialized Engine Rebuilding Tools

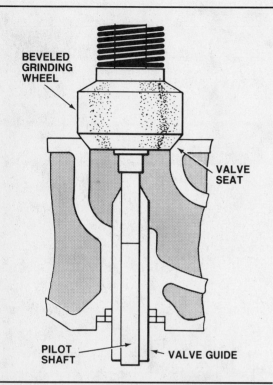

Figure 3-14. The concentric valve seat grinder grinds the surface of the seat in one motion.

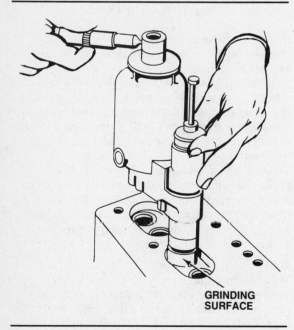

Figure 3-15. The eccentric valve seat grinder rotates on a different axis than that of the valve seat.

Valve seat cutter
Carbide cutters are also used in valve seat reconditioning, figure 3-17. The cutters are located by a pilot, similar to the pilots used on

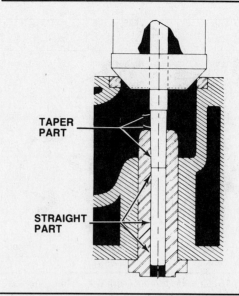

Figure 3-16. The stones for a hard-seat grinder are mounted on a hollow-centered holder.

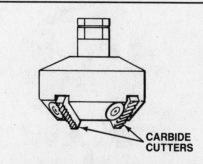

Figure 3-17. Carbide cutters mount a tool bit onto a holder, instead of using a conical grinding stone.

hard seat grinders. The big difference is that you turn the cutters by hand, or with a special motor that turns at about 60 rpm, figure 3-18. This is extremely slow when compared with the hard seat grinders, which turn at about 1800 rpm.

Because the carbide cutters are very hard and turn slowly, they will last much longer than grinding stones. The grinding stones on a hard seat grinder may last through only about 30 valve seats, and must be dressed before beginning each job, and usually at least once during a valve job on a typical V8 engine. Blueprint engine builders might dress the stone before each seat.

Valve grinding machine
A valve grinding machine has a chuck that accepts one valve at a time, figure 3-19. The chuck is mounted on a sliding table controlled manually with a lever. When the chuck is engaged with the motor, the valve rotates. The

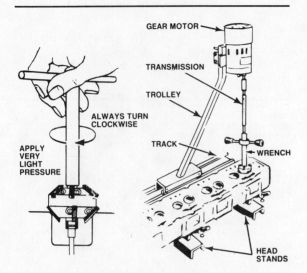

Figure 3-18. Carbide cutters work with a hand-turned arbor, or at extremely slow cutting speeds.

Figure 3-19. Valve-grinding machines use a chuck mounted on a sliding table.

Figure 3-20. You can use the second wheel on some grinding machines for grinding rocker arms or refinishing valve tips.

operator then uses the lever to move the table so the face of the valve moves across a high-speed grinding wheel. The operator controls the amount of metal removed from the face of the valve. Ordinarily, only enough metal is taken off to make the valve face smooth and concentric.

The valve face angle is adjusted where the chuck is mounted on the table. Degree marks indicate whether the valve will be ground at 30 degrees, 45 degrees, or any other angle the operator selects.

Some valve machines have two grinding wheels. The second wheel is used for grinding the grooves out of rocker arms, or refinishing worn valve tips, figure 3-20. Special chucks or holders ensure that the grinding is perpendicular to the part.

Some larger rebuilding shops have a machine that will refinish the bottom of a tappet, and put a slight convexity on it at the same time. The convexity helps to make the tappet spin in its bore while the engine is running, and spreads out the wear.

Valve guide renewing

Valve stem metal is very hard, but valve guides are soft cast iron, bronze, or silicon bronze. Because of this, a valve guide will wear faster than the stem. When doing a valve job, check the guides for wear. You must renew or replace the guides if they are out of specification.

Valve guide bore gauge

You can check valve guide clearance with a valve guide gauge, figure 3-21. The valve guide gauge consists of a dial indicator with a slim probe that you can insert into the valve

Power and Specialized Engine Rebuilding Tools 35

Figure 3-21. Valve guide gauges check the valve guide clearance.

Figure 3-22. A knurling bit resembles a forcing tap, in that it rolls the metal into threaded ridges, reducing the valve guide internal diameter.

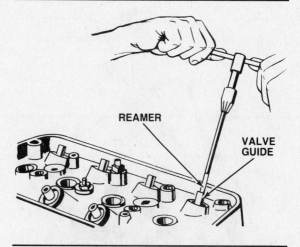

Figure 3-23. The reamer opens the knurled metal up to the correct diameter.

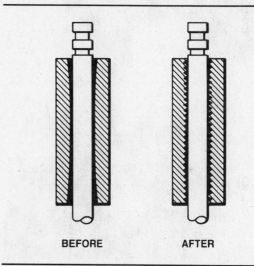

Figure 3-24. If the knurling is successful, the newly-made threads will support the valve stem through its entire length.

guide. The gauge gives the diameter of the guide, but will also show any taper or out-of-round condition.

Knurling

There are many methods of renewing valve guides. The easiest method is knurling. A knurl is a special bit that is inserted into the valve guide and turned with a drill motor that has a drive reduction, figure 3-22. The knurl rolls a thread into the valve guide and raises the metal. The points of the thread will reduce the hole size so that it is smaller than the valve stem.

The next step in knurling is to run a reamer through the hole to take the tops off the threads and make the hole fit the valve stem, figure 3-23. The remainder of the threads will support the valve stem figure 3-24.

Guide insert

Another method of valve guide repair uses a bronze spiral coil. A special tap cuts an oversize thread in the guide, figure 3-25. Then a piece of spiral bronze is threaded into the guide. A swaging tool forces the bronze spiral into the guide. Finally, a reamer is used to bring the guide out to size.

Another guide renewal method employs a thin wall bronze sleeve, similar to a bushing, but thinner. The guide is first reamed to accept the sleeve. Then the sleeve is pressed or driven in. A swaging tool is run through the

Figure 3-25. First use a special tap to cut the thread in the guide before threading in the bronze coil.

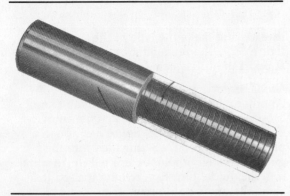

Figure 3-26. You can also drive in a special thin-walled tube to replace integral valve guides.

sleeve to force it into contact with the guide. Lastly, the sleeve is reamed to the correct size.

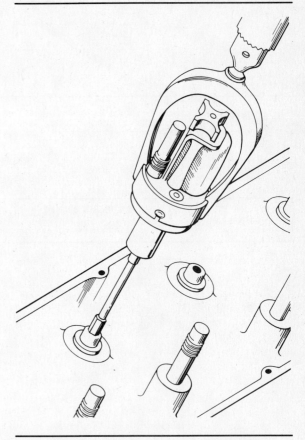

Figure 3-27. A guide hone gives you much better control over the inner clearances.

When the original guides are the integral type, that is, a part of the cylinder head casting, you can use a special thin-walled tube, figure 3-26. The original integral guide is bored out, and the new insert driven in. The final step is to ream the guide to size.

When the original guides are the insert type, simply drive them out and press in and ream the replacements.

Although reaming has been used to size guides for many years, honing allows the machinist to finish a guide to a specification between normal reamer sizes, figure 3-27. Normally, reamers come in sizes 0.001 inch apart. If the size you need ends in an even thousandth, such as 0.3120-inch, reaming with the correct tool will work fine. However, honing allows you to finish guides to a specification of, perhaps, 0.3126-inch. This is very handy if you are blueprinting an engine, and the nearest-thousandth reamer size is not what you want.

You can accomplish all of these operations with handtools or portable electric tools. In addition, there are valve guide machines for production shops. The machine speeds up the

Power and Specialized Engine Rebuilding Tools

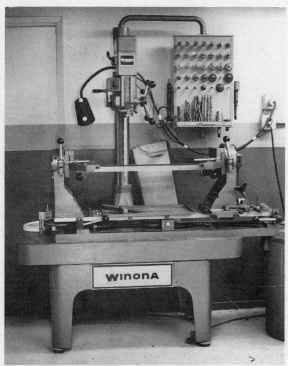

Figure 3-28. The head machine resembles a drill press, but is much more specialized.

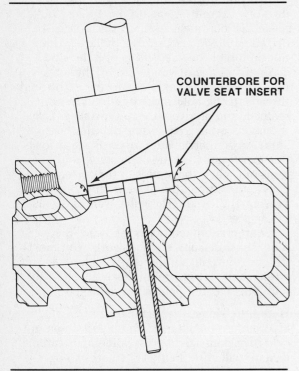

Figure 3-29. The valve seat fits tightly into a counterbored area in the port.

Figure 3-30. Counterboring requires a special tool bit.

floats on a cushion of air to make it easily movable. The head machine looks like an oversize drill press, figure 3-28.

Seat insertion tool or machine
After several valve jobs, the valve seat may have been ground so far that it should be replaced. If the seat is an insert type, it is knocked out, and a new, oversize insert is installed. New inserts usually come in a slightly larger oversize diameter than the original seats. This means that you must enlarge the counterbore, or the cut-out area in which the seat inserts are placed for the oversize insert, figure 3-29. You can also install inserts in heads that originally had integral seats. The head material that formed the original seat is cut away until there is a counterbore to accept the replacement insert.

Machine the counterbore with a special tool, figure 3-30, driven with an ordinary drill motor. You can also cut counterbores on a valve guide machine.

When the counterbore is machined to fit the new insert, a driver is used to install the insert. Hammer blows to the driver will knock the insert into position and seat it in the counterbore. You must guide the driver by the valve guide machine or by the tool that clamps to the cylinder head. Using the driver without a guide is not recommended because it is easy to cock the seat and shave metal from the side of the counterbore. This metal then ends up under the seat and makes it sit at an angle. The result is poor heat transfer and eventual valve failure from overheating.

Since the insert is not made of the same material as the head, it won't expand at the same rate. To keep the inserts in position, machinists install them with a press fit, usually 0.006 to 0.008 inches (0.150 to 0.200 mm) for inserts in aluminum heads, and 0.004 to 0.005 inches (0.100 to 0.125 mm) in cast iron heads. Sometimes machinists stake the head metal around the seat to keep it from working loose. If the seats are installed with a good interference fit, staking the seat is not necessary.

work because the head is shifted on a special table to bring the guide under the drilling or reaming chuck. In some machines, the table

Figure 3-31. Positive valve stem seals prevent leaks.

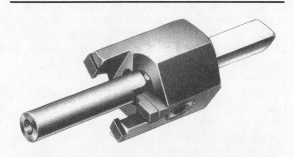

Figure 3-32. This special tool lets you cut valve guides for seals on engines that were never equipped with them.

Tool for positive valve seal installation
You must machine the valve guides to install positive guide seals on an engine that did not have them originally. Machining the guide ensures that the seal will be concentric with the valve stem. It also means that the seal will be a tight fit on the guide and prevent leaks, figure 3-31.

The tool that machines the valve guide consists of a pilot and cutters, figure 3-32. The pilot is inserted into the guide and turned with a slow speed drill motor. As the tool is pushed down over the guide, it cuts the outside of the guide to the right diameter for the seal. If the head is mounted in a head machine, the tool can be operated by the machine.

The valve is inserted into the guide before installing the seal. A plastic cap goes over the end of the valve to prevent damaging the seal. A tool is then used to push the seal over the valve stem and onto the guide, figure 3-33. Some engines come with positive valve seals as original equipment. If so, it is not necessary

Figure 3-33. Using this tool to push the valve seal over the guide prevents damage to it.

to do any machining as long as the replacement seals are the same size as the original equipment.

Rocker stud puller
Rocker arm studs also wear out, especially if the valve train geometry changes after a valve job, and allows the rocker to hit the stud.

If the rocker wears a groove in the stud, you should pull the old stud out, and drive in a new one. Commercial stud drivers and pullers make the job much easier. A stack of washers on the stud will also act as a puller. Tighten a nut on top of the washers to pull the stud.

A new stud is driven in after tightening two nuts on its end to prevent damage to the stud threads. If the hole is damaged, it can be reamed, and an oversize stud installed. Drive the new stud in to the same depth as the old stud. A commercial stud driver is made for this job, figure 3-34. It is adjustable and bottoms out when the stud is at the correct depth. Check and correct the valve train geometry to prevent stud wear. The valve train is discussed in Chapter 14.

If a stud is broken off in the head, the techniques just discussed won't work. You must drill the stud out until there is nothing left but a thin shell. Then carefully take the shell out with pliers so the hole is not damaged.

Hydraulic valve lifter remover
Hydraulic valve lifters slide up and down inside precision-sized bores. Sometimes carbonized oil will stick them in the bores so tightly that you can't remove them by hand. Hydraulic valve lifter removers, figure 3-35, operate like a small slide hammer to pull the lifters with repeated sharp blows.

To use them, you remove the engine castings to get access to the tops of the lifter bores. Slide the tips of the removal tool into the top of the lifter and rotate the jaw attachment to

Power and Specialized Engine Rebuilding Tools

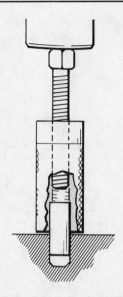

Figure 3-34. Stud drivers position the stud at exactly the correct depth.

Figure 3-35. You can remove stuck valve lifters with this inexpensive tool.

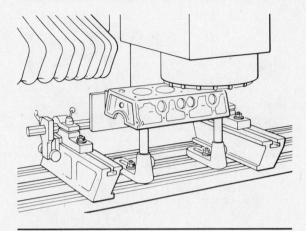

Figure 3-36. Head resurfacing in a milling machine is mostly automatic.

Figure 3-37. When the cutters are mounted below the table, the machine is called a broach.

spread the jaws, clamping them into the lifter. Then tap the weight against the top of the tool to pull the lifter.

You can also use some removal tools with screw-type pullers to extract hydraulic valve lifters that are really stuck.

Head surfacing machines

The machined surface of a cylinder head often becomes distorted after long periods of service. Aluminum cylinder heads are especially prone to warping. High and low spots may develop and cause leaks. Whenever rebuilding an engine, check the head with a straightedge before assembly, as described in Chapter 14. If the head is distorted, you will have to resurface it.

Milling machines and broaches

A milling machine uses one or more large rotating cutting tools, figure 3-36. The cylinder head is mounted on a table. Either the table or the cutters are moved so that metal is taken off the gasket surface. This makes the head perfectly flat. The depth of the cut is regulated by the operator. Ordinarily, you would take off only enough metal to be sure the cutters reach all the low spots.

There are many different milling machine designs. When the cutters are mounted below the table, the machine is called a broach, figure 3-37. The cutters rotate in a horizontal plane, and the head moves from one end of the machine to the other.

Grinder

The grinder uses a large diameter grinding wheel to resurface a head, figure 3-38. The head is mounted horizontally on a table with

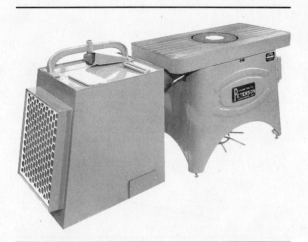

Figure 3-38. Grinders are probably the most common head resurfacing tool.

Figure 3-39. This particular tool not only cuts the seat wider in the head, but also cuts the valve guide to allow using another spring inside the main spring.

the gasket surface up. The grinding wheel is mounted so the flat side surfaces the head. The wheel is mounted on a sliding table that travels the full length of the head.

The depth of cut on the grinder is controlled by raising or lowering the table on which the head is mounted. When grinding, the wheel may not take off all the metal on its first pass. It is always allowed to make a return pass over the head before increasing the depth of cut. Automatic trippers will reverse the direction of the grinding table and send it on the return pass.

The maximum amount of metal removed with each round trip pass should be 0.002-inch (0.05-mm). Make the final pass without changing the wheel height, to be sure that the entire surface is cleaned up.

One of the big differences between the grinder and the mill is that the mill cuts dry. The grinder must have a constant stream of coolant splashing on the head to prevent overheating. The coolant is recirculated by a pump in the bottom of the machine.

Spotfacer for enlarging spring seat

To keep valves from floating, you can sometimes install stronger springs which may be larger in diameter than the original equipment springs. In this case, you must cut the spring seat in the cylinder head so the spring will fit.

Camshaft manufacturers have a special spring seat cutting tool called a spotfacer. The tool is driven by a drill motor or head machine, figure 3-39. A pilot centers the tool concentric with the valve guide. The tools are normally not adjustable, but are made for a particular spring diameter.

Flycutter

Occasionally an owner specifies oversize valves or a camshaft with higher lift or different timing from stock. Whenever this is done, you should check the assembled engine for interference between the piston and valves as the pistons sweep over top dead center. If the valves do strike the piston, you can usually correct the problem by enlarging the valve reliefs in the piston crown.

You can do this easily with a flycutter, figure 3-40. The flycutter consists of one or more tool bits mounted in a bar or cylinder. The cutting surfaces protrude ahead of the tool, so that it cuts a circular path as you feed it into the workpiece.

Be very careful whenever you enlarge valve reliefs. You must be sure that both the diameter and the depth of the relief matches that required by the valve diameter and lift, but you must not cut the relief so deep that you weaken the piston crown or cut into the back of the top ring groove. If you can't deepen the relief enough to keep the necessary clearance, you can sometimes avoid interference by advancing the camshaft a few degrees.

Block Service Tools and Equipment

You use engine block tools during major overhauls or complete rebuilds.

Camshaft bearing removal tool

Camshaft bearing surfaces wear slowly, and often are still in good condition even when the

Power and Specialized Engine Rebuilding Tools

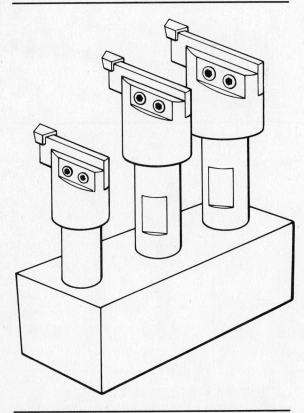

Figure 3-40. Flycutters make modifying piston crowns easy.

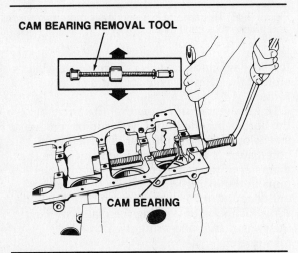

Figure 3-41. The camshaft bearing removal tool works like an expandable drift.

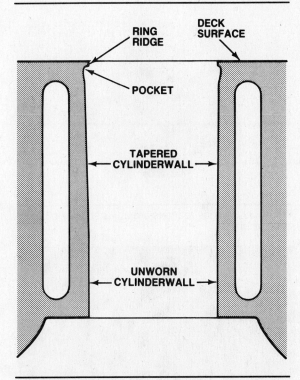

Figure 3-42. The ridge at the top of the bore is worn by the top compression ring.

rest of the engine is completely worn out. During a simple overhaul, you can often leave them in place. You should remove them for a thorough rebuild, however, if only to make sure that the oil passages behind them are thoroughly clean.

Camshaft bearings in OHC engines are usually simply precision bores through the cylinder head, or through the joint between a cylinder head and a bolt-on saddle. These bearings usually require no service. In many OHV engines, however, the bearings are a separate set of inserts that you must remove in order to clean the oil passages or galleries behind them. They are seldom expensive, and as a general practice you should replace them during block overhaul.

The camshaft bearing removal tool, figure 3-41, enables you to extract these bearings from inside the block without damaging the precision bores into which the shells are pressed. You can usually use the same tool to both remove and install the bearings.

Ridge reamer

As the piston rocks to top dead center, the uppermost piston ring slides almost to the top of the bore, and then descends again. Over many miles, the normal wear from the ring results in a small step at the top of the bore, figure 3-42. Unless removed, this ridge will catch on the piston rings if you try to remove the pistons through the top of the bore, and can damage the ring lands. In addition, if you do not plan to rebore the cylinders during an overhaul, you must be sure to remove the ridge or it will interfere with the new, unworn piston rings.

Figure 3-43. Ridge reamers are adjustable for different cylinder sizes.

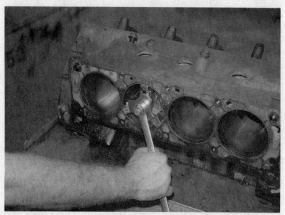

Figure 3-44. Cut only enough metal to remove the ridge.

Figure 3-45. Set the self-aligning bore gauge to zero after you place it in the bore.

Figure 3-46. This style of bore gauge is more precise and much more convenient to use.

Ridge reamers, figure 3-43, let you cut this ridge away without enlarging the bore. Most are hand-operated lathe-type tools that you center in the bore and then turn with a wrench, figure 3-44. Typically, they remove the ridge completely in just a few turns.

Production shops often do not bother with ridge reaming. Their attitude is that if the ridges are worn deep enough to prevent the pistons from being removed, then the cylinder should be rebored and fitted with new pistons anyway. They just knock the pistons straight out from beneath and throw them away.

Cylinder bore gauge

Using a cylinder or dial bore gauge to measure cylinder bore diameter is much faster than using an inside micrometer.

There are two types of bore gauges. The sled type, figure 3-45, is self-aligning. You insert it into the bore and the plunger spring pushes the sled against the cylinder wall. Then set the dial to zero. Sliding the sled up and down in the bore will indicate taper. Turning the sled around the cylinder will show if it is out-of-round.

To get the instrument within the range of the bore being measured, you add or remove extensions from the plunger. The sled-type bore gauge is not used to directly measure the diameter of the cylinder, but to measure variations in the diameter to determine taper and out-of-round. However, you can determine the actual diameter by zeroing the dial, removing the gauge, and measuring it with a micrometer.

The rocking-type bore gauge, figure 3-46, is more precise than the sled type, and is designed to be used with a honing machine. It has two aligning slides that center a plunger in the bore. To get the diameter of the bore, the instrument is rocked up and down until the smallest reading is shown on the dial. The sled-type gauge usually reads to 0.0005-inch, while the rocking-type reads to 0.0001-inch.

Power and Specialized Engine Rebuilding Tools 43

Figure 3-47. The setting micrometer makes using the dial bore gauge much easier.

A big advantage of the rocking-type bore gauge is the setting micrometer available for it, figure 3-47. The setting micrometer instantly aligns the plunger with the micrometer faces. You then zero the gauge dial at the standard bore size, so that the gauge reads *deviations* from the target size that you need. Inserting the gauge shows if the cylinder is oversize, and moving the gauge around shows if it is tapered or out-of-round.

Boring machines
The automotive boring machine is used to bore cylinders to the next available piston size. The cylinder is bored to the required size minus about two thousandths of an inch for honing. Honing removes the tool marks left by the boring machine, brings the cylinder out to final size, and gives the cylinder walls a proper surface for ring seating.

Older shops may still have boring bars somewhere on the premises, but they are seldom used any more. These boring bars simply bolted to the block using the cylinder head bolt holes, and depended on a good deck surface to position the tool correctly. If you use one of these tools, you *must* resurface the head first, or the bar may cut off-center.

The modern boring machine is a heavy piece of equipment that contains a long rigid bar, figure 3-48. It has a revolving cutting tool on the end, and centering shoes which expand to center the bar. You bolt the block to the tool using steel spacers held by the main bearing saddles. These tools will cut a correctly-aligned hole no matter what the condition of the deck is.

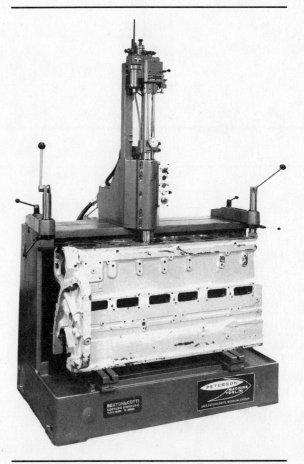

Figure 3-48. Boring bars cut a rough-finished cylinder close to the final size you need.

The machine has a feed mechanism that moves the cutting tool through the cylinder as it rotates. You can take light cuts in one pass. If you must remove a lot of metal, you should take several passes.

Most boring machines have shutoff switches that turn off the motor when the bar stops cutting at the bottom of the cylinder. This prevents the bar from continuing to rotate and perhaps hitting the main bearing webs.

Boring machines all have limitations on the sizes of cylinders they are able to bore. You can bore small cylinders with a small boring machine. Larger cylinders require a larger boring machine, figure 3-49. Air-cooled engines usually have individual cylinders. Since there is no block surface on which to bolt the block, you must use a jig, figure 3-50. The jig is a heavy plate with a hole in the middle. You bolt the cylinder to the bottom of the plate, and then bolt it to an appropriate boring machine. You can then bore the cylinder just as if the boring machine were bolted to the cylinder

Figure 3-49. Larger engines with multicylinder castings require a heavier tool.

Figure 3-50. Air-cooled cylinders require you to use a jig to set the tool up.

block. You will probably not bore many air-cooled engines because the cost of the machine work is often more than the cost of a new cylinder.

Cylinder Hone

The cylinder hone uses abrasive stones to remove metal from cylinder walls. Traditional designs have four "wings" that stick out from the body of the hone, figure 3-51. Two of the wings, opposite each other, hold the stones; the other two hold the centering shoes.

A nut at the top of the hone expands the shoes and stones until they press against the cylinder wall. The hone is turned by a heavy-duty drill motor and moved slowly up and down in the cylinder.

Honing stones are available in different grits or degrees of coarseness. The coarse stones cut faster, the fine stones more slowly. The fine stones produce the type of finish required for good ring seating. For the best finish, and to avoid loading up the stones with metal particles, all fine honing stones should be used with honing oil. Coarse grit stones may be used dry.

The second design is called a Flex-Hone®. Flex-Hones resemble a large bottle brush with small, round grinding stones set at the ends of the bristles, figure 3-52. These tools are easy to use, and can produce a good ring-seating surface all the way to the very top of the cylinder bore.

Glaze breaker

The glaze breaker is a set of four stones mounted on a spring-loaded shaft, figure 3-53. A glaze breaker is used when re-ringing an engine. It removes the glaze from the cylinder walls and provides a fresh crosshatch surface.

Line boring equipment

After long use, the expansions and contractions that the engine block goes through may cause the main bearing bores to become misaligned. When out of alignment, the main bearings put bending forces on the crankshaft. Main bearing bores that are out of alignment will cause the bearings to wear out much faster than those in alignment.

Power and Specialized Engine Rebuilding Tools

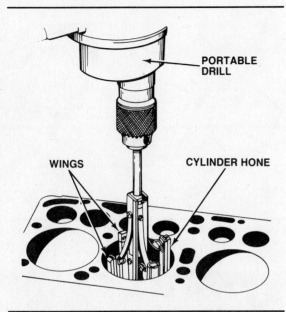

Figure 3-51. Hand hones center the abrasive stones in the bore with centering shoes.

Figure 3-52. Flex-Hones are easy to use correctly.

Figure 3-53. Use a glaze breaker to prepare the engine for new rings when wear does not require a rebore.

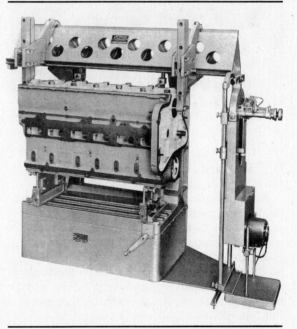

Figure 3-54. Main bearing bores can only be recut with a line boring machine.

Line boring equipment consists of a boring bar that can be centered in the two end main bearing webs, figure 3-54. To use it, you must first remove the main bearing caps and grind their bases to remove about 0.002 inch (0.05 mm). This reduces the size of the web hole slightly, so that the machine can bore back to the original standard size.

The disadvantage of this method is that the location of the crankshaft is changed. The crankshaft then sits slightly higher in the block. However, if you keep the cap grinding to a minimum, the difference in crankshaft location may be only a few thousandths of an inch, which should have no effect on engine operation.

Avoid excessive relocation of the crankshaft. It changes compression ratio, causes slack in the timing chain, and moves the crankshaft out of line with the seal in the front cover. It can also cause misalignment between the crankshaft and the transmission.

Align Honing
Align honing is a technique for correcting minor misalignments in the main bearing bores of engine blocks. Instead of a rigid boring bar, as used in line boring, an align hone uses a

Figure 3-55. The align-honing arbor uses spring-loaded stones to cut all the main bearing bores at once.

Figure 3-56. The machine rotates the cutting arbor automatically, but you must stroke it back and forth by hand.

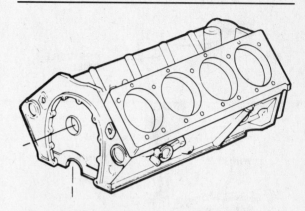

Figure 3-57. Torque plates bolt onto the cylinder block to simulate the distortion caused by a real cylinder head.

Figure 3-58. Torque plates reproduce the normal dimensional changes of an assembled engine during machining.

long arbor with spring-loaded honing stones set along its length, figure 3-55.

To align hone a block, you first grind a *small* amount of metal from the bases of each main cap with the same tool you use to grind connecting rod caps. Then you bolt the caps in place, insert the align honing arbor into the bores, turn on the machine, and stroke the arbor back and forth by hand, figure 3-56. Align honing removes only small amounts of metal, but typically that's all that is necessary.

Torque plates for boring and honing
A torque plate is a metal slab about two inches thick, and the same length and width as a cylinder head, figure 3-57. When bolted to the block, it sets up the same stresses and dimensional changes within the cylinders that will exist when a real cylinder head is attached to the engine. The torque plate has holes above each of the cylinders so that you can perform all the usual engine boring and honing operations with it bolted to the block, figure 3-58.

For many years, torque plates were used exclusively by high-performance and racing engine builders. Boring and honing with a torque plate was considered an excellent practice, but not really necessary for production engines. However, many modern thin-wall castings *require* a torque plate to be used with all boring and honing operations. These engines do not have the stiffness that heavier engines possess and so require a torque plate.

The torque plate is installed just like a cylinder head. The better torque plates have bolt towers — extensions around the bolt holes — that permit you to use the same head bolts or studs that you will install on the engine. You must use every head bolt, and tighten them to the same torque value as on the cylinder head.

Power and Specialized Engine Rebuilding Tools

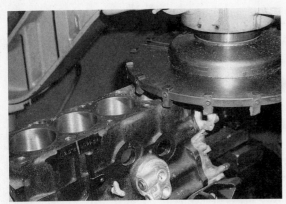

Figure 3-59. You can use the same grinder for both heads and blocks, if it fits into the machine.

Figure 3-60. If there is the slightest chance that you will reuse the harmonic balancer, use only the correct puller for removing it from the crankshaft snout.

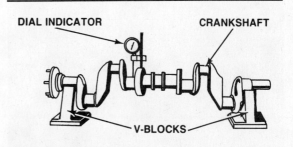

Figure 3-61. Set the crank up on V-blocks in order to test it for straightness.

Without this precaution, the bores will distort when the head is installed, and the rings won't seat against the cylinder walls.

Block surface mill and grinder
You can use the same equipment for surfacing blocks that you used to surface the cylinder heads, if the block fits into the machine, figure 3-59. The machines come with supports strong enough for the heaviest block that will fit in the machine, so support is not a problem.

Block resurfacing is similar to head resurfacing. Position the block so that you take the cut parallel to the crankshaft, and square with the cylinders. If the cut slants in any direction, it may cause trouble in assembling the engine. If the slant is extreme, it can cause interference between valves and pistons, and varying compression ratios between cylinders. Proper setup ensures that the new surface will be aligned with the cylinders and crankshaft.

Crankshaft Service Tools and Equipment

The crankshaft, after the block and cylinder heads, is probably the largest, heaviest and most expensive part of the engine. The various tools and equipment used to straighten and restore the crankshaft are described in this section.

Harmonic balancer puller
Harmonic balancers fit tightly on the crankshaft and require a special puller for removal, figure 3-60. Because of the rubber ring between the hub and outside steel ring, you will pull the balancer apart if you use a jaw type puller on the outer edge. To prevent this, the hub of the balancer has tapped bolt holes in it. Capscrews are threaded into the holes, and then you tighten the puller to pull on the capscrews while pushing on the end of the crankshaft. Some pullers handle both removal and installation.

Always use the correct puller to remove or install a harmonic balancer. It turns at crankshaft speed, and if the outer inertia ring separates from the inner hub, the damage to the engine and car will be impressive.

V-block and dial indicator
The first step in restoring a crankshaft is to check it for straightness. To do this you use a dial indicator and two V-blocks, figure 3-61. The V-blocks must be mounted firmly and high enough so that you can rotate the crankshaft.

The crank is supported at each end main journal by the V-blocks. The dial indicator is positioned near the center on one of the other main journals, and the crank is rotated. If the dial indicator moves as you turn the crank, it

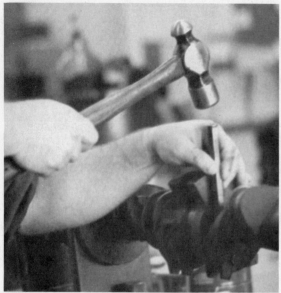

Figure 3-62. You can straighten a misaligned crank with some very careful blows against the fillets. Use a *dull* chisel.

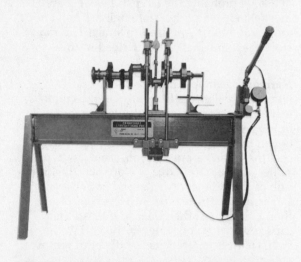

Figure 3-63. You can also use a specialized press to straighten a bent crankshaft.

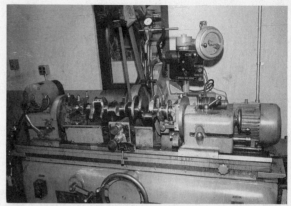

Figure 3-64. Crankshaft grinders rework the journals to an undersize or remove metal built up by welding.

indicates an out-of-round main journal or a bent crankshaft. Measuring the main crankshaft journal with a micrometer will quickly show if it is out-of-round. If the journal is good, then the crank is bent, and must be straightened, if possible.

You can straighten bent cranks in two ways. The simplest method is to use a round nose chisel and hammer, figure 3-62. A blow in the fillet radius, next to the main journal will straighten the crankshaft slightly. Succeeding blows will bring it into alignment.

You can also straighten the crankshaft in a press, figure 3-63. Either method can crack or break the crank if not done carefully. After straightening, grind the crankshaft to the next available bearing undersize, if necessary.

Crankshaft grinder
Rod and main journals will become worn after many miles of use or through neglect. The most common form of neglect is failure to change the oil. Particles of hard carbon and metal worn from other parts of the engine will damage the bearings. This is especially true if the filter is not changed also. It can become clogged and the oil will bypass it.

The crankshaft grinder restores worn journals by grinding away metal until the journal is perfectly round and smooth, figure 3-64. The grinder is actually a large lathe with special mountings for the crank. When grinding a main journal, the crank rotates around the shaft centers. A large grinding wheel spins at high speed. It is moved against the turning crankshaft until the shaft is ground to the next available undersize bearing.

Whenever you grind rod journals, you must offset the center of the crank so the entire shaft rotates around the center of the rod journal. Because each rod journal is in a different position, you must reposition the crank to a different mounting location for each one.

Hand grinders
After grinding the new journal surfaces, you can use a high speed grinder to put a slight radius on the oil hole in the journal, figure 3-65. This prevents the sharp edge of the journal from digging into the bearing.

Crankshaft balancing machine
Crankshafts turn at very high rpms, and the slightest out-of-balance condition can cause the entire assembly to vibrate itself to pieces. Crankshafts for two, three, four, five, six, and eight-cylinder engines must all be balanced in

Power and Specialized Engine Rebuilding Tools

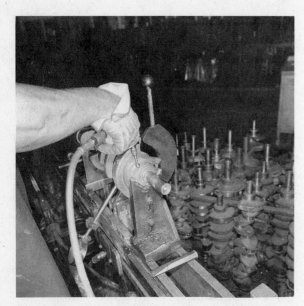

Figure 3-65. After grinding the journals, break the sharp edge of the oil holes by chamfering with a high-speed grinder.

Figure 3-67. A double-pan, double-reading scale lets you balance each end of the connecting rod separately and at the same time.

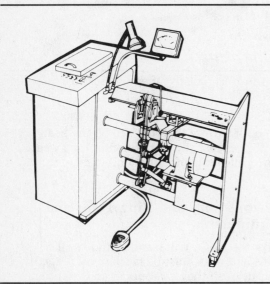

Figure 3-66. You can balance any of the rotating parts on a versatile crankshaft balancer, not just the crankshaft.

different ways in order to keep them running smoothly at the rpm of interest to the driver.

Balancing a crankshaft requires that you pay attention to both reciprocating and rotating weight. Reciprocating weight is that which moves up and down in the cylinder, such as the piston. Rotating weight is that which turns around the crankshaft axis, such as the crankshaft itself. The connecting rod both rotates around the crankshaft (at the big end) and reciprocates up and down the cylinder with the piston (at the little end). For practical purposes, you can consider that about half of the connecting rod weight reciprocates, and about half rotates.

A crankshaft balancing machine, figure 3-66, spins the crankshaft with weights attached to the journal equal to one half the weight of the reciprocating parts. Crankshaft imbalance is indicated on a meter or by a light which tells where the shaft is heavy. The operator then drills a hole in the counterweight to lighten the heavy side. If he has to add weight to the light side, he drills a hole and inserts a heavy metal plug in the hole. Lead plugs are the most common, but you can also use tungsten because it is slightly heavier than lead.

Balancing scale
Balancing the crankshaft requires that you know the specific rod and piston weight. All the pistons and rod assemblies must weigh the same. Different piston or rod weights would make it difficult, if not impossible, to balance the crankshaft.

The pistons are usually weighed to find the lightest one. Then you shave metal off the others in a lathe until they are all equal. Shave the pistons on the inside, around the pin bosses or on a heavy part of the skirt.

The easiest way to weigh rods is to use a double-pan scale which weighs each end of the rod separately, figure 3-67. You can use a single-pan scale if you support the other end of the rod in a sling so that the rod beam is at the same height as the pan. To balance the rod, grind metal from the ends until all the small ends weigh the same and all the large ends weigh the same.

These scales are very sensitive. They usually measure in tenths or hundredths of a gram, or to $1/32$ of an ounce.

Figure 3-68. Be careful when you polish crankshaft journals to avoid wearing flat spots in the surface.

Figure 3-70. Piston pin drifts come in many different sizes to accommodate different piston designs.

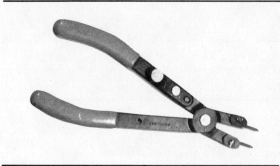

Figure 3-71. Snap ring pliers will remove most piston pin circlips.

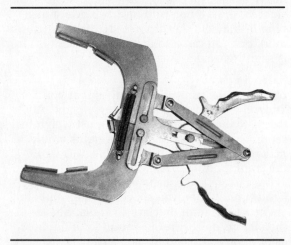

Figure 3-69. A piston ring expander prevents the rings from bending permanently by expanding them from the ends.

Crankshaft polisher

The surface left by the crankshaft grinder is too rough for long bearing life, and must be polished. The polisher rotates the crank, while the operator applies a belt sander to the journals, figure 3-68. The polisher removes very little metal. Its only purpose is to create a smooth surface that will run on the bearings with the least amount of friction and wear. The polisher is mounted on the crankshaft grinder.

Piston, Rod, and Ring Tools and Equipment

Like the block, crank and cylinders, the pistons and rods must be machined to exact specifications so they can operate most efficiently. These machining procedures use a number of different machines.

Ring expander

A piston ring expander, figure 3-69, is a tool that will spread the piston rings enough to allow them to slip over the head of the piston and into the grooves. Using this tool avoids deforming or breaking the rings.

Piston-pin drift

Piston pins can be either free floating or a press-fit into the small end of the rod. For pins that are a press-fit in the rod, a piston pin drift, figure 3-70, is used to remove them. The drift may also be part of the pressing fixture. Some pistons have snapring pin retainers that fit into grooves in the piston pin bore. Remove these with snapring pliers, figure 3-71.

Power and Specialized Engine Rebuilding Tools

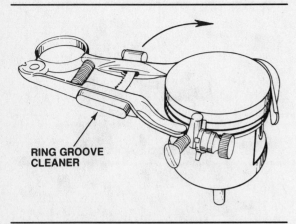

Figure 3-72. Use the ring groove cleaner very carefully to avoid damaging the groove surfaces.

More elaborate tool sets for pressing out piston pins include a fixture that mounts to the table of a hydraulic press, and includes various sizes of drifts and supporting inserts for handling a variety of piston pin sizes.

Ring-groove cleaner

You will often reuse pistons during an engine overhaul. Carbon collects in the ring grooves behind the rings and you must remove it or the new rings won't fit properly into the grooves. The piston ring groove cleaner, figure 3-72, is a hand tool with scrapers to fit different sizes of grooves.

In the past, engine rebuilders often cleaned piston ring grooves with a sharp piece of broken piston ring. This is *not* a satisfactory technique for a piston in good condition, because the sharp edges will scratch the inner surfaces of the ring grooves. These scratches will prevent the ring from forming a gas-tight seal during engine operation, and will reduce compression and increase blowby.

Piston knurler

A high-mileage engine may have considerable wear on the pistons, resulting in too much piston clearance in the bore, yet not be so worn that an overbore is absolutely required. The excessive clearance lets the piston rock in the bore, which will wear out the new rings, and cause piston slap. If you are rebuilding the engine, an overbore with new pistons will solve the problem. However, if the customer's budget requires you to reuse the pistons, you can bring them up to size with a knurler, figure 3-73. Situations such as this sometimes arise with expensive autos. For instance, new pistons for an older Porsche might cost $150 each. If the engine is already apart, but the old pistons are not quite worn out, knurling is certainly an alternative.

Figure 3-73. Knurlers can bring an undersized piston up a few thousandths to fit a slightly oversize bore.

A knurler is a special tool that expands the effective diameter of a piston skirt. To use it, you chuck the piston into the tool, and slowly turn it while forcing the hardened wheels on the knurler into the skirt of the piston. The wheels roll a series of small ridges into the piston skirt, which make the piston fit more tightly into the bore.

You can also knurl new pistons if they do not fit the cylinder bore. Knurling usually takes up a maximum of 0.006 inch (0.15 mm) on the diameter. The knurler will raise the metal more than that, but the fine points of the projections wear off quickly.

Knurling should be done with one pass of the roller. If the roller is passed over the skirt a second time, it may go on top of the previously raised metal and push it down again. After knurling, you can tailor the pistons to the proper clearance in each cylinder by lightly passing a file over the knurling. After each filing, use a micrometer to check piston diameter.

Remember that knurling a piston is an economy measure, and will never restore the performance or engine life as will a rebore with new pistons. Knurled pistons will retain their clearance for perhaps 50,000 miles (80,000 km), and after that the small ridges will be worn away. If you knurl pistons to fit cylinders that truly should be bored instead, the knurling will last only a few thousand miles.

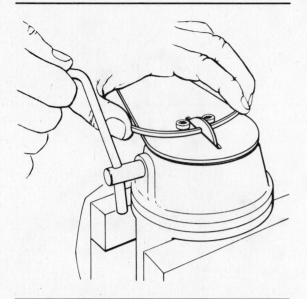

Figure 3-74. Ring end-gap grinders are quick and less likely to damage the ring.

Piston ring grooves also wear, and the cost of reconditioning the skirts and ring grooves so as to reuse a piston will eventually exceed the cost of new pistons and boring. When you realize that the price difference between engine rebuild kits with and without pistons is often only about $75, you can see why many shops do not recommend piston knurling.

Ring end-gap grinder

Piston ring manufacturers supply their rings in sizes that fit standard bores and the usual oversizes. These rings fit within the cylinder bore with a very small gap between the ends of the rings. If this gap is too narrow, increase it by filing the ends of the ring. While you can do this by hand, a ring end gap grinder does a much quicker and more precise job, figure 3-74. To use this tool, you simply place the ring flat on the grinding table and squeeze the gap closed onto the grinding wheel. Turning the grinding wheel with a small crank or a drill motor will quickly remove metal from the ends of the ring without damage. Some ring end-gap grinders also tilt up to 45° to let you grind beveled-gap rings correctly.

Rod honing machine

A rod honing machine consists of a motor that rotates a hone or fine abrasive stone, figure 3-75. The operator moves the piston or connecting rod back and forth the length of the hone. Periodically, he checks the diameter of the piston pin hole and rod on a gauge mounted on the machine. The hone removes such small amounts of metal with each pass that the

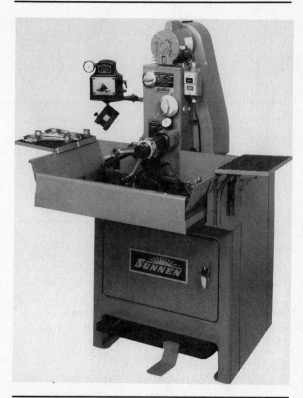

Figure 3-75. Rod honing machines recondition both the little and the big end of the connecting rod.

operator can make the hole fit the pin within a few ten-thousandths of an inch or hundredths of a millimeter.

Honing is done with lubricant flowing over the part and hone at all times. A tray under the hone catches the honing oil. The oil drains into the machine and is recirculated by a pump. The hone is operated by a foot switch, so the operator can use both hands to hold the part being honed. More elaborate honing machines have power stroking. Instead of the operator moving the part over the hone, the machine moves it.

Rod aligner

If a rod is bent or twisted, the piston won't move correctly in the cylinder. A bend in the rod that cocks the piston pin will cock the piston in the cylinder. This will put stress on the pin and the connecting rod bearing that will cause early wear and failure.

If a rod is twisted, the rotation of the rod around the piston pin or in the piston will force the piston into the cylinder wall, figure 3-76. This will cause early ring failure or piston wear.

Straighten bent and twisted rods on a rod aligning fixture, figure 3-77. A mandrel expands in the big end of the rod and holds it

Power and Specialized Engine Rebuilding Tools

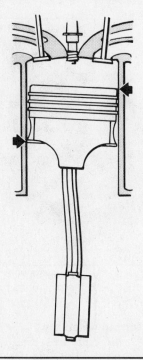

Figure 3-76. If the rod is twisted, it will force the piston into the cylinder wall as the crankshaft turns.

firmly. The small end of the rod must have either the pin or the piston installed. If only the pin is installed, the aligner checks the alignment of the piston skirt.

Checking the alignment is done by sighting between the pin or piston and a flat plate or bar, figure 3-78. Straighten the rod by inserting a bar into the hole in the piston pin and bending or twisting the rod. The fixture is strong enough so that you can bend the rod without removing it.

Some rod aligners also check offset. An offset rod is one which is bent in two directions. The two bends cancel each other so that no bend or twist shows up. But the piston is offset from the bottom of the rod. This defect can result in the big end of the rod rubbing against the sides of the crank. Offset is difficult to correct. An offset rod usually must be replaced.

Rod cap grinder

The inside of the large end of the connecting rod can be damaged if a bearing insert comes loose and spins. If an engine is subject to heavy use, the hole in the big end of the rod can stretch to an oval shape. In either case, you must bring the rod back to the standard size, so it will fit a normal bearing insert.

Do this by removing the bolts from the rod, and grinding about 0.002 inches (0.05 mm) from both the rod and cap parting faces to

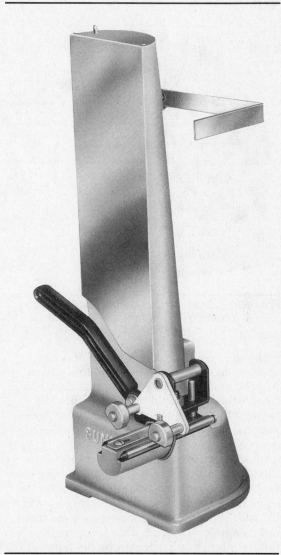

Figure 3-77. The rod aligner lets you check the connecting rod pin, or the piston skirt, depending upon which item is installed into the fixture.

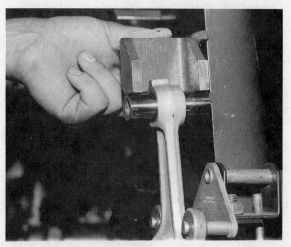

Figure 3-78. Check alignment by sighting between the pin or piston and a flat plate.

Figure 3-79. Grind the rod and cap parting faces to make the opening in the connecting rod smaller.

Figure 3-80. The rod heater heats only enough of the rod to allow you to push the pin out by hand.

make the hole smaller. Use a rod and cap grinder, figure 3-79, and then follow with a hone to enlarge the hole to the standard size. Center the hone so that it just touches the sides of the rod at the parting line, but hones material off the top and bottom of the hole to restore it.

Ideally, you should remove the bolts from the rods so that you can grind the mating surfaces of both the rods and caps. Some shops leave the bolts in, and take the entire cut off the cap. When they assemble and hone the rod, the hole moves up. This shortens the rod's center to center length by the amount cut from the cap, as much as 0.004 inches (0.10 mm). For a standard rebuild, this won't affect how the engine runs. But if a rod is rebuilt several times, the compression ratio of that cylinder will become progressively lower. Remember also, that each rebuild removes part of the bearing tang groove.

When both the rod and cap are ground to make the hole smaller, the rod is still shortened, but not as much. Since an equal amount of metal is removed from the cap and the rod, the hole moves up just half as much.

Rod heater for pin assembly

Because of the difficulty in setting up a hydraulic press, and the possible damage to the pistons, you will sometimes use rod heaters for assembling press-fit pins. The heater only heats the small end of the rod, figure 3-80. It heats it just enough to allow you to push the pin into the rod by hand. A small assembly jig ensures that you can't push the pin in too far.

After the pin is inserted into the rod and piston, it takes only a few seconds for the rod to cool and seize onto the pin. Then you can handle the rod and piston assembly without fear of dislodging the pin. Some rod heaters have two heat units. One rod can be heating while you assemble the other. This makes the job go faster.

Arbor press

Engines that use full floating piston pins usually have a bushing in the end of the connecting rod. If a hydraulic press is not available to push the old bushings out and the new ones in, you can use an arbor press, figure 3-81. The arbor press must be large enough to do the job. Some small arbor presses do not have enough leverage because their handles are too short.

Power and Specialized Engine Rebuilding Tools

Figure 3-81. Use an arbor press to push out old bushings and install new ones.

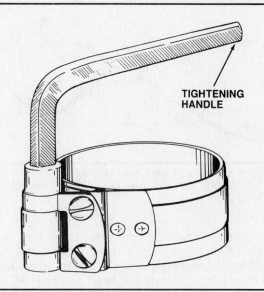

Figure 3-83. The most common piston ring compressors use wide spring-steel bands to compress the rings into the grooves.

Hydraulic press

A small hydraulic press, mounted on a table, is used to assemble pistons to press-fit connecting rods, figure 3-82. You can also use it to take them apart. The press usually has a capacity of about ten tons. A hydraulic ram does the actual pressing. The ram can be manually operated with a lever, or by an electric motor that drives a hydraulic pump. Pilots guide the pin into position.

Ring compressor

In order to install the pistons in the cylinders, you must compress the rings. A ring compressor squeezes the rings into their grooves. When you push the piston into the cylinder the rings pass into the cylinder in their compressed position.

Ring compressors come in a wide variety of designs. The traditional-style tools use a wide spring-steel band that you tighten around the entire piston with a key, figure 3-83. These tools come in various lengths so that you can cover all piston rings at once in a variety of piston sizes.

A second popular ring compressor design resembles an oil filter wrench, figure 3-84. These clamp around the piston ring using a narrower metal band, and are much quicker to open and close than the full-length styles.

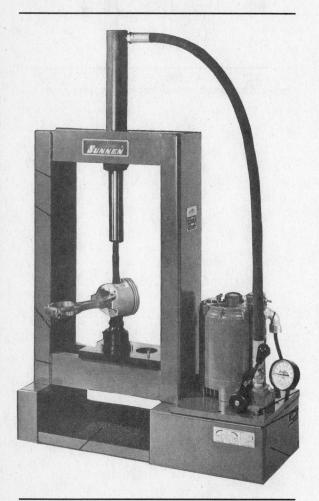

Figure 3-82. You can assemble and disassemble pistons and connecting rods with press-fit piston pins using a special-purpose hydraulic press.

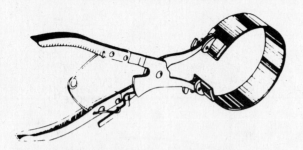

Figure 3-84. These quick-acting piston ring compressors open and close like an overcentering pair of pliers.

A final piston ring compressor is the simplest of all, figure 3-85. These are just metal collars that you slide over the piston before inserting it into the cylinder bore. The hole inside the collar is tapered, so that each piston ring is squeezed into its groove as you push the piston through the collar and into the block. The drawback of these tools is that you must have a separate collar for each bore size that you plan to work with.

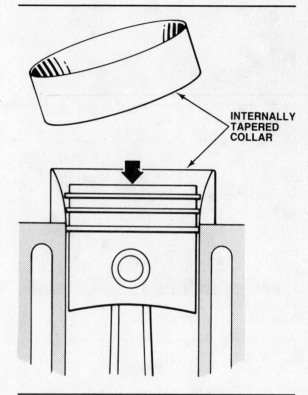

Figure 3-85. Professionals often use these simple metal collars, because they have no moving parts.

Chapter 4
Measurement and Measuring Equipment

If you intend to rebuild or repair an engine and have it last more than a few minutes after you first start it, then you must understand how to measure dimensions and parts accurately and precisely. You will frequently need to decide whether a part can be reused or not based on measurement differences as small as 0.0001 inch (0.002 mm). Your machining and assembly accuracy must often be accurate to 0.001 inch (0.02 mm). And you will most likely need to understand and work with these differences in both the traditional U.S. Customary and the Metric systems of measurement.

In this chapter, we cover measurement systems and the tools you will typically use in engine work. This information is general. More information on using these tools is in the individual chapters covering those phases of repair and rebuilding.

THE METRIC SYSTEM

The International Metric System is here to stay, and as an auto technician, you must be able to use it in order to service modern automobiles. In the past, service publications often gave specifications in both metric and U.S. Customary units, as a convenience to technicians who were still learning the metric system. While this is still generally true, it is no longer universal. Just as working on modern cars requires an investment in metric sockets, wrenches, and taps and dies, you as an automotive technician must invest the time necessary to understand those aspects of metric measurement that are important to your job.

Luckily, you need to know only five or six metric measures to work on almost any system. We discuss those you will need in engine repair and rebuilding in the next sections.

How The Metric System Works

The metric system is easy to use because of two main features designed into it:

- The prefix of the unit tells you how large it is.
- All units change magnitudes in multiples of 10.

This is similar to the way we count money in the United States: ten cents to a dime, ten dimes to a dollar, etc. For example, the basic unit of length or distance in the metric system is the meter, which is a distance slightly over three feet long. Dividing the meter into ten equal parts gives a unit of measurement called the decimeter, from the Latin *deci*, meaning one-tenth. Dividing a meter into one hundred

| STANDARD METRIC PREFIXES AND ABBREVIATIONS ||||
Prefix	Magnitude		Abbreviation
Giga	1,000,000,000	or billion	G
Mega	1,000,000	or million	M
Kilo	1,000	or thousand	k
Hecto	100	or hundred	h
Deka	10	or ten	da
Deci	1/10	or one-tenth	d
Centi	1/100	or one one-hundredth	c
Milli	1/1,000	or one one-thousandth	m
Micro	1/1,000,000	or one one-millionth	μ
Nano	1/1,000,000,000	or one one-billionth	n

Figure 4-1. All metric units use the same prefixes.

equal parts gives a unit called the centimeter, from *centi*, meaning one-one hundredth. Finally, dividing the meter into one thousand parts gives the millimeter, from the Latin *milli*, meaning one-one thousandth. Ten millimeters make one centimeter, just as ten centimeters make one decimeter, and ten decimeters make one meter. You will rarely see decimeters — the most common divisions are the centimeter and millimeter.

Latin prefixes indicate the subdivisions of a basic unit. Greek prefixes indicate multiples of that unit. For instance, the Greek prefixes for 10, 100, and 1000 are deka, hecto, and kilo. A distance of ten meters is one dekameter, one hundred meters is one hectometer, and one thousand meters is one kilometer. Only meter and kilometer are common.

The metric system uses these same prefixes, figure 4-1, to indicate the same basic divisions or multiples of all metric units of distance, weight, and time, as well as force, torque, power, volts, amps, etc.

For example, one kilometer (km) is a distance of one thousand meters (m). One milliamp (mA) is a current of one one-thousandth of an amp (A). One megaohm (MΩ) is a resistance of one million ohms (Ω).

Once you get used to it, the metric system's units of ten are easy to use. In the U.S. Customary system, we always had to remember units that changed in units of 4, 12, 16, 5280, and so on. To measure small quantities, we had to keep track of units that changed in units of 1/4, 1/8, 1/16, 1/32, and 1/64. In the metric system, working with any quantity is as simple as making change from a dollar bill.

Length, Volume, and Weight

Weights and measures are particularly convenient, because the metric system usually defines them in terms of each other. Remember that the basic unit of length in the metric system is the meter (m). The meter corresponds roughly to 39.37 inches. An *exact* relationship (by definition) is that one inch is exactly 2.54 centimeters (cm) long. As already mentioned, a millimeter (mm) is one-tenth of a centimeter, so there are exactly 25.4 mm in one inch.

The basic unit of volume is the liter, the space occupied by a cube one-tenth of a meter to each side. Dividing a liter into one thousand parts gives a milliliter (ml) or the equivalent volume, one cubic centimeter (cc). One quart is slightly less (0.946) than one liter. In Canada, motor oil is sold in liter containers that are slightly larger than the American quart cans or plastic bottles. One fluid ounce (a measure of volume, not weight) equals about 29.6 milliliters.

The basic unit of weight is the gram (g). One thousand grams make up the most commonly used next largest unit, the kilogram (kg). One one-thousandth of a gram is one milligram (mg). A kilogram is about 2.2 pounds. One ounce weighs about 28.3 grams.

Converting between the U.S. Customary and the metric system is straightforward, figure 4-2.

Torque

The metric measure of torque is the Newton-meter (Nm), which corresponds to 8.85 inch-pounds (in-lb) or 0.74 foot-pounds (ft-lb). To convert inch-pounds to foot-pounds, simply divide or multiply by 12: one foot-pound is the same as twelve inch-pounds. Many auto manufacturers now give torque specifications in both inch-pounds, foot-pounds, and Newton-meters.

Occasionally you will see torque specifications given in kilogram-meters, or even kilogram-centimeters. These are old metric ways of expressing torque. One kilogram-meter equals 1.5 ft-lb, and one kilogram-meter equals 100 kilogram-centimeters.

Measurement and Measuring Equipment

STANDARD CONVERSIONS FROM THE U.S. CUSTOMARY SYSTEM TO THE METRIC SYSTEM			
Description	**Unit**	**Multiply By:**	**To Get:**
length	inch	25.4	mm
	foot	0.3048	m
	yard	0.9144	m
	mile	1.609	km
volume	gallon	3.7854	l
	fluid ounce	29.57	ml
weight	pound	0.4536	kg
	ounce	28.3495	g
pressure	psi	6.895	kPa
vacuum	in-Hg	25.4	mm-Hg
torque	in-lb	0.11298	Nm
	ft-lb	1.3558	Nm
velocity	mph	1.6093	km/h
fuel performance	mpg	0.4251	km/l

STANDARD CONVERSIONS FROM THE METRIC SYSTEM TO THE U.S. CUSTOMARY SYSTEM			
Description	**Unit**	**Multiply By:**	**To Get:**
length	mm	0.0394	inch
	m	3.281	foot
	m	1.094	yard
	km	0.6215	mile
volume	liter	0.2642	gallon
	ml	0.0338	fluid ounce
weight	kg	2.2046	pound
	g	0.0353	ounce
pressure	kPa	0.1450	psi
vacuum	mm-Hg	0.0394	in-Hg
torque	Nm	8.8511	in-lb
	Nm	0.7376	ft-lb
velocity	km/h	0.6214	mph
fuel performance	km/l	2.3524	mpg

Figure 4-2. Conversions between the metric and the U.S. system can be read from this table.

Temperature

In America, temperature is measured in degrees Fahrenheit (°F), while in most other countries it is measured in degrees Celsius (°C). In the Fahrenheit system, water freezes at 32°F, and boils at 212°F. In the Celsius system (also called the Centigrade system in some older manuals), water freezes at 0°C and boils at 100°C. Converting is easy:

To convert Fahrenheit temperatures to Celsius:

$0.556 \times (°F - 32) =$ Degrees Celsius (°C)

For example, 45°F corresponds to 7.222°C, because:

$0.556 \times (45 - 32) = 7.222$

To convert Celsius temperatures to Fahrenheit:

$(1.8 \times °C) + 32 =$ Degrees Fahrenheit (°F)

For example, 10°C corresponds to 50°F, because:

$(1.8 \times 10) + 32 = 50$

Comparing U.S. Customary and Metric Measurements

When measuring parts, you can use either inch or metric measuring tools, but you may have to convert from one system to the other in order to use the manufacturer's specifications. The simplest way to handle metric specifications is to use metric-graduated tools, just like the metric wrenches in your toolbox. This avoids the problems of converting tolerance specifications.

Converting tolerance specifications

If the specification or sizes are given in metric units, and you use inch tools to do the measuring, you must convert your measurements

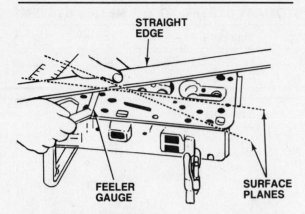

Figure 4-3. Use a straightedge with a feeler gauge to test flatness in engine castings.

to metric units. Because millimeters are so much smaller than inches, you must remember that tolerance specifications to the same number of decimal places are *not* interchangeable between inches and millimeters. One one-hundredth of a millimeter is 25 times smaller than one one-hundredth of an inch.

When you convert metric specifications to inches so that you can use your inch measuring tools, always keep one more decimal place in the inch value than you had in the millimeter value. Measuring to that level of accuracy will keep you out of trouble. When you convert inch specifications to millimeters so that you can use metric measuring tools, you can drop one decimal place. For example, these measurements are equivalent:

Metric	Inch
28.05 mm	1.104 inch
30.78 mm	1.212 inch
10.2 mm	0.40 inch
17.9 mm	0.71 inch
11 mm	0.4 inch
17 mm	0.7 inch

You can avoid these cautions by simply using metric measuring tools for metric specifications, which is by far the simplest, quickest, and most fool-proof solution. Manufacturers often give their specifications in tables that include the correct specifications for both systems, but that is not always the case, and it will only become more uncommon with time.

GENERAL MEASUREMENT TOOLS

Engine repair and rebuilding requires tools to measure dimensions, clearances, depths, and runout. Some are quite inexpensive, but a few necessities are fairly costly.

Straightedges

Cylinder heads and blocks heat up and cool every time the engine is started and then stopped. Over time, this cycle can warp the head and block mating surfaces, especially if the customer ever let the engine overheat. You can also cause these parts to warp if you disassemble the engine while it is still warm.

Always check head and block gasket surfaces for warping with a straightedge when rebuilding an engine. A straightedge is a thick, flat steel tool with at least one edge ground perfectly straight. Place this edge on the surface to be checked, and try to slip a thin feeler gauge under the straightedge wherever possible, figure 4-3. The size of the feeler gauge you can get under the straightedge tells you how much the cylinder head is warped. If the warping is severe, you must grind or mill it to make it flat.

Straightedges come in lengths from one to six feet. The least expensive are simply precision-ground stiff metal bars, but deluxe versions also come with a beveled and graduated edge.

Steel Rules

The simplest way to measure an object is with a ruler, and quality steel rules are available with graduations as fine as one one-hundredth of an inch, and one-half millimeter. You can't measure to any great accuracy with them, but they are useful for helping you select the correct size micrometer, for rough measurements of valve or piston diameter, valve spring length or installed height, or even the width of valve cutouts in a piston crown.

Steel rules range from one inch (2.5 cm) to 12 feet (3.7 m) long. One of the most useful versions is the small 6-inch (15-cm) pocket rules. They are machine-graduated and carefully surfaced to make the small graduations easier to read.

You can buy steel rules in virtually any combination of graduations you want. In U.S. Customary inch units, you can specify rules graduated in 12ths, 14ths, 16ths, 24ths, 28ths, 32nds, 48ths, and 64ths, in addition to the usual half, quarter, and one-eighth-inch tick marks. You can also obtain rules graduated in 10ths, 20ths, 50ths, and 100ths of an inch. In practical use, the one-hundredths scale is about as fine as you can read without a hand lens. Finer measurement than this requires a vernier or dial caliper.

Metric scales are usually graduated in half-millimeters, millimeters, centimeters, and meters.

Measurement and Measuring Equipment

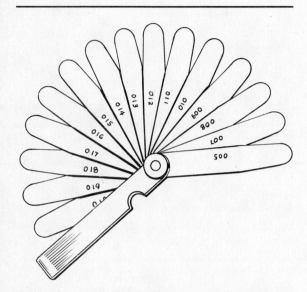

Figure 4-4. Blade-type feeler gauges are inexpensive and versatile.

Figure 4-5. Go-no-go gauges are especially useful for adjusting mechanical valve lifters.

Feeler Gauges

You use feeler gauges to measure clearances in a wide variety of situations, from piston cylinder wall clearance to crankshaft endplay. They can supply fairly exact measurements in the absence of other precision tools.

Blade-type

The most familiar feeler gauge is the blade-type, figure 4-4, in which a stack of thin metal leaves with precise thicknesses are bound into a protective metal handle. To use a feeler gauge, you insert various blades into the gap to be measured until you find the one that just fits. The thickness of the gauge then indicates the size of the gap.

To use a feeler gauge to *measure* a clearance, begin with a blade that you know is smaller than the gap, and work up to thicker sizes. The correct blade will have a light, sliding fit, one with which you feel a slight drag on the blade as you slide it back and forth in the gap. This is the technique you use to measure crankshaft endplay, or the existing valve clearance in a shim-and-bucket OHC engine.

When using a feeler gauge to *adjust* a clearance, such as a contact point set or the valve clearance on many OHV solid-lifter engines, you use a slightly different technique. Choose the correct size blade first, loosen any locking screws, and slip the blade into the gap. Then, while sliding the blade back and forth, you decrease the clearance with the adjuster mechanism until you feel a slight drag on the blade.

At that point, the clearance is correct and you can tighten the locking mechanism.

Feeler gauges in U.S. Customary styles commonly range in thickness from 0.001-inch to 0.025-inch in 0.001-inch increments. Metric feeler gauges typically range from 0.05 mm to 1.00 mm in 0.05-mm increments, and many gauges are marked in both systems. Thicker gauges of both types are available for special purposes, but you can usually just stack thinner blades to obtain the necessary measurement.

Go-no-go

Because using feeler gauges requires a certain skill in judging the amount of drag on the blade, some technicians prefer to use special versions, called go-no-go gauges, when adjusting clearances. These gauges have blades with a step separating two different thicknesses along the length of the blade, such as 0.007 and 0.009, 0.008 and 0.010, and so on, figure 4-5.

To use the tool, you begin with a thin blade, and insert progressively thicker ones until you find one in which the thinner tip can enter the gap easily, but the thicker base can't. The gap is then equal to the dimension between the two thicknesses of the blade.

Feeler gauge stock

You can buy feeler gauge stock in rolls up to 25 feet long for use in situations where you wear out the blade quickly, such as adjusting solid lifters in a running engine. This gauge stock is also useful for precise shimming of various components during assembly, such as when the camshaft support towers in an OHC engine must be raised slightly to compensate for resurfacing the cylinder head.

Figure 4-6. Non-magnetic feeler gauges let you measure clearances around magnetic components in distributors.

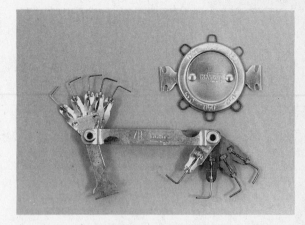

Figure 4-8. Wire-type feeler gauges for spark plugs come in several different styles.

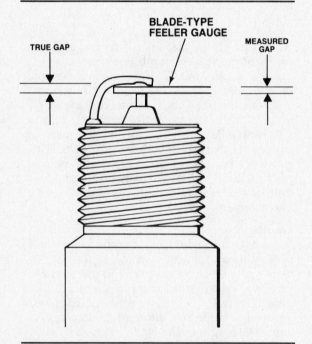

Figure 4-7. Blade-type feeler gauges read too low when you try to use them on spark plugs.

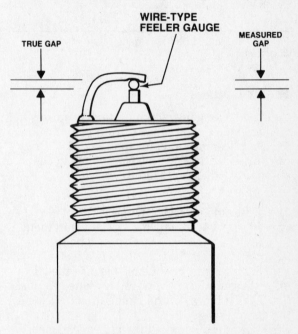

Figure 4-9. The wire-type feeler gauge fits snugly into the eroded gap between spark plug electrodes.

Non-magnetic

Sometimes you must measure a clearance between components that are magnetic, such as the air gap between an iron reluctor and a magnet inside an inductive distributor. An ordinary steel feeler gauge will be attracted to the magnet and will drag no matter what the gap is. For these situations you can use feeler gauges made of non-magnetic hard brass, figure 4-6.

Round-wire

Spark plugs require precise gaps between the center and the side electrode in order to prevent misfires. An ordinary blade-type feeler gauge can't be used for spark plugs, because the wide blade won't fit into the depression eroded into the side electrode, figure 4-7.

Wire-type feeler gauges, figure 4-8, use short loops or hooks of wire in precise sizes to fit into the confined space of a spark plug gap without giving readings that are too large, figure 4-9. They come in a variety of styles, and usually also have small notched bending attachments that you can use to bend the electrode while adjusting the gap. Some even have small wire brushes to let you clean the electrodes.

Measurement and Measuring Equipment 63

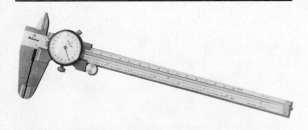

Figure 4-10. Dial or vernier calipers give you rapid measurements accurate to about 0.001 inch (0.02 mm).

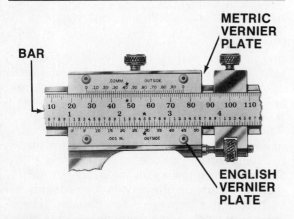

Figure 4-11. The vernier scale lets you measure to the final decimal place in a caliper — usually 0.001 inch or 0.02 mm.

Calipers

The next most precise tool for measuring dimensions after a machinist's ruler is the dial or vernier caliper. These tools aren't precise enough for measuring critical engine parts, but for many general purposes they are quite satisfactory.

Calipers are simple. They have a graduated steel beam with a fixed jaw, and a movable slide with another jaw and a vernier or dial readout, figure 4-10. Extensions on the opposite ends of the jaws let you use the same tool for either internal or external measurements. On many calipers, opening the jaws also extends a narrow depth measuring rod from the end of the beam. Most quality calipers read to an accuracy of 0.001 inch, or to 0.05 or 0.02 mm. Many also come with both U.S. Customary and metric scales, so you can use a single tool for measurements in either system.

Calipers come in several sizes and styles for specialized purposes. The usual sizes for most routine work are the 8-inch (200-mm) and 12-inch (300-mm) versions, but you can buy them

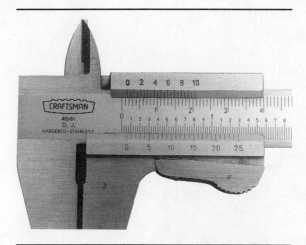

Figure 4-12. One of the graduations on the vernier scale will always align with one of the graduations on the beam.

up to several feet long. Typical uses for calipers include measuring pistons, valve springs, push rods, connecting rods, approximate bore sizes, or any task where accuracy to the nearest one-thousandth is sufficient.

You can also purchase electronic calipers with a digital display. These are the quickest tools to use, but they are also quite expensive.

Vernier calipers

This basic caliper style is named because it uses a vernier scale, figure 4-11. Vernier scales let you use a basic inch or millimeter scale and a carefully graduated slide to measure to a precision or perhaps ten to twenty-five times as fine as the basic scale. Vernier caliper beams are often marked directly to 0.025 inch, with the vernier scale providing increased precision to 0.001. Metric versions typically read to millimeters directly, and use the vernier to read to the nearest 0.02 mm.

You read a vernier caliper scale in the same way as the vernier scale described in the section on micrometers. The left edge of the vernier scale on the slide has a zero mark that aligns with the basic scale on the beam to give you the base measurement. If the zero mark aligns *exactly* with one of the divisions on the beam, then the last digit of the measurement will be zero. If not, one of the graduations on the vernier scale will align with one of the graduations on the beam, figure 4-12. Add that number from the vernier scale to the base measurement to get the total measurement.

Dial calipers

Dial calipers are much quicker to use than vernier calipers, and can give you the same precision. Their construction is similar, except that

Figure 4-13. Dial calipers can display the last one or two decimal places using an easy-to-read pointer and dial face.

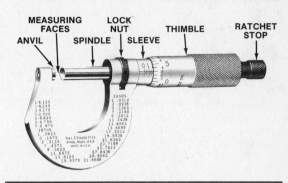

Figure 4-14. You will use outside micrometers regularly in all phases of engine rebuilding.

instead of a vernier scale they have a dial and pointer, similar to a clock face, figure 4-13. When you use the calipers, you read the inch and tenths of an inch directly from the beam. A delicate rack-and-pinion mechanism spins the pointer around the dial as you open and close the calipers, and the pointer gives you the thousandths measurement directly. One complete revolution of the pointer corresponds to precisely 0.10 or 0.20 inch. Metric tools may have the beam graduated in 1-mm divisions, with the dial tick marks every 0.02 mm.

Because they are so much quicker to use, dial calipers have essentially replaced vernier calipers as the standard tool for medium-accuracy measurement in engine repair and rebuilding. When you purchase one, buy a model that has the rack and pinion mechanism sealed. Open-rack calipers can jam on small metal particles lodged in the gears.

Micrometers

For precise measurements to finer levels of accuracy, you will use a micrometer. The standard micrometers you will usually have in the shop measure to 0.001 inch and to 0.01 mm. You can buy both U.S. and metric-reading micrometers that read to the next decimal place (0.0001 inch, or 0.001 mm), and you will need to if your work includes crankshaft measurements. They are considerably more delicate and expensive. These tools will give you the capacity to measure all of the parts you will find in an engine to the usual level of precision. These include critical measurements such as valve stems, bearings and journals, pistons and piston pins, camshaft bearing bores, lifter bores, and shims and thrust bearings.

Because they are essential in so many phases of engine work, micrometers and micrometer-like tools come in a wide variety of styles. There are micrometers for measuring inside dimensions, outside dimensions, curved surfaces such as engine bearings, screw threads, tubes, and many more. Some micrometers have interchangeable anvils so that you can use them for different purposes. Micrometer heads are available separately for situations in which a machine shop needs to build its own measuring tools for a specific purpose.

You will use outside micrometers for many measurements in engine work. Outside micrometers, figure 4-14, come in one-inch ranges. For example, a one-inch micrometer can measure objects from 0 to 1 inch in size, while a two-inch micrometer measures objects between one and two inches. Metric micrometers are sold in steps of 25 millimeters, such as 0 to 25 mm, 25 to 50 mm, and so on. The standard outside micrometers have anvils and spindles with flat surfaces. To measure the thickness of a bearing shell, you will need a micrometer with a ball attachment that can fit within the curvature of the shell, figure 4-15.

You can buy outside micrometers in sizes ranging from one to six inches (25 mm to 150 mm), but if you intend to work professionally in engine rebuilding, you should purchase a set containing all of these. You will use the smaller micrometers almost daily, and the two largest sizes when you work with larger engines, such as domestic big-blocks and diesels.

If you can't afford a complete set of micrometers, then you should probably purchase a one-inch micrometer first, and then a four- or

Measurement and Measuring Equipment

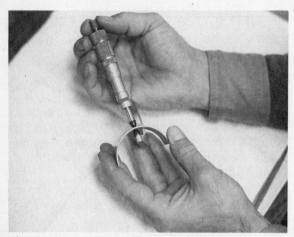

Figure 4-15. Ball attachments give the micrometer spindle a convex surface, which lets it fit into the hollow of a bearing shell.

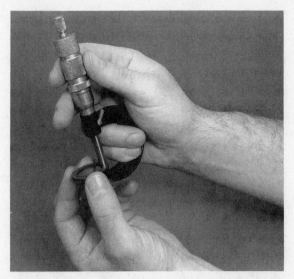

Figure 4-17. Rest the micrometer frame in the palm of your hand and rotate the thimble with your fingertips.

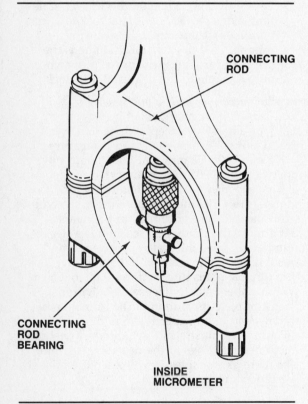

Figure 4-16. Inside micrometers resemble an outside micrometer without the frame.

five-inch micrometer to use in measuring pistons. After these two, buy the micrometer that lets you measure the crankshaft journals of the engines you usually work with.

You may be tempted to purchase a single one-size-fits-all micrometer that uses interchangeable anvils to measure any object between zero and six inches in size. These may seem like a bargain, but be aware that they are clumsy to use for the smaller measurements that are the most common. They also take more time to set up than does simply reaching for a different size.

Inside micrometers use the same measuring thimble and sleeve system as the outside micrometers just described, figure 4-16. Inside micrometers are useful for measuring cylinder, valve guide, and lifter bores, as well as the insides of crankshaft and camshaft bearings.

Using a micrometer

To use an outside micrometer, place the object to be measured between the anvil and spindle, figure 4-17. Turn the thimble to move the spindle into contact with the object being measured until you feel a certain pressure. This amount of pressure is important, and you must learn the right feel for it by practice.

Some micrometers have a ratchet or a friction stop built into them to provide the proper amount of pressure. To use these tools, turn the smaller end knob until you can feel it slip, or until the ratchet clicks. This is an easy way to maintain consistent accuracy in a production shop, especially if the tools will be used by different machinists, with different feels. Many professionals consider the feel you develop in your hands to be more accurate than a friction or a ratchet stop, however, and adjust their own personal micrometers to their own hands. If you don't use the friction or the ratchet stop, you can overtighten the tool and bend the frame.

Quality micrometers read to 0.0001, or the nearest ten-thousandth of an inch. By long-standing tradition, machinists in the United States refer to measurements made to ten-

Figure 4-18. The vernier scale on an inch micrometer reads to the 0.0001-range, or "tenths".

thousandths accuracy as so-many *thousands* plus so-many *tenths*. For example, if the micrometer reading is 0.2817 inches, a typical machinist would read this out loud as "two-hundred eighty-one *thousands*, and seven *tenths*".

Whenever you use a micrometer, don't forget to add the inch or millimeter capacity of the tool to the reading you get. For example, a two-inch micrometer that reads 0.7862 inches really indicates 1.7862 inches.

Reading a vernier inch micrometer
A vernier inch micrometer reads directly to 0.001 inch, and the more expensive versions use the vernier to read to the "tenths", or the nearest 0.0001 inch, figure 4-18. The sleeve has a graduated scale, numbered every 0.100 inches. The smaller lines indicate 0.025 inch divisions between them. The thimble graduations read directly from 0.001 to 0.025, with numbers marked every five divisions. Each complete revolution of the thimble moves the measuring face exactly 0.025 inch.

Figure 4-19 shows three different micrometer settings. Use these as examples as you learn how to read the micrometer. Follow this simple procedure:

1. Close the micrometer gently on the object to be measured. Lock the spindle in place and carefully remove the micrometer from the workpiece.
2. Inspect the numbered lines on the graduated scale of the sleeve to find the last complete 0.100-inch unit. It will be a number between 0 and 0.9 inches.
3. Inspect the tick marks between the numbered lines to find the last complete 0.025-inch unit. It will be either zero, 0.025, 0.050, or 0.075 inches.
4. Inspect the scale on the thimble to find the last complete 0.001-inch unit. It will be a number between zero and 0.025.
5. Add them all together. The sum is the micrometer reading to the last complete 0.001-inch.

If the micrometer has tenths capacity, you must now use the vernier scale to obtain the last 0.0001-inch, or "tenths" unit.

6. Locate the vernier scale on the sleeve. It is numbered from zero to ten, or perhaps from zero through nine to zero again.
7. Locate the single line on the vernier scale that aligns perfectly with any of the lines on the 0.001 through 0.025 scale on the thimble. The number for that line is the last digit for your measurement, which is now accurate to the nearest 0.0001 inch.

Reading a vernier metric micrometer
You read vernier metric micrometers in basically the same way except for the different graduations. Vernier micrometers measure to 0.001 mm or 0.002 mm, but you should buy one with the 0.001 mm accuracy.

A metric micrometer reads directly to 0.01 mm, figure 4-20, and the more expensive versions use the vernier to read to the nearest 0.001 mm. The graduated scale on the sleeve is numbered above the scale every 5 mm, and marked below the scale with 0.5 mm divisions. The thimble reads directly from 0.01 to 0.50, with numbers marked every five divisions. Each complete revolution of the thimble moves the measuring face exactly 0.5 mm.

Figure 4-21 illustrates three different metric micrometer settings to the nearest 0.001 mm. Use them as examples as you learn the following procedure:

1. Close the micrometer on the object to be measured. Lock the spindle in place and carefully remove the micrometer from the workpiece.
2. Inspect the numbered lines on the graduated scale of the sleeve to find the last complete millimeter unit.
3. Inspect the tick marks below the graduated scale to find the last complete 0.500-mm unit: either zero, or 0.500.
4. Inspect the scale on the thimble to find the last complete 0.01-mm unit. It will be between zero and 0.50.

Measurement and Measuring Equipment

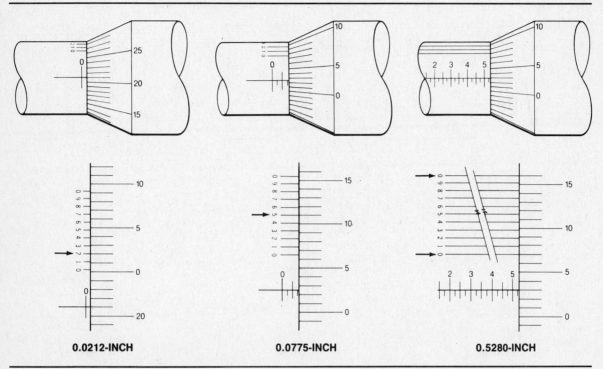

Figure 4-19. These three micrometer settings correspond to 0.0212 inch, 0.0775 inch, and 0.5280 inch.

Figure 4-20. This metric micrometer lets you measure to the nearest 0.01 mm.

5. Add them all together. The sum is the micrometer reading to the last complete 0.01-mm.

If the micrometer has thousandths capacity, you must now use the vernier scale to obtain the last 0.001-mm unit.

6. Locate the vernier scale on the sleeve. It is numbered from zero through ten, or perhaps zero through nine to zero again.
7. Locate the single line on the vernier scale that matches any of the lines on the 0.01 through 0.50 scale on the thimble. That number is the last digit for your measurement, which is now accurate to the nearest 0.001 mm.

Digital micrometers

Digital micrometers can be quicker to use than vernier micrometers. They use a digital display similar to an odometer to read the first digits of your measurement, figure 4-22, although you must still read a vernier scale for the final decimal place.

Be sure to check a digital micrometer regularly to make sure that the display reads accurately.

Caring for Your Micrometers

Always remember that your micrometers are delicate, precision instruments, and that they can be permanently ruined if you drop or overtighten them. If you do drop one, inspect it right away for damage. Sometimes you can readjust them, but if not, you should return it to the manufacturer for repair.

Keep your micrometers clean and lightly oiled with a high-grade instrument oil. When you purchase the micrometer, be sure to get a hard, protective case. Most micrometer sets already come boxed. Keep the micrometer in the case unless you are using it. If you have to set it down outside the box, place it on a rag or some other soft surface.

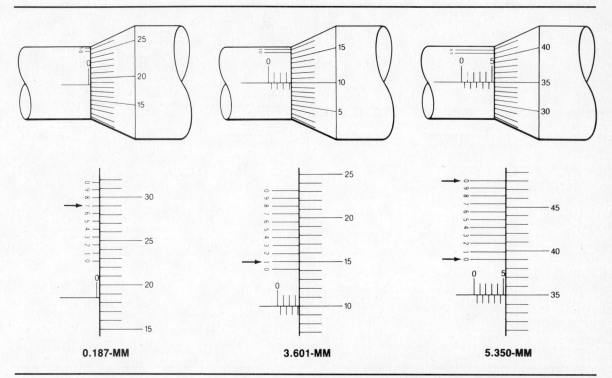

Figure 4-21. These three micrometer settings correspond to 0.187 mm, 0.601 mm, and 0.350 mm.

Figure 4-22. Digital micrometers use a readout to display the first digits of your measurement.

Inspect your micrometers regularly to make sure they are accurate. Zero-to-one inch and zero-to-25 mm micrometers are easy to check:

1. Clean the faces of the spindle and anvil by closing the micrometer lightly on a clean piece of paper, such as a business card, and gently pulling the card out.
2. Close the micrometer jaws with normal pressure and inspect the reading. It should be zero; that is, the zero mark on the thimble should line up with the graduated scale on the sleeve. If not, follow the manufacturer's directions to adjust it to zero.
3. Open the micrometer all the way, and insert a one-inch (or a 25-mm) setting standard.
4. Close the micrometer jaws, and inspect the reading again. It should be exactly one

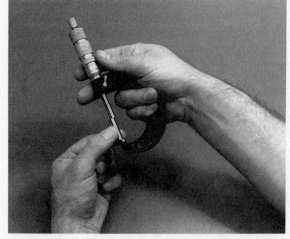

Figure 4-23. Setting standards let you test the accuracy of micrometers with jaws that do not meet.

inch (or 25 mm). If the micrometer has been overtightened so that the frame is bent out, the reading will be more than that. If the reading is less, the micrometer frame is bent inward, probably from being dropped.

Larger micrometers come with larger setting standards that you can use in the same way to test their accuracy, figure 4-23. Treat these standards with the same care that you give the micrometer itself.

Measurement and Measuring Equipment

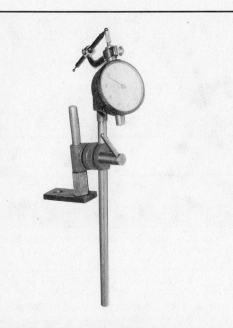

Figure 4-24. This dial indicator measures the travel of a plunger that contacts a moving part.

Figure 4-25. The dial indicator must be firmly attached to the workpiece. Any slack or movement will cause inaccuracy.

Dial Indicators

You may not use your dial indicator as often as you use your micrometers, but you can't do without this essential tool in precision engine work. Dial indicators measure movement, and are essential for checking valve lift, valve clearance, valve-to-valve guide clearance, crankshaft and camshaft run-out, crankshaft end play, and a host of other dimensions.

Dial indicators make it easy to locate exact top dead center if you need to replace a harmonic balancer and want to check the timing marks. You will use a dial indicator extensively if you install a high-performance camshaft in

Figure 4-26. Set the indicator to zero by rotating the face.

one of your customer's engines, not only to determine top dead center, but to degree the camshaft, and to check for valve-to-piston interference.

Finally, you will use the dial indicator repeatedly during engine machine operations, such as boring, honing, and deck refinishing, both to set up the workpiece and to check your work.

Dial indicators are very simple, figure 4-24. They consist of a dial face with a pointer that is moved around the dial by a spring-loaded plunger or a small toggle lever. The amount of movement of the plunger or lever is indicated on the dial in either inches or millimeters. Toggle-style dial indicators are most often used with lathes. The plunger style of dial indicator is the kind you will commonly use.

To use a dial indicator, you attach it to its stand and fix the stand firmly to the workpiece, or to a rigid bracket or table holding the workpiece, figure 4-25. You then carefully adjust the stand so that the plunger or lever touches the workpiece. Most dial indicators have a movable face so that you can rotate the zero marker to the exact tip of the pointer, figure 4-26. At this point, any movement of the workpiece or the dial indicator will register exactly on the dial face as the plunger or lever follows the surface of the workpiece.

Dial indicators differ in graduations and plunger travel. One of the most useful measures in thousandths of an inch, with one revolution over one inch of plunger travel. Others have ranges of several inches. A metric dial indicator might register a single millimeter over an entire revolution of the pointer. You can purchase dial indicator faces that read continuously around the dial, or that begin at zero in both clockwise and counterclockwise directions, figure 4-27. The latter are very useful for measuring runout.

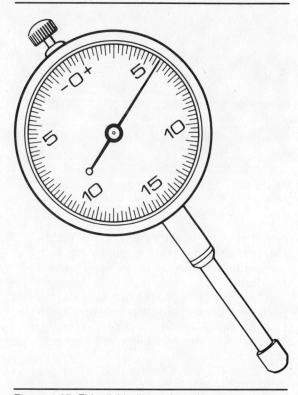

Figure 4-27. This dial indicator faceplate measures deviations by reading both ways.

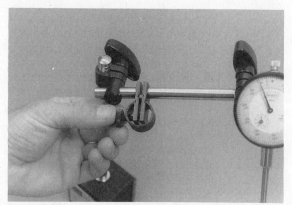

Figure 4-28. Adjusting this thumbscrew moves the dial indicator slightly to help you zero the needle.

Dial indicator stands
For a dial indicator to read accurately, it must be held rigidly in place. Most indicators have a small lug projecting from the back of the housing so that you can attach it to a stand, but many different backs are available for different purposes.

Dial indicator stands come in many forms. Some clamp in position using a small C-clamp or specially-made locking pliers. Others use

Figure 4-29. Degree wheels are essential for accurate valve timing.

permanent magnets with on-off levers to hold them to steel or iron pieces. The stands will often have flexible joints of some sort so that you can position the dial indicator in different positions, and will often have a micro-adjuster of some kind for fine changes, figure 4-28.

Dial indicator accessories
Dial indicators are versatile tools, and there are different accessories for almost any purpose you can think of. You can buy plunger extensions, replacement or exotic-shaped contact points, different faceplates, and many other parts. One measuring-tool manufacturer lists almost five hundred separate part numbers in over thirty pages of its catalog, just for dial indicators and accessories alone.

Degree wheels
Degree wheels are essential for precision valve train assembly. They consist of a simple metal or plastic disk, figure 4-29, graduated in degrees around the circumference. To use a degree wheel, you bolt it to the end of the crankshaft snout, set the engine at top dead center, and then attach a pointer to the engine so that it points exactly at the zero mark on the degree wheel, figure 4-30. You can then rotate the engine and read the exact degrees of rotation from the scale on the wheel.

You use a degree wheel primarily to check valve timing during engine assembly or blueprinting, and usually in combination with a dial indicator. For instance, if the camshaft specifications call for the intake camshaft lobe to reach maximum lift at 111 degrees after top dead center, you can turn the engine to that position using the degree wheel, and check the lift with a dial indicator on the lifter.

Measurement and Measuring Equipment

71

Figure 4-30. Position the degree wheel so that zero is near your pointer, and then bend the pointer to adjust it exactly.

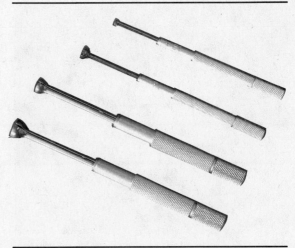

Figure 4-32. Small-hole gauges use a split-ball mechanism to contact the inside of the hole.

Figure 4-31. Telescoping or snap gauges must be measured with an outside micrometer to achieve the final measurement.

Telescoping Gauges and Small-Hole Gauges

Telescoping gauges, figure 4-31, are sometimes called snap gauges because the contacts are spring loaded and will snap out when released. They are less expensive than inside micrometers, but serve the same purpose. To use them, you retract the measuring faces and insert them into the hole to be measured. You then loosen the lock screw, and a spring extends the faces against the work. Lock the gauge by turning the handle, and rock it to make sure it is at the largest diameter. If not, you must loosen the handle and retighten the contacts. Then remove the gauge in the locked position and measure it with an outside micrometer. Measurements made this way are called transfer measurements.

Although telescoping gauges can be used to measure cylinders several inches in diameter, they are most useful for measuring small cylinders in which your hand holding an inside micrometer won't fit. You can buy sets of telescoping gauges that can measure inside diameters between 5/16 and 6 inches (7.9 to 150 mm).

Small-hole, or ball gauges are similar to telescoping gauges in function, but use a different mechanism to provide an inside measurement, figure 4-32. These tools can measure holes even smaller than those that telescoping gauges can fit. Typical sets of small hole gauges can measure holes ranging between 0.125 and 0.500 inches (3.2 to 12.7 mm). These are useful for measuring valve guide inside diameters.

A small-hole gauge consists of a split ball, or two convex, spoon-shaped blades, mounted on a narrow handle. To use it, you insert the tool into the hole to be measured, and then turn a knurled knob on the top of the tool. This pulls a small wedge up between the two halves of the split ball, and expands it into the sides of the bore. Like a telescoping gauge, you must then remove the tool from the work and measure it separately with a micrometer. Both telescoping gauges and small-hole gauges require two separate operations to make the transfer measurement: one to adjust the gauge, and another to measure them with a second tool. They are more prone to errors than inside micrometers unless you use them very carefully.

FASTENER TORQUE

You will often find that a nut or bolt must be tightened a specific amount. This is true of *all* fasteners, but it is particularly important for critical ones such as cylinder head bolts or connecting rod nuts. On any part, tightening beyond the specifications can stretch the fastener beyond its elastic limit, also called its yield point. This can break the fastener, strip the threads, or distort the parts you want to fasten together.

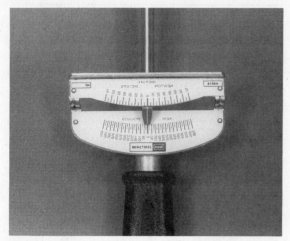

Figure 4-33. Modern torque wrench scales usually list inch-pounds or foot-pounds, as well as Newton-meters.

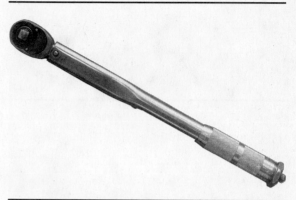

Figure 4-34. Click-type torque wrenches are easy to use, but must be recalibrated periodically by the manufacturer.

When you tighten a bolt to the correct torque setting, what you are actually doing is stretching the bolt an exact amount. Metal is elastic — you can stretch it a little, and it will spring back to its original length. If you stretch it too far, however, you will exceed its yield point and the bolt will never spring back completely. It will also be weaker than it was before.

Manufacturers keep these limits in mind when they specify torque settings. The correct torque settings are high enough to stretch the fasteners, but not so high that you permanently distort them. By tightening to the correct torque setting, you get the maximum clamping power from the fastener, and you can safely use the fastener several times.

You can tell when you overtighten a fastener. As you gradually tighten the bolt or nut with the wrench, the torque that you need to keep turning it increases evenly. As soon as you exceed the yield point, however, the effort you need to turn the wrench stops rising, and the wrench becomes easier to turn. Any more tightening weakens the bolt, and will eventually break it.

Torque Settings

Manufacturers publish correct torque values for virtually all the fasteners on the engine. The chances are that if you have to tighten the fastener with anything besides your hand, there is a torque value for it. When the manufacturer doesn't specify a specific torque value for a bolt, then you can find the maximum torque value by looking at the bolt itself. Manufacturers grade bolts according to their tensile strength, and all but the lowest quality fasteners have marks on the bolt head to tell you what those grades are. We explain these marks in the *Classroom Manual*. A particular bolt diameter within a particular grade has a certain standard torque setting. This brings up another issue with torque settings. If you replace the original fasteners on an engine with new ones, be sure to use fasteners of the same or higher grade as the original. If you use low grade bolts to replace high grade originals, then you will overtighten the bolts if you use the original torque setting, or undertighten the joint if you use the correct torque setting for the cheaper bolts.

Always look up the torque settings before you assemble an engine, and refer to them frequently as you work.

Torque Wrenches

You obtain proper fastener tightness by using a torque wrench. A torque wrench is one of the measuring tools that you will use over and over in engine repair and rebuilding. Most have a scale on them that indicates how tight you are twisting the nut or bolt, figure 4-33. The scale may be graduated in inch-pounds, foot-pounds, Newton-meters, kilogram-meters, or a combination of these values depending on the wrench. The more expensive click-type torque wrenches are different, figure 4-34. To use these, you rotate a collar on the wrench handle until the zero line on the collar lines up with the correct number on a scale. When you use the wrench, it "clicks" when you reach that torque setting.

Measurement and Measuring Equipment

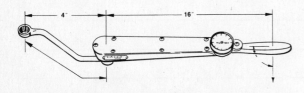

Figure 4-35. Anything that makes the wrench longer between the fastener and handle makes the wrench read too low.

You can also use torque wrenches with extensions or adapters to get around obstructions. Any combination of adapters that changes the length of the drive-end of the wrench will make the wrench read incorrectly. They will increase the leverage, figure 4-35, and apply more torque than the wrench indicates. Whenever you use an adapter that lengthens the drive end of the wrench, use this formula to determine what the reading should be for the actual torque you want:

$$\frac{\text{Wrench}}{\text{Reading}} = \frac{\text{Torque at Fastener} \times \text{Wrench Length}}{\text{Wrench Length} + \text{Adapter Length}}$$

If, for example, you wanted to tighten a bolt to 30 foot-pounds using a 16-inch wrench with a 4-inch adapter, you would calculate it this way:

$$\frac{\text{Wrench}}{\text{Reading}} = \frac{30 \text{ foot-pounds} \times 16 \text{ inches}}{16 \text{ inches} + 4 \text{ inches}}$$

or

$$24 = \frac{480}{20}$$

Therefore, to get 30 foot-pounds on the fastener using a 16-inch torque wrench with a 4-inch extension, you would pull on the wrench until it indicated 24 foot-pounds. At that point the fastener would be tightened to 30 foot-pounds.

Extensions that simply increase the height of the wrench above the bolt or nut won't affect the torque reading, *only if they are strong enough not to twist*. A long, thin 3/8-inch extension acts very much like a torsion bar spring, and will absorb some of the torque from the wrench without transferring it to the fastener.

You can test this for yourself quite easily. Take a long extension from your tool box, and clamp the end of it in a bench vise. Then put your torque wrench on the extension and try to turn it. Depending on the length and thickness of the extension, you can get a reading on the torque wrench of as much as ten or twenty ft-lbs (15 to 30 Nm), without having moved the bottom of the extension at all. If the torque wrench had been attached to a cylinder head bolt instead of the bench vise, you would have undertightened the bolt by the amount the extension deflected.

When you must use an extension to increase the height of the wrench above the fastener, try to use a half-inch drive extension, and keep it as short as you can. Some tool manufacturers can supply charts that give you correction factors to use with their torque wrenches when you have to use a long extension.

Torque-to-Yield Fasteners

Torque-to-yield fasteners are very different. You do not tighten these bolts to a certain torque setting and stop. Instead, you tighten them to a specific intermediate value, then turn the wrench through an additional angle to get the correct clamping force. Torque-to-yield cylinder head bolts are common in modern engines.

Up to about 25 ft-lb (35 Nm) of torque, there is little bolt-to-bolt variation in clamping force between the bolts in a set. Beyond this amount, tiny variations between bolts and threaded holes become magnified because of worn or damaged threads, small dirt particles, or oil left in some holes and not others. The actual clamping force exerted by bolts tightened to the same apparent torque wrench settings can vary by as much as 200 percent, figure 4-36. However, the yield points of a set of identical bolts doesn't vary nearly as much. After you tighten a bolt to the yield point, you can rotate the bolt an entire turn or more without increasing the clamping force. By tightening a set of bolts to a tension above their yield points, you can make sure that the clamping power is much more even.

You must tighten torque-to-yield bolts using the exact method the manufacturer specified. Typically, you tighten the bolts in three steps, up to the recommended torque in the shop manual (the yield point). Then you carefully tighten each bolt an additional amount, such as 1/4 turn, or 90°, while ignoring the torque wrench reading. Some import manufacturers are very specific. One requires an additional turn of 114°, which means you must use a special protractor that fits on top of the torque wrench.

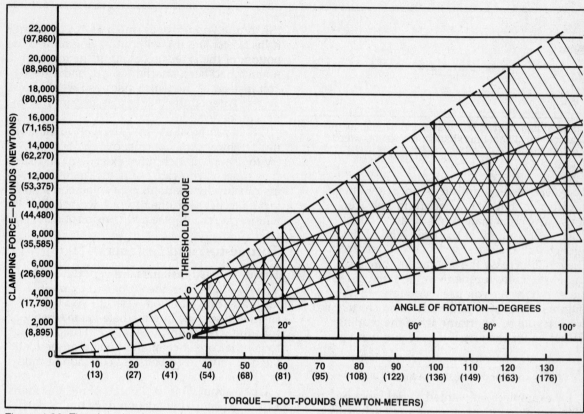

Figure 4-36. The clamping force exerted by bolts tightened to the same apparent torque wrench settings varies substantially.

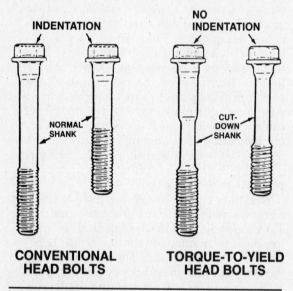

Figure 4-37. Torque-to-yield bolts sometimes look different from standard bolts.

Follow the manufacturer's recommendations for replacing these bolts. Some manufacturers allow you to reuse the bolts once, but others state that you must replace the bolts any time you remove them. Check the shop manual very carefully, and never reuse a head bolt if the manufacturer says not to do so. If you have any doubt, replace the bolt.

If you are unsure whether the engine uses torque-to-yield bolts, here are some clues to look for. Normal cylinder head bolts often have an indentation in the head. Torque-to-yield head bolts do not, figure 4-37. Some torque-to-yield head bolts have a cut-down shank, although many do not. Finally, some torque-to-yield bolts have a spiral pattern cut into the bolt shank. Again, if you are unsure, check the shop manual.

Chapter 5

Obtaining Specifications, Manuals, Engines, and Kits

Engine repair and rebuilding requires more than hand tools and new parts — you also need information. Before you can even buy replacement parts, you must identify the engine correctly, learn the correct disassembly sequence, and obtain the specifications for the parts you will inspect.

Once you do begin work, you must then decide just what parts the engine needs to return it to satisfactory condition. For instance, will new rings restore compression, or should you install oversize pistons? Your answer to this question dictates whether you should obtain a gasket kit and rings, or buy an entire crank kit with all new reciprocating parts. You might decide that the best solution is to skip repair entirely and install a short block, long block, or a complete remanufactured assembly.

In this chapter, we will explain where you get the information to help you make these decisions, how to use it to guide your planning, and how to obtain parts once you begin.

MANUFACTURER'S SPECIFICATIONS

One of the most important tools that you will use when repairing or rebuilding an engine is your ability to locate and read manuals and specifications. There are many different printed information sources. We will begin with the basics, and move on to more specific information later in this section.

Identifying the Engine

First things first. Before you can start, you must know exactly what kind of engine you have. If your customer drives the car into your shop and tells you that the engine has never been worked on before, then your job is easy. You can identify the engine quickly by interpreting the VIN and engine codes. Often, however, the engines that you rebuild will arrive in the back of a pickup truck, and you might have no idea what year or model car the engine came from. Sometimes the engine will already have been rebuilt once, and the heads, manifolds, block, and inner components may have come from different cars. For these engines, you may have to interpret casting marks and numbers on the heads or block.

Using VIN codes
Engines differ, and sometimes the differences are not obvious. For example, the camshaft bearings for a 1956 small block Chevrolet won't work in a 1957 model, and the crankshaft seals for the 1985 small block won't fit the 1986. Parts for a Ford 351 Windsor won't

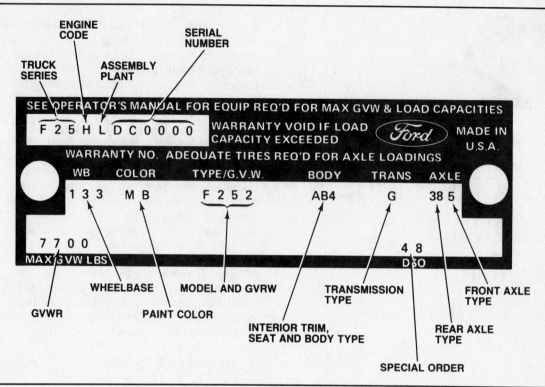

Figure 5-1. Once decoded, the VIN provides you with much detailed information about the vehicle and its equipment.

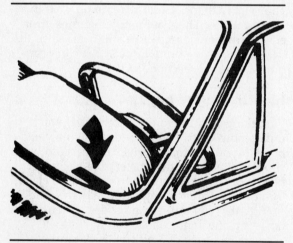

Figure 5-2. You can always find the VIN on a modern car by looking through the windshield at the dashboard above the steering wheel.

fit a Ford 351 Cleveland. The rings you buy for a 1335 cc Honda won't fit the pistons in the 1342 cc engine.

Bearing and piston clearances, valve grinding information, and many other specifications differ between engines that appear very similar. Unless you can positively identify the engine you work on, you may easily waste time and money on the wrong parts, or spoil an otherwise satisfactory rebuild by making an incorrect adjustment. To keep this from happening, manufacturers always supply VIN and engine codes. With these, you can conclusively identify the engine in order to use the correct specifications and purchase the correct parts.

The VIN is the Vehicle Identification Number. Manufacturers assign every car they build a unique number that you can decode to get specific information about that car, figure 5-1. This number usually appears on a small plate on the top left corner of the dashboard, where you can read it through the windshield, and in some spot under the hood or inside a door jamb as well, figure 5-2. Domestic carmakers use the same number for engine and transmission serial numbers, and stamp it onto the engine block and transmission during assembly. Many manufacturers now attach a bar code containing the VIN to the same plate.

The most important information you can extract from the VIN is the model year, and often the engine code, discussed below. All the manufacturers use different coding systems, however, so you will need a shop manual, an aftermarket specification guide, or sometimes just the owner's manual to decode them. The engine code is a combination of numbers or letters (sometimes a single number or letter) that identifies the number of cylinders, the en-

Obtaining Specifications, Manuals, Engines, and Kits

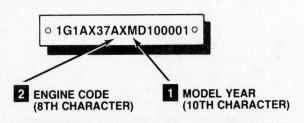

Figure 5-3. The VIN often includes the engine code in a specific position.

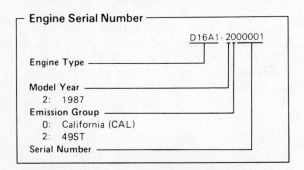

Figure 5-4. Many import manufacturers use a separate numbering series for their engines, with the engine code embedded in it.

Figure 5-5. The metal tag identifies this engine as a 1972 Ford 351 CID V8, manufactured in Cleveland, Ohio.

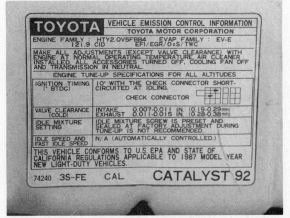

Figure 5-6. The VECI sticker will be placed in some highly visible spot underneath the hood.

gine displacement, the fuel system type, and the horsepower.

Engine and calibration codes

All domestic and many import car manufacturers insert the engine code into a specific position of the VIN, figure 5-3. Some import manufacturers, however, use a separate engine numbering sequence, with the engine code in a specific position in the engine number, figure 5-4. The engine number may be stamped on the engine.

Some manufacturers also attach tags or stickers directly to the engine. Ford Motor Company and AMC formerly attached a metal identification tag to the engine block, manifold, or head that listed the engine code, date of manufacture, and other information, figure 5-5. Modern cars often display a plastic sticker containing even more information, and including the calibration code.

The calibration code identifies the engine and original transmission type, and exactly what exhaust and emission-related equipment should be installed. The calibration code also lets you look up the correct vacuum diagrams for the engine.

The unit number is another code that you will find on GM engines. It appears on a plastic tag, usually stuck to the oil pan, and is used by the assembly plant to track the engine as it is being built, and to ensure that it is built with the correct components. The unit number is also used to keep track of parts on an engine. For instance, if a certain camshaft begins to fail too frequently, the manufacturers can use the unit number to identify which engines were equipped with that camshaft for service bulletins or recalls.

Look for these stampings, tags, or stickers near the front or the sides of the engine. You can find stampings on machined pads on the engine block, but tags and stickers can be attached to the block, valve, or timing covers, or to the oil pan. Do not destroy them — reattach them to the engine upon reassembly. Many state emission testing facilities use these codes for periodic emission inspection. If you lose the tag, you may lose a future customer, because he will rightly blame his registration headaches on you.

VECI

The Vehicle Emission Control Information sticker gives you additional information, figure 5-6. Manufacturers tailor this underhood stick-

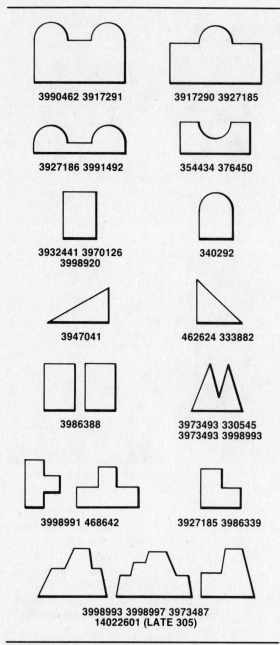

Figure 5-7. Casting marks on Chevrolet cylinder heads identify the basic casting.

- Fast-idle speed
- Enriched idle speed
- Exhaust emission limits
- Initial timing setting
- Breaker-point dwell
- Spark plug type and gap
- Carburetor adjustment instructions
- Vacuum routing diagram, including location of vacuum control valves, vacuum ports, and EGR valves
- Valve lash
- Emissions calibration code.

Much of this information is tune-up related, but many accessory and some emissions parts wear out along with the engine block and heads, and you may have to replace them at the same time. Often, the VECI sticker is the *only* source for this information to let you identify the correct parts.

Interpreting casting numbers and marks
Casting numbers and marks are helpful for identifying blocks, cylinder heads, and manifolds, as well as crankshafts and camshafts. Any given casting might have several different types of marks on it, and they might give you very different kinds of information. They are *not* necessarily part numbers, and so you must be careful when you interpret them. For instance, a row of Chevrolet engine blocks with the same casting number reading 14022801 may all look about the same. However, the cars they can come from range from 1980 to 1986 models, and the engine capacity could be either 276 or 305 CID. They may have been machined for different starter motor mounts, clutch bellcrank mounting pads, transmission bolt holes, and exhaust manifold bolt patterns.

Casting marks
With some manufacturers, when the foundry workers construct the molds for a given casting, they mark the mold so that the castings have external marks that identify the general series. These marks may be numbers, or they might be triangles, half circles, rectangles, or other shapes that correspond to one or a few types of castings, figure 5-7. Often, castings with the same mark end up being further modified or machined for use in a variety of other applications.

You can often use these marks to identify an assembled engine, even if the more specific casting numbers are hidden inside or behind covers. They can usually tell you the manufacturer, division, the general engine displacement, and a broad range of production years. You will need a manufacturer's or an aftermarket manual to interpret them.

er to each car on the assembly line to provide a permanent record of emissions equipment, tune-up settings, and up-to-the-minute running changes that may not make it into the specifications book until the following model year. The VECI sticker may contain:

- Engine identification, including model year and whether the engine is a California or Federal version
- Slow-idle speed

Obtaining Specifications, Manuals, Engines, and Kits

Figure 5-8. Some blocks have casting numbers that are visible from the outside without disassembly.

Figure 5-9. This casting number (A314) indicates that this manifold was cast on January 31, 1974. (A = January, 31 = 31st, 4 = 1974).

Casting numbers

Casting numbers are more specific than casting marks. On a cylinder head, for example, the casting numbers might enable you to identify the volume of the combustion chamber, and the intake and exhaust port sizes. On a cylinder block, the casting numbers might indicate the size of the main bearings, and whether the block has two-bolt or four-bolt main bearing caps. Sometimes these numbers are visible without disassembling the engine, such as the casting numbers on the left rear of small-block Chevrolet engine blocks, figure 5-8. Often however, the numbers are inside the assembled engine or behind covers. They still do not tell you conclusively whether a given part will fit your engine. For instance, small-block Chevrolet cylinder heads with a casting number reading 462464 could come with either large or small intake and exhaust valves, and were used on various 5.7 liter engines between 1977 and 1988. They may or may not be drilled for the accessory mounts or temperature sensors on your particular engine.

Miscellaneous marks and labels

In addition to casting numbers and marks, the foundries often mark the castings with the date and even time of manufacture, figure 5-9. Various other marks or numbers cast or stamped into the casting might tell you which alloys were used in manufacturing.

Some manufacturers, such as Ford, cast a highly specific production number into the parts which decodes to division, model, model year, basic part number, etc. These numbers can tell you much more than the simpler casting mark/casting number method.

As we discussed before, remember that the VIN will also be stamped into major castings. This is the same number that appears on the dashboard and on the vehicle's registration.

Recognizing oversize/undersize codes

Manufacturers produce their engines on a carefully calculated budget. These budgets allow for a certain scrap rate to cover parts with unacceptable core shift or damage from machining errors, but the more castings that can be salvaged, the more money the manufacturer saves.

Engine blocks with out-of-spec or damaged lifter bores can sometimes be returned to the assembly line by installing oversize hydraulic lifters. Crankshafts with damaged journals can also be ground undersize, and then fitted with undersize connecting rod or mainbearing inserts. Manufacturers often mark blocks and other components built with non-standard parts in some way. Chrysler, Dodge, and Plymouth formerly marked their V8 engines with a diamond-shaped stamp on a small pad on the right front of the block to indicate 0.001, 0.008, or 0.030 oversize lifters. A maltese cross in the same spot indicated undersize rod or main journal diameters, figure 5-10. Oldsmobile at one time stamped the oversize figure on the side of the lifter bore, and etched the same figure into the lifter itself.

Some manufacturers produce entire coding systems to indicate whether an engine contains off-size parts. For example, some Chrysler 238-cid (3.9-liter) engines are stamped on both the crankshaft and cylinder block:

Crankshaft stampings on the #6 counterweight:

- An "R" or "M" combined with numbers indicates that *some* rod or main journals are 0.001-inch (0.025-mm) undersize. For example, "R-2-3" indicates that rod journals

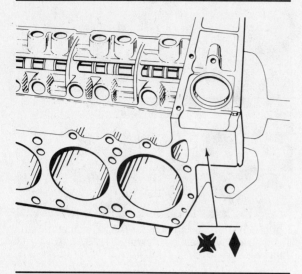

Figure 5-10. Crosses and diamonds stamped into an older Chrysler engine block indicate one or more non-standard sized parts installed by the factory.

Figure 5-11. All manufacturers provide extensive series of service manuals for each car or engine that they produce.

#2 and #3 are undersize, and "M-1-4" indicates that main journals #2 and #3 are undersized.
- An "RX" or "MX" indicates that *all* rod or main journals, respectively, are 0.010-inch (0.250-mm) undersize.

Cylinder block stampings:
- An "A" after the engine serial number indicates 0.020-inch (0.500-mm) oversize lifter bores.
- A 3/8-inch (10-mm) diamond stamped at a lifter bore indicates that it is 0.008-inch (0.200-mm) oversize. The same diamond will be stamped on the top pad at the front of the engine.
- An "X" stamped on a milled pad of the cylinder head indicates 0.005-inch (0.125-mm) oversize valve stems.

As another example, AMC often used select-fit bearing inserts on their 4.2 liter six-cylinder engine. This means that it's not particularly uncommon to find a standard-size bearing shell paired with a 0.001 or 0.002 *undersize* bearing shell on the same journal. AMC used a set of color-coded marks on the crankshaft and block to let rebuilders in on the situation. Installing standard-size replacements will create excessive oil clearances, and the pressure in the rebuilt engine will be too low.

Aftermarket engine rebuilders commonly regrind crankshafts, and will stamp the undersize information on the webbing or an end flange of the crank.

The various oversize and undersize marks may or may not be explained in the manufacturer's service manuals for the engine. If you discover marks similar to these on an engine block, make sure you measure parts as you remove them. Then purchase the correct over- or undersize parts for reassembly.

Obtaining Specifications

Specifications for the engines you work on will include torque settings, clearances, surface finishes, valve machining angles, maximum wear limits, and a variety of other essential information. There are several sources for engine specifications.

Manufacturers' service manuals

Because of their focus on a single vehicle line, manufacturer's service manuals are the most detailed sources of engine specifications, figure 5-11. However, these manuals generally cover only a single model year, and it can often be difficult and time consuming to locate a specific piece of information among the many repair procedures. In addition, all the manufacturers organize their manuals differently. If you work on more than one brand of engine you will find locating similar information in the different manuals to be extremely frustrating. Finally, the manufacturer's service manuals are expensive, and you must usually buy new ones for each model, each year.

Manufacturers' service bulletins

Car manufacturers regularly publish service bulletins for the technicians in their dealers'

Obtaining Specifications, Manuals, Engines, and Kits

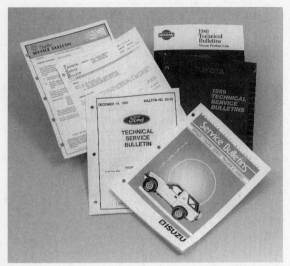

Figure 5-12. Service bulletins can keep you up-to-date on mid-year changes and field problems that come to the manufacturer's attention.

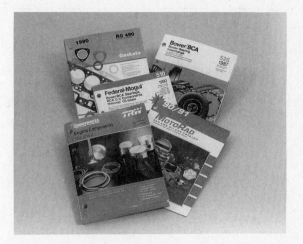

Figure 5-13. Engine component manufacturers also publish manuals and other literature with specifications, procedures, and miscellaneous information.

service departments, figure 5-12. These bulletins contain updated specifications, solutions to common problems, and various other recommendations that may not have appeared in the service manuals.

Unfortunately, service bulletins can be hard to obtain unless you work in a dealer's service department. Many manufacturers are reluctant to make them generally available, because they prefer this information to benefit their own dealers and service departments. However, some aftermarket companies collect this information and republish it for a fee to subscribers. If you regularly work on a particular brand or specific category of vehicle, you may be able to obtain specific service bulletins that would be useful. Look for advertisements for service bulletin subscriptions in monthly trade magazines such as *Motor Service*, *Service Station and Garage Management*, *Super Service Station*, *Brake & Front End*, and *Automotive Rebuilder*, to name a few.

Aftermarket service manuals
Several publishers print consumer-level aftermarket service manuals for automobiles. These contain summaries of the information in the manufacturer's service manuals, as well as occasional hints and tips for accomplishing tasks without factory tools. Because the publishers intend these books primarily for the do-it-yourself market, they must usually cover a range of models and years, and can't offer as much information as the manufacturer's service manuals.

Other companies publish yearly manuals which include specifications and repair procedures. These books are quick and easy to use, and contain exact service and rebuilding specifications for many engines.

Aftermarket component suppliers
Many engine component manufacturers publish parts catalogs, interchange manuals, training manuals, installation booklets, specification tables, and newsletters for the types of products they sell, figure 5-13. These publications contain a wealth of specific information about engine capacities, standard bore sizes, and bottom-end and top-end component sizes, clearances, and interchanges. They are much easier to keep on hand and use in a shop than an extensive library of factory publications.

REPLACEMENT PARTS, SUBASSEMBLIES, AND ENGINES

How do you decide which parts you must purchase in repairing or rebuilding an engine? In the course of your work you will routinely need to obtain crankshafts, rods, pistons, valves, springs, and seats, as well as rings, bearings and seals. Sometimes the customers will supply the necessary parts themselves, especially if the rebuild involves high-performance modifications. At other times, the customer will simply bring an engine to your shop and expect you to supply whatever is needed. If you work in a production engine rebuilding shop, you will need to obtain fairly large quantities of these parts regularly and economically. In the following sections, we outline what you need to know to purchase engine parts easily and economically.

Figure 5-14. Short blocks are economical and fairly quick to install.

Overhaul versus Rebuild

In a complete engine rebuild, you will replace or recondition all the worn parts inside the engine. You will bring all wear surfaces and clearances back to as-new specifications. You will thoroughly inspect any old parts that you retain, and reject any with flaws that could compromise performance or service life. The result will be an engine that is in new, or better-than-new condition throughout.

An overhaul is a different matter. When you overhaul an engine, you will reinstall many parts that are worn, but still within the service limits. This means that the original pistons might be reinstalled with new rings, or that bearings in good condition might be reinstalled.

Engine Families and Interchangeability

Engine designers frequently develop an entire series of engines from the same basic design, often from almost identical castings. These groups of engines are called engine families, and they frequently share parts. The small-block Chevrolet engine family, for instance, began with an engine in 1955 displacing 265 cid (4.34 liters). Over the next thirty-five years Chevrolet marketed this engine in displacements ranging from 262 cid to 350 cid (4.3 to 5.7 liters), with a closely related casting displacing 400 cid (6.6 liters).

Engines within a family can interchange many of their internal and external parts, including crankshafts, rods, and pistons, as well as intake and exhaust manifolds, cylinder heads, oil pan, and valve covers. Not all components fit all other engines, but enough interchange to make parts shopping a good deal easier.

Other engine families include big-block Chevrolets, and small, intermediate, and big-block Fords. Engines from Oriental manufacturers evolve in a similar way, such as the older 2T, 2TC, and 3TC four-cylinder inline engines from Toyota, superceded by the various 4A-series engines. One example from Nissan includes the older A-12, A-14, and A-15 four-cylinder engines, which evolved into the E-15, E-16S, E-16I, and GA-16 engine series.

You can learn the various engine families in a number of ways. If you work in automotive repair and rebuilding, you will learn them through the experience you get every day. You can also purchase aftermarket repair and rebuilding manuals that explain the history of a particular engine family in detail. Because the engines within a family also share clearance, adjustment, and tune-up specifications, you will find your work easier and more accurate the more you learn.

Service Replacement Engines — Pros and Cons

The service departments at new car dealerships seldom perform major rebuilds on a customer's engine. Rebuilding the engine correctly requires time-consuming and meticulous work, and also requires a major investment in space and tooling that a dealer emphasizing new car sales can ill-afford. Most, therefore, prefer to install service replacement engines, or farm the cylinder head and block work out to local machine shops.

You can purchase service replacement engines from a dealer or from local rebuilding companies as short blocks, long blocks, or entire engine assemblies. Dealer assemblies are often more complete than those you buy from an independent rebuilder, because the rebuilders assume that you will reinstall some components from the original engine, such as oil pans and valve covers. An assembled short block, figure 5-14, is a complete engine block assembly with core plugs and dowel pins installed. The crankshaft, rods, pistons, and

Obtaining Specifications, Manuals, Engines, and Kits

Figure 5-15. Long blocks also include the cylinder head or heads.

bearings are also installed. Some also include an installed camshaft and timing chain or gear set.

An assembled long block, figure 5-15, includes everything in the short block, as well as the cylinder heads, correctly torqued, with all the valve gear in place. The long blocks also include the oil pump and pickup and the timing set and camshaft (if they weren't in the short block). Some long blocks also include the harmonic balancer, and the flywheel or flexplate.

Complete service replacement engine assemblies include the entire engine, but what the manufacturer considers an entire engine varies. The General Motors Mr. Goodwrench service replacement engine (formerly called Target Master) is a 350 cid (5.7 liter) V8 assembly less the manifolds, damper, water pump, distributor, flywheel, and clutch — essentially a complete long block. The exact components delivered with the engine vary between manufacturers. They may or may not include the engine manifolds or accesories.

Manufacturers build short blocks, long blocks, and service replacement engines so that they can be installed easily into a customer's car by reusing components still in good condition. For instance, an engine that has thrown a rod may not need new cylinder heads, while a high-mileage, abused car may need a complete replacement engine. In addition, an individual chassis may require unique exhaust manifolds, induction equipment, oil pans, and other components that still use the same basic long block or service replacement assembly. Customers balk at purchasing a replacement engine with incorrect parts already installed.

Under some circumstances, your customers may specifically *not* want to replace the original engine with a newer version, even if it will cost them more money. Sometimes original equipment is important to the value of car. For example, a 1967 Corvette Sting Ray with the original, matching VIN engine block is worth about $10,000 more than it would be with another engine identical in every way except for the serial numbers.

Another example includes high-performance car owners. Older engines, especially those from the 1960s, were manufactured when casting technology was not as advanced as it is today, and when light weight was less important. They frequently have thicker decks, cylinder walls, and heavier casting in critical areas of the cylinder heads. Modern engines are manufactured under stricter economy, and use thin-wall castings to save weight, at the expense of rigidity and resistance to overheating damage. Your customer may prefer to rebuild the heavier block, rather than purchase a newer, lightweight engine.

Lastly, rebuilding an older engine always has the advantage of letting you start with a seasoned block. A new engine built with a green casting is full of areas under stress, and its shape is not quite permanent. During the countless warm-ups and cool-downs of the engine's service life, these stresses relax, and the castings settle into a constant, final shape. Unfortunately, all the machining has already been done. When the engine is broken in, the cylinder walls may no longer be exactly round, the deck may be slightly warped, the crankshaft saddles may be a little out of line, etc. An older engine has warmed up and cooled off thousands of times, and you can be pretty confident that if you remachine these seasoned castings to original specifications, then the engine will probably last longer and perform better during its second lifetime than it did during its first. Race car builders limited to production castings often prefer to start with worn-out, hundred-thousand-mile pickup truck motors to build up for their cars.

Engine Kits and Gasket Sets

Engine kits are usually the simplest and most economical way for you to purchase parts for an engine you will rebuild. Typically these kits include both new and remanufactured parts (most often the crankshaft and rods, if they're included). Crankshaft kits usually include a reground crankshaft, pistons, piston rings, and the rod and main bearings. A crankshaft and camshaft kit often includes the same components as a crankshaft kit, along with a camshaft, lifters, timing gears, and chain.

Of course, you can also purchase sets of pistons, rods, bearings, and rings separately. Remember, though, that the larger kits are usually a much better value for the money.

Gasket sets are packages including all the gaskets (and often seals and O-rings) you will need for a certain task on a certain engine. For instance, top-end overhaul gasket kits include intake and exhaust manifold gaskets, head gaskets, and sometimes carburetor base gaskets and valve seals. Master engine overhaul gasket sets include the same parts, but add oil pan gaskets and seals, water pump gaskets, and timing cover gaskets and seals.

Used Engines

A final alternative you should consider when shopping for parts for an engine overhaul or rebuild is to skip the repairs entirely, and simply install a used engine. The prices of used engines you buy from a dismantler vary with condition, but you can often buy them in excellent condition.

Japanese engines in particular are often more economical to purchase used than to rebuild. In Japan, motor vehicles over three years old must pass an extremely rigorous and expensive inspection program every two years — every year for cars over ten years old. The *typical* fees for an average subcompact car inspection total upwards of $700. As a result, automobiles in Japan are often scrapped with the engines still in excellent condition, often with no more than 30,000 to 50,000 miles (48,000 to 80,000 km). In recent years, importers have begun purchasing these engines in Japan and shipping them in bulk to the United States for resale. In many areas, buying a low mileage Japanese engine costs less than purchasing the parts needed to rebuild it.

Domestic engines, on the other hand, benefit from a long-established tradition of rebuilding, and factory and aftermarket parts are inexpensive and readily available. Nonetheless, you may find it more economical to purchase a used domestic replacement engine than to repair or rebuild one.

PART TWO

Diagnosis, In-Car Repair, and Engine Removal

Chapter Six
Engine Testing and Diagnosis

Chapter Seven
Engine Repair in the Vehicle

Chapter Eight
Engine Removal

Chapter Nine
Disassembly, General Inspection, and Cleaning

Chapter Ten
Repairing Cracks and Damaged Threads

Chapter 6

Engine Testing and Diagnosis

Complete engine rebuilding is not always necessary, or desirable. Sometimes all the engine needs is specific repair to one system, such as grinding the valves and seats, or replacing a head gasket with internal leaks. Sometimes the customer wants an overhaul, rather than a complete rebuild. Your skills in evaluating engine condition *before* the teardown will prevent much unnecessary work and expense, and will help keep your customers. Remember, nobody likes to hear that a repair estimate for their car must be revised upwards, or that the repair they just paid for failed to solve the problem.

In this chapter, we explain what you need to know, and what equipment you use to identify common mechanical problems in an engine before you disassemble it. Obviously, if the engine is running and still in the car, your task is much easier than if the customer brings you the engine in the back of a truck. In cases like those, you must question the owner very carefully and avoid making promises about the engine that you cannot keep once you have had a chance to inspect it internally.

OIL CONSUMPTION

All engines burn oil during normal operation. In the past, engine oil consumption followed a pattern. Oil use was always higher for a brand new engine during break-in, settled down to a long period of low consumption during the normal service life, and then began to increase as the engine finally wore out. Modern engines, with closer piston ring clearances, higher quality ring material, and carefully-prepared cylinder bores, require almost no break-in procedures at all, and you can expect them to use no more oil during their first service interval than during the remainder of their service life.

But how much oil consumption is too much? This is always a judgment call, but in general, if the engine consumes more than about one quart of oil every 500 hundred miles, then it needs attention. At that level, the oil being burned in the combustion chamber will form significant deposits on the intake valve, piston crown, cylinder head, and spark plug. Performance, fuel economy, and emissions will likely all be unsatisfactory.

Unless you can actually see the engine smoking, be sure to check whether the engine is really burning the oil and if any unusual circumstances exist. For instance, the engine may have severe leaks, and not burn oil at all. Ask whether the customer drives continuously at highway speeds, or tows a trailer. In both these situations, oil consumption will normally increase. Check to see what oil the customer

Engine Testing and Diagnosis

Figure 6-1. This spark plug has a layer of ash caused by burning engine oil in the combustion chamber.

Figure 6-2. Oil-wetted plugs are a sure sign that the engine is a severe oil burner.

Figure 6-3. Varnish can stick a ring firmly in the groove, preventing it from sealing against the cylinder wall at all.

has been using in the engine. If the oil viscosity is lighter than what the manufacturer recommends, then the engine will burn oil even if in good condition. As a last check, pull a few spark plugs to inspect their condition. If the engine burns oil, the spark plugs will have whitish or gray ash deposits, figure 6-1. If the engine is burning a lot of oil, the plugs will still be covered in liquid oil when you remove them, figure 6-2. Oil normally enters the combustion chamber either past piston rings, or past valve guides. One quick check that sometimes helps you distinguish between the two is to look at the spark plug deposits. Often, an engine pulling oil through the valve guides will have deposits built up only on one side of the plug electrodes — that side nearest the intake port. If the engine pulls oil up past the piston rings, then spark plug deposits will likely be more even all over the end of the plug.

Oil Passing the Piston Rings

When the piston rings operate correctly, not more than one or two *thousandths* of a drop of oil is left behind on the cylinder wall when the piston descends. This very small amount of oil increases substantially if the rings, pistons, and cylinder walls are not in perfect condition, or if there are other problems:

- Ring and cylinder bore wear or cylinder scoring
- Stuck or broken rings
- Rings installed incorrectly
- Using oil with too low a viscosity, or diluted with fuel
- Excessive crankcase pressure
- Excessive oil thrown onto cylinder walls.

Worn, stuck, or broken rings

Over time, the piston rings and cylinder bores wear, and can no longer keep a gastight seal around the combustion chamber. When this happens, engine vacuum and ring movement tends to pump oil up past the piston into the cylinder, while simultaneously the worn oil scraper ring fails to remove enough oil film from the cylinder wall.

When oil reaches the hot upper ring lands, it begins to char. This forms varnish deposits in the ring grooves that can stick a ring firmly in place, preventing it from sealing against the cylinder wall at all, figure 6-3. Both compression rings and oil control rings can stick, dramatically reducing compression and increasing oil consumption. Non-detergent oils are much likelier to stick a ring than detergent oils. A stuck ring is usually permanently stuck, but

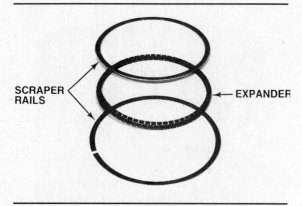

Figure 6-4. Oil rings often come as a multi-piece assembly, with two scraper rails and a central expander.

Figure 6-5. If the ends of the expander ring overlap instead of butting together, the ring will damage the cylinder wall immediately.

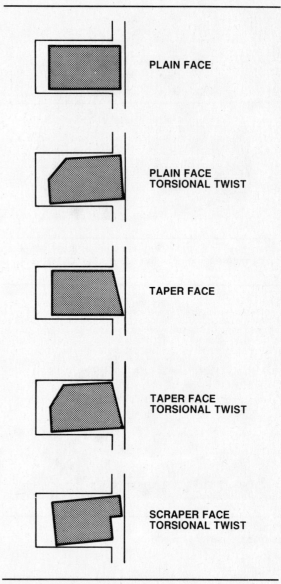

Figure 6-6. Compression rings come in a variety of shapes so that they will twist when installed and contact the cylinder wall with a sharp edge.

sometimes an overnight soaking in penetrating oil can loosen it enough to free up when you start the engine.

Broken rings cause the engine to begin consuming oil immediately, instead of slowly increasing as occurs with normal wear. The broken rings can also score the cylinders, causing damage that even new rings cannot seal.

You will detect stuck, broken, or worn-out compression rings during a compression test. Broken rings will smoke through the tailpipe immediately. You can assume that if the compression rings are in poor shape, the oil control ring is probably equally worn.

Incorrectly installed rings

Incorrectly installed rings also can cause excessive oil consumption. Multi-piece oil control rings, figure 6-4. have a central expander ring between the two scraper rails. The ends of the expander should butt against each other, figure 6-5. If they overlap instead, the ring will fail and score the cylinder as soon as you start the engine.

You can also install compression rings incorrectly. These rings appear to be simple hoops, but they can actually be quite complex, figure 6-6. Their shape helps the oil control ring keep oil out of the combustion chamber by scraping

Engine Testing and Diagnosis

Figure 6-7. Piston ring installers protect both the ring and the piston from damage during assembly.

and pumping oil *down* as the piston moves. If you install them upside down, they scrape and pump oil *up* instead, and will increase oil consumption substantially.

You can damage either kind of piston ring fairly easily during installation. Always use a piston ring expander to install rings, figure 6-7, or open them up *with your fingertips at the ring ends*. Avoid "screwing" the rings down into the ring groove. This can bend the ring, preventing it from rotating or sealing to the bottom of the groove.

Oil viscosity too low

Engine oil viscosity problems can result from running an oil that is too thin for the engine's operating conditions. Manufacturers always specify the correct oil to use in their engines, and these recommendations often specify a lighter oil for colder weather. For instance, an engine designed to run on 10W-30 oil may require a switch to 5W-30 or even 5W-20 if operated in very cold weather.

If you run an engine oil that is too thin, its low viscosity prevents the rings from removing it effectively from the cylinder walls. The oil is left behind in the cylinder and burned during the power stroke. Oil with a greater viscosity is easier for the rings to keep scraped off the cylinder wall because it can not easily squeeze past the rings.

Engine oil can also become thin because of fuel dilution. Any condition that encourages raw, liquid gasoline or diesel fuel to collect on the piston crown, cylinder walls, and ring surfaces will promote oil dilution. The liquid fuel drips around the rings or collects on the cylinder bore, to be scraped into the sump by the descending piston rings. Oil dilution has a variety of causes, including:

- Choke plate stuck closed or adjusted incorrectly
- Coolant thermostat stuck open
- Flooding carburetor or leaking injectors
- Mechanical fuel pump with an internal leak
- Short trips in cold weather
- Excessive blowby.

Excessive crankcase pressure

Excessive crankcase pressure, such as that caused by a clogged PCV valve, upsets the normal vacuum balance inside the engine. Crankcase gases then tend to flow out along seams and seals, taking engine oil with them. The pressure also makes it harder for the rings to do their job. The rings must overcome the crankcase pressure to retain the oil in the oil ring groove and scrape it through the slots in the piston.

Excessive oil on cylinder walls

Excessive oil splashed onto the cylinder walls also causes oil to pass the piston rings, mainly because there can simply be more oil than the rings were designed to remove. Some amount of oil on the cylinder walls is necessary for piston and ring lubrication, but any left behind as the piston rings descend is burned. Oil splash can never be eliminated, as the oil inside a running engine at high RPM is distributed throughout the crankcase in ropes, drops, floating balls, and vertical or inverted puddles. However, *excessive* cylinder wall oiling can easily occur if the oil level is too high. An oil pressure relief valve that is stuck open can cause the same result, as oil pressure rises excessively and overloads the piston rings.

Oil Passing the Valve Guides

Valve stems and valve guides also wear out with high mileage. Although valve stems need a small amount of clearance in the guide to prevent them from galling or seizing inside the guide, too much clearance lets oil flow down the valve stems into the ports. Normally the valve seals help prevent excessive oil from entering this clearance space, but if they are in poor condition, they compound the problem.

Oil flows down the valve guides because of manifold vacuum. The space at the top of an intake valve guide is always at atmospheric pressure, but the portion inside the intake port is exposed to manifold vacuum whenever the valve is open. When this occurs, any oil that was splashed onto the top of the guide or onto the upper valve stem is likely to be drawn into the intake port. Some of it forms carbon deposits on the back of the intake valve, but the remainder passes into the cylinder to be burned.

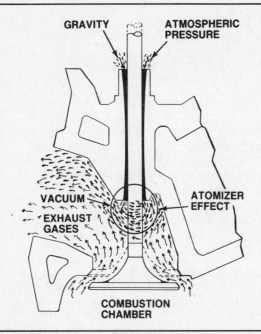

Figure 6-8. Exhaust valve guides leak oil when they wear, too.

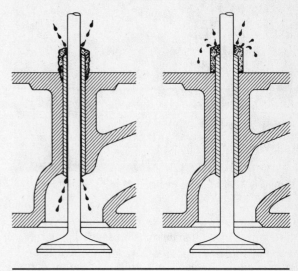

Figure 6-9. New valve seals minimize the oil flow down the guides.

Exhaust valves also draw some oil down their valve stems — without oil, the valves would seize to their guides as soon as the engine warmed up. The amount of oil is much less, though, because the exhaust port vacuum is never as high as in an intake port. Most of the oil is drawn down the exhaust valve by the slight vacuum created by the moving exhaust gases, figure 6-8, in the same way the early road-draft systems ventilated crankcases before PCV systems were invented.

The oil flow down the valve stems can be minimized by fitting new valve seals, which can be done without disassembling the engine, figure 6-9. This procedure is discussed in a later chapter. If the compression test showed that the piston rings are still holding reasonable compression, and your customer is on a tight budget, then this is an option you can suggest. This just fixes the symptoms, however, not the problem. Worn valve guides prevent the valve head from properly fitting into the seat, and this will eventually need to be repaired.

Remember that any engine condition that results in too much oil washing around under the valve cover can increase oil consumption through the valve guides, even if there is nothing wrong with the guides or seals. Both clogged oil drain back passages and worn pushrod tips (for engines that pump oil through the pushrods) can increase the amount of oil sloshing around the valve stems.

For some engines, simply running the engine at higher rpm, such as on highway trips, loads the valve area with enough oil to increase oil consumption substantially.

External Oil Leaks

Engines can use up considerable oil through leaks. If undetected, a leak can fool you into thinking that the engine is burning the oil.

Most leaks are obvious, and caused by deteriorated or incorrectly installed gaskets or seals, or fittings such as oil pressure sensors. Sometimes oil can leak through cracks in the block or cylinder head.

One important oil leak is that caused by oil mist blowing through the PCV intake hose, as will happen if the engine has excessive blowby or a clogged PCV valve. The oil mist can blow through the PCV fresh air intake into the intake manifold, or out the dipstick tube, especially at wide-open-throttle or high rpm.

Internal Oil Leaks

V-type engines can draw oil through leaks in the intake manifold gasket, where it attaches to the cylinder heads, figure 6-10. The symptoms of a gasket leak here are the same as that of poor valve seals, except that replacing the seals will fail to fix the leak. This problem also causes a vacuum leak, and can create engine smoke. See the section later in this chapter for a way to locate this problem using a vacuum gauge.

Engine Testing and Diagnosis

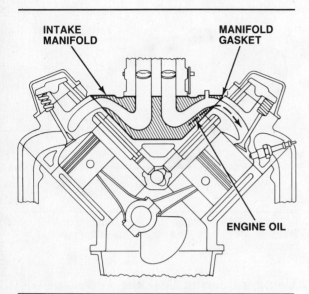

Figure 6-10. Intake manifold vacuum can draw oil into the cylinder around a defective or poorly installed gasket.

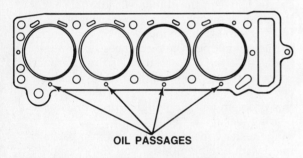

Figure 6-11. These small holes contain the pressurized oil that flows to this overhead camshaft.

Most OHC engines have head gaskets with passages in them to direct pressurized oil to the camshaft, figure 6-11. If the gasket leaks oil into a combustion chamber, the plugs will foul repeatedly, just as if a ring were broken. If it leaks into a coolant passage instead, then you will likely see oil blobs floating in the coolant, and water drops on the dipstick.

Running an engine for only a short time with coolant or water in the oil turns the mixture into an unmistakable milky, chocolate-colored emulsion. Operating the engine with oil that looks like this will cause serious damage.

Cracks in the block or cylinder head can also cause internal oil leaks. We discuss these in the next section.

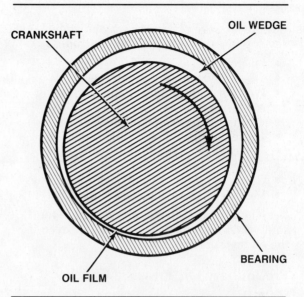

Figure 6-12. The oil wedge in the clearance space between the journal and the bearing depends on oil pressure to feed it lubricant as the oil leaks out of the bearings.

ENGINE OIL PRESSURE

Oil pressure is critical to proper engine operation and life. If the oil pressure goes, so does the engine. With low pressure, insufficient oil will be forced into the clearance spaces in the camshaft, connecting rod, and main bearings, making them wear out faster, figure 6-12. If there is no pressure at all, metal-to-metal contact will quickly heat the bearing shells until they weld to the journal and spin. This can seize the engine if you are lucky, or throw a rod through the side of the crankcase if you are not.

You can detect low oil pressure from the dashboard oil pressure gauge, or from a flickering or steadily burning oil warning light. The actual pressures from factory oil pressure gauges are often fairly inaccurate, but you can still use them to detect changes from normal conditions. The oil warning light might stay on for too long after the engine starts, come on whenever the engine returns to idle, or stay on all the time.

Low oil pressure causes various noises. Engines with hydraulic lifters will tick when the oil pressure is low, because the lifters are unable to take up the valve clearance without a steady flow. The noise is the same as the one that hydraulic lifters make when you first start a cold engine, especially after you change the oil. Low oil pressure also causes the crankshaft bearings to knock. We discuss engine noises more thoroughly later in this chapter.

There are several reasons why an engine will develop low oil pressure. These include:
- Low oil level
- Oil diluted with unburned fuel
- Worn or damaged main and connecting rod bearings
- Worn or defective oil pump
- Loose or plugged oil pickup screen
- Defective oil pressure relief valve
- Clogged oil passages
- Cracked, porous, or plugged, oil galleries
- Missing or loose oil galley plugs.

Keep in mind that the oil pressure light for some older engines will flicker at idle even if the oil pressure is normal.

Preliminary Oil Pressure Checks

If you suspect low pressure, first make sure that the oil level and idle speed are correct. Next, check for oil dilution. Crankcase oil can become diluted with unburned fuel from a plugged PCV system, flooding, a stuck choke, or a coolant thermostat stuck open. Pull the dipstick and watch how fast the oil drips from the end. Heavily diluted oil will run off the end of the dipstick noticeably faster than fresh oil. In severe cases, you can even drip the engine oil from the dipstick onto the ground and watch the gasoline in it evaporate. If it is obviously diluted, change the oil and filter and correct whatever problem caused the dilution in the first place.

If the oil condition and idle speed seem okay, hook up a pressure gauge as explained in the next section.

How to Use an Oil Pressure Gauge

Attach an external oil pressure gauge and measure the pressure directly, figure 6-13. You connect it to the engine by removing an oil passage plug or the oil pressure sending unit and attaching the gauge or its hose fitting. The gauge will register the oil pressure in actual pounds per square inch or kPa. If the problem only appears during driving, hook the gauge up with a long hose so you can read it while driving the car.

Oil pressure specifications vary with different engines, and you should look up the correct values in the manufacturer's service manual or in an equivalent aftermarket manual. As a rule of thumb, however, typical pressures for many warmed-up cars at idle speeds are in the region of 5 to 20 psi (35 to 140 kPa), and pressures at 2000 rpm should be in the region of 30 to 60 psi (200 to 400 kPa).

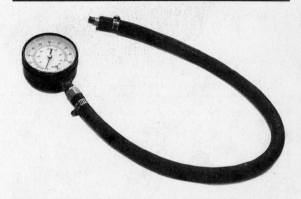

Figure 6-13. Oil pressure gauges connect to the sending unit port, and let you measure the pressure directly.

Using a Bearing Leakage Tester to Confirm Wear

When the engine is running, the main resistance that the oil pump meets is the narrow spaces between the crankshaft journals and the main and connecting rod bearing shells. In a high-mileage engine, wear in the bearing surfaces will open the clearances up enough to decrease the oil pressure. As the clearance space increases, so does the rate at which oil can escape from the bearing. You can confirm this problem using an engine prelubricator as a bearing leakage tester.

The tester consists of a small tank, partially filled with engine oil, that you can pressurize with compressed air, figure 6-14. It is sometimes called an engine prelubricator because it can also be used to lubricate the engine's pressure-fed bearings before starting.

You must remove the oil pan to observe the oil flow directly. Connect the oil hose from the tank to an oil passage on the engine block. The tapped hole for the oil pressure sending unit is usually the most accessible place. When you connect the tank, the pressurized oil will flow from the crankshaft bearings and out of the oil pump screen. Any bearing that drips at least 20 drops of oil per minute has acceptable clearance. If there is no dripping at all, the bearing is too tight, or the oil hole is blocked. If the oil comes out in a stream, the clearance is excessive.

The engine prelubricator will also tell you whether the engine oil galleries are cracked, porous, or plugged, or whether a gallery plug is missing. If oil runs from some point other than the bearings or the oil pump screen, then investigate further to locate the leak.

Engine Testing and Diagnosis

Figure 6-14. Leakage testers let you tell right away whether the bearings or seals are doing their jobs.

ALTITUDE	INCHES OF VACUUM
Sea Level to 1,000 Ft.	18 to 22
1,000 Ft. to 2,000 Ft.	17 to 21
2,000 Ft. to 3,000 Ft.	16 to 20
3,000 Ft. to 4,000 Ft.	15 to 19
4,000 Ft. to 5,000 Ft.	14 to 18
5,000 Ft. to 6,000 Ft.	13 to 17

Figure 6-15. Elevation has a strong effect on the vacuum an engine can develop.

If low oil pressure is caused by a faulty oil pump, pickup screen, or pressure relief valve, then you will have to perform further disassembly in order to inspect the parts.

BASIC DIAGNOSTIC TESTS

Aside from the basic considerations of oil consumption and low oil pressure, there are several basic tests you should run to locate other engine mechanical problems. The tests that you will perform on a running engine include:

- Manifold vacuum tests
- Engine compression test
- Cylinder leakage test
- Power balance test.

What a Vacuum Gauge Can Tell You

In general, the higher and steadier the vacuum reading, the better the engine condition is. Normal engine vacuum at idle is from 15 to 21 in-Hg (375 to 525 mm-Hg). Vacuum decreases as the throttle opens. At steady part-throttle cruising, engine vacuum should run between 10 to 15 in-Hg (250 to 375 mm-Hg). When the throttle plates are fully open, as in WOT acceleration, normal engine vacuum is nearly zero.

All vacuum readings decrease with increases in elevation, figure 6-15. As a rule of thumb, subtract one inch or 25 mm from the manufacturer's specifications for every 1000 feet (300 m) above sea level. For instance, if the specifications call for 18 in-Hg (460 mm-Hg) of vacuum at idle, expect an engine in good condition to have only 13 in-Hg (330 mm-Hg) of vacuum when tested at 5000 feet (1500 m).

Normal vacuum readings also vary between different engines because of the differences in cam timing and valve lift. Older engines will idle with several inches more vacuum than modern emission-controlled engines, and their lower valve overlap also gives them a steadier reading. Engines with high-performance camshafts will idle with lower vacuum for the same reasons.

Even with these differences, a vacuum reading is invaluable when comparing engines of the similar year, make, and design, or in making comparisons with a given engine before and after adjustments. A low vacuum reading could be caused by something as simple as retarded timing, but it could be a much more serious problem and should be investigated and corrected. Vacuum gauge readings can pinpoint:

- PCV system problems
- Intake manifold gasket and vacuum line leaks
- Intake or exhaust valve and valve guide problems
- Retarded ignition and valve timing
- Exhaust system restrictions
- Poor combustion chamber sealing.

In some cases, such as with exhaust system restrictions, the vacuum gauge reading is the only test that can pinpoint the problem. In other cases, such as ignition timing, the vacuum reading is an inconclusive test. Other tests will indicate the problem exactly, but the vacuum test gives you a general idea of the fault.

A vacuum gauge can be a simple dial with an attached hose, figure 6-16, or a more complex pump-type instrument that also lets you

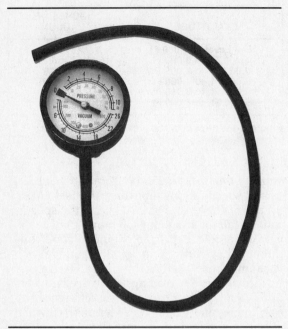

Figure 6-16. This simple vacuum gauge reads both vacuum and pressure.

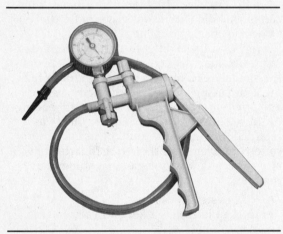

Figure 6-17. More versatile gauges let you create vacuum with a hand pump to test vacuum motors or other diaphragms.

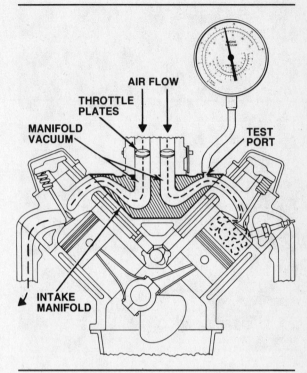

Figure 6-18. Intake manifolds often have a special vacuum port screwed into the casting.

pump down a vacuum inside a hose, figure 6-17. These better tools are extremely useful for many other troubleshooting and tune-up tasks, but for a basic manifold vacuum check the simpler gauge will work as well.

All the following tests require you to measure manifold vacuum. Look for a vacuum port installed into the intake manifold, figure 6-18, or into a special manifold in the base of the carburetor. Cars with automatic transmissions may have an accessible vacuum modulator hose leading from the intake manifold. You can simply attach the vacuum gauge to the manifold fitting, or you can cut into a vacuum line and install a tee-fitting. If you do install a special fitting, be sure to remove it or cap the opening when you are through testing.

Avoid tapping into any of the vacuum lines attached to the distributor, or to the exhaust gas recirculation valve, if there is one. These lines carry smaller vacuum signals, which are much less than manifold vacuum. Also avoid tapping the vacuum lines leading to the vapor recovery canister. If you must tap into a vacuum line containing a vacuum control valve or restrictor, be sure to tap into the line upstream from them. Many transmission modulator vacuum lines have control valves in them for emissions control.

How to test the PCV system

You can test the PCV system using a simple vacuum gauge or an infrared engine exhaust gas analyzer. We discuss both methods below.

With a vacuum gauge
1. Connect the vacuum gauge to the intake manifold.
2. Disable the ignition system.

- For electronic ignition systems, disconnect the wiring harness connector from the distributor or coil pack, or disconnect the wire from the negative (–) coil primary terminal. When disabling an

Engine Testing and Diagnosis

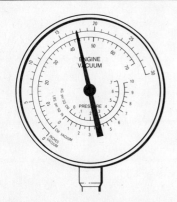

Figure 6-19. An engine in good condition will read steady at 15 to 21 in-Hg (375 to 525 mm-Hg) at idle.

Figure 6-20. A steady reading three to nine inches (75 to 225 mm) below normal may indicate internal leakage around the piston rings or late timing.

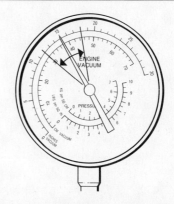

Figure 6-21. A gauge reading with the needle fluctuating three to nine inches (75 to 225 mm) below normal, indicates a vacuum leak in the intake system.

electronic ignition, tape the wire ends so that the disconnected wires do not ground.
- For point-type ignition systems, simply disconnect the wire connecting the distributor with the negative coil primary terminal.

3. Crank the engine and note the vacuum gauge reading.
4. Pinch the PCV hose closed with a pair of pliers. and again note the vacuum reading while cranking the engine. Compare the readings taken in steps 3 and 4. The step 3 reading should be greater than the step 4 reading if the PCV system is working properly.

With an infrared exhaust gas analyzer
1. Bring the engine to normal operating temperature (upper radiator hose hot) and run at normal idle.
2. Connect an infrared exhaust gas analyzer according to the manufacturer's instructions.
3. Remove the PCV valve from the engine, not from the intake manifold or the carburetor.
4. Plug the open end of the valve and read the CO meter. The meter should show a large increase in the CO reading. If the reading remains the same or only increases slightly, the valve is not working. Check the valve or hose for a restriction and then retest.

How to spot vacuum leaks
Use this test to find a leak in the engine valve system, intake system, cylinder head and manifold gaskets, or in the vacuum hose circuit.

1. Connect the vacuum gauge to a manifold vacuum source.
2. Connect a tachometer to the engine.
3. Start the engine and run until normal operating temperature (upper radiator hose hot) is reached.
4. With the engine running at its correct idle speed, note the gauge reading and needle action.

- The gauge needle should hold a steady, normal reading between 15 and 21 in-Hg (375 and 525 mm-Hg), figure 6-19.
- A steady reading three to nine inches (75 to 225 mm) below that value at idle may indicate internal leakage around the piston rings, or late ignition or valve timing, figure 6-20.
- A gauge reading with the needle fluctuating three to nine inches (75 to 225 mm) below normal, indicates a vacuum leak in the intake system, figure 6-21.

Vacuum gauge readings can also indicate head gasket leakage. Leaking head gaskets will cause the gauge needle to vibrate as it floats between a low and a high reading, figure 6-22. Do a compression test or power balance test to determine where the leak is.

Defective manifold gaskets and carburetor or throttle body mounting gaskets are the most common sources of intake vacuum leaks. Throttle plates that are not fully seated when in the closed position will result in low vacuum at idle. You can check for vacuum leaks at the manifold and carburetor or throttle body

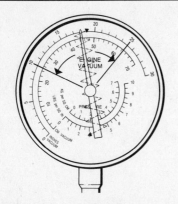

Figure 6-22. A leaking head gasket will cause the needle to vibrate as it floats through a range from below to above normal.

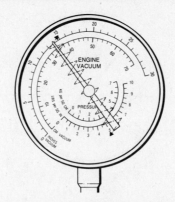

Figure 6-23. An oscillating needle one to two inches (25 to 50 mm) below normal could indicate an incorrect fuel mixture.

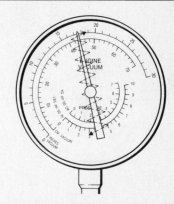

Figure 6-24. A rapidly vibrating needle at idle that steadies as engine speed increases indicates worn valve guides.

gaskets by squirting noncombustible cleaning solvent or tune-up solvent along the gasket joints with the engine running. If the idle speed stabilizes and vacuum increases for a few seconds, you have located the leak. If the leak is large, you may be able to see the solvent drawn into the gasket joint.

Checking for gasket vacuum leaks with solvent will not tell you that the gaskets are 100 percent leak-free. The lower edges of intake manifold gaskets on V-type (and some inline) engines are not accessible for checking. The joints between the lower edges of the intake manifold and the heads of V-type engines are inside the engine valley, and these are common locations for vacuum leaks. A sharp technician can spot this problem by hooking the vacuum gauge to the manifold and testing the vacuum while cranking. Normal vacuum is 3 to 7 in-Hg (75 to 175 mm) for a cranking engine with the throttle plates *closed* and all vacuum lines plugged. Anything less than this probably indicates an internal vacuum leak.

How to pinpoint problems with the valves
With a vacuum gauge, you can also detect leaking valves, incorrect valve timing, sticking valves, and weak or broken valve springs. Before proceeding, make sure that the ignition and fuel systems are in good working order. A spark plug misfire can produce vacuum gauge readings similar to those resulting from internal valve problems. An incorrect fuel mixture can cause the needle to oscillate one to two inches (25 to 50 mm) below a normal reading, figure 6-23.

Worn and leaking valve guides can be detected using a vacuum gauge. Valve guide wear is indicated by a rapidly vibrating needle at idle. As engine speed is increased the needle will steady, figure 6-24. You can double-check valve guide condition by removing the engine valve covers and squirting motor oil at the tops of the valve guides with the engine running. If a large cloud of blue oil smoke appears in the exhaust and the vacuum gauge reading increases, the valve guides are worn.

If the needle intermittently drops one to two inches (25 to 50 mm) from the normal reading, one of the engine valves is not seating properly, figure 6-25. Each time the valve fails to seat, the needle fluctuates or flutters. If the valve is actually burned, then the needle may drop as much as 9 inches (225 mm) from the normal reading. Do a compression test or a cylinder power balance test to locate the leaking valves.

An irregular drop of one or two inches (25 or 50 mm) from a normal reading indicates sticking valves, figure 6-26. Needle action is similar to that of a leaking valve, but not as consistent. If you are unsure whether the valve is sticking or leaking, remove the valve cover and run penetrating oil down the valve stems to lubricate the valve guides. Retest and compare your reading. Penetrating oil often temporarily prevents a valve from sticking, but has no effect on one that is not seating properly.

To check for weak or broken valve springs, increase the engine idle speed to about 2,000 rpm. If the needle fluctuates rapidly between 12 and 24 inches (300 and 600 mm), and the fluctuations increase in speed as rpm increase, the valve springs are weak, figure 6-27. If the fluctuation is rapid but irregular, a valve spring is probably broken. The rapid fluctuation occurs each time the valve tries to close.

Engine Testing and Diagnosis

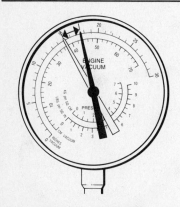

Figure 6-25. If the needle regularly drops one to two inches (25 to 50 mm) from the normal reading, one of the engine valves is burned or not seating properly.

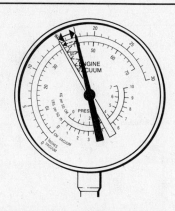

Figure 6-26. An irregular drop of one or two inches (25 or 50 mm) from a normal reading indicates sticking valves.

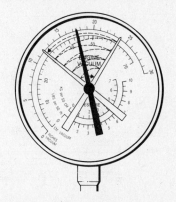

Figure 6-27. Weak valve springs will produce a normal reading at idle, as engine speed increases the needle will fluctuate rapidly between 12 and 24 inches (300 and 600 mm).

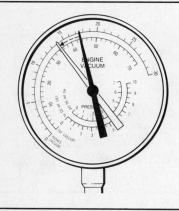

Figure 6-28. A steady reading that drops two to three inches (50 to 75 mm) when the engine is brought off idle indicates the ignition timing is retarded.

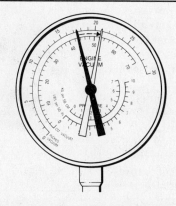

Figure 6-29. A steady reading that rises two to three inches (50 to 75 mm) when the engine is brought off idle indicates the ignition timing is advanced.

Figure 6-30. A needle that drops to near zero at high rpm then rises slightly to a reading below that of a normal idle indicates an exhaust restriction.

Ignition timing test
Vacuum gauges can also tell you whether the timing is incorrect, assuming you know what the reading for the engine ought to be. If the gauge needle remains steady at idle, but drops two or three inches (50 or 75 mm) if engine rpm is increased slightly, figure 6-28, late ignition timing may be the problem. Early ignition timing is shown by a steady gauge reading that increases two or three inches (50 or 75 mm) above normal when brought off idle, figure 6-29.

How to spot an exhaust restriction
With the vacuum gauge and engine tachometer connected as in the vacuum leak test, slowly open the throttle until engine speed reaches 2,000 rpm. The needle should rise quickly from the normal idle vacuum reading. Hold the throttle at that position and watch the gauge. If the exhaust is restricted, the vacuum reading will decrease to near zero, and then rise and level off at a reading below that of a normal idle, figure 6-30.

With some large-displacement engines, you may need to do this test while actually driving the car. Connect the vacuum gauge hose to a tee fitting installed in a manifold vacuum line. Run the vacuum hose under the hood and through the car window. This lets you watch the gauge action while driving the car through the speed range where performance starts to fall off.

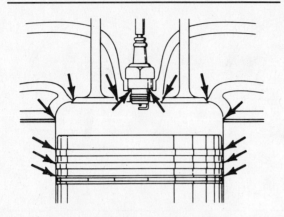

Figure 6-31. Good compression depends on proper cylinder sealing at many different points.

Figure 6-32. Good quality compression testers screw into the spark plug hole with a hose and adapter.

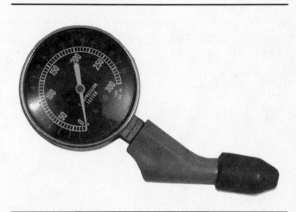

Figure 6-33. Inexpensive compression testers have a simple rubber cone that you hold in the spark plug hole with your hand.

How to check compression with a vacuum gauge

Before testing for poor compression, make sure the engine has normal readings on all other tests. The crankcase oil also must be in good condition. Diluted oil causes an incorrect reading which indicates a loss of compression when there is no such loss.

Connect the gauge and engine tachometer just as you did for the vacuum leak test. Open the throttle quickly and hold until engine speed reaches 2,000 rpm. Quickly close the throttle. A reading increase of less than five inches (less than 125 mm) indicates a compression loss. Be sure to do a compression test to verify the vacuum gauge findings, and to be sure there are no restrictions in the exhaust.

Engine Compression Testing

The power output of any automotive engine depends on the compression in its cylinders. This compression depends on how well each cylinder is sealed by the piston rings, valves, cylinder head gasket, and also by the spark plugs, figure 6-31. If any of these points are not properly sealed, compression is lost, causing a drop in power output.

You measure the compression pressure in each cylinder and variations in pressure among the cylinders by performing a compression test. By measuring how well each cylinder will hold its compression, any pressure loss can be located. Compression pressure is measured in pounds per square inch (psi) or kilopascals (kPa) using a compression gauge.

A compression gauge measures the amount of air pressure that a cylinder is capable of producing. Gauges come in two designs: those that thread into the spark plug hole and have a flexible rubber hose, figure 6-32, and those with a rubber cone that you simply hold tightly into the spark plug hole, figure 6-33. In both types, a valve in the gauge holds the pressure reading until a vent valve is pressed. Each automaker prints compression specifications for its engines. You can find these in the engine specification section of the factory shop manual and in many independent service manuals.

Specifications are usually stated in one of two ways. Compression specifications may require the lowest-reading cylinder to be within a certain percentage (generally 75 percent) of the highest cylinder. If the highest cylinder reading is 160 psi (1100 kPa), for example, the lowest cylinder reading must be at least 120 psi (830 kPa). In other cases, compression specifications are sometimes stated as a minimum value, with a certain allowable variation between the cylinders. If the minimum is 130 psi (900 kPa), and the variation is 30 psi (200 kPa),

Engine Testing and Diagnosis

99

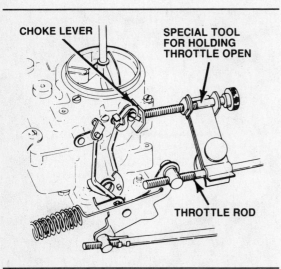

Figure 6-34. The choke and throttle must be held open during a compression test.

Figure 6-35. Remote starter switches will save you many trips back to the driver's seat.

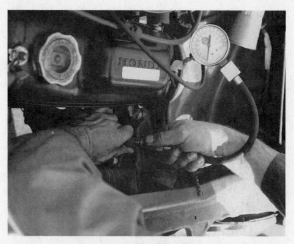

Figure 6-36. For screw-in testers, thread the hose into the spark plug socket first, then attach the gauge.

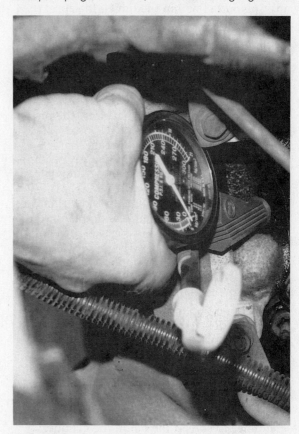

Figure 6-37. For press-in testers, firmly hold the tip of the gauge in the hole.

the compression in each cylinder is satisfactory as long as it is at least 130 psi (900 kPa) but no more than 160 psi (1100 kPa).

How to test engine compression
Use this procedure to test engine compression:

1. Place the transmission in Neutral or Park and set the parking brake. Start the engine and run at fast idle for several minutes to bring it to its normal operating temperature (upper radiator hose hot). Shut off the engine.
2. Disable the ignition system as described in the earlier PCV testing section.
3. Remove the air cleaner from air injection system or carburetor. Block the throttle linkage to hold the choke and throttle wide open, figure 6-34.
4. Remove the cables from all spark plugs. Loosen each plug one turn. Blow the dirt from around the spark plugs with compressed air, then remove the spark plugs.
5. Connect a remote starter switch according to the manufacturer's instructions for switch use, figure 6-35.
6. Thread the compression gauge adapter into a spark plug opening finger-tight. Connect the tester to the adapter, figure 6-36. If the tester has a tapered rubber tip instead of screw threads, insert the tip firmly into the spark plug opening and hold it in place during engine cranking, figure 6-37.
7. Crank the engine for at least five complete compression strokes. This will give you

the highest possible reading. Note and record the tester reading on the *first* and the *last* strokes.
8. Release the compression gauge pressure to return the gauge to zero. Disconnect the tester from the adapter and remove the adapter from the cylinder.
9. Repeat steps 5 through 8 for each remaining cylinder.

Compression test results
Compare the highest, or fourth stroke, reading for each cylinder with those specified by the manufacturer. The compression pressure should be within the maker's specifications for each cylinder. When limits are given for pressure differences between cylinders, the reading should fall within the limits. Interpret the gauge readings as follows:

- Compression is normal when the tester gauge shows a steady rise to the specified value with each compression stroke.
- If the compression is low on the first stroke and builds up with each succeeding stroke, but not to specifications, the piston rings are probably worn.
- A low compression reading on the first stroke that builds up only slightly on the following strokes indicates sticking or burned valves.
- When two adjacent (side-by-side) cylinders have equally low compression readings, there is probably a head gasket leak between the two.
- A higher than normal compression reading usually means excessive carbon deposits in the combustion chamber.

When you find low compression in one or more cylinders, there are a few additional tests you should make. Squirt about one tablespoon of engine oil through the spark plug opening in each low-reading cylinder. The oil acts as a temporary seal between the cylinder wall and the rings, but has no effect on leaking valves or a blown head gasket. Crank the engine several times to spread the oil on the cylinder walls and rings. Then recheck the compression for each of these cylinders.

When compression increases on the wet retest, it means rings or cylinder walls are worn. Oil should increase compression about 5 percent. If there is no compression increase, bad valves or a leaking head gasket probably cause the low compression.

Cylinder Leakage Test

A cylinder leakage tester, also called a leak-

Figure 6-38. A cylinder leakage tester tells you where and how much leakage you have from a combustion chamber.

down tester, figure 6-38, gives even more detailed results than a compression test. A leakage test can tell you:

- The exact location of a compression leak
- How serious the leak is in terms of percentage of the cylinder's total compression.

The cylinder being tested must have the piston at top dead center on the compression stroke, so that both valves are closed. The tester forces air from an ordinary air compressor into the sealed combustion chamber through the spark plug hole with the gauge installed in the line. The gauge indicates how much pressure leaks out of the combustion chamber.

The gauge scale is marked from 0 to 100 percent. A reading of 0 percent, physically impossible to obtain, indicates a perfectly sealed chamber with absolutely no leaks. A reading of 100 percent indicates that no pressure at all is being held within the chamber. You must calibrate the gauge before each test, following the equipment manufacturer's instructions.

Cylinder leakage testers differ, and you should be sure to follow the manufacturer's instructions for the model that you use. In general, however, this procedure will work for most equipment:

1. Bring the engine to normal operating temperature (upper radiator hose hot) and run at normal idle.

Engine Testing and Diagnosis

2. Turn off the engine and disable the ignition system as described in the earlier section on testing the PCV system.
3. Number the spark plug cables for reassembly, and remove all spark plugs and plug gaskets or tubes.
4. Remove the air cleaner. Block the carburetor or fuel injection linkage so that the choke plates or throttle plates are held completely open.
5. Disconnect the PCV hose from the crankcase.
6. Remove the radiator pressure cap, or be sure you can see the liquid in the cooling reservoir.
7. Carefully mark the spark plug wire positions on the base of the distributor with a piece of chalk or a crayon. Remove the distributor cap. If the engine does not have a distributor, you will need to position the pistons at TDC on the compression stroke visually, or in some other manner.
8. Calibrate the leakage tester according to the manufacturer's instructions.
9. Use a wrench on the crankshaft pulley nut to turn the engine by hand. Turn the engine until the distributor rotor tip points to the mark on the distributor base corresponding to the number 1 spark plug tower. The number 1 piston is now near top dead center on the compression stroke.
10. Install the tester adapter in the spark plug opening. Connect the tester to the adapter.
11. Connect the compressed-air hose to the tester, figure 6-39, following the equipment manufacturer's instructions and precautions.
12. Pressurize the cylinder.
13. Note the percentage reading on the tester's scale. Use the following table to interpret the reading:

0-10%	Good
10-20%	Fair
20-30%	Poor
30% and above	Dead!

14. If the cylinder has more than 20-percent leakage, pinpoint the cause of the leaks as follows:

    - Listen for air escaping through the carburetor, indicating a leaking intake valve.
    - Listen for air escaping through the exhaust pipe, indicating a leaking exhaust valve.
    - Listen for air escaping through the crankcase and PCV system, indicating

Figure 6-39. Connect the compressed-air hose to the tester.

   worn or damaged piston rings, worn cylinder walls, or a worn or cracked piston.
    - Watch for air bubbles in the coolant, indicating a leaking head gasket or a crack in the block or head.
    - If two adjacent cylinders have high leakage readings, the head gasket is leaking between them or the head or block is cracked.

15. Shut off the compressed air source.
16. Disconnect the tester hose from the adapter.
17. Turn the engine by hand until the distributor rotor tip points to the next mark on the distributor base. The next cylinder *in the firing order* is now near top dead center.
18. Remove the tester adapter from the cylinder just tested and install it in the cylinder to be tested.
19. Repeat steps 11 through 18 until all cylinders have been tested.

You can save time in spotting problems by using a few plastic bags and some rubber bands. Attach a plastic bag over the radiator filler neck opening, the tailpipe, and the carburetor mouth or fuel injection intake. If compressed air escapes from these places during the test, it will inflate the bag, and let you *see* that the system leaks.

Power Balance Tests

The power balance test shows you if an individual cylinder, or a group of cylinders, is not producing its share of power. During the test, you short out the suspected spark plug or plugs so that there are no power strokes from the cylinder or cylinders being tested. If an engine is in good condition, all of its cylinders provide the same amount of horsepower. Shorting out a cylinder should cause the same horsepower loss as shorting out any other cylinder. Various conditions, including engine rpm and manifold vacuum, change when cylinders fail to fire. The changes that occur tell you whether that cylinder or group of cylinders has been doing its share of the engine's work.

You measure these changes in terms of engine rpm drop, manifold vacuum drop, or a combination of these factors. The changes that you choose to observe during a power balance test depend both on the test equipment you are using and the particular engine you are testing.

Remember that the ignition and fuel systems must be working correctly. The power balance test can not distinguish between a dirty injector, a bad spark plug, and a cylinder with worn rings. All cause the cylinder to have lower output.

Power balance test equipment

To perform an engine power balance test, you usually use an engine analyzer that lets you disconnect the spark plugs with pushbuttons, figure 6-40. The numbers on the pushbuttons usually refer to cylinder *firing order*, not to the cylinder numbering. For example, the number 5 pushbutton controls the fifth cylinder in the engine's firing order, not cylinder number 5. Some consoles have specially marked scales that make it easier to measure the changes that occur.

Some older-model consoles have a single knob that controls the individual spark plugs. This type of console may not be safe for use with a modern electronic ignition system. Be sure that your equipment is compatible with the car you are testing.

Power balance test precautions

If the car you are testing has an EGR valve with a movable valve, disconnect it during the

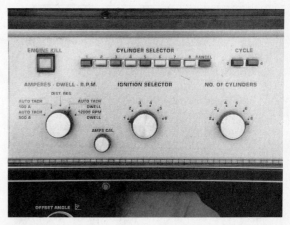

Figure 6-40. To do a power balance test easily, you should have a tester like this one that lets you disconnect a given cylinder with a switch or push button.

power balance tests. Otherwise, the cycling of the EGR system will affect the engine rpm and interfere with your test results.

If the car you are testing has a catalytic converter, you must try to limit the amount of unburned fuel reaching the converter. Various test equipment makers have different methods of testing a converter-equipped car:

- One recommendation is to short each plug for no more than 15 seconds and then let the engine idle normally for 30 seconds between tests. This keeps the unburned fuel from building up in the system.
- Another recommendation is to do the test as quickly as possible without pauses. This is to keep the unburned fuel from building up.
- Engines with feedback carburetors or injection systems that have exhaust gas oxygen (EGO) sensors should be in open-loop mode for accurate power balance testing. When you kill a cylinder, excess (unburned) oxygen enters the exhaust and creates a false signal. This can drive the fuel system fully rich and cause inaccurate power balance results. Follow the carmaker's instructions to do a power balance test on an electronically controlled engine.
- If the engine has an electronic idle speed control system, you cannot test it accurately at idle. The idle speed control system will allow a momentary decrease in speed and then automatically return it to normal. Some carmakers recommend not doing a power balance test on such an engine or doing it at 1,500 to 2,000 rpm. Always follow the manufacturer's directions.

The following paragraphs give basic procedures for power balance testing, but you must

Engine Testing and Diagnosis

modify these in accordance with carmakers' specific instructions for late-model vehicles.

Performing a cylinder power balance test
In this test, you short out individual cylinders one-by-one while the rest of the cylinders continue to fire.

1. Connect the engine analyzer according to the equipment maker's instructions.
2. If the engine has an EGR system, disconnect and plug the vacuum line from the EGR valve.
3. If the engine has electronic controls, place it in open-loop operation according to the carmaker's directions. On many systems, this is done by disconnecting a specified sensor.
4. Start the engine and bring it to normal operating temperature (upper radiator hose should be hot).
5. Run the engine at fast idle (about 1,000 to 1,500 rpm) or at another test speed as specified.
6. Press the button to kill one cylinder. Note and record the engine rpm drop and manifold vacuum drop. Release the button.
7. Repeat step 6 for each remaining cylinder.

After testing all cylinders, compare the results. If the changes in engine rpm and manifold vacuum are about the same for each cylinder, the engine is in good mechanical condition. On the other hand, if the changes for one or more cylinders are noticeably different, the engine has a problem. The fault may be mechanical, or it may be in the ignition or fuel systems. You will have to make further tests to pinpoint the problem. Some of the remaining engine tests may help you to find the fault.

Interpreting Power Balance, Compression and Leakage Test Results

Compare the results of the power balance test, the compression test, and the cylinder leakage test to narrow the list of possible engine problems. The following paragraphs describe some typical combinations and their meanings.

Good balance, good compression, poor leakage
If an engine has even power balance, compression within specifications, but heavy leakage, it is probably a worn, high mileage engine. Symptoms include excessive blowby, lack of power, and poor economy. Combustion chamber deposits can account for relatively good compression.

Poor balance, good compression, good leakage
Many problems can cause poor power balance test results on an engine that has good compression and low (good) cylinder leakage. Ignition problems in one or more cylinders are one common cause. If no ignition problems exist, the cause is something outside of the combustion chamber, such as one or more:

- Broken or bent pushrods
- Broken rocker arms
- Worn camshaft lobes
- Collapsed hydraulic valve lifters
- Leaking intake manifold gaskets
- Leaking valve guides
- Defective multi-port type fuel injectors.

Poor balance, poor compression, good leakage
A cylinder with low (good) leakage but poor compression and poor power balance has a valve-related problem. For any one of a number of reasons, one valve may fail to open at all, not open all the way, or open at the wrong time. If all the engine's cylinders show low leakage but poor compression, the camshaft could be out of time. Or, the wrong camshaft, pistons or crankshaft could have been installed in the engine.

ENGINE NOISES

Engine noises are difficult to locate and pinpoint, because so many things in the engine make noise, because these noises tend to travel through the engine, and because they can come and go with temperature, engine load, rpm, and other factors. They are also very difficult to describe. Some auto manufacturers distribute training videotapes with actual recordings of engines with various knocks and rumbles. These are good tools, and help you become familiar with sounds not often heard.

However, remember that virtually any moving engine part can make noise if it wears out, is damaged, or interferes with some other part, and that noises from unusual sources can sound just like the common ones in troubleshooting charts. And remember this: the very same problem in two different types of engines can sound very different, just as very different problems in different types of engines can sound pretty much the same. Something as simple as whether the engine has a cast aluminum or a stamped steel valve cover can make a radical difference in the way a noise sounds to your ears.

Be sure to eliminate sounds that seem to

come from the engine, but might be the normal rythym of an accessory. It takes only a minute to loosen a belt and see if the rattle comes from the alternator, power steering pump, or other engine driven accessory.

We will discuss some common engine noises in this section, but remember that you should not rely only on your ears. Try other tests, too, such as a compression check, vacuum measurement, or a cylinder power balance test. Sometimes the only way to identify the noise is to tear down the engine to find and replace the broken parts. And remember that there is no substitute for experience in diagnosing engine noises. When you hear a noise, remember what it sounds like. When you discover the cause, file it away in your memory with the noise so that you can use the information the next time.

Ask the Customer

Be sure to ask the customer for as much information as possible about the noise. You may not get a useful description of the noise, but you might find out something critical about when the customer hears it:

- Does it happen all the time, only just after the engine starts, or only after the engine has warmed up?
- When did the noise first start?
- Does it happen at idle, at higher rpm, or both?
- Does it only occur with the car moving?
- What makes the noise get louder or go away?

The answers to questions like these can help you identify problems that are related to transmission, drivetrain, or the wheels, as well as other mysterious noises that have nothing to do with the engine. Lastly, ask the customer to point out the noise as you both listen to the car. You might not hear it right away, and talking to the customer can save you time spent looking over the engine when the problem is a loose tailpipe.

Using a Stethoscope

If the engine condition is pretty bad, you will be able to listen to the noise just by standing next to the car. To isolate more subtle noises, you can use a mechanic's stethoscope. These tools are also great for isolating and interpreting specific engine noise. The tool works like an ordinary medical stethoscope — it has a thin diaphragm inside a housing that magni-

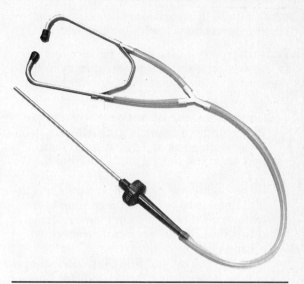

Figure 6-41. Mechanic's stethoscopes are constructed like a medical stethoscope, but have a long probe to reach into restricted areas.

fies sounds and sends them up the tubes to your ears. The difference is that the mechanic's stethoscope has a long needle attached to the diaphragm that you can use to probe individual engine parts, figure 6-41. If you do not have a stethoscope, you can use a long screwdriver, with the handle held against the bone under your ear. Be careful not to touch the screwdriver to a moving engine part while you hold it up to your ear. Once you use a stethoscope, though, you will not want to go back to the screwdriver.

Top or Bottom End?

The first decision you must make when you listen to engine noise is to decide where the noise is coming from — the top end or the bottom end. Often (but not always), bottom end knock and rumbling noises occur with high frequencies, at crankshaft speed. Top end noises, on the other hand, have frequencies one-half that of crankshaft speed. You can also simply search with the stethoscope on the cylinder head or on the engine block, and see where the noise is loudest.

You can use a timing light to determine whether a noise is from the top or bottom end of the engine, and even which cylinder it is coming from. Hook the timing light up to a spark plug wire and see if the engine noise cycles in time with the flashing light, or if it cycles at half that speed. Obviously, if the noise is in time with the valve gear, then it is

Engine Testing and Diagnosis

related to the top end of the engine. You can move the timing light pickup from wire to wire to pinpoint the noise to a particular cylinder.

Stay away from the distributor, water pump, or fuel pump, because those parts make their own normal noises, and abnormal noises can be quite loud. On front-wheel-drive cars, try to stay away from the transaxle assembly, too. Disconnect any suspect accessories and listen again before you make a final diagnosis.

Valve train noises
First listen to the top end. Run the engine long enough to let the idle speed drop down so that the frequency is as low as possible. Then get the stethoscope and start listening. Avoid placing the stethoscope on the valve cover directly — the thin metal vibrates like a drum and picks up engine noise from all over the top of the engine. Also avoid placing it on a cylinder head bolt, because the bolt will transmit noises from the engine block. Instead, touch the probe to the bolts or studs holding the cover on. You will not be able to identify exactly which valve makes which click, but you will be able to decide which bank of a V-type engine contains a faulty part, or if a noise is actually coming from the valve train.

Normal top end sounds
When you listen to the top end of a healthy engine, expect to hear a continuous, high-pitched whirring noise with a very rapid and much fainter sewing-machine clicking coming from the valves. The whirring is the combined, averaged sound of all the moving parts in the engine. The more valves the engine has, and the higher the idle speed, the more the individual clicks will blend into a constant drone.

The clicking will be louder if the engine has solid (mechanical) lifters, or (sometimes) if it has a roller camshaft. However, it should be even, and not loud enough to hear easily without the stethoscope. You should not hear regular louder clicks, or hear regular gaps in the clicking.

Abnormal top end sounds
Most abnormal valve train noises have a regular, repeated rhythm, either a noise that repeats itself at rapid but regular intervals, or a rattle that cycles in and out at intervals a few seconds apart. However, a camshaft with excessive endplay makes a clacking, knocking noise with an irregular rhythm, as the shaft, gear, or pulley bangs against the block or the timing cover as the camshaft rotates.

Another irregular noise coming from the top of the engine is that caused by a loose timing belt. The slack side of the belt flaps and slaps the side of the cover or the pulleys, making an irregular thumping that comes from the very front of the engine. Sometimes this noise can even sound like a diesel engine at idle. A tight timing belt makes a whirring, whining, hum that rises and falls in pitch with rpm.

Mechanical problems that cause noise in the valve train include:

- Worn-out valve guides
- Worn-out camshaft lobes
- Worn or bent pushrods
- Collapsed lifters
- Worn rockers
- Loose rocker arm studs, shafts, or locknuts
- Sticking valves
- Broken valve springs
- Too much runout of valve seats or valve faces.

Many of these noises are difficult to distinguish, and require some more investigation. For example, both a collapsed lifter and a broken valve spring will make a single, clear clack whenever the valve opens. They sound almost exactly the same, but you can locate either easily by removing the valve cover and inspecting the valve hardware.

Both rocker arm pivot wear and valve guide wear can cause a louder, cycling, valve rattle that you can hear over the normal valve noise. The rattle is caused as the rocker arm or the valve repeatedly strikes the rocker stud or valve guide each time the valve opens.

In addition, loud valve train noise might be caused by oil problems. If the oil level in the crankcase is too low, then hydraulic lifters can not get enough oil pressure to take up the valve clearance. The loose valves will clack against their seats, even though there may be nothing wrong in the valvetrain itself. Low pressure or restricted oil flow to the cylinder head will also cause excessive top end noise. In this instance, the frequency and rhythm of valve train noise might be regular and even, but the volume will be much louder than normal.

Too *high* an oil level can cause a similar problem in some engines. If it is high enough to contact the spinning crankshaft, then the crank will beat it into foam. This aerated oil is difficult for the oil pump to pick up and circulate, and when the bubbles reach the lifter galleries, they allow the valve springs to collapse the lifter. This also causes many problems besides noisy hydraulic lifters.

Under some conditions, the oil becomes aerated and causes valve noise only at high engine speeds, and idling the engine will make the valve noise disappear. The same symptoms can appear if the oil pump pickup tube or the pump itself develops a leak that lets it draw in air along with the oil.

Last, remember that noisy valves may simply need adjusting. This is more common with mechanical valves, but even the hydraulic lifters in pushrod engines may need an adjustment after very high mileage, or if the valves have been adjusted incorrectly. This occurs because a hydraulic lifter has a limited range of adjustment within which it can increase or decrease valve lash to the necessary setting. With high mileage, the valve train may wear so far that the lifter can no longer take up all the slack. When that happens, any further wear will cause valve noise to increase. This is easily corrected by tightening the rocker stud adjustment locknut to depress the plunger assembly the correct distance into the lifter body once more.

Bottom end noises

Noises from the bottom end of the engine include those made by the crankshaft and the piston and connecting rod assemblies. When you want to listen to the bottom end of the engine, put the stethoscope probe onto an accessible part of the block casting, such as below a cylinder head or down next to the oil pan rail. Sometimes you must put the car on a hoist to get at the bottom of the engine, figure 6-42.

Depending on where exactly you put the stethoscope, you may or may not hear the valves. Try to find a location on the block where the droning from the valves is faintest.

At the bottom end of a healthy engine, expect to hear the same rapid whirring as before, and if everything is well, nothing else. The oil pump should operate silently, and you should hear no rumbling or pounding noises.

Knocking or thumping noises are signs that something is wrong. In general, bottom end noise can be caused by:

- Worn main bearings
- Worn connecting rod bearings
- Too much crankshaft endplay
- Crankshaft main or rod journals worn out-of-round
- Too much piston pin clearance
- Bent connecting rod
- Piston slap from worn cylinder bores
- Insufficient piston to bore clearance
- Cracked piston skirt.

Noises from the bottom end are often less distinct and deeper in tone than the noises from

Figure 6-42. From underneath the car, the stethoscope is used to isolate noises in the bottom end of the engine.

the valve gear, although it can be much louder. Except for crankshaft endplay and connecting rod clatter, bottom end noises have a regular rhythm and change with engine rpm.

Crankshaft endplay

Too much crankshaft endplay produces an irregular knock similar to a camshaft with excessive endplay, and is also most noticeable at idle. You can make the noise louder or fainter by playing with the clutch pedal.

Connecting rod bearings

The sharp, clattering knock from bad connecting rod bearings may be continuous at idle, or may only appear as you close the throttle suddenly. You can make bad rod bearings knock by floating the throttle back and forth at intermediate rpm. The knock will often disappear while the engine speed is rising, then return as the speed drops or steadies.

A cylinder power balance machine also lets you check a noise you suspect may be a rod bearing. Ground each cylinder in turn and listen to the noise. Rod bearing noise decreases as you ground the noisy cylinder, because you remove the load from the rod.

Piston slap

Piston slap is a common noise that is loudest with a cold engine, and decreases and may go away completely as the engine warms. It is caused by too much clearance between the pistons and the cylinder bores. The piston rocks back and forth at bottom and top dead center

Engine Testing and Diagnosis

as the connecting rod's big end swings sideways. The sound is a hollow metallic clatter that sometimes sounds exactly like the noise a diesel truck engine makes at idle.

To confirm piston slap, retard the spark and listen again. Reducing the pressure on the piston crown this way will reduce the slap. One time-honored test is to squirt a tablespoon or so of 40- or 50-weight motor oil into the suspect cylinders, and restart the engine. The heavy oil should reduce the piston noise for a few seconds.

Grounding an affected cylinder will often make piston slap louder, because the piston does not have the cushioning of the extra gas pressure holding it down at top dead center. This makes the slap at the top of the cylinder louder.

Wrist pin noise

Bad wrist pins may or may not give off the classic "double rap" noise that they are famous for. The actual noise is a sharp knocking that stands out most at idle, caused by the piston pin striking the top and the bottom of the pin bores as the piston changes direction at top and bottom dead center. If clearances are very large, the noise will sound like a very loud sewing machine at higher rpm.

Grounding a cylinder increases the racket from noisy wrist pins, because the pin then taps at top dead center as well as bottom dead center. This may also be audible as the "double rap." Retarding the spark decreases wrist pin noise.

Main bearings

Main bearing noise is a rapid, steady dull pounding which you may not hear at idle, but which increases under load, such as under acceleration or at higher speeds. You can sometimes hear a dull pounding from main bearings with excessive clearance when you first start the engine, especially after changing the oil.

Other noises

Noises that seem to come from the engine bottom end can also be caused by other problems:

- Loose or broken harmonic balancer or pulleys
- Loose torque converter bolts
- Cracked flywheel or flex plate
- Flywheel hitting splash shield
- Oil pump gear rotor wear.

Never rule out anything when you diagnose noises.

Chapter 7

Engine Repair in the Vehicle

Some of the repair and rebuilding jobs you will perform can be started and finished without ever removing the engine from the car. Cylinder heads are usually fairly easy to remove and install right in the engine compartment. Timing chains or gears are also easy to remove and install. In some engines, you can inspect and replace the main and connecting rod bearings after dropping the oil pan.

Working on the engine in the car is not as convenient as when the engine is in a stand or on a bench, but it can save you time for certain repairs. Be sure to plan ahead, though, before choosing this route. If you are going to work on both the top and bottom end, you will probably save more time in the long run if you go ahead and pull the engine out so that it is easier to work on.

In this chapter, we will explain some of the basic engine mechanical tasks that you can perform with the engine still in the car, including repairs to the top and bottom end, camshaft drives, and routine valve service.

SERVICING THE CAMSHAFT DRIVE

Both OHC and OHV engines drive the camshaft through a belt, chain, or gear set, accessible for repairs on the front of the engine, figure 7-1. For many years, most designers specified chain or gear drive to rotate the camshaft. The camshaft in a pushrod engine is mounted at a fixed distance from the crank, with no intervening gasket surfaces to disrupt gear mesh or alter chain tension. The gears or chain and sprockets are usually just pressed onto keyways in the shaft ends, figure 7-2. These systems are simple and reliable, and normally last the life of the engine.

Many overhead-camshaft engines use composite camshaft drive belts. Timing belts and sprockets reduce weight, require no lubrication, and produce less noise than chains and gears, figure 7-3. Belts are often mounted outside the engine behind a simple shroud, figure 7-4. Chain and gear drives require oil lubrication and are always concealed behind a cover, figure 7-5. Manufacturers who use belts in their cars usually specify a recommended service inspection or replacement interval, chain drives do not require periodic service.

Valve Interference

When you install timing chains, belts, or gears, all of the timing marks on the engine

Engine Repair in the Vehicle

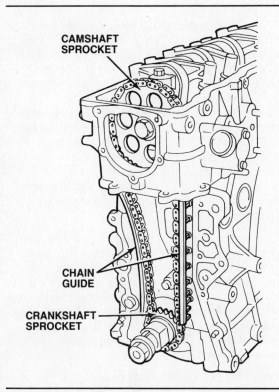

Figure 7-1. This Nissan engine takes the camshaft drive off the front of the crank using a single roller chain.

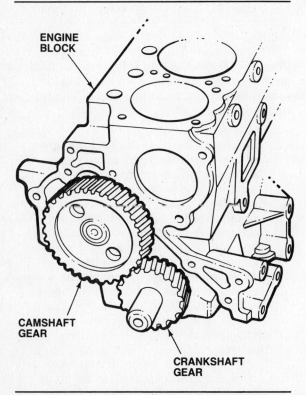

Figure 7-2. Geared camshaft drives use a direct connection between the crankshaft and the camshaft.

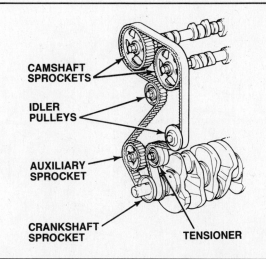

Figure 7-3. Timing belts are quieter than chains or gears, but require more frequent maintenance.

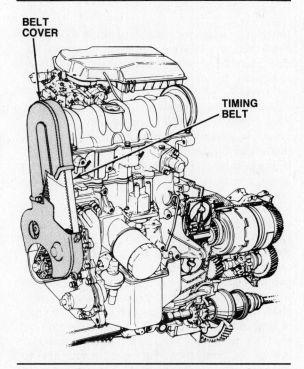

Figure 7-4. The timing belt on this transverse OHC engine is protected by a simple plastic shroud.

must be in alignment. The initial timing position for most engines is with number 1 cylinder at TDC of the compression stroke. There are exceptions. Some European engines are timed with the number 4 cylinder at TDC of the compression stroke. Whenever you are going to remove timing components, it is a good

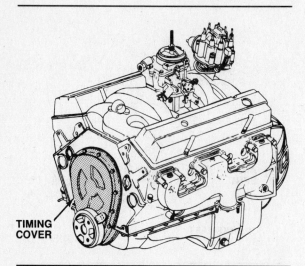

Figure 7-5. The oval shaped cover on the front of this engine seals the camshaft drive (a chain on this engine).

practice to make rotating the engine to align the timing marks your first step. You want to avoid having to rotate the crankshaft or camshaft with the drive chain, belt, or gears removed.

Many engines are an interference design. This means the combination of cam timing, valve lift, and high compression allows the pistons and valves to sweep the same volume in the combustion chamber at different times in the engine cycle. If you rotate an interference engine with the timing mechanisim disconnected, there will be valve-to-valve or valve-to-piston contact. Contact of this nature can easily bend valve stems to create bigger problems. The engine will not run properly. Compression is lost because the valves cannot properly seat. Severe contact can also damage the piston crowns. If you must rotate the camshaft more than about 1/8-turn to position the timing marks, be sure to *first* rotate the crankshaft about 90° off TDC in the normal direction of rotation (clockwise for most engines) to give the valves clearance for opening. Once the camshaft is in position, rotate the crankshaft backward to TDC and finish the job.

Timing Chains and Gears

Timing chains are common in OHV engines. Chain drives are also used on some OHC engines, although composition rubber belts are currently more common. Chain and gear designs vary from engine to engine. Single or double roller chains may run over steel, cast iron, metal-nylon, or metal-composition gears or sprockets, refer to Chapter 9 of the *Classroom Manual*.

Some OHV engines use gear-to-gear setups to drive the camshaft, gear sets are normally not used on OHC engines because a precise mesh between the gears is very critical. Changing the distance between the centerline of the crankshaft and camshaft centerline by as little as 0.002 inch (0.05 mm) can upset the clearance between timing gears in a two-gear setup. This means that routine head resurfacing or head gasket replacement could alter the timing gear interface and cause severe damage.

How to service OHV timing chains and gears
Use these procedures for servicing the standard OHV timing chain and gears. Removing and replacing a geared timing set is more difficult than removing and replacing an OHV timing chain, because in many cases the camshaft must be removed from the car to press the gear on or off. This may not be possible with the engine still in the car.

Removing the OHV timing cover
An OHV pushrod engine might have a cast cover similar to those of most OHC engines, or it may have one made of stamped steel. If the engine uses a bolt-on cast cover, refer to the section explaining how to remove the cover from an OHC engine, and follow those procedures. The following directions explain how to remove a stamped steel OHV timing cover. Removing them is easy, as they seldom hold anything other than the crankshaft seal, and sometimes a tag with timing marks. Follow this procedure:

1. Disconnect the battery.
2. Drain the radiator. Disconnect the upper and lower radiator hoses at the engine. If the vehicle has an automatic transmission, disconnect and plug the cooling lines at the radiator.
3. The radiator, fan, and fan shroud all need to be removed. Designs will vary. Some allow you to un-bolt the fan and remove the radiator and shroud as a unit, on others you may have to remove the shroud and then the radiator before you can get to the fan bolts.
4. Remove all accessory drive belts, and any accessories or brackets that interfere with the cover.
5. Remove the water pump, if necessary. V-type engines often use a water pump that straddles the timing cover. On an inline engine, you may not need to disturb the water pump at all.

Engine Repair in the Vehicle

6. Remove the crankshaft pulley. If the engine has a harmonic balancer, remove it using the proper puller, figure 7-6. Do *not* hammer or pry on the outer ring of a rubber-bonded harmonic balancer.
7. Determine what other parts need to be removed before you take off the cover, and remove them.
8. Sometimes the lower part of the front cover is held on by the oil pan. If so, loosen the front oil pan bolts.
9. Unscrew the bolts holding the cover to the front of the block, and remove the timing tag, if necessary.
10. Gently pry the timing cover off the block, figure 7-7.

Figure 7-6. Pulling a harmonic balancer requires a specific puller. A puller that hooks over the rim of a rubber-bonded balancer will destroy the balancer.

Inspecting, removing, and reinstalling the OHV timing gears

Camshaft and crankshaft timing gears might be pressed onto their shafts and require a puller to remove, although most are a slip fit. With some engines, you must remove the camshaft from the engine to remove the gear; with others, you can remove the camshaft gear without removing the cam. With either type, you can inspect the timing gears for excess backlash and runout, and the camshaft for excessive endplay while they are still installed on the engine:

- If there is too much backlash, then you must replace the gears.
- If there is too much endplay, you should check to see whether the camshaft spacers or thrustplate are missing or worn.
- If there is too much runout, check to see if a particle of metal is caught between the gear and the shoulder of the shaft.

Follow this general procedure for inspecting the camshaft drive:

1. Remove the timing cover as already explained.
2. Remove enough of the front radiator grille and bodywork to give you clear access to the front of the engine (you should have already removed the radiator).
3. Label the spark plug wires and remove the spark plugs.
4. Remove the valve cover as explained elsewhere in this chapter.
5. You must remove the valve spring load from the camshaft before you can measure endplay, backlash, and runout. Loosen the locknuts holding the rocker arms in place and twist them to the side. You do not have to remove them from the cylinder head. If the engine has shaft-mounted

Figure 7-7. Remove the cover by lifting around the edge to break the gasket seal.

rockers, remove the bolts holding the rocker shaft in place and remove it from the head. Be sure to use the factory sequence for removing the rocker shaft, and loosen the bolts in several steps.

6. *To measure camshaft runout*, set up a dial indicator on the engine block with the plunger resting on the face of the camshaft gear just inside the gear teeth, figure 7-8. Zero the dial indicator.

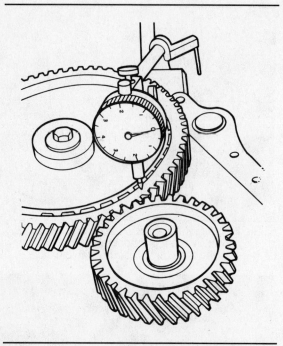

Figure 7-8. Set up the dial indicator on a flat, machined surface of the camshaft gear or sprocket to measure runout.

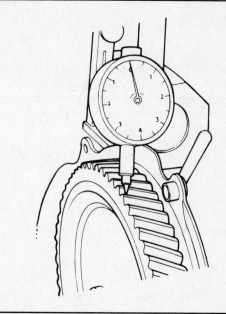

Figure 7-9. Set up the dial indicator on the edge of a tooth to measure backlash.

7. Rotate the camshaft one complete revolution while watching the dial indicator. Use a wrench on the crankshaft pulley bolt to turn the engine over, motion should be smooth and slow. Add the dial indicator deviations for both directions to calculate maximum runout.
8. Compare the runout to the specifications for that engine. Not all manufacturers specify runout, but a typical number is 0.004 to 0.005 inches (0.100 to 0.125 mm) for the camshaft gear, and 0.003 to 0.005 inches (0.075 to 0.125 mm) for the crankshaft gear, measured in the same manner with the dial indicator set up on the crankshaft gear.
9. *To measure camshaft backlash*, set the dial indicator up again with the plunger tip resting on the edge of a camshaft gear tooth, in line with (tangential to) the edge of the gear, figure 7-9.
10. Take all the backlash out of the gear mesh by turning the wrench on the crankshaft bolt slightly, and zero the dial.
11. Turn the wrench the other way to take up all the backlash in the other direction, read and record the total backlash on dial indicator.
12. Repeat on at least six equally-spaced gear teeth around the camshaft gear, and average the result.
13. Compare the backlash to the specifications for that engine. You can expect backlash to be in the 0.002 to 0.006 inch (0.05 to 0.15 mm) range.

Note: Backlash can also be measured using a feeler gauge, however, readings will not be as accurate as with a dial indicator. To use a feeler gauge, take the backlash out of the gear mesh and insert a feeler gauge between the gear teeth of the slack side. Backlash will be equivalent to the thickness of the largest feeler gauge blade that will easily fit between the gears. Repeat the procedure at several points on the gear and average your findings.

14. *To measure camshaft endplay*, set up the dial indicator with the plunger resting directly on the end of the camshaft center fixing bolt, or on a machined surface near the center of the gear, figure 7-10.
15. Push the camshaft towards the engine to take up all the clearance, and zero the dial.
16. Pull the camshaft away from the engine by wedging two large screwdrivers under opposite sides of the gear and gently prying out and then releasing.
17. Measure the total endplay by reading the dial indicator, and compare it to the specifications for that engine. The acceptable endplay varies between engines. Typical numbers are between 0.001 to 0.007 inches (0.025 to 0.175 mm). If the endplay is outside limits, check the camshaft spacers or thrustplate for wear or damage.

Engine Repair in the Vehicle

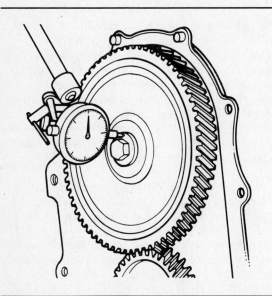

Figure 7-10. Set up the dial indicator near the center of the gear. To measure end play, gently pry out on and then release the gear.

Figure 7-11. Remove the bolts holding the camshaft retainer in position.

If endplay is out of specification, you will have to remove the camshaft gear to replace the thrustplate or spacers. If the backlash or runout are out of specification, you will have to remove and replace both the camshaft and crankshaft gears. Both gears are generally machined as a matched set and should be replaced as a unit. Follow this general procedure:

1. Remove the spark plugs and rotate the engine until the timing marks on the timing gears align.
2. Check the service manual for the engine you are working on to determine the method used to remove the camshaft gear. Some designs allow you to simply remove the gear using a jaw-type puller. For others you will have to remove the gear and camshaft from the engine block as an assembly. The gear is then separated from the shaft using an arbor press. If you need to remove the cam, follow the additional steps. Some front body panels may have to be removed. You will need enough free access in front of the engine to allow you to pull the camshaft straight out.
3. Remove the pushrods and mark them so that they can be replaced in their exact original positions. A simple way is to push them through numbered holes in a piece of cardboard.
4. For an inline engine, access and remove the lifter inspection cover or covers from the side of the engine. Reach into the lifter cavity and remove the lifters. Label the lifters for reassembly in the same bore.
5. For a V6 or V8 engine, you will have to remove the intake manifold in order to remove the lifters. Be sure to keep them in the correct order for reassembly.
6. Remove the camshaft retainer or thrustplate bolts, figure 7-11.
7. Carefully pull the camshaft straight out of the front of the engine, figure 7-12. The camshaft bearings are fragile and can easily be damaged by the lobes and journals of the camshaft.
8. Put the camshaft in a press and carefully press the camshaft gear off the shaft.
9. Use a puller to remove the crankshaft timing gear, if necessary.

To reinstall the camshaft and crankshaft gears, follow this general procedure:

1. Inspect any woodruff keys and keyways in the camshaft, crankshaft, and gears for damage. Replace the key if it is damaged, and carefully remove any burrs at the edge of the keyways with a fine oilstone. You can lightly polish the shaft snouts with crocus cloth to help the gears slip on snugly.
2. Be sure any spacers or thrustplate components are correctly installed.
3. For press-fit camshaft gears, heat the gear gradually in an oven or hot oil bath to expand it and make it easier to press on. If the camshaft is out of the engine, press the camshaft gear onto the shaft in an arbor press. If the cam is still in the engine, use a gear installing tool to pull the gear onto the camshaft.

CAUTION: Using a hammer and drift to drive

Figure 7-12. Remove the camshaft by sliding it straight out, be carefull not to damage the bearings with the cam lobes or journals.

the camshaft gear onto its shaft can destroy the thrust surfaces and force you to repeat the whole job. On some engines, the crankshaft gear can be tapped into place with a large open drift and a hammer. Check the engine manufacturer's service manual for the proper procedure.

5. For slip-fit camshaft gears, just slip the gear over the end of the cam.
6. Lubricate and install the camshaft, if necessary, and reinstall any retaining bolts securing the camshaft gear.
7. Position the crankshaft gear onto the crankshaft snout, and make sure that the timing mark aligns correctly with the camshaft mark. Then use the gear installing tool to pull the crankshaft gear onto the shaft, or tap it into place with an open drift and a hammer.
8. Reinstall the lifters, pushrods, inspection covers or intake manifold, and spark plugs.
9. Lubricate the gears with engine oil before installing the timing cover.

Removing, inspecting, and reinstalling the OHV timing chain

Heavy-duty and truck OHV engines with timing chains often have all-metal sets with roller element chains. These are strong and long-lasting. Many domestic OHV passenger car engines with timing chains use the metal-nylon sprocket sets because they run more quietly than all-metal arrangements. Unfortunately, the nylon teeth wear fairly rapidly, especially in hot weather or under heavy-duty use. When the wear becomes bad enough, the chain can jump a tooth and retard the valve timing. Sometimes the nylon disintegrates catastrophically, which lets the camshaft freewheel and litters the sump with plastic and metal fragments. Aluminum gears can also fail dramatically. Always inspect a nylon or composition set very carefully, and replace them if they show significant wear.

Another problem caused by wear is called chain stretch. The metal chain will not actually stretch, but does become longer as the pins and the pivot holes in the plates wear, and joints loosen up. By itself, chain stretch is usually not a problem, as the chain can lengthen considerably before sprocket problems develop. What does matter is that the slack causes the camshaft sprocket to lag behind the crank sprocket as the engine runs, retarding the valve timing. In the small block Chevrolet, for example, every 0.020 inch (0.5 mm) that the chain elongates retards the timing about one degree from the intended value. Retarded valve timing can cause poor idle and low rpm performance.

This problem is so common that many aftermarket camshaft manufacturers purposely position the keyways or locating slots on their camshafts about 4° advanced from the ''straight-up'' position. This built-in camshaft advance ensures that the valve timing will remain closer to the correct specifications through the normal life of the camshaft as the chain stretches.

Most OHV engines do not use a chain tensioner because the distance between the crank and camshaft centerlines never changes. Check slack in these engines this way:

1. Use a wrench to rotate the crankshaft counterclockwise just enough to take all slack out of the left side of the timing chain. Avoid turning the camshaft. All you want to do is reposition the slack in the chain.
2. Mark or establish a reference point on the block face in line with the midpoint of the tensioned (left) side of the chain. Measure the distance from the reference point to the chain with a machinist's scale, figure 7-13, and record your findings.
3. Without turning the camshaft, rotate the crankshaft clockwise just enough to transfer the tension to the opposite side of the timing chain. All the chain slack should now be on the left side of the chain, when viewed from the front.

Engine Repair in the Vehicle

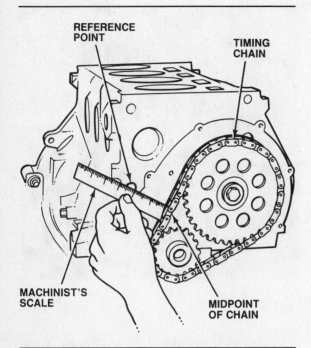

Figure 7-13. With the slack on the opposite side of the chain, measure from a reference point on the block to the midpoint of the chain.

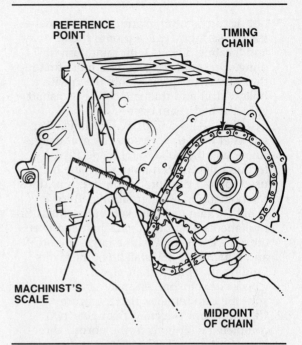

Figure 7-14. Transfer chain slack to the other side of the chain, force the chain out with your finger, and measure from the reference point to the chain.

4. Use your finger to force the slack portion of the chain towards the reference point, measure and record the distance from the reference point to the chain with a machinist's scale, figure 7-14.
5. Total chain deflection is the difference between your two measurements. Compare this number to the manufacturer's specifications for that engine. In practice, one-half inch (1 cm) is the maximum slack you should accept. Even this amount will seriously affect valve timing, and is a normal amount of wear to find after 50,000 miles (80,000 km). If wear exceeds specifications, replace the timing chains and sprockets.

To remove and replace the timing chain, follow this general procedure:

6. Remove the spark plugs. Rotate the engine, using a wrench on the crankshaft end-bolt, until the timing marks on the sprockets align, figure 7-15. If you removed the end-bolt in order to pull off the harmonic balancer, refit it temporarily. *Never turn the engine with the camshaft bolt.*
7. The camshaft sprocket may or may not be pressed in place, and then secured by a bolt. Remove the bolt and gently tap the edge of the sprocket to loosen it. The crankshaft sprocket is sometimes pressed in and must be removed with a puller.

Figure 7-15. The timing marks on sprockets are often just punch marks that you must align.

8. Inspect the Woodruff key and the keyway in the crankshaft snout. Replace the key if it is damaged, and carefully remove any burrs at the edge of the keyway with a fine oilstone. You can lightly polish the crankshaft snout with crocus cloth to help the new sprocket slip on snugly.
9. Arrange the new timing chain and sprockets so that the timing marks align, and then slip the sprockets over the camshaft and crankshaft snouts. If the engine was

not disturbed after aligning and removing the old set, they will slip onto the keyway and locating dowels with minimum fiddling. On some engines, you must first tap the crankshaft sprocket into place with a hollow drift and then attach the camshaft sprocket and chain over it.

10. Install the bolts and draw the camshaft sprocket onto the camshaft by gradually tightening the bolts to their specified torque value. When more than one bolt is used to hold the sprocket to the cam, draw the bolts up in an alternating pattern to avoid cocking the sprocket on the shaft. Simultaneously tap the crankshaft sprocket onto the crankshaft with a large open drift and a hammer. Reinstall any other bolts holding the sprockets in place.
11. Check the camshaft endplay, and make sure the cam retainer plate is secure.
12. Carefully turn the crankshaft gear two complete revolutions in the normal direction of rotation, TDC to TDC, and check that the timing marks on the crankshaft gear and the camshaft gear line up again. If the marks on either gear fail to align exactly after exactly two revolutions, remove and reposition the gears.
13. Lubricate the new chain and sprockets with engine oil before installing the timing cover.

Reinstalling the OHV timing cover

If the cover is made of cast aluminum, refer to the section explaining how to install the cast covers on OHC engines, and follow those directions. If the cover is made of stamped steel, follow the directions below. Installing an overhead valve timing cover is simple:

1. Tap the old oil seal out of the cover.
2. Clean the cover in solvent and inspect it for damage. You may find that the cover has been scarred by contact with a loose timing chain. Excessive camshaft endplay will let the camshaft walk forward into the back of the cover, and either distort it or grind through it. If the damage is not repairable, replace the cover. The gasket sealing surface may have been damaged when prying the cover off during disassembly. Remove any burrs with a fine file or emery cloth, clean the gasket area of any old gasket and sealing materials.
3. Lightly oil the new oil seal, and position its open end to the inside of the cover. Carefully press or tap it into place with a seal driver, figure 7-16.

Figure 7-16. Use a suitable tool to press the new seal into the timing cover.

4. Clean all traces of old gasket and sealer from the block face.
5. Use a thin film of non-hardening gasket sealer, or narrow bead of silicone sealer, to attach the cover gasket to the engine block. You may choose not to use sealer on the cover itself to make later removal easy.
6. Place the cover onto the block and run in all the cover bolts, snug the bolts up in an alternating pattern to avoid distorting the cover. If you removed the timing tag with the cover, be sure to position it correctly. Torque the bolts to the manufacturer's recommended specification, usually only 6 to 8 ft-lbs (8 to 11 Nm).
7. Reinstall any components you removed earlier.
8. Reinstall the radiator, transmission cooling lines, and radiator hoses. Fill the radiator.
9. Connect the battery and start the engine.
10. Check the automatic transmission fluid, engine oil, and coolant levels, and top off as necessary.
11. Run the engine and check for leaks.

How to service an overhead cam timing chain

Although composition belts are more common in late model OHC engines, many older engines from Mazda, Isuzu, Datsun, Mitsubishi, and Toyota use chain drive. More recent OHC designs such as Nissan's Sentra and Infiniti Q45, and the General Motors Quad 4 engine also use a timing chain. European manufacturers, such as Alfa Romeo, BMW, and Mercedes-Benz, have always used timing chains for their OHC engines.

Engine Repair in the Vehicle

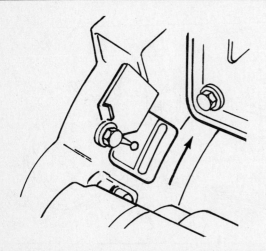

Figure 7-17. These are typical marks on the flywheel that you inspect through a hole in the rear cover plate.

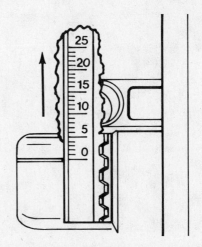

Figure 7-18. You will occasionally find numbered scales on the flywheel, as on this Saab.

Removing an OHC chain timing cover
Engines with a chain-driven overhead camshaft typically use a cast front cover to protect the timing components and to retain oil. Because the chain requires no periodic inspection or replacement, as do belts, the cover is not designed to be removed and installed quickly. It often serves several other functions such as, providing a place to locate a crankshaft seal, housing the water pump, and providing an attachment point for alternator, power steering pump, or other engine driven accessory brackets. Some engines also use the cover to house the distributor, or the oil pump. Determine what other parts have to be removed before attempting to remove the cover.

Some timing covers seal against the underside of the head, the front of the engine block, and the front part of the oil pan. Others are simpler, and use a larger cover over the front of the whole engine which has its own simple gasket. Depending on the situation, you might have to raise the car to remove an under-engine splash shield or an engine mount, and support the engine on a jack or hoist. Although some manufacturers recommend that you remove the oil pan, this is not always necessary if you are careful.

If the car is rear-wheel-drive, use the procedures listed previously to remove the radiator, shroud, and fan. The following step-by-step cover removal procedure is typical of a transverse, front-wheel-drive engine:

1. Remove all accessory drive belts, and any accessories or brackets that attach directly to the cover. The bolts that hold the brackets on may also hold the cover to the block or head. Pay attention to the position of

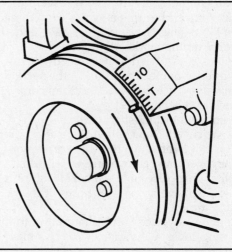

Figure 7-19. This Mazda times with a notch on the crankshaft pulley and degree marks on the engine block.

the bolts as you remove them, often several different length bolts are used.
2. Remove the spark plugs and rotate the engine to set the number one piston at TDC on compression, using the engine timing marks. Timing marks for engines with OHC chains differ. You may have to align:
   - Notches or marks on the flywheel, figure 7-17, with a hole in the engine rear cover plate.
   - A numbered scale on the flywheel, figure 7-18, with a pin on the cylinder block casting.
   - Grooves or marks on the crankshaft pulley, figure 7-19, with a pointer on the engine block or timing cover.

Figure 7-20. You must remove any water pump bolts that also support a bracket.

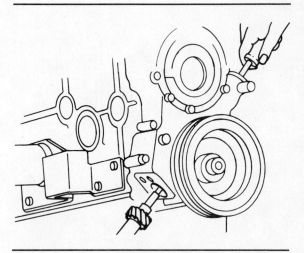

Figure 7-21. Push the oil pump/distributor drive out through the hole.

Figure 7-22. Remove any other bolts holding the cover to the engine block or cylinder head.

3. You will normally need to remove the valve cover, but this may not be necessary on all engines. Most engines have additional timing marks on the camshaft sprocket that need to be aligned with a reference point. If so, skip to the section on valve cover removal and then come back to this point.
4. Remove the crankshaft pulley.
5. Some covers also contain the water pump. You can often leave the water pump on the cover, but remove any water pump bolts which also hold a bracket, figure 7-20. Remove any heater hoses at the same time.
6. Remove any accessories that mount to the timing cover. On various cars, this may include the:
    - Distributor
    - Fuel pump
    - Chain tensioner
    - Oil pump.
7. From under the vehicle, remove any oil pan bolts that connect the cover and pan. Loosen but do not remove the bolts attaching the oil pan to the engine block on both sides of the engine. This allows you to remove and reinstall the cover without damaging the pan gasket.
8. If the oil pump is externally mounted to the cover, remove the pump and the pump drive shaft. It may be necessary to use a screwdriver to remove the pump drive. Insert the screwdriver into the distributor drive hole to push the shaft out with a slight twisting motion, figure 7-21.
9. Remove the remaining timing cover bolts, including any bolts that secure the cover to the head, figure 7-22.
10. Tap the sides of the cover gently with a soft hammer to break the seal. Remove the cover by sliding it forward off the front of the engine. Avoid prying if at all possible. Most covers are aluminum alloy that can be easily damaged.

CAUTION: When removing the cover, break the upper and lower seals very carefully. The head and oil pan gaskets have to be reused when you reinstall the cover. If one of the gaskets is stuck to the cover, use a thin gasket scraper to carefully break the seal without tearing the gasket.

How to remove, inspect, and replace the timing chain
Inspecting the timing chain, tensioners, slippers, and guides is easier after you remove

them from the engine. Timing chain inspection and replacement procedures differ from engine to engine, especially between OHC and DOHC engines. Be sure to check in the shop manual for the details.

Before you remove the timing chain, check the camshaft and timing chain alignment marks. Manufacturers often mark OHC timing chains with plated links that line up with punch marks on the sprockets during reassembly, figure 7-23. Others specify a certain number of chain links between the camshaft and crankshaft. If there is any possibility that you may reuse the timing chain, you can make your own marks on both the chain and each sprocket with paint so that you can easily reassemble the set. If the timing setup involves two chains and an intermediate sprocket, mark each sprocket against a location on each chain.

On some DOHC engines, you must rotate the engine by turning the crankshaft with a wrench to align a mark on the crankshaft sprocket with a mark on the lower cylinder block. Next, insert alignment dowels though a hole in each of the camshaft sprockets to hold the camshafts in the correct position. Single cam engines may also need to be locked into position by inserting a dowel into the sprocket or fitting a key into a notch cut into the back end of the camshaft. Always check the service manual for any special tools you will need before you begin tearing an engine down.

This is an ideal time to check the camshaft endplay. If you need to install a new thrustplate or spacers, do it after you remove the sprocket. The procedure is the same as that described earlier for OHV engines.

Once you align the camshaft and timing chain marks, follow this general procedure to remove the chain:

1. Loosen the chain tensioner, if not previously removed, and remove any chain guides or slippers.
2. Hold the camshaft in position using a suitable tool and remove the bolt(s), or locknut holding the camshaft sprocket to the camshaft, figure 7-24. If you do not have a tool for holding the camshaft, lock it in position by holding the crankshaft with a breaker bar and socket while loosening the camshaft sprocket. An impact wrench will generally loosen the camshaft sprocket without turning the shaft.
3. Remove the camshaft sprocket from the camshaft. It may simply slip off a keyway on the camshaft by hand, or you may have

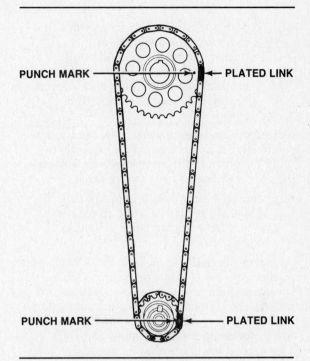

Figure 7-23. Some chains have specially-plated links that you use to position the timing chain.

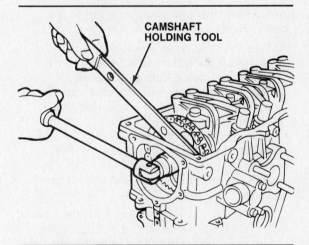

Figure 7-24. Lock the camshaft to keep it from turning, and remove any bolts or nuts holding the sprocket in place.

to use a puller to remove it. If you have to use a puller, you may be able to thread a fixing bolt partway back into the camshaft to give the forcing screw a firm footing to push against.
4. Lower the chain and sprocket and slip the chain off the crankshaft sprocket to remove it from the engine.

- If there are two chains and an intermediate sprocket, you may have to slide

both sprockets off their keyways first to remove the lower chain.
- If the timing sprocket will not fit through a chain tunnel in the head or block, remove the chain and lower it through the tunnel without the sprocket.

Note: You can be extra certain that you reassemble the chain and sprockets correctly by tying the chain to the sprockets with a short piece of wire before you remove them.

5. Remove the crankshaft sprocket from the crankshaft. It may be a slip fit over its keyway, or you may have to remove it with a puller.

Inspect the chain, sprockets, tensioner, and guides for wear or damage, figure 7-25. Inspect the chain carefully for cracked or worn rollers. The slippers or chain guides may show slight scoring — this is normal and is not reason to condemn the parts unless the wear is excessive. If any parts show damage or excessive wear, replace them all as a set. If you decided earlier to correct camshaft endplay by installing a new thrustplate or spacers, do so now. Follow this procedure to reinstall the timing chain:

1. First loop the chain around the sprockets, lining up the factory timing marks or the marks you made during disassembly.
2. Slip the sprockets with the chain attached over the camshaft and crankshaft noses, aligning any keys in the shafts with the keyways in the sprockets. It may be easier to slip the crankshaft sprocket on first, then loop the chain over it while you install the upper sprocket. Watch out for these exceptions:
   - On DOHC engines in which the camshaft sprockets stayed on the engine, loop the timing chain over the sprockets, making sure that the sprockets are still lined up with the dowel pins you installed during disassembly, and that the crankshaft alignment marks are still in position. You may have to remove one of the dowel pins temporarily to loop the chain over both camshaft sprockets. You must be able to reinsert the pin easily after attaching the chain. Be sure to remove the dowel pins after the chain is attached.
   - If there are two chains and an intermediate sprocket, fit the top chain first, then work all three sprockets and the lower chain onto their shafts simultaneously, figure 7-26.

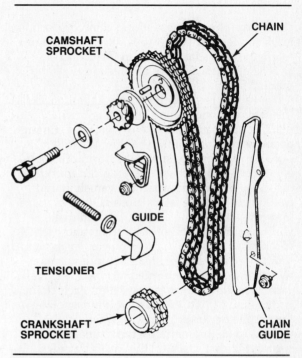

Figure 7-25. Inspect all of the timing components for wear or damage.

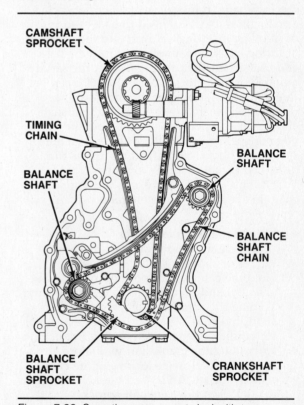

Figure 7-26. Sometimes you must deal with two chains, and several sprockets.

Engine Repair in the Vehicle

- If the timing sprocket will not fit through a chain tunnel in the head or block, figure 7-27, remove the sprocket and raise the chain through the tunnel without it. Then fit the sprocket onto the chain and slip both sprockets over their shafts.

3. If you have not disturbed the engine since removing the timing chain and sprockets, all the keys, keyways, and bolt holes should align. If not, carefully rotate the camshaft or crankshaft until the sprockets will slip into place with all the marks aligned.
4. Install the bolt(s), or locknut holding the camshaft sprocket in place, and tighten to the correct torque setting.
5. Install any chain guides, slippers, and tensioners that mount to the cylinder head or block. Some chain tensioners must be pre-loaded, while others simply snap against the chain under fixed spring tension.
6. Carefully turn the crankshaft sprocket two complete revolutions in the normal direction of rotation, TDC to TDC, and check that all of the timing marks on the crankshaft sprocket and the camshaft sprocket line up again with the pointers or marks on the engine, timing tag, or cylinder head. If the crankshaft and camshaft sprocket marks fail to align exactly after two revolutions, remove the chain and reposition it.

Note: Disregard the alignment marks on the chain itself — depending on the number of links in the chain, they may not line up again exactly for hundreds of revolutions, even when the timing is correct. The longer the chain, the more revolutions it will be before the same teeth in the sprockets match the same rollers in the chain.

7. Lubricate the new chain and sprockets with engine oil before installing the timing cover.
8. Reinstall the timing chain cover and any other components you had to remove, as explained earlier.

How to install the OHC timing cover
Be very careful when installing the front cover on engines with an overhead camshaft. Sometimes it will be difficult to get the cover in between the bottom of the head and the top of the oil pan, without destroying the seal. On all timing covers, the greatest source of problems is improper sealing that causes oil leaks. You can avoid it by following these steps:

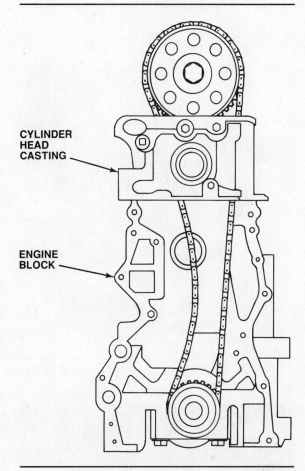

Figure 7-27. Some engines have a casting around the chain that forces you to attach the sprocket after pulling the chain up through the casting.

1. Ensure that the gasket surfaces on the timing cover and the engine are clean and dry.
2. Attach the timing cover gasket to the cover with gasket sealer. A good weatherstrip cement such as 3M's works well, as do many of the silicone sealers. Apply a thin film of sealer on the timing cover flange and on the gasket (timing cover side), figure 7-28. Let the sealer on both surfaces set-up for about a half minute then press the gasket into position on the cover. Make sure that the holes in the gasket align perfectly with the holes in the timing cover.
3. If the oil pan gasket tore when you removed the cover, cut off the gasket cleanly where the pan and block meet, figure 7-29. Replace this section with a piece of gasket cut from a new oil pan gasket.
4. To install the new crankshaft seal, thoroughly clean the seal area of the cover.

Chapter Seven

Figure 7-28. Apply a thin bead of gasket sealer to the flange of the timing cover and the gasket.

Figure 7-29. Cut the torn gasket where the pan and the block meet.

Then press in the seal with a seal installation tool. This will prevent distortion which could result in leakage. Never drive in an unsupported seal with a hammer.

5. Run a bead of silicone sealant in the four corners where the block meets the head and oil pan.
6. Carefully slide the timing cover into place, figure 7-30, making certain the gaskets do not bunch up or shift.

Figure 7-30. Carefully slide the timing cover into place on the front of the engine.

7. Loosely install the timing cover bolts that will not interfere with the installation of other parts. Draw the bolts up evenly in an alternating pattern and torque to the manufacturer's recommended specification.
8. Reinstall all the accessories and brackets that you removed. Generally, it is best to reinstall them in the reverse order in which they were removed.
9. Reinstall the radiator. Reconnect the cooling lines from the transmission. Connect the radiator hoses and fill the radiator.
10. Connect the battery and start the engine. After the engine reaches operating temperature check and adjust the ignition timing if you removed the distributor.
11. Check the automatic transmission fluid, engine oil, and coolant levels and top off if necessary.
12. Check the engine for leaks.

Timing Belts

Timing belt construction and reliability have greatly improved since their introduction, however, most still require routine inspection, adjustment, and replacement. Early belts had a projected service life of about 30,000 miles (48,000 km), while later model engines call for replacement at about double that mileage.

Timing belts are installed under high tension, operate at high rpm, and make several tight bends — creating stress from all directions. Modern alloy engines are designed to dissipate heat rapidly, and the heat released

Engine Repair in the Vehicle

from the front of the engine is trapped behind the timing shroud or cover, increasing the already high temperatures generated by the friction of the belt and components. Coolant or oil leaking onto the belt will also shorten service life. With age, belts dry out and become brittle. Continued operation can shear teeth or even cause the belt to break. When failure occurs, camshaft to crankshaft synchronization is lost. At best, the engine will not run. At worst, it may destroy itself.

How to service an OHC timing belt

Belt construction is unique to each engine design. Belts vary in length, width, tooth profile, installation procedures, and tensioning requirements. Always make certain the replacement belt meets original specifications and the teeth fit the sprockets properly. Installing some belts requires special tools, and all belts must be properly tensioned to provide adequate service life. Consult the appropriate factory repair manual to guarantee a correct installation.

Removing the OHC belt timing cover

Timing belts are usually fairly easy to expose, although a tight engine compartment might make it more awkward than you would like. Engine designs are so different that there really is no single technique that will substitute for common sense and careful work.

In general, first remove any components that interfere with access to the timing cover or shroud. On many rear-wheel-drive cars, you must remove the fan, spacer, and various pulleys and belts. Front-wheel-drive cars are often easier, because the fan is usually mounted on the radiator. In some restricted engine compartments, you must also remove under-car shields, alternator and brackets, and even an engine mount. Occasionally the valve cover must come off, too.

Next remove the cover or shroud. It may be a fairly large cover over the front of the engine, or simply a partial inspection cover that you remove to expose the top of the belt. The easiest to remove are the simple one-piece covers, which might be cast, stamped metal, or plastic. Once you have the cover removed, look it over for damage. Then inspect the timing belt itself.

Inspecting the timing belt

Once you can see the timing belt, examine it for wear or damage. If the engine has a partial cover, rotate the crankshaft with a wrench so that you can inspect every inch of belt and sprocket. Look the belt over carefully even if you plan to replace it, because its condition can reveal problems that might wear out the new belt prematurely.

Be careful with the belt. Although they can take a lot of tension in normal use, they are actually quite fragile. Never twist the belt more than 90° (with the teeth out), or bend it sharply to inspect the underside. Watch out for:

- Dirt, coolant, and oil-soaked areas. You can gently clean off dirt by wiping with a rag (never use solvent), but replace any belt soaked with oil or coolant.
- Hardened or cracked outer surface. You should be able to dent the surface of the belt with a fingernail.
- Separating cloth and rubber layers.
- Worn, cracked, or missing teeth, figure 7-31.
- Wear or cracks on the side or back of the belt, figure 7-32.

Be sure to correct any engine problems revealed by the belt condition before you install a new belt:

- Coolant or oil-soaked belts indicate a gasket problem. A poor-fitting shroud can let water or dirt damage the belt.
- Broken belts may have been installed incorrectly, damaged before installation, or can indicate a faulty belt tensioner. Never crimp a belt, or twist it more than 90 degrees.
- Cracked or missing teeth can indicate a locked or binding water pump or camshaft. Look for foreign materials jammed between the sprocket teeth, too.
- Wear or damage to the outer belt face might be caused by a jammed or nicked idler pulley, interference with the belt cover or shroud, or engine overheating.
- Wear or damage on only one side of the belt might indicate a misaligned or damaged belt guide or sprocket.

Also inspect the crankshaft and camshaft sprockets, the tensioner, and any idler pulleys for wear or damage.

Removing and replacing the timing belt

If the timing belt is worn or damaged, or if it is simply time for a routine replacement, then use the general procedure in this section. If you are removing the timing belt to get at some other engine component behind it, and plan to reinstall the belt, be sure you mark it so that you can put it back in the same position as when you removed it. You can make it easy on yourself at reassembly if you make a

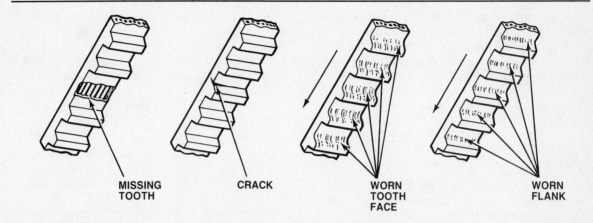

Figure 7-31. Timing belts deteriorate in several ways. Check internal surfaces for worn, cracked, or missing teeth.

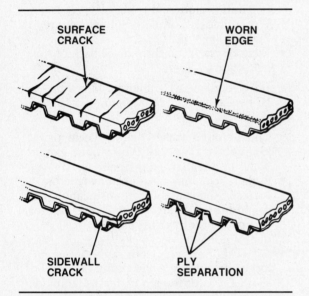

Figure 7-32. Check the timing belts for wear or cracks on the sides and external surfaces.

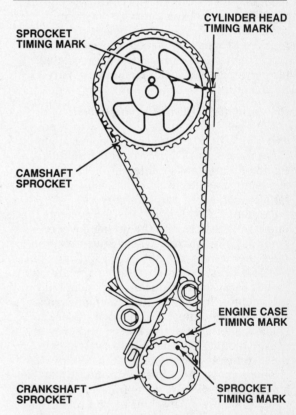

Figure 7-33. On some engines you have to align punchmarks on the sprockets and the engine case.

small mark on the belt opposite the timing marks on each belt sprocket. Because the timing belt is an inexpensive wear item, it is good practice to replace it with a new one whenever you have to remove it.

1. Remove the spark plugs, and use a wrench to turn the crankshaft until the timing marks show that the number one piston is at TDC on the compression stroke. Timing marks differ. You may have to align:
   - A groove on the pulley with a timing scale, figure 7-19.
   - A pointer on the block with a mark on the flywheel or flexplate, figure 7-18.
   - A punchmark on the crankshaft timing belt sprocket with another mark on the engine case, figure 7-33.

2. Follow the procedure explained earlier to remove the timing belt shroud or cover. If the cover is two-piece, you must remove both the upper and lower sections. This may be fairly complicated, involving removal of an engine mount, dipstick tube, and crankshaft pulley. On some engines

Engine Repair in the Vehicle

you must then either raise or lower the front of the engine slightly to get the necessary clearance.

3. This is an ideal time to check the camshaft endplay, figure 7-34. If you need to install a new thrustplate or spacers, do it after you remove the sprocket. The procedure is the same as that described earlier for OHV engines.
4. Check the timing marks on the camshaft sprocket to make sure that they line up — if they do not, you may have the engine 180° out of position. This is rare, but possible if the engine pulleys have extra timing marks for valve adjustment. The timing marks let you position the camshaft correctly with respect to the crankshaft when you install the new belt, and let you verify that the current valve timing is correct. The marks differ between engines. Look for:
    - A small hole in the camshaft sprocket that lines up exactly with a mark on the cylinder head, figure 7-35.
    - A punch mark on the sprocket that lines up with a pointer on the cylinder head, figure 7-33. Marks on the camshaft sprocket that align with the cylinder head upper surface, when the word "UP" is up, figure 7-36. If the engine has dual overhead camshafts, it will have marks on both sprockets. Check to see that they both line up with their respective pointers.
5. Remove or loosen any timing belt guides that align the belt with the sprockets.
6. Loosen the timing belt tensioner — it may be held in place with a central bolt, or by several bolts on a bracket — and push it as far away as you can to loosen the belt. Tighten the bolts to hold the adjuster in that position temporarily. Some tensioners require a special tool to release and hold them, check the service manual for the particular engine you are working on before you tear it down. Remove the timing belt, it should slip off the sprockets easily.
7. Inspect the belt tensioner for wear, it should spin freely and quietly, if not replace it.

Installing the belt

Installing a timing belt is basically the reverse of removing it, with a few extra things to keep in mind:

1. Line up the crankshaft marks as explained in the previous section so that the number

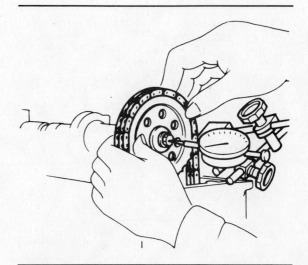

Figure 7-34. Check the camshaft endplay while the cover and belt are removed.

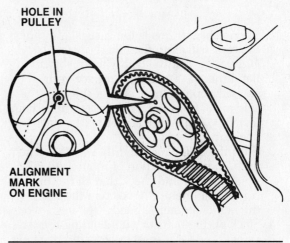

Figure 7-35. This Toyota system is a small hole in the camshaft sprocket that you align with a mark on the cylinder head.

one piston is at TDC on compression. Without the belt in position the crankshaft will turn easily, often just removing the belt will move the crank, make sure it is in position. Line up the camshaft marks, too. If you are re-using a good, low mileage belt, you can align it using the marks you made on the belt and the pulleys during disassembly. Some new belts will have reference marks on them for easier installation.

2. Slip the belt over the notches in the crankshaft sprocket, then fit it onto the camshaft sprocket(s). Since the engine was positioned with a worn belt, the teeth of the new belt may not line up perfectly with

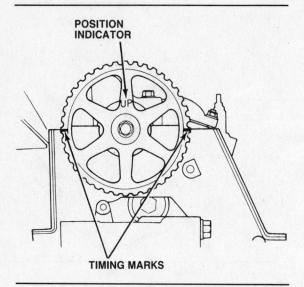

Figure 7-36. This Honda engine has two marks on the pulley that you line up with the cylinder head surface.

the notches. If this is the case, slightly turn the camshaft until they do. *Never pry or force a belt on*. If the belt will not slip on by hand there is a problem, retrace your steps. Keep the drive side of the belt (the side away from the tensioner) tight as you install it. Make sure all the marks stay lined up.

3. Turn the crankshaft slightly with a wrench in the normal direction of rotation to take up any slack, and check that all the timing marks still line up. Reposition the belt if the marks are not aligned.
4. Slack off the bolts or nuts holding the tensioner, and let it snap into position against the belt.
5. Attach a wrench to the crankshaft and slowly turn the engine two complete revolutions in the normal direction of rotation. Check to make sure all of the timing marks line up in the TDC position, if not, remove and reposition the belt.

Note: Disregard the alignment marks you made on a used belt — they will not align exactly with the marks on the sprockets again for many more engine revolutions — possibly hundreds. The longer the belt, the more revolutions it will be before the same notches in the pulleys match the same teeth in the belt.

6. Tighten the bracket bolts or center fixing bolt that hold the tensioner in position, then check the belt tension. On some engines, the pulley is not spring loaded, and you should check for a specific amount of slack at a particular point in the belt run.
7. Reinstall any belt guides that you had to remove.
8. Again, carefully turn the crankshaft sprocket two complete revolutions in the normal direction of rotation, TDC to TDC, and check that the timing marks on the crankshaft and the camshaft line up properly. If the crankshaft and camshaft sprocket marks fail to align exactly after two revolutions, remove the belt and reposition it.
9. Reinstall the timing belt cover or shroud and any other components you had to remove, as explained earlier.

VALVE TRAIN ADJUSTMENT AND REPAIR

Because the valve train contains the majority of moving parts in an engine, you will often have to adjust and repair the various components. Many valve train repairs can be performed while the engine remains in the chassis. Many routine maintainence procedures as well as mechanical repairs are performed after removing the valve cover. Other more complex valve train repairs, such as camshaft replacement, will require a combination of operations detailed in various sections of this chapter. In this section, we will cover some of the top-end tasks you will be performing:

- Removing and replacing the valve cover
- Checking and adjusting valve clearances
- Replacing the valve springs and seals.

These are really the most common valve train repairs you will make with the engine still in the car. The remaining tasks are explained later under engine disassembly in Chapter 9 of this *Shop Manual*.

How to Remove and Replace the Valve Cover

Valve covers are usually made of cast aluminum or stamped from mild steel sheet. Some newer engines use valve covers that are molded in plastic or a composite material. Plastic valve covers, along with other plastic engine components, may become more common in the future as designers attempt to trim weight. The OHC engine valve covers are often made of aluminum, OHV engines usually have covers made of stamped steel. With either type of engine, the cover is generally fastened to the head with bolts in one of two manners. Two or more bolts located along the centerline of the cover attach it to the head, or a number of

Engine Repair in the Vehicle

bolts are located around the edge of the cover hold it in place. With either design, you might find the cover secured by studs and nuts in place of the bolts. To remove a valve cover from an engine, use the following steps:

1. Remove all accessories and brackets that interfere with the removal of the valve cover. These may include emissions equipment hoses, fuel lines, or air cleaner support brackets. Some engines require the removal of the intake manifold runners to get at the valve cover.
2. If the spark plug wires are routed across the valve cover, mark them with numbered tags to ease reassembly, then remove them at the spark plugs, figure 7-37.
3. Remove all of the fasteners holding the cover to the head, figure 7-38.
4. If the cover will not lift off the head easily, gently tap it with a soft hammer to break the gasket seal, then remove it, figure 7-39.

Installing a valve cover is quite simple:

1. Clean the valve cover in solvent and scrape off all gasket material from its flange surface, figure 7-40. Check cast covers for cracks.
2. Check a steel valve cover for dimpled bolt holes, where over-torquing has raised the gasket rail around the bolt hole. Lay the gasket rail across the top of a vice, a block of wood, or other flat surface and tap around the bolt holes with a ballpeen hammer to level the surface, figure 7-41. This will allow even compression of the new gasket and reduce the possibility of leaks.
3. Remove all old gasket material and sealer from the mating surface on the head. The surface must be clean and dry.
4. Many late model engines use RTV silicone in place of gaskets to seal the cover to the head. When installing this type of cover, be sure to apply the silicone evenly around the entire sealing surface. Covers that do use a gasket can generally be installed without sealer. Should you decide to use a sealer, use it sparingly. Spread a thin layer of gasket sealer along the gasket surface of the valve cover and press the gasket into position. Be sure to align any locating tabs in the gasket with the cutouts in the cover.
5. Reinstall the valve cover on the engine. Tighten the fasteners to the manufacturer's specified torque, normally no more than 6 to 8 ft-lbs (8 to 11 Nm). Do not over-tighten.
6. Replace any spark plug wires, brackets, and accessories that you removed.

Figure 7-37. Remove the spark plug wires at the plugs, mark them, and move them aside.

Figure 7-38. Remove all the fasteners holding the valve cover to the cylinder head.

Figure 7-39. Tap the cover with a soft hammer to break the gasket seal, and then carefully lift it off.

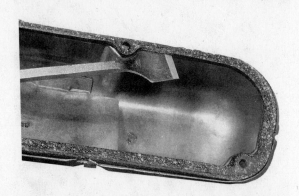

Figure 7-40. Scrape all the gasket material off the flange of the valve cover.

7. Start the engine and check for leaks.

Valve Clearances

Ideally, the valves should operate with zero clearance, or lash, so that the valve movement will follow the profile of the camshaft lobe exactly. Hydraulic lifters do operate with zero lash, once the engine has oil pressure, because the lifters automatically take up all slack in the valve train. Hydraulic lifters compensate for wear in the valve, valve seat, pushrods, and rocker arms. They also automatically readjust the clearance to zero as the valve train and engine castings expand when the engine warms up. Hydraulic valve lifters do not require routine adjustment.

Mechanical lifters, on the other hand, must usually be set with a precise amount of clearance so that the valves operate with very close to zero lash once the engine warms up, figure 7-42. In some engines, the valves actually operate with more than a minimal amount of clearance in order to work correctly with specially designed clearance ramps on the camshaft lobes. Adjust the mechanism yourself periodically to compensate for wear in the valve train. Adjusting the valves to operate correctly when the engine is warmed up, will cause too much valve clearance when the engine is cold, which will create more noise.

Maintaining the correct clearance is important. If you set the lifters with too much clearance, the camshaft cannot open the valves fully. In addition, you will shorten its effective duration, opening the valves later, and closing them earlier. Too much clearance also hammers all the valve train parts together, and can wear them out prematurely.

If you set the lifters with too little clearance, then the reverse happens. The valves open too far, and effective duration is increased — the

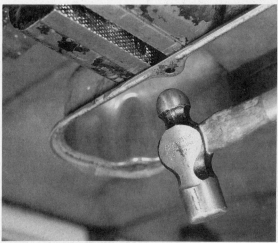

Figure 7-41. Carefully tap the metal around the bolt holes in the valve cover to flatten them again.

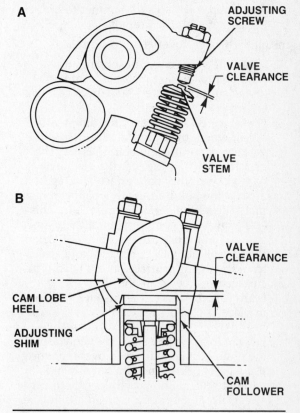

Figure 7-42. Static clearances in the valve train allow the metal parts to expand without leaving too much or too little clearance during operation. You measure the gap between the rocker arm and the valve stem (A), or between the heel of the cam lobe and the tappet or cam follower (B).

valves open earlier and close later. Once the engine warms up, the valves might not even be able to close completely, in which case the engine will run poorly, backfire through the intake, overheat, and burn exhaust valves.

Engine Repair in the Vehicle

Hydraulic valves seldom need attention, unless you removed the rockers for service or the engine has a very high mileage. Most domestic vehicles converted to hydraulic lifters by the mid-1970s. Some exceptions that retained mechanical lifters include:

- Chrysler: 3.7 liter slant 6-cylinder engines through 1980
- Chrysler: captive import 1.6 liter, 1.7 liter, and 2.6 liter engines used through 1986, 1983, and 1987, respectively
- Ford: 2.0 liter and 2.4 liter diesel engines, 1984-87
- General Motors: 1.8 liter Chevette diesel, 1981-86
- General Motors: captive import engines (Sprint, Spectrum, and Nova), 1985-88.

Many OHC engines have mechanical, or solid, valve lifters which need periodic service. Until recently, nearly all imported cars relied on mechanical valve lifters. In the mid-1980s manufacturers developed hydraulic lifters that were compatible with the high-rpm use of small displacement engines. Some of the popular imports such as Honda, Mazda, Mercedes-Benz, Mitsubishi, Nissan, Saab, Subaru, Toyota, and Volkswagen use hydraulic lifters in most of their engines. Mechanical lifters are still used by European manufacturers such as Audi, BMW and Volvo and Japanese manufacturers such as Hyundai and Isuzu for some applications.

Adjusting mechanical valve clearances
Typical clearance specifications for intake valves range from 0.004 inch (0.10 mm) to 0.025 inch (0.64 mm). Exhaust valve clearances range from 0.004 inch (0.10 mm) to 0.030 inch (0.76 mm). You will usually hear the inch measurements called simply thousandths. For example, valve clearances range from 4 thousandths to 30 thousandths.

Service manuals list valve clearances either as hot or cold specifications. Some manufacturers will provide both a hot and cold specification, however only one is the preferred condition. Cold clearances are generally larger that hot clearances for the same engine, to allow for heat expansion during operation. If the valves are to be set cold, the engine should be just that, cold, or at least not run for several hours. For cold setting, manufacturers often provide a maximum coolant or oil temperature at which the valves can be adjusted or a minimum waiting time after the engine has been run. As a general guideline, if you have run the engine prior to adjusting the valves, then check the coolant temperature. It needs to be about the same as the air temperature, and the valve cover should feel cool to the touch. To use a hot specification, you must have the engine warmed to its normal operating temperature. If you are servicing the valves with the engine hot but not running, you may have to reinstall the valve covers halfway through the job and run the engine for a few minutes to bring it back up to temperature. Valves and rocker arms can cool rapidly with the valve covers off, even though the coolant temperature stays high.

There are three common valve adjustment mechanisms:

- An adjustment nut holding the rocker to the rocker stud
- An adjustment screw on the rocker arm
- Replaceable adjustment shims (or discs) between the camshaft lobes and cam followers.

Older domestic pushrod engines often used mechanical lifters based on a hydraulic design. Tightening or loosening the locknut on the rocker stud will decrease or increase valve lash.

Most other pushrod engines have adjustment screws on the rocker arms themselves. The screw can be at the center pivot of the rocker arm, figure 7-43, or it might be at the end of the rocker arm, figure 7-44. Some OHC engines have rocker arms and adjustment screws as well, figure 7-45. The adjustment screws can be either of two types:

- A single, interference-fit capscrew or jamscrew that you adjust with a single wrench, figure 7-46. This type of screw will stay in whatever position you leave it in.
- An adjusting screw with a separate locknut, figure 7-47. To adjust, first loosen the locknut with a wrench. Then use a screwdriver or wrench to turn the adjusting screw. When the adjustment is correct, hold the screw in position while tightening the locknut. Some manufacturers make special tools that combine the adjusting and locking functions.

The Chevrolet Vega and Porsche 924 overhead-cam engines have a system that uses a tapered adjustment screw in the cam follower, or tappet. By inserting a hex key through a hole in the follower, you thread the screw in or out to adjust the gap.

Most DOHC and some SOHC engines have replaceable adjustment shims (or discs) between the camshaft lobes and cam followers,

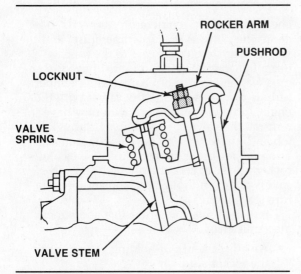

Figure 7-43. Sometimes the locknut holding the rocker arm to the stud also serves as the clearance adjuster.

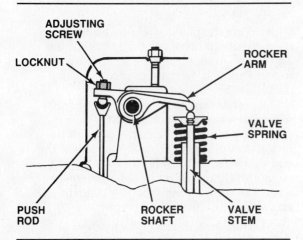

Figure 7-44. Shaft-mounted rockers usually have a separate adjusting screw and locknut on one end of the rocker arm.

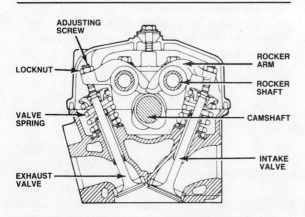

Figure 7-45. Some SOHC engines use adjustment screws and locknuts on the rocker arms.

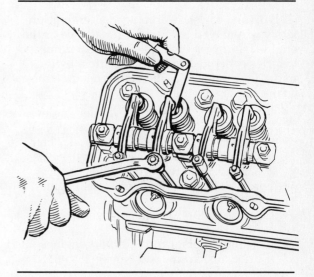

Figure 7-46. You only need one wrench if the rocker arm has interference-fit screws or nuts.

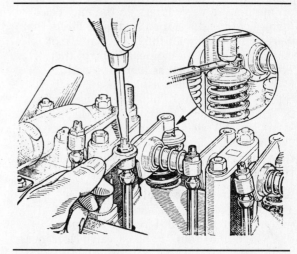

Figure 7-47. This rocker arm has a locknut holding the adjusting screw to the rocker.

figure 7-48. If the clearance must be adjusted, you use a special tool to press the follower down, figure 7-49. You then remove the shim with a magnet or a prying tool, and install a thicker or thinner shim to decrease or increase the gap. Some engine manufacturers, such as Alfa Romeo and Jaguar, use a smaller shim that fits directly on top of the valve stem underneath the cam follower. On this type of engine, clearance is also adjusted by replacing the shims. However, to get to the shim you first have to remove the camshaft and the followers.

Use a flat feeler gauge to measure the valve clearance gap on mechanical lifters. Make sure the gauge is clean and accurate. Feeler gauges do wear after extensive use. You might have to

Engine Repair in the Vehicle

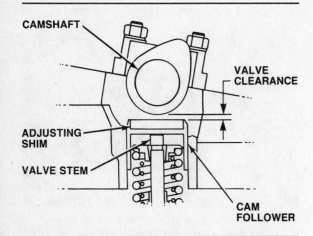

Figure 7-48. Many OHC designs use replaceable shims between the camshaft and the valve lifters.

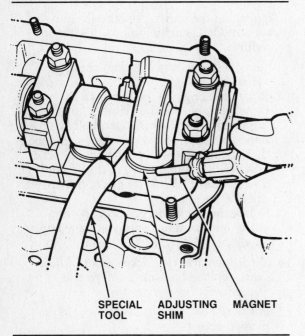

Figure 7-49. You will use a special tool to press the cam follower down so that you can replace the shim.

service the valves with the engine off and cold, off and warmed-up, or with it running at normal operating temperature. These methods are explained in the next sections.

Different feeler gauge designs are available for valve work. The gauge may be bent at an angle to make it easier for you to check valves near the hot exhaust manifold. A go, no-go gauge, with two different thicknesses on one blade, can be convenient in some cases. The thin section should fit a particular gap, but the thick section should not fit into the same gap.

If the valves still seem noisy after adjustment, or if you cannot get consistent measurements when the engine is running, the rocker arm may be worn above the valve stem, figure 7-50. In this situation, the rocker arms will have to be removed and resurfaced or replaced.

Adjusting clearances with the engine running
With many OHV engines, you can adjust the valves easily and accurately with the engine running at its normal operating temperature. You generally cannot use this method with OHC designs. Since the rocker arms are lubricated by engine oil pressure, running an engine with the valve cover off can create quite a mess. If you do a running adjustment, an inexpensive set of oil deflectors that clip onto the rocker arms to prevent oil from splashing is a wise investment. Look up the proper hot valve clearance specifications for the engine you are working on and have the necessary feeler gauges and wrenches at hand.

Follow a specific pattern so that you do not miss any valves. Work from one end of the engine to the other. Set all of the intake valves or all of the exhaust valves first. The general procedure for valve adjustment with the cover removed is as follows:

1. Make sure the engine is at its slowest idle speed and its normal operating temperature.
2. Insert a feeler gauge of the correct thickness between the rocker arm and valve stem, figure 7-51. The gauge should pass through the gap with a slow, steady drag:
   - If you have to force the gauge in, or if the engine starts missing when the gauge is inserted, the clearance is too tight.
   - If the gauge slips through too easily, or if there is a choppy, jerking feel as you pass the gauge through, the clearance is too loose. Double check by inserting a gauge that is one or two thousandths thicker than specified, or by using a go, no-go gauge.
3. If the clearance needs adjustment, turn the adjusting screw in or out as required. If the adjusting screw is in the pushrod end of the rocker arm, keep the wrench extension or screwdriver in line with the pushrod, figure 7-52, to minimize any whipping action. You may want to use a swivel socket, to help cushion the rocker arm movement.
4. Recheck the clearance after adjustment. If a separate locknut is used, check the clearance after the locknut is tightened.

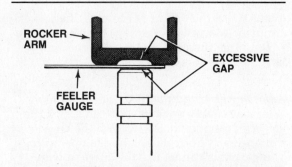

Figure 7-50. When the rocker arm is worn, a feeler gauge will not let you get an accurate measurement.

Figure 7-51. Use the feeler gauge at the same time that you tighten or loosen the valve clearance.

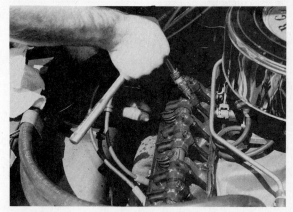

Figure 7-52. You can minimize the disturbance while you adjust the valve by keeping the extension in line with the pushrod, a swivel socket will help maintain alignment by cushioning movement.

NOTE: Feeler gauges are pretty delicate items, and the hammering they take adjusting valve clearances in a running engine eventually pounds them thinner. Higher-quality feeler gauge sets use a nut and bolt to hold the blades in place, so that you can replace individual blades that become flattened. Change the blade every three or four times you use it to adjust valves on a running engine.

Incidentally, if you are reassembling an engine and need to set the valves up initially so that you can start it up, most manufacturers will supply a cold lash specification in addition to the hot lash. Use the *cold lash* specification with the engine stopped, to get the valve clearance in the ballpark, as explained in the next section. Then start the engine and adjust them to the correct hot specifications.

Adjusting clearances with the engine off
You can also adjust the valves on many engines with the engine off. Overhead cam engines are often much easier than OHV engines, because you can actually see the camshaft contacting the lifter, and there is no question as to when the valve is closed.

To position the valves on either a shim or rocker arm OHC engine, follow this general procedure:

1. Remove the valve cover and choose a cylinder.
2. Rotate the engine while you watch the camshaft lobes for the cylinder you are going to adjust. When the camshaft has rotated so that the base circle of a cam lobe is directly in line with its follower or rocker arm pad, then that valve is closed, and you can insert the feeler gauge to check the clearance.
3. Rotate the engine to position another valve, and continue until you have adjusted all the valves.

To adjust the valve clearances in an OHC engine with replaceable shims, follow this general procedure:

1. Position a valve for measurement as explained above.
2. Measure the clearance between the base circle of the camshaft lobe and the surface of the follower. Record the clearance.
3. Rotate the engine and measure clearances for all the remaining valves, and record the values.
4. If any clearances are outside specifications, you must replace the shim for that lifter with a thicker or a thinner one.
5. Rotate the engine until the nose of the camshaft lobe of the first valve to be adjusted points directly away from the follower. The valve should be closed, with the camshaft base circle against the follower.
6. You will need one and possibly two special tools to compress the valve spring and to hold the lifter away from the cam lobe.

Engine Repair in the Vehicle

7. Compress the follower and hold it down using the special tools, figure 7-49.
8. Pry the shim loose with a small screwdriver, or other suitable tool, and remove it with a magnet. Some manufacturers make special pliers with jaws that fit around the camshaft to lift the shim off the follower.
9. Measure the thickness of the shim with a micrometer, figure 7-53.
10. To calculate the thickness of the new shim to install, use a pencil and paper or a hand calculator:
    a. Choose the midpoint of the manufacturer's valve clearance range.
    b. Subtract that amount from the measured valve clearance.
    c. Add the result to the thickness of the shim that you just measured with the micrometer.
    d. The final result is the thickness of the new shim you must install to set the new clearance squarely in the middle of the manufacturer's range. Choose a shim that is as close to that thickness as you possibly can.
11. Install the new shim in the recess in the top of the lifter, make sure it is firmly seated in place. Shims often are hardened on only one side, install this type shim with the hard face toward the cam lobe.
12. Release the follower by removing any holding and compressing tools.
13. Recheck the valve clearance.

Often, the engine manufacturer will make it simpler for you to choose the correct new shim during valve adjustment by printing a selection chart that has the math already built in. Some shims will have their thickness stamped or printed on the lower face, this should only be used for reference. Always measure shims with a micrometer, they will wear after being run in an engine.

If you are assembling an engine after performing a valve job, you may want to set valve clearance closer to the highest recommended tolerance. Newly ground valves and valve seats will settle in, the clearance will decrease slightly as the engine is run. When the car comes back for the initial service, you will often find the valve clearance will have closed up to be near the preferred mid-point of the recommended setting range.

Adjusting hydraulic valve clearances

You will not normally need to adjust the valve clearance of hydraulic lifters. If a vehicle is routinely serviced, lifters typically run for the

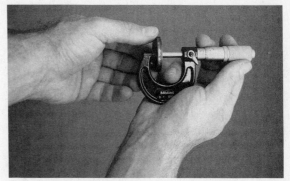
Figure 7-53. Use a micrometer to measure the thickness of the existing shim.

life of the engine with no attention. You will need to tighten them only after very high mileage causes wear to exceed the lifter's adjustment range, or to reset them after engine disassembly.

There are two adjustment techniques for engines with stud-mounted hydraulic rockers, depending on the engine. In the first system, used in most Ford OHV engines, you tighten the rocker locknut until it bottoms, and ensure that the lifters are set within their operating range by using longer or shorter pushrods. In the second, common to General Motors, Chrysler, and other Ford OHV systems, you continue to tighten the rocker locknut to set the lifter within its operating range.

You can also find shaft-mounted hydraulic rockers on many General Motors and Chrysler engines, and on some Ford engines. Typically, you adjust the valves on these engines as you do those with non-adjustable stud-mounted rocker arms, by fitting longer or shorter pushrods.

On an OHV engine, you can not see the camshaft lobes, so the procedure for making sure the valves are closed is more complicated. Start by rotating the engine to TDC on the compression stroke of number one cylinder, with both valves closed. Since valve clearance is not as critical with hydraulic lifters as it is with solid lifters, some manufacturers recommend checking and adjusting the lash for the valves of more than one cylinder at a time. The valves that can be adjusted together varies for each engine, therefore you must know the sequence for the individual engine you are working on. If this information is unavailable, there are several methods you can use to make sure each cylinder is at TDC.

The simplest method is to mark the crankshaft pulley for each cylinder. To do this you will need to know the firing interval of the engine. Most V8 engines fire at 90 degree intervals so you will need three marks, in addition to TDC on the pulley, figure 7-54. Inline 6-cylinder engines, and some V6 engines, fire at 120 degree intervals so you need two additional marks, one at 120 degrees BTDC and one at 120 degrees ATDC, on the pulley, figure 7-55. The common firing interval for a 4-cylinder engine is 180 degrees, so you need only one mark directly opposite from TDC, figure 7-56. After setting the valves for the number 1 cylinder, turn the crankshaft in the normal direction of rotation to align your next mark with the TDC indicator on the timing cover. Now the next cylinder in the firing order will be at the top of its compression stroke and the valves can be adjusted. Continue in this process until all the valves have been adjusted.

Another way to position the engine for adjusting valve lash is called the "running mates" method:

1. Write down the firing order. For example:

   1 5 4 2 6 3 7 8

2. Write the second half of the firing order directly beneath the first half:

   1 5 4 2
   6 3 7 8

3. The cylinder numbers that are in vertical pairs are called running mates or companion cylinders. In our example, the running mates are 1 and 6, 5 and 3, 4 and 7, and 2 and 8. When the exhaust valve of one cylinder is just closed, its companion cylinder will be just past top dead center on its power stroke. Both the intake and the exhaust valves for that cylinder will be closed and in position for valve adjustment. You can tell when the valve has just closed because the pushrod will turn freely and the rocker will be slightly loose.

4. To adjust the number 1 cylinder valves, turn the engine until the number 6 cylinder exhaust valve just closes. Next, turn the engine until the exhaust valve for number 3 has just closed, and adjust both valves for number 5. Continue for each cylinder, finishing by adjusting number 8 with number 2's exhaust valve just closed.

The running mates method works for all even-firing 8-, 6- and 4-cylinder engines, domestic and imported, because in each of these engines two pistons move up and down together, 180-degrees opposite in firing order. Positioning one cylinder with the exhaust valve just closed means that it will have just

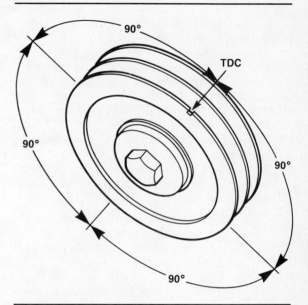

Figure 7-54. For a V8 engine, mark the crankshaft pulley at 90 degree intervals.

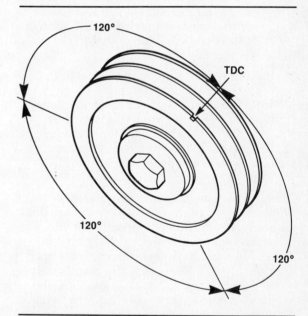

Figure 7-55. Most inline 6-cylinder and some V6 engines fire at 120 degree intervals. You will need to make two marks in addition to TDC on the pulley.

passed TDC, beginning its intake stroke. Its running mate will be just past TDC, beginning its power stroke, and in the correct position for valve adjustment. The system cannot work for uneven-firing V6 engines because no two pistons are ever exactly 180 degrees apart.

Adjusting clearances with non-adjustable rocker arms
Engines with this type of valve gear use zero-lash hydraulic lifters, but you center the

Engine Repair in the Vehicle

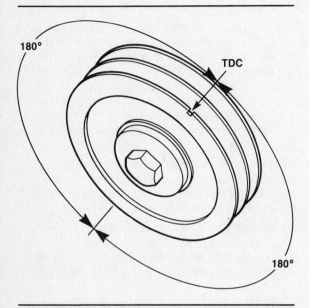

Figure 7-56. Four-cylinder engines normally fire at 180 degree intervals so you only need one additional mark on the pulley.

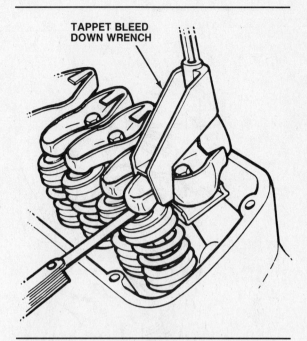

Figure 7-57. To check clearance on non-adjustable rocker arms, use a tappet bleed down wrench to depress the lifter and bottom out the plunger.

plunger inside the lifter body by installing shorter or longer pushrods. Adjust the valve clearance on non-adjustable rocker arms this way:

1. Use one of the methods explained previously to position a cylinder at TDC on its firing stroke.
2. Tighten the rocker arm locknut to the correct torque specification.
3. Press on the pushrod end of the rocker arm with a tappet bleed down wrench, a large screwdriver, or other suitable tool to bottom the plunger in the lifter, figure 7-57.
4. Hold the end of the rocker arm down, and measure the clearance between the valve stem tip and the rocker.
5. Compare this value to the specification range for that engine. If the clearance is too great, install a longer pushrod. If it is too little, install a shorter pushrod.
6. Rotate the engine so that the next cylinder is at TDC on the power stroke, and repeat.

Ford OHV engines often required this technique for adjusting valves. Pushrods for these engines are available in the nominal lengths, and 0.060 inches (1.5 mm) oversize and undersize. One of these three choices should bring the valve clearance within the specification range. In fact, the standard pushrods are usually fine, unless the head and block mating surfaces have been resurfaced, or the valve seats have been ground several times.

Adjusting clearances with adjustable rocker arms

Stud-mounted adjustable rocker arms first appeared on the 265-cid (4.3-liter) small-block Chevrolet in 1955. Many engineers were skeptical of their reliability at first, but the design proved quite durable. Since that time the system has become common throughout General Motors engines, and also among many other manufacturers. Adjusting clearance on engines with adjustable rocker arms is easy. To set them with the engine not running, follow this procedure:

1. Use one of the methods explained earlier to position a cylinder at TDC on its power stroke.
2. Slowly tighten the rocker arm locknut, without compressing the hydraulic lifter, until you remove all the slack from the valve train. You can tell when the slack is gone when the pushrod no longer rotates freely and the rocker can not wiggle from side to side, figure 7-58.
3. Slowly tighten the rocker arm locknut another 3/4 to 1 1/2 full turns (depending on the engine) to position the plunger in the center of its travel inside the lifter. You must tighten it slowly. If the lifter is primed with oil, it will take a few seconds

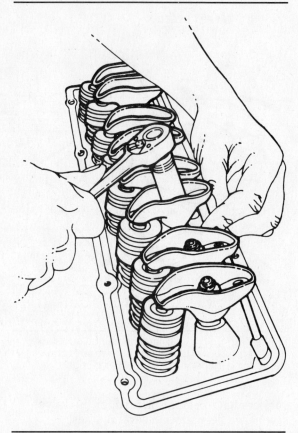

Figure 7-58. Slowly tighten the locknut until the pushrod no longer rotates freely to remove slack from the valve train.

to bleed down each time you turn the locknut.
4. Rotate the engine to position the next cylinder, and repeat the procedure.
5. Continue until you have adjusted all the valves.

If you have not disassembled the valve mechanism, and just want to reset the clearance on an engine that already runs, follow this procedure:

1. Start the engine. When it has warmed up, stop it, and remove the valve cover.
2. Start the engine again, and let it idle with the cover off. Attach oil deflector clips to the rocker arms to minimize oil splash.
3. Starting with any valve, back off the rocker arm locknut until the valve starts to clatter. At this point, the valve has too much clearance.
4. Slowly tighten the nut until the clatter just stops. You have barely removed all lash, but have not yet compressed the lifter plunger into the lifter body. If you stop now, the lifter will have no reserve travel left to compensate for wear in the future.
5. Slowly tighten the locknut $3/4$ to $1 1/2$ full turns, in $1/4$ turn steps, waiting about 10 seconds between steps. You must do this slowly to give the lifter a chance to bleed down between steps. The amount you tighten the locknut differs between engines.

CAUTION: Be sure to tighten the nut slowly, and wait for the lifter to bleed down each time. If you try to hurry, you can easily extend the valve so far that the piston will hit it.

6. Repeat for all the other valves.
7. Stop the engine, clean the gasket surfaces, remove the oil deflector clips, and reinstall the valve cover.

How to Replace the Valve Springs and Seals

It is possible to replace valve springs and seals on an engine in a vehicle with the cylinder head in place. In order to do this, both valves must be held firmly against their seats. You can hold the valves by using compressed air in the cylinder.

Position the piston in one cylinder at TDC on the firing stroke. Insert an air fitting into the spark plug hole and attach an air line to it. The air pressure will hold the valve against its seat so that keepers and spring can be removed without the valve falling into the cylinder. The piston must be exactly at TDC, or the air pressure will rotate the crankshaft.

You can use a special valve spring compressor to replace broken springs, collars, and keepers, and install new valve guide seals, figure 7-59.

How to Replace a Camshaft

If either an OHC or OHV camshaft is worn, you can often replace it without removing the engine. Most overhead camshafts mount in saddles bolted to or cast in one piece with the cylinder head, and once the valve train is detached, can be simply unbolted and lifted out. The camshafts in OHV engines are more complicated to remove, but the job seldom requires that you pull the engine. Follow the procedures explained in this chapter on removing the timing gear and timing cover, and then skip to the sections under engine disassembly in Chapter 9 of this *Shop Manual*.

Engine Repair in the Vehicle

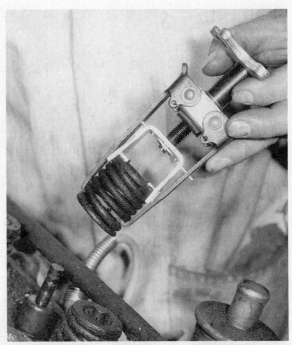

Figure 7-59. This set up allows you to service valve springs and seals without removing the cylinder head. Air pressure holds the valve in place, attach the spring compressor and remove the keepers, then you can lift the tool off along with the spring.

REMOVING AND REPLACING THE CYLINDER HEAD

Cylinder head removal procedures vary greatly from engine to engine depending on whether the engine is a V-type, inline, or flat, and whether it is an OHV or OHC design.

In all cases, first disconnect the battery, then remove all accessories, hoses and wires that may interfere with removal. Drain the coolant and oil from the engine.

On overhead cam engines, refer to the previous sections in this chapter for information about removing timing chains and belts. Some engines allow you to remove the head without completely disassembling the timing mechanism, however you may have to loosen or block up the chain tensioner to prevent it from fully extending. Check the manufacturer's service literature for the specific engine you are working on. Remove the timing chain or belt and the upper sprocket or pulley, and unbolt the timing cover from the cylinder head.

When your diagnosis indicates cylinder head gasket leakage, keep in mind that head gasket failure is generally the result of another problem. Overheating is usually the culprit, and you will have to locate and repair the cause to do the job properly. If the engine got hot enough to blow the gasket there may be additional damage that is not clearly visible, even with the cylinder head off. Both the cylinder head and the block should be checked for warpage, cracking, and other conditions that may have led to gasket failure. The procedures for inspecting can be found in *Chapter 14 of this Shop Manual*.

Using torque-to-yield, or stretch, bolts to fasten the cylinder head to the block is gaining in popularity. *Whenever you remove torque-to-yield bolts they must be replaced*, reusing them will not provide enough clamping force to hold the engine together. For information on how these bolts work and how to recognize them see *Chapter 14 of the Classroom Manual*.

CYLINDER HEAD GASKET REPLACEMENT

The following procedure details underhood steps to remove the cylinder head of General Motors front-wheel drive V-6 engine. The sequence takes into account that the battery has been disconnected, the engine oil and coolant have been drained, the exhaust crossover has been removed, and the exhaust pipe has been disconnected from the manifold. This is a general procedure, some steps will vary depending on the engine displacement and year and model of the vehicle. Cylinder head removal for many engines will be similar.

CYLINDER HEAD GASKET REPLACEMENT

1. Remove the air cleaner, disconnect the spark plug wires at the plugs, and remove the distributor cap with the plug leads.

2. Prepare the intake manifold for removal by disconnecting all hoses and wires that couple it to the body and engine block.

3. Disconnect the front torque arm and remove it.

4. Remove the air conditioning compressor bracket to access the valve cover bolts, do not disconnect any air conditioning lines or hoses.

5. Remove the cap screws holding the front valve cover in place.

6. Lightly tap the valve cover with a soft hammer to break the seal. If the seal does not want to break, carefully pry the cover to break it loose, then lift it off.

7. Disconnect the accelerator linkage from the carburetor and disconnect the cable from the bracket on the rear valve cover.

8. Disconnect the Pulsair bracket at the valve cover and swing it under the Pulsair housing. Do not remove the Pulsair unit.

9. Remove the bolts holding the deceleration valve bracket to the valve cover, swing the valve and bracket out of the way leaving the hoses attached.

Engine Repair in the Vehicle

CYLINDER HEAD GASKET REPLACEMENT

10. Remove any remaining bolts holding the valve cover to the cylinder head, tap the cover with a soft hammer to break the seal, and lift it off.

11. Mark the position of the rotor on the base of the distributor, remove the distributor holddown clamp, and take out the distributor.

12. Prepare the intake manifold for removal by disconnecting all hoses and wires that join it to the engine block or body.

13. Disconnect and plug the fuel lines from the carburetor. Use two wrenches to loosen the inlet line to avoid distorting the fitting.

14. Loosen and remove the intake manifold retaining bolts.

15. Lift off the intake manifold assembly. You may have to use a prybar to break it loose; avoid prying on the sealing surface.

16. Loosen the rocker arms just far enough to free the push rods.

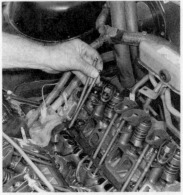

17. Lift the pushrods out. Be sure to keep them in order and return them to their same position on reassembly.

18. Break the cylinder head bolts loose in the reverse order of the specified tightening sequence. Once all the bolts have been loosened, remove them.

CYLINDER HEAD GASKET REPLACEMENT

19. Break the cylinder head loose from the block with a prybar; do not pry on the sealing surface.

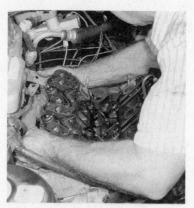

20. Carefully lift the cylinder head off the engine block.

21. Clean and inspect the sealing surfaces of the cylinder head and engine block. Be sure to clear out any debris that falls into the engine and the head bolt holes.

22. Place the new head gasket on the block. It must fit on the dowels with the word "front", on the gasket, at the front of the engine.

23. Carefully lower the head into position on the new gasket so that the alignment dowels fit securely into the holes in the head.

24. Reinstall the head bolts and bring them up to torque in stages following the manufacturer's specified sequence.

25. Reinstall the push rods to their original location and position the new intake manifold gaskets on the cylinder heads.

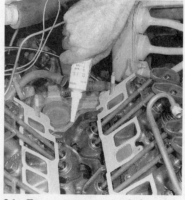

26. Run a quarter-inch bead of RTV silicone sealant at each end of the valley.

27. Carefully lower the intake manifold into position, avoid dislodging the sealant and gaskets. Reinstall the bolts and torque in the proper sequence.

Engine Repair in the Vehicle

CYLINDER HEAD GASKET REPLACEMENT

28. Tighten the rocker arm nuts to obtain the correct hydraulic lifter lash.

29. Apply a bead of RTV silicone sealant around the valve cover surfaces on the cylinder heads.

30. Reinstall the valve covers and tighten to the specified torque value. Replace any other items that were removed on disassembly.

SERVICING THE BOTTOM END

Many technicians automatically assume that any bottom end work requires that you remove the engine from the car. While this is certainly true for opposed engines, and for any engine with permanent chassis or driveline components in the way, you can do bottom end work on many cars after simply dropping the oil pan. You must first look over the chassis to determine if the oil pan can be removed with the engine in the car.

Up to the 1950s, mechanics could perform most engine work with the engine still in the car, including cylinder boring and honing. With special equipment, a mechanic could even regrind the connecting rod journals with the engine still in place. Current engines require more precision than you can reasonably achieve with tools like this, but main and connecting rod bearing replacement, as well as oil pump service, can often be performed with the engine still in the car.

With the oil pan off it is possible to replace the rod and main bearings without removing the engine. However, since there is not adequate clearance to accurately inspect and measure the crankshaft, connecting rods, and block, this is not a recommended procedure. In addition, most bearing failures result from some other related problem and this procedure cures the symptom and not the cause. If bearings are worn due to high mileage, other components are also worn and engine removal and disassembly is required to do the job properly.

How to Remove and Replace the Oil Pan

You must usually remove the oil pan to service the oil pump, to replace rod and main bearings, to re-ring the pistons, or to replace main seals or the oil pan gasket.

Most oil pans are made of sheet metal, although you will occasionally see them constructed of cast aluminum. You install a cork or composite gasket to seal the pan to the engine. Most oil pan removal and replacement procedures are similar, so we will cover the basic procedures for a typical removal and installation.

1. Drain the engine oil.
2. Remove any under-car shields that prevent you from reaching the pan. If there is a dust cover over the clutch or torque converter, remove it, too.
3. In most cases you must unbolt the engine mounts and raise the engine to provide clearance for the oil pan. If this is true of the vehicle on which you are working, check the top side of the engine to determine how much lifting room you have before the engine, transmission, or an accessory strikes the firewall or binds.
4. Remove any exhaust pipes that cross underneath the pan.
5. Remove any suspension pieces that interfere with the oil pan removal, such as drag links or center links. For most vehicles with parallelogram steering, unbolting the idler arm at the frame will allow the components to drop far enough out of the way.

6. When you have provided adequate clearance for the oil pan to be removed, remove the oil pan bolts, figure 7-60.
7. Then lower the pan and carefully remove it from the car. If the oil pump or pickup keeps the pan from clearing the engine, reach inside and unbolt the pump pickup, figure 7-61, or the pump itself, then remove the pan.

Reinstalling the oil pan is as simple as taking it off. Follow the steps below:

1. Clean the oil pan and inspect it for dents, gouges, or other damage. On a stamped steel pan these can often be pressed or tapped out from the inside. If the damage is serious you should replace the pan.
2. Rest the flange on a flat surface and tap the area around the bolt holes with a ballpeen hammer to restore it to the same level as the rest of the gasket surface. This will enable the gasket to evenly seal around the entire flange.
3. Clean the oil pump pickup, replace the screen if damaged, and prime the pump with fresh oil.
4. Position the new gasket on the pan. Gasket sealer can be used but is generally not required. The gasket can be held in place by tying it to the pan using specially designed clips, or small pieces of thin wire run through several of the bolt holes.
5. Hold the oil pan up to the engine and reinstall the oil pump, using a new gasket on the pickup. Never use sealer on an oil pump gasket, a small piece of sealer that finds its way into the pump will cause serious damage. Start several of the pan bolts to hold the pan and gasket in position, remove the clips or tie wires, and install the rest of the bolts. Tighten the oil pan bolts to the specified torque in an alternating pattern. Avoid over-tightening. Pan gaskets are easily distorted by over-tightening and a leak results.
6. Reinstall the engine mounts, center link, exhaust pipe, and any other pieces you removed in order to take off the pan.
7. Replace the oil filter, and refill the engine with fresh oil.
8. Disable the ignition or fuel system and crank the engine over to circulate oil through the engine. Restore to running condition, then start the engine and check for oil pressure.
9. Run the engine to operating temperature, and check for oil leaks. Add oil as needed to bring level to the full mark on the dipstick.

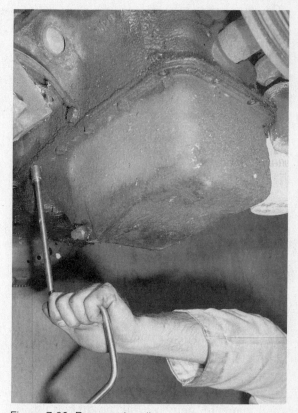

Figure 7-60. Remove the oil pan bolts.

Figure 7-61. Unbolt the oil pickup, if necessary, to get clearance to drop the pan.

How to Replace the Rod Bearings

Rod bearing service with the engine in the car should only be considered as a last resort. If the situation warrants an in-chassis bearing replacement, begin by raising the car to a comfortable working height, so that your arms can move freely and not tire as you work. Follow this general procedure with the oil pan removed:

Engine Repair in the Vehicle

1. Remove the spark plugs to relieve compression.
2. Rotate the crankshaft with a wrench until the front connecting rod journal is at the lowest point of its travel. Replace the bearings one rod at a time starting at the front of the engine.
3. Loosen the rod cap nuts on the first rod and run them down to the end of the bolts.
4. Hold the cap firmly against the crankshaft journal and push the rod and piston up into the cylinder by tapping evenly and alternately on the bolts with a plastic hammer. When the bolts are all the way up into the cap, remove the nuts, separate the cap from the bolts, and remove it.
5. Slip small sections of plastic tubing or fuel line over the rod bolts to prevent the rods from nicking the crankshaft.
6. Push the piston up into the bore till the rod bolts clear the crankshaft, rotate the crankshaft to gain clear access to the rod.
7. Remove the bearing shell from the connecting rod.
8. Remove the bearing shell from the rod cap.
9. Inspect both bearing shells for excessive wear or damage that might indicate problems with the crankshaft journal.
10. Clean the crankshaft journal and inspect it for wear or damage. You can check the journal for taper or out-of-round with a micrometer from beneath the car. If the crankshaft is out of tolerance on any journal, or if any journal is damaged, you will have to remove the crankshaft to repair or replace it.
11. Clean the rod and cap bores and parting surfaces, and install new bearing shells. Be sure the locating tangs fit into the notches correctly.
12. Check the clearance of the new bearing with an inside micrometer or Plastic-Gauge®, see Chapter 16 of this *Shop Manual*.
13. Lubricate the bearing shells and the journal with clean motor oil.
14. Rotate the crankshaft until the crank pin is in the position in which you removed the cap, gently slip the connecting rod down until it fits over the crankpin, and remove the pieces of protective hose.
15. Make sure the cap is facing in the proper direction, then slide the cap onto the rod bolts. Push the cap gently into place, and start the retaining nuts on the bolts.
16. Tighten the rod nuts evenly to the correct torque setting.
17. Rotate the crankshaft until the next connecting rod is accessible, and repeat the procedure until all the bearings have been replaced.

How to Replace the Crankshaft Main Bearings

You can replace the crankshaft main bearing inserts at the same time that you replace the rod inserts. To do this correctly, you need a special tool called a rollout pin. The pin is a small tool that fits into an oil hole in the crankshaft, and lets you push the upper bearing shell out by rotating the crank. Follow this procedure with the oil pan, oil pump, and spark plugs removed:

1. Break all the main bearing caps loose, do not remove the bolts. This will lower the crankshaft slightly, and make it easier to roll the bearing inserts out.
2. Starting at one end of the crankshaft, remove the main bearing cap and take out the insert, figure 7-62.
3. Place the rollout pin in the hole in the main bearing journal. Turn the crankshaft in the normal direction of rotation until the pin is against the side of the bearing insert opposite the locking tang. Check to be sure that the pin is positioned correctly. It must be lower in height than the bearing shell, or it will score the saddles in the block as you rotate the crankshaft. Continue to turn the crankshaft in the same direction, the rollout pin will push the bearing out. When the insert is all the way out of the web, remove it from the journal.
4. Clean the crankshaft main journal, and inspect it for scoring or other damage. There is no way to actually measure the journal because it will still be in position in the saddle, but you can check to see if anything is obviously wrong. Clearances of the new bearings can be checked using Plastic-Gauge®, as detailed in Chapter 16 of this *Shop Manual*.
5. Lubricate the journal, and turn the crankshaft so the rollout pin is not visible. Lubricate the new bearing insert and place it on the crank journal.
6. While holding the bearing against the crank journal, turn the crank in the opposite direction of normal rotation to roll the new shell into place. The rollout pin must push against the end of the bearing with the tang. Make sure that the locking tang

Figure 7-62. Remove the main bearing cap and the lower bearing shell insert.

is properly aligned and seated in the groove on the engine block.

7. Rotate the crankshaft back and remove the rollout pin.
8. Install the other half of the insert in the bearing cap, lubricate it, and reinstall it loosely.

9. Proceed to the next main bearing and replace the shell. Continue in this manner until all inserts are replaced. See the notes below about replacing the thrust bearing.
10. Torque the cap bolts as specified by the manufacturer.

If the thrust bearing cap was one of the ones that you removed, the procedure for replacing it is a little different:

1. Install the thrust bearing cap with the bolts finger-tight.
2. Use a stout screwdriver or prybar to pry the crankshaft forward against the thrust surface of the upper half of the bearing.
3. Hold the crankshaft in this position, and pry the thrust bearing cap to the rear. This will align both halves of the thrust bearing.
4. Keep holding the crankshaft with forward pressure while tightening the cap to the correct torque setting.

Oil Pump

An in-pan oil pump is easy to replace once you remove the oil pan. Remove the pan, as already described, then remove the bolts holding the oil pump in position. There may be a separate bolt holding the pickup tube. Once the bolts are out, remove the pump.

When reinstalling the pump, be sure that you first prime it by filling it with fresh motor oil. Slip it into place, and be sure that the drive mechanism (a tang or rod) fits correctly into the bottom of the distributor shaft. If you have any difficulty, remove the pump, reposition it, and try again.

Chapter

8
Engine Removal

There is no single, detailed procedure that you can use to remove the engine from all cars. Engine removal procedures vary with make, model, year, options, and accessories. Describing the exact removal and replacement procedures for every engine is neither possible nor necessary. There are certain procedures common to removing all engines. In this chapter we will discuss these so that with a little care and thought you will be able to adapt them and be able to remove the engine from almost any vehicle.

On most rear-wheel-drive cars, you can easily remove the engine through the top of the engine compartment with an overhead hoist. On other cars, such as air-cooled Porsches and Volkswagens, you must remove the engine from below, supported on a floor jack. Possibly the most difficult engines to remove are those installed in RWD vans. Depending on how the particular model and engine are constructed, you might be able to slide the engine out through the front after removing the radiator and grille, lift it out through the passenger compartment side door, or lower it onto a floor jack underneath the car.

Front-wheel-drive cars have a few special considerations. Often the transaxle must come out with the engine. You may have to remove the wheels and disconnect the front suspension in order to remove the drive axles. Some engine transaxle assemblies lift out the top of the car, others drop out the bottom of the chassis. If you remove the engine without the transmission attached, you often have to support the transmission with a special jig.

PREPARATION

Even though removal procedures are unique for each individual vehicle, there are some common guidelines that can be applied to removing all engines. Read over these sections before you begin.

Clean the Engine First

Start by cleaning the engine and engine compartment with a high pressure washer or steam cleaner. A clean engine is not only easier to work on, but also less likely to slip out of your hands as you maneuver it out of the chassis. Dirt and grease can also hide wires and fasteners that must be removed or disconnected before the engine can be taken out.

If a steam cleaner is not available and the car can be driven, take it to a specialty company and let them do the cleaning for you. These companies are equipped with all the tools, safety equipment, and waste disposal facilities that high pressure steam cleaning requires.

One thing you should remember: think carefully before you steam clean an engine while it is still in the car. Any domestic or import engine manufactured after about 1980 is likely to have computer-controlled ignition, fuel injection, and emission equipment. Unless very carefully handled, a steam cleaner can turn delicate electronic components into junk. Earlier engines up through the 1970s can usually be steamed cleaned without trouble, but you must still stay away from electrical components.

Steam cleaning, as well as other methods of engine and parts cleaning, are discussed in more detail in Chapter 9 of this *Shop Manual*.

Plan Ahead

Before removing anything from the vehicle, first spend a few minutes planning. Determine exactly what you intend to do so you will know what to include in the engine removal. For example, if you do not plan to work on the transmission, it is often simpler to remove the engine and leave the transmission in the car. In other cases, it may be best to remove the engine and transmission together, especially if the firewall makes it difficult to get at the engine-to-transmission bolts.

Have a notebook on hand so that you can write down a list of parts that will need to be repaired or replaced when you reinstall the engine. This includes belts, hoses, engine mounts, gaskets, hose fittings, filters, and any damaged components you discover when you remove the engine. You do not want the job spoiled because you reused a part that should have been replaced. The fuel, ignition, and cooling systems must work perfectly when the engine is reinstalled. If not, the customer will be back, and they will not be happy.

You should also get a roll of masking tape and a felt-tipped pen to use in labeling vacuum hoses, electrical connectors, and any wiring harnesses that you will remove. There may not seem to be very many of these on the engine, but there are almost always too many to remember. Save yourself time in reassembly and label *everything* before you remove or disconnect it.

Step back and survey the installation before you begin removing parts. Think ahead to avoid removing unnecessary items. Removing assemblies rather than individual parts will save time and make reassembly easier. Most engines have engine electrical harnesses with only several connectors joining them to the body, there is no need to disconnect all of the wires on the engine. When removing the air cleaner, leave the ductwork and hoses attached to it. Mark the hoses and disconnect them from the engine. The hoses will hold their shape making installation easier and you have only one end to connect. If you have to remove the distributor, take it off as a unit along with the cap and spark plug wires. Air conditioning compressors, power steering pumps, and clutch slave cylinders can often be unbolted from their mounting brackets and moved aside without disconnecting fluid lines or pressure hoses. Be sure to securely tie hydraulic units out of your way. Never let them hang by their lines or hoses.

To help you plan, look over the following checklists. You might not have to perform all listed operations on every car. Check the following items to make sure that the removal goes smoothly.

Remove these items from *above*, through the engine compartment:

- Air cleaner assembly
- Battery, including cables, brackets, and tray
- Hood
- Fan and fan shroud
- Radiator and radiator hoses
- Heater hoses
- Throttle and kickdown linkage
- Fuel lines, both feed and return
- All accessible wiring
- External accessories: power steering pump, air conditioning compressor, air injection pump, etc.

Remove these items from *below*, with the car raised, if necessary:

- Engine oil and coolant
- Splash pans and shrouds
- Starter motor
- Exhaust components
- Engine mounts
- Anti-roll struts
- All accessible wiring.

Preliminary Operations

To remove the engine without the transmission, you will also detach the torque converter fasteners (automatics), clutch linkage and transmission bolts (manuals), and any fasteners or linkage that hold the engine to the transmission. For cars in which the engine and transmission come out together, you will remove the driveshaft, transmission mounts,

Engine Removal

transmission linkage, speedometer cable, and any electrical or other connection between the transmission and body.

Removing wiring and accessories
Be very careful about removing wiring harnesses and electronic components under the hood. Modern cars usually have computer-controlled fuel, ignition, and emission systems, and these parts and their connections are delicate. Label anything you remove, and make an entry in your notebook should you encounter anything unusual.

Be sure to *remove the battery*, do not simply disconnect it — you will be busy enough removing the engine without worrying about dragging the oil pan or a manifold across the terminals.

Air conditioning systems
Removing the air conditioning system takes some special considerations. Try not to disconnect any hose or pipe fittings at all. The system is full of pressurized R-12 refrigerant (also called Freon®), and if you open the system without carefully discharging it, it will spray out. This is *dangerous,* as R-12 can freeze skin on contact, and can blind you if it sprays in your eyes.

The best way to get the air conditioning compressor out of your way is to unbolt and slide it out from the bracket that holds it to the engine. Then tie it securely off to the side, figure 8-1. Refrigerant leaks and electrical circuit failure can result if you let the compressor hang from the hoses or wiring. Simply leave the compressor in the car while you remove the engine. You may also have to disconnect the condenser and tie it back in the same way when you remove the radiator.

If you cannot avoid disconnecting the air conditioning system, make sure that you know the exact procedures. Use the manufacturer's service manual, or talk to someone familiar with the particular system in the car.

Be aware, also, that discharging R-12 into the atmosphere is illegal in most areas. R-12 is one of several man-made gases destroying the ozone layer in the upper atmosphere. The ozone layer protects you from cancer-causing ultraviolet light, and every drop of R-12 that you can keep out of the air helps keep the ozone layer in good shape. Many professionals now use sealed recycling machines so that they not only keep the refrigerant out of the air, but get to reuse it when they charge up another car, figure 8-2.

Relieving fuel system pressure
Since fuel injection systems are pressurized at

Figure 8-1. Avoid disconnecting air conditioning lines or hoses unless you absolutely have to. Just tie the compressor out of the way.

Figure 8-2. If you have to discharge the air conditioner use the proper equipment, releasing refrigerant into the atmosphere is illegal in most areas.

all times, the pressure must be properly relieved before any fuel lines or hoses are disconnected. A residual pressure of 10 to 50 psi (7 to 350 kPa) may be present in the system even while the engine is off, this ensures quick, easy, starts. Opening any fuel injection system connection without first relieving system pressure will cause gasoline to spray out on your hands and engine components.

There are several methods for relieving fuel pressure. Most work in one of these ways:

- If the system has a fuel pressure regulator with an accessible vacuum nipple, connect a vacuum pump and apply 10 to 15 in-Hg (250 to 400 mm-Hg) of vacuum to release fuel pressure through the fuel return line.

- If the system has a Schrader valve on the fuel rail or throttle body, connect a hose with a valve depressor and release the fuel into a suitable container.
- Remove the electric fuel pump fuse and start the engine. Run the engine until it runs out of fuel and stalls. Then crank the engine for 3 to 5 seconds to ensure that all pressure is released.

Any car, both injected and carbureted, with a non-vented fuel tank can build up pressure in the fuel tank. Built up pressure can cause fuel to spray out when you loosen fittings. Take a moment to remove and replace the fuel filler cap to vent any possible pressure before you loosen the fuel lines.

DISCONNECTING THE ENGINE

There are certain procedures you need to follow when removing any engine, you will be working under the hood as well as under the car. Begin by setting the vehicle up on a hoist, use jack stands if a hoist is not available. Make sure the car is firmly supported, jack stands must be on level ground. As the engine comes out a lot of weight is being removed, if the car is not properly supported it could shift position. Protect the body parts and grill work near the engine compartment with fender covers to prevent damage to the paint. The following general steps apply to most any engine removal.

1. Drain the engine oil and replace the drain plug. Leave the oil filter on the engine, this will prevent any residual oil from dripping into your work area.
2. Drain the cooling system, remove both the upper and lower radiator hoses, and disconnect the heater hoses.
3. Disconnect and remove the battery, hold down, and tray. On some cars you will also have to remove the battery support that is under the tray.
4. If the engine is to be removed out the top, the hood will have to come off. Mark the position of the hood hinges with a marking pen, figure 8-3, this will speed up reinstallation and alignment. Have an assistant hold the hood as you remove the bolts.

CAUTION: When you remove the bolts from an opened hood, remove the *highest* bolts first so that the lower bolts can still hold the hood securely to the hinge while you loosen them. If the lower bolts are removed first, the hinge can tip down and bend the sheet metal as you loosen the top bolts.

The hood should be laid flat for storage, a

Figure 8-3. Marking the position of the hinges before you remove the bolts will make installation easier.

good place is upside down on the roof of the vehicle. Use a fender cover or blanket to protect the paint. You can also hang the hood on the wall. Never rest the hood standing up on the floor; the corners will easily chip or bend and the hood can fall over. Once the engine is out of the car, you can set the hood back in place.

5. Remove the air cleaner assembly. Leave any hoses or valves attached to the housing, disconnect and tag them at the engine end, figure 8-4.
6. Disconnect all engine electrical connections, most engines use a wiring harness that can be disconnected at a multi-plug and removed with the engine, figure 8-5. You will also need to remove the engine to body ground strap. Survey the engine compartment and disconnect any wires, such as the alternator power feed, cooling fan thermo-switch, and ignition coil leads that are not part of the engine harness, figure 8-6.
7. Remove the radiator, fan shroud, and fan, figure 8-7. On cars with an automatic transmission you will have to disconnect and plug the transmission cooler lines before removing the radiator. If the air conditioning condenser is attached to the radiator, unbolt it and leave it in the chassis if possible.
8. Disconnect and tag all vacuum hoses that attach the body to the engine. If several hoses run to a control box mounted to the body, leave the hoses attached to the box and unfasten the box from the body, then lay the unit on top of the engine, figure 8-8.

Engine Removal

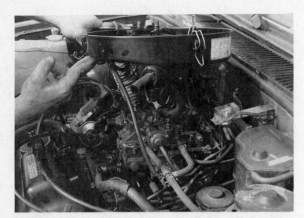

Figure 8-4. Leave all the ductwork, vacuum lines, and hoses attached to the air cleaner assembly, disconnect them at the engine and remove the assembly as a unit.

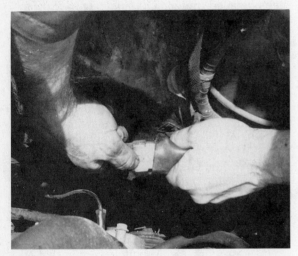

Figure 8-5. Most engines are wired with a harness, disconnect at the multiplugs and remove the harness with the engine.

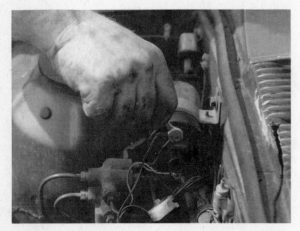

Figure 8-6. Some electrical connections, such as the ignition coil, alternator power lead, and thermo fan switch, will have to be disconnected separately from the engine harness.

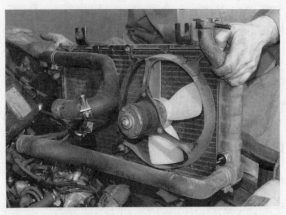

Figure 8-7. The radiator, fan, and fan shrouds can often be removed from the chassis as a unit.

Figure 8-8. If vacuum hoses run to a control box on the body, disconnect the box from the body, lay the unit on top of the engine, and remove the assembly with the engine.

9. Relieve fuel pressure, then disconnect both the fuel feed and return lines from the engine. Plug the fuel lines to prevent spillage.
10. Disconnect the throttle cable or linkage. You may also have to disconnect the choke cable or the kickdown linkage for the automatic transmission.
11. Unbolt the air conditioning compressor from its mounting bracket, swing it off to the side, and tie it out of the way.
12. Power steering pumps can often be unbolted from their mounting bracket and swung off to the side, similar to air conditioning compressors, figure 8-9. If you do have to remove the pump, drain the system into a container. Plug all of the lines and the fittings you disconnect to keep contaminants out of the system.

Figure 8-9. Disconnect the power steering pump, these can often be loosened from the bracket and swung out of the way without disconnecting the fluid lines.

Removing the Engine

At this point everything that attaches the engine to the chassis, with the exception of the engine mounts and transmission, should be disconnected. Take a moment to look things over to be sure nothing has been overlooked. Remove any brackets or other protruding parts that might catch, snag, or be damaged as the engine is hoisted out.

What remains to be done will vary greatly with the specific vehicle and engine you are working on.

Hoisting the engine from a RWD chassis

Removing the engine from a RWD chassis is generally a straight forward operation that varies little from manufacturer to manufacturer. There are some exceptions, British car makers were fond of placing engine to transmission bolts in inaccessible locations. As a result, the transmission has to be removed along with the engine and some additional steps are required. The transmission will have to be drained, the drive shaft and mounting brackets removed, speedometer cable, linkages, and wiring disconnected. The lifting equipment to be used is another consideration, the hoist must be able to raise the engine and transmission high enough to clear the body. Fortunately, most RWD engines can be removed by leaving the transmission in the chassis and the procedures are similar to the one outlined below.

The following sequence provides step-by-step procedures for removing a Ford 4-cylinder 2.3L OHC from a RWD chassis. The sequence takes into consideration that the preliminary steps detailed under *"Disconnecting the Engine"* have been taken.

Before you begin, take a moment to check that you have disconnected everything that couples the engine to the body. Keep in mind, you will be lifting a considerable amount of weight. If you overlook only a single electrical wire, the result could be disastrous. As the engine comes out, it is hanging on a chain and free to move like a pendulum. Should anything snag on the way out, then suddenly break free, the engine can jump violently and begin to swing out of control. Several hundred pounds whipping around on the end of a chain is not only a threat to your safety, it can also tear through the car like a wrecking ball. Avoiding a situation like this is simple, double check your work. Be absolutely certain that all wires, hoses, and cables are disconnected and the engine has a clear path on the way out.

Engine Removal

RWD ENGINE REMOVAL

1. Mark the position of the hinges on the hood. Remove the top bolts, then remove the lower bolts and lift off the hood.

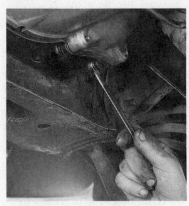

2. Disconnect the starter electrical wiring, remove the mounting bolts, and lift off the starter.

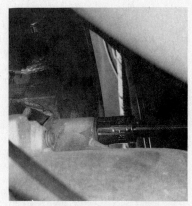

3. Remove the bolts attaching the engine to the top of the bellhousing.

4. For automatic transmission models, remove the bolts attaching the torque converter to the flex plate.

5. Disconnect the exhaust pipe at the manifold, loosen brackets and hangers if necessary, and pull the exhaust pipe down and out of the way.

6. Disconnect the right and left engine mounts from their brackets.

7. Disconnect the clutch cable, linkage, or slave cylinder.

8. Disconnect any transmission linkage that couples the transmission to the engine block.

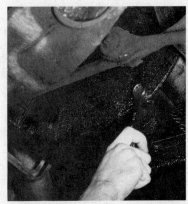

9. Remove the lower bellhousing bolts and any other fasteners that connect the engine and transmission.

RWD ENGINE REMOVAL

10. Attach a short piece of chain to the lifting hooks on the engine, if no hooks are provided attach the chain to secure studs or bolts at opposite corners of the engine.

11. You may have to remove the exhaust manifold in order to obtain a secure fitting for the lifting chain.

12. Support the transmission with a floor jack and position the lifting hoist over the center of the engine.

13. Connect the hoist, raise it just enough to take the tension off of the engine mounts, double check to make sure everything is disconnected from the engine.

14. Break the transmission free from the engine, you may have to gently pry it loose, then raise the hoist as you guide the engine out.

15. Once the engine is out, bolt it in an engine stand. Continue your disassembly by removing the manifolds and any accessories still attached to the engine.

Hoisting the Engine and Transaxle From a FWD chassis

Many front wheel drive cars require that the engine and transaxle be hoisted out the top of the chassis as a unit. Some suspension components, as well as the axles, have to be disconnected before the assembly can be removed.

Three common methods of disconnecting the suspension to free the axles are:

- Disconnect the tie rod end from the steering knuckle and separate the lower control arm from the strut. You will have to remove the cotter pins and lock nuts, then unseat the ball joint. With the steering disconnected, swing the knuckle and strut out of the way, then slide the axle out of the transaxle.
- Some manufacturers provide a way to separate the strut from the lower control arm, two bolts generally hold the knuckle and strut together. Remove the bolts and the strut can be moved aside without unseating the ball joint.
- The lower ball joint of some vehicles is separated by removing a single bolt that fits through a flange on the strut and a notch in the ball stud of the joint. Once the bolt is out, the control arm can be pried down to separate the connection.

Following are the step-by-step procedures for lifting the engine and transaxle from a Honda Civic. The sequences take into consideration that the previous steps described under DISCONNECTING THE ENGINE in this chapter have been taken.

Engine Removal

FWD ENGINE AND TRANSAXLE REMOVAL

1. Drain the engine oil, coolant, and transaxle fluid.

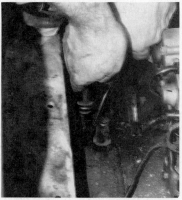

2. Slide back the rubber boot, remove the lock-pin, and disconnect the speedometer cable.

3. Disconnect the clutch cable at the transaxle, swing the cable out of the way and tie it back.

4. Remove the front wheels. Remove the cotter pin and nut, then separate the tie rod from the steering knuckle. Repeat for the other side.

5. Remove the cotter pin and nut from the lower ball joint. Pry down on the control arm, and pivot the strut free of the arm. Repeat for the other side.

6. Gently pry the inner constant velocity joint away from the transaxle. Slide the axle free of the transaxle and tie it out of the way. Repeat for the other side.

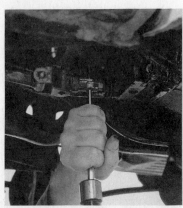

7. Slide the transmission shift linkage clip back to expose the roll pin, drift the pin out with a punch, and separate the linkage.

8. Remove the bolt holding the torque rod to the transaxle and pry the rod and bushing off its mounting.

9. Disconnect the exhaust pipe from the manifold, drop it down, and position it out of the way.

Chapter Eight

FWD ENGINE AND TRANSAXLE REMOVAL

10. Attach a sling chain to the lifting hooks on the top of transaxle and the front of the cylinder head.

11. Connect a lifting hoist to the sling and raise the hoist just enough to take the tension off of the engine mounts.

12. Remove the front engine mount to the cylinder head bolts. Remove the center bolt of the mount, then slide the mount into the body cavity.

13. Remove the nuts from the bottom of both the front and rear engine mounts.

14. Remove the bolts that connect the front and rear torque rods to the engine, loosen the bolts that attach the rods to the body.

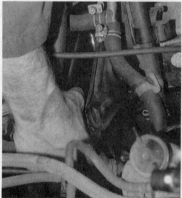

15. Swing the torque rods out of the way by pivoting them on the loosened bolts. Push them as far away from the engine as possible.

16. Check to make sure both engine and transaxle are free of the chassis, raise the engine hoist about six inches, and recheck to see that everything has been disconnected.

17. Have an assistant slowly raise the hoist as you guide the engine and transaxle out, you may have to tilt the assembly to clear the bodywork.

18. When clear, roll the hoist with the engine and transaxle away from the car.

Engine Removal

Hoisting the engine from a FWD chassis

Some front wheel drive transverse engines can be hoisted out from the top while the transaxle is left in the chassis. Chrysler engines are often removed in this manner. The procedure for this operation is a combination of the two previous removal methods. There are, however, some unique considerations.

All items that link the engine to the body will have to be disconnected and the transmission separated from the engine block. Since space is restricted in a transverse engine chassis, some additional equipment may have to be removed from the engine to allow sufficient clearance. Because both the engine and the transaxle are supported in the chassis by the same mounts, a special holding fixture must be used to support the transmission when removing the engine. The amount of side travel is also limited by the design. Additional body panels may have to be removed and lifting the engine out must be done slowly and carefully.

Dropping the engine and transaxle from a FWD chassis

Another method of removing a front-drive engine is to drop it out the bottom of the chassis along with the transaxle. The Ford Escort and General Motors N-Body vehicles: Buick Skylark, Oldsmobile Calais, and Pontiac Grand Am, are examples that require this type of engine removal. The procedure is similar to lifting the engine and transaxle out the top, however some specialized equipment may be required.

Suspension components need to be disconnected since the transaxle is being removed. Ford requires that the axles be removed from the chassis. General Motors designs allow the axles to come out with the transaxle. You disconnect the steering knuckle from the strut and slip the axle out of the hub.

A special dolly, designed to support and lower the engine transaxle assembly, is required for a General Motors car. The dolly is securely attached before the engine to body mounts are disconnected. With the dolly in place: remove the engine mounts, check and disconnect any remaining attachments, then slowly raise the vehicle off of the dolly and drivetrain assembly.

Ford recommends using a standard engine hoist to lower the assembly. A special support bracket must be installed temporarily to hold the engine in position while the drivetrain to chassis components are disconnected. Install the support bracket, then disconnect all engine to body connections except the right side engine mount. Connect the lifting hoist and raise just enough to relieve tension. Remove the support bracket and right side engine mount. Slowly lower the lifting hoist until the engine and transaxle rest on the floor, disconnect the hoist and roll it out of the way, then raise the vehicle.

LIFTING THE ENGINE

To lift the engine, you will usually use a hoist and a lifting sling of some sort.

Hoists

There are two basic types of hoists used for engine removal, overhead and portable. Overhead units might be a simple stationary chainfall, figure 8-10, or a chainfall mounted to an overhead track system that allows you to move the engine around the shop, figure 8-11. Overhead track systems are often self-powered, either hydraulically, or by an electric motor, figure 8-12.

Stationary chainfalls are the least convenient to use tools for removing an engine. To use one, you must position the car underneath the hoist, and be careful not to damage the paint with the chains as you lift the engine. In addition, you must reposition the car, not the engine, as removal continues, and finally roll the car out from underneath it once the engine is raised. Now you will have to find some way of moving the engine to where you will disassemble it.

Track-mounted hoists (especially the powered ones) are easier to use because you can lift the engine and move it away from the vehicle without having to reposition the car. The most elaborate units let you maneuver the engine anywhere in the shop. This is a real advantage during block machining operations, because you can transfer the block from station to station quickly and easily.

Portable engine hoists, commonly called cherry-pickers, use a lever-operated hydraulic cylinder to lift an engine. They consist of a boom that reaches over the engine compartment and legs on wheels that roll under the vehicle, figure 8-13. With one of these units, you can pull an engine anywhere that you can move the cherry-picker to. When you must pull an especially dirty engine and want to avoid making a mess inside the shop, this is an advantage. You can also lower the engine almost to the floor, and then use the cherry-picker to roll it around the shop. However, lifting and lowering the engine hydraulically is

Figure 8-10. One type of overhead hoist is a stationary chainfall mounted to the roof of the shop.

Figure 8-11. Overhead track systems allow you to move the engine around the shop on the hoist after you remove it from the vehicle.

Figure 8-12. A self-powered hoist is definitely top-of-the-line lifting equipment.

Figure 8-13. Cherry-picker engine hoists are compact, mobile, and fairly inexpensive.

harder to control, and the legs of the hoist might interfere with the car wheels or jack stands.

For any hoist you use, make a careful inspection of it to ensure that it is in proper working order before proceeding. The hoist you use must be rated to support the amount of weight you will be lifting. Read over the safety section in *Chapter 1* of this Shop Manual before you actually begin.

Lifting Slings

Slings provide the engine to hoist connection and can be simple or elaborate. A simple length of chain is commonly used for a sling, figure 8-14. More complicated crank-operated units that let you tilt the engine while it is supported by the hook are also available. Shops that do a lot of the engines often have lifting plates that bolt to the intake manifold and hook on the hoist. Be sure to use sturdy attachment points on the engine. Most engines have lifting brackets bolted onto them somewhere. These will usually let the engine balance correctly for removal. If there are no hooks or loops attached to the engine, you can use any sturdy bolt or stud to attach the sling, figure 8-15. You can also insert bolts into tapped holes in the block or cylinder head to attach the chain. Be sure to use large washers over them to prevent the chain from slipping off the end. The last thing you want to do is drop the engine on top of yourself, a co-worker, or the car while lifting it out.

Engine Removal

Figure 8-14. A short length of heavy chain and some sturdy fasteners will let you pull most engines easily, and without damage.

Figure 8-15. If there are no factory lifting brackets, you can use any convenient and sturdy stud or bolt as an attaching point for the sling.

Pulling the Engine

Get someone to help you pull the engine, lifting an engine is not something that you want to do alone. An engine on a hoist can swing out of position easily and unexpectedly to cause injury and damage. Exact methods vary and over a period of time you will develop your own style.

Removing the engine with the transmission disconnected

If you are leaving the transmission behind, remember that the engine must come straight forward a few inches before it will be free of the transmission input shaft. Strain on the input shaft can damage the transmission, have the transmission firmly supported. With automatic transmissions, bringing the torque converter out with the engine can damage the transmission. Make sure the converter stays with the transmission as you slip the engine forward. You can ensure this by holding the converter in place with a large screwdriver as you remove the engine. Tie the converter securely to the transmission as soon as the engine is clear.

Removing the engine with the transmission attached

If you are bringing the transmission out with the engine from a front-engine, rear-wheel-drive car, you will find it easiest to tip the back of the engine down so that the transmission can slip out from under the car without dragging on the floor and firewall. Front-wheel-drive engine/transaxles assemblies may simply come straight up, but most often will have to be slightly tilted or twisted to clear the body panels.

Hoisting the engine

Follow this procedure to pull the engine out with a hoist:

1. Make a last check to ensure that nothing connects the engine to the car except the loosened engine mounting bolts. Be sure to remove any engine anti-roll struts or vibration dampers that connect the engine or transmission to the chassis or sheet metal. Mark them before removal so that you can replace them in the same position.
2. Attach the sling diagonally across the engine, the ends should be far enough apart so that the engine will be stable and come up straight, figure 8-16. If you use bolts to hold the chain, tighten them snugly, but not too tight, to reduce the chance of them bending or breaking. Be sure to use washers that are large enough to prevent the chain link from slipping over the head of the bolt.
3. Center the lifting hook over the engine. This prevents the engine from suddenly swinging to one side as it clears the mounts. If you use a collapsible cherry-

Figure 8-16. Attach the chain across the engine so that it is balanced as you lift it up.

Figure 8-17. You can continue disassembly with the engine attached to a stand.

picker, make sure the legs are fully extended. This will keep the hoist from tipping as you raise the engine.
4. Lift the engine slightly to take the weight off the engine mounts. Then remove the engine mount bolts. At this point, there should be nothing else holding the engine to the car.
5. Lift the engine out of the engine compartment, slowly. Raise the engine several inches and then make a final check for any wires, hoses, or brackets that still hold it to the car. Have one person operate the lift, and another steady the engine as it comes out. Never lift an engine any higher than absolutely necessary.
6. Once the engine is clear of the compartment, move the hoist or the car until you can lower the engine. This makes moving it easier, and is *essential* if you use a cherry-picker. Portable hoists are unstable with a heavy load more than a few feet off the ground and tip easily. Always lower an engine slowly. This will prevent damage to the engine, and keep the load under your control at all times.
7. Bolt the engine to an engine stand, figure 8-17, or set it on a sturdy bench for disassembly.

Chapter 9

Disassembly, General Inspection, and Cleaning

Now that you have the engine out of the car, you can begin taking it apart. Ideally, the engine has already been steam-cleaned. If not, it should be cleaned now. Disassembling a clean engine is easier and considerably more pleasant. You are also less likely to overlook nuts and bolts hidden by a caked on accumulation of grease, oil, and dirt. If you do not have the facilities to clean the engine you might consider disassembling it outside in order to keep the mess out of the shop.

In this chapter, we will cover the procedure to break-down an engine into its major components. From this point until final assembly, you will be working with individual components — not a complete engine. Keep this in mind, and work in an organized way. Make sure that you have enough room to lay out parts for inspection during tear-down, keep a collection of small cans, boxes, or other containers to hold fasteners and small parts you remove and must be reinstalled when the engine goes back together.

Remember that the procedures in this chapter are general. There are many different types of engines. Some engines have unique features that will not be explained here. Refer to the manufacturer's service manual, or another reliable source that details the specific engine you are working on. You will need accurate specifications to guide you through tear-down and assembly.

INSPECT AS YOU GO

Keep your notebook handy. Disassembly can go quite rapidly, but you will be much better off if you take time and inspect parts as you go. The condition of the parts you will be removing will help you decide whether to rebuild the engine with all new parts, or just to overhaul it reusing some of the pieces. For instance, if you plan a total rebuild, you might just knock the pistons straight out of the bore without removing the ring ridge. You may discover later that the pistons could have been reused, saving you time and your customer money. If you have decided to overhaul instead of rebuild the engine, the camshaft bearings can remain in the block, and the pistons can be left attached to the connecting rods. Through experience, you will be able to read the wear patterns to determine what areas of the engine will need special attention. The most important rule to remember is to never compromise. Never reuse a questionable part, it can come back to haunt you.

Even if you are familiar with the engine you are working on, take notes about unusual things you find as you disassemble it. This might not be routine or even possible in a production shop, but taking the extra time is always a good idea. Watch out for:

- Damaged or worn-out parts
- Asymmetric wear patterns
- Non-stock components and internal marks
- Mistakes made during previous engine work

Major damage or excessive wear will be obvious as you take the engine down. However, minor damage, that is often overlooked, can provide clues as to why the engine failed. Asymmetric wear patterns on bearings and pistons are especially helpful for diagnosing warped blocks, bent or twisted connecting rods, and bent crankshafts.

Engines might be fitted with oversize or undersize bearings, lifters, pistons, valve stems, or other components from the factory, as explained in Chapter 5 of this *Shop Manual*. In addition, the engine might have been worked on internally by someone before you. They might have added non-stock components that you must return to the same position or replace with identical items. Engines might have oversize valves, a high-performance camshaft, or higher-ratio valve lifters. There may also have been some custom machine work done. Stock pistons might have been fly-cut to provide more valve-to-piston clearance than would be needed in a stock engine. The depth of this cut might be very slight, but it will be essential to duplicate it on the replacement pistons if you reuse the other non-original parts. You might also discover a non-stock oil pump, camshaft drive mechanism, or a variety of other changes.

Finally, watch out for mistakes someone else made inside the engine before delivering it to you. Many engines end up being rebuilt twice: first by an inexperienced owner or mechanic, and then soon after by a professional who is supposed to fix the problems created during the first rebuild. You might or might not be told what was done to the engine before it came to you. You might discover missing or incorrect thrust bearings, fasteners, mismatched rods and caps, or mismatched main bearing caps. Parts might be numbered or punch-marked incorrectly, or not at all. If the customer tells you that someone worked on the engine in the past, inspect it very closely while you disassemble it.

TOP END DISASSEMBLY AND INSPECTION

Start your disassembly right by working around the engine bolts (especially the exhaust manifolds) with a can of penetrating oil. No matter how careful you are, you are bound to experience broken fasteners at some point. Although we explain how to fix them in Chapter 10 of this *Shop Manual*, there is no reason to make them any more numerous than absolutely necessary. Make sure that the engine oil and coolant are drained, it is much easier to do it now than later. Also keep in mind that no matter how long you let the crankcase drain there will still be residual oil in the engine. Keep a drip pan under the engine as you disassemble it.

Remove Accessories and Manifolds

Follow this general procedure to remove, disassemble, and inspect the accessories and manifolds:

1. Remove any bolt-on engine accessories, such as the water pump, distributor, and fuel pump, that are still on the engine, figure 9-1. Any remaining motor mounts and brackets should also be removed at this time.
2. Remove the valve covers and set them aside.
3. Remove the bolts holding the intake manifold to the cylinder head or heads. Loosen the bolts sequentially, in reverse order of the torquing sequence, a turn or so at a time to prevent the manifold from warping. Look for hidden bolts, some manifolds have bolts that are easily overlooked. Note any bolts that are loose so that you can check under them for evidence of leaks or stripped threads after the manifold is off. On some older inline engines, the intake manifold is cast in one piece with the cylinder head. The manifolds on V-type engines will be bolted to both cylinder heads. Some fuel injected engines have two-piece manifolds that are disassembled by first removing the top casting and then the lower casting.

    Some intake (and exhaust) manifolds use different-size or different-length bolts in the different bolt holes, and putting the wrong bolt in a hole can result in coolant leaks or damage to internal parts. If you

Disassembly, General Inspection, and Cleaning

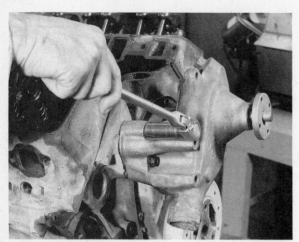

Figure 9-1. Remove the fuel pump, water pump, and any other accessories that are still bolted to the engine.

Figure 9-2. Gently pry the manifold off the engine block.

are working on an unfamiliar engine you can keep track of bolt positions by pushing them through holes in a piece of cardboard in spots that correspond to their position on the manifold.

4. Remove the intake manifold from the engine. You might have to break the manifold loose by taping it with a soft hammer, or by using a block of wood as a punch, then lift it up with a prybar, figure 9-2. If the engine has a separate lifter valley cover, remove it too.

Inspect the mating surfaces of the intake manifold for torn or misplaced gaskets, and warped flanges that might have permitted oil or coolant leaks into the combustion chambers. Now that the manifold is off, you can peer down the intake and exhaust ports with a light to see if the valves have excessive carbonized oil deposits. If any valve has significantly more or less carbon on it than the others, inspect that port's gasket surface for oil and coolant leakage. Coolant leaking into a port or combustion chamber tends to clean off carbon deposits. Coolant leaking along cylinder head or deck surfaces will often etch the metal, and leave behind a trail of discoloration to mark its path.

The manifolds for V-type engines often have a heat crossover passage. If the passage is clogged with carbon, expect to find excessive engine wear due to over-choking and oil dilution.

On V-type engines, look over the lifter valley and the underside of the manifold or lifter valley cover to gauge the care the engine received:

- If the valley is relatively clean, then the engine probably received fairly frequent oil changes, and many components are probably reusable.
- If the area is covered by a thick layer of oil sludge, you can expect to find cylinder wall, bearing, and journal wear. Sludge is caused by infrequent oil changes, or engine operation below normal operating temperature on a regular basis.
- If the region is full of hard-baked or charred oil deposits, then the engine was severely overheated, some of the castings (especially cylinder heads) could be warped or cracked, pistons may be collapsed, and cylinder bores damaged.
- If you find a light-brown milky foam or fluid, then coolant has been leaking into the oil. A cracked cylinder head or leaking head gasket is the probable cause.

5. Remove the exhaust manifold nuts or bolts, and remove the exhaust manifold. Take care when removing exhaust fasteners. Because the high exhaust heat has speeded up corrosion, exhaust nuts and bolts may be very difficult to break loose. Use penetrating oil, heat, or any other means available to avoid breaking a fastener. Once the manifold is off, look over the mating surfaces of both the manifold and the cylinder head, figure 9-3. If the exhaust manifold gaskets were leaking for an extensive time, then the hot gases might have eroded a groove in one of the mating surfaces. If surface damage will not allow a

Figure 9-3. Remove the exhaust manifold and inspect the mating surface for cracks or signs of leaks.

new gasket to seal, you will have to resurface or replace the head, manifold, or both.

Remove the Cylinder Heads

You are now ready to remove the cylinder head or heads. This is a straightforward operation on many engines, but on others it can be fairly complicated. If the engine has an overhead cam, the drive belt or chain will have to be disassembled, return to Chapter 7 in this *Shop Manual*. Disassemble the camshaft drive mechanism as explained in the section titled, SERVICING THE CAMSHAFT DRIVE. Then turn back to this chapter and continue with Cylinder head removal. Otherwise, follow the OHV rocker arm disassembly procedure.

OHV rocker arm disassembly

1. Loosen the locknuts holding stud-mounted rocker arms in place, then twist the arms to the side to free the push rods. This type of rocker can be left attached to the cylinder head. If the engine has shaft-mounted rocker arms, the arms and shafts are normally removed as an assembly. Remove the bolts holding the rocker assembly in place, loosen the bolts in several steps. Be sure to use the factory sequence for loosening and removing the rocker shaft. Note the position of any bolts that are drilled or relieved to supply oil. These will have to be reinstalled in the same position on assembly.
2. Take the pushrods out of the engine one at a time. Wipe any oil off the pushrod with a rag, and inspect for damage:

- Pushrod tips, whether convex or cupped, should be smooth and round. There should be no visible galling or flat spots on the ends.
- Look at the oil drill holes in the ends — they should be round, not misshapen or worn oversize.
- You can also check for straightness by rolling the pushrod on a surface plate, thick piece of glass, or other perfectly flat surface, figure 9-4.

If the pushrods are to be reused, keep them in order. You must match pushrods to the same lifter, and fit them in the same bore they were removed from.

Cylinder head removal

1. Locate the correct cylinder head bolt tightening pattern for the engine you are working on. There are many different specific patterns, depending on the number of cylinders, valve configuration, and casting materials. Typically, the head tightening sequence will start with the two middle bolts, then spiral out toward the ends of the block. *Loosening* sequence is normally the exact opposite of the tightening sequence. Sequence is *important*, especially with aluminum cylinder heads. Loosening in the wrong sequence, even on a cold engine, can warp the head.

   Loosen the bolts in several stages to relieve the clamping force slowly, then remove them. Watch out for:

- Hidden or inconspicuous bolts. Many engines have head bolts in inconspicuous locations. Trying to pry the cylinder head off without removing them can cause serious damage. For instance, small-block Chevrolet engines have nine obvious bolts underneath the valve cover, and eight less-obvious bolts outside, near the spark plugs. Some General Motors V-6 diesels have three bolts inside the intake ports under the valve cover. These head bolts are concealed behind Allen-headed plugs.
- Bolts of different lengths. Many engines have more than one size cylinder head bolts. Be sure to note which bolts went where as you remove them, they must go back in the same place.
- Torque-to-yield bolts, described in chapter 4 of the *Classroom Manual*. This type of bolt usually *cannot be reused*. Discard torque-to-yield bolts on disas-

Disassembly, General Inspection, and Cleaning

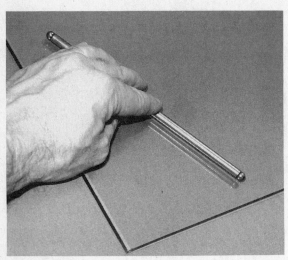

Figure 9-4. Check push rods for straightness by rolling them across a surface plate, piece of glass, or other flat surface.

Figure 9-5. Carefully lift the cylinder head off the block.

sembly and use new ones when the engine goes back together.

2. The cylinder head is now ready to be lifted off. Adhesion of the cylinder head gasket to the mating surfaces, may hold the two castings quite firmly together. Gently prying up on the cylinder head should break it free. Look for a prying pad cast into the cylinder head and use a prybar to lift it straight up. If there is no pad, insert the prybar into an intake port — never pry on a machined surface. It should not take much force to break the gasket loose, problems in removing the head might be caused by an overlooked bolt. Remember, many engines use dowel pins to locate the head securely in position on the block, and working the head from the side will not break the seal. Once the head is broken loose, lift it off the block, figure 9-5. For V-type and opposed engines, repeat the procedure for the other head.

3. Look over the cylinder head and its gasket for evidence of leaks. Check for leaking between adjacent cylinders, as well as between the combustion chambers and coolant or oil passages. Gas leaks often show up as trails of carbon stuck to the block or head surface. If you see anything like that inspect the combustion chambers, the deposits in the head can tell you a lot:

   - Thick, black carbon deposits that leave a wet oil smear on your fingers are caused by oil leaking into the combustion chamber.
   - Dry, sooty, black deposits indicate either an ignition misfire or a fuel mixture that is too rich.
   - A clean, shiny metal surface in a combustion chamber indicates that engine coolant has been leaking into the chamber. This indicates head gasket seepage or porosity in the castings due to metal fatigue.
   - The ideal conditions, not normally seen on an engine in for a rebuild, are a very light, fairly even layer of carbon covering the combustion chamber surface and intake valve face. The exhaust valve, because it operates at a much higher temperature, will often be colored a light, whitish gray.

4. This final step applies only to diesel engines. Because diesels have such high compression ratios, often 20 to 22 to 1, there is very little clearance between the top of the piston and the cylinder head. Resurfacing the block and the cylinder head can increase the compression ratio or decrease piston-to-head clearance beyond specifications.

   To avoid problems on reassembly, measure the deck clearance of each cylinder after you remove the cylinder heads. Measuring deck clearance is explained in more detail in Chapter 16 of this *Shop Manual*.

Set the cylinder heads aside for now, and continue with bottom end disassembly. The remaining steps of cylinder head disassembly and thorough inspection are covered in Chapters 14 and 15 of this *Shop Manual*.

BOTTOM END DISASSEMBLY AND INSPECTION

Continue with the bottom end disassembly exercising the same care you used removing the top end of the engine. Make sure you have enough room to lay out parts for inspection. There is still a considerable amount of oil, and possibly coolant, trapped in the block casting. If you are working on an engine stand, keep a drain pan under the engine. A good tear-down bench will have a lip around the top edge to retain fluids. The top may also be angled slightly to provide drainage. Keep your work area as clean as possible.

Remove the Harmonic Balancer and Flywheel

Remove these components before you do any further disassembly. They can be awkward to remove later:

1. Remove the bolt and washer at the front of the crankshaft and remove the harmonic balancer or pulley. Always use the proper tool for removing a balancer, figure 9-6. You can use a locking tool on the flywheel to prevent the crankshaft from turning.

   Inspect the harmonic balancer for damage to the keyway, and for wear to the hub caused by the timing cover seal. If the groove is too deep, you can usually press on a repair sleeve available from gasket manufacturers, figure 9-7.

   If the balancer has a rubber bonding ring, then look it over carefully. These rings eventually deteriorate. The first signs of deterioration are cracks and pieces of the ring breaking loose. When enough of the rubber is loose, then the outer ring can begin to walk around the inner hub. The immediate result is that the timing marks are no longer accurate. Eventually, the outer ring can come loose completely. Since it turns at crankshaft speed, the damage is immediate and spectacular.

2. If the engine has a clutch, use a punch to mark the pressure plate in relation to the flywheel. If you reuse these components, you will want to reinstall them in the same position to maintain correct engine balance.

   To remove the pressure plate, loosen the bolts a little at a time in an alternating pattern. This keeps bolts from stripping and the pressure plate from distorting. Keep the bolts together — they are often specially-designed for this application. Remove the clutch assembly.

3. Use a punch to mark the flexplate or flywheel in relation to the rear crankshaft flange, and remove it from the crankshaft,

Figure 9-6. Use only the correct tool to remove a crankshaft damper or harmonic balancer.

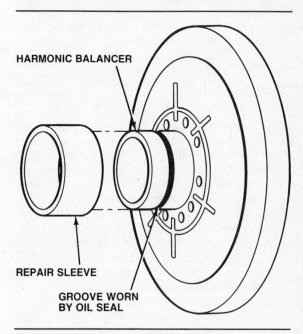

Figure 9-7. These thin repair sleeves provide a completely new surface for the stock type seal to contact.

figure 9-8. Marking the flexplate or flywheel is unnecessary on some engines, because the bolt holes may be arranged asymmetrically, or the flange may have a dowel hole that you line up to find the correct position.

To remove the bolts, you will probably have to hold the crankshaft using the same tool that you used to remove the balancer. Mark the bolts you remove from the crankshaft flange. You must use these bolts, or identical replacements, when you reassemble the engine.

Inspect the ring gear for broken or worn-out teeth, figure 9-9, and the flywheel or flexplate for cracks around the hub area.

Disassembly, General Inspection, and Cleaning

Figure 9-8. Unbolt the flywheel or flexplate after marking it with a center punch.

Figure 9-10. Remove the hydraulic lifters from their bores.

Figure 9-9. Look for worn, chipped, or missing teeth on the ring gear.

Figure 9-11. Valve lifter removers make short work of removing stuck lifters.

You can sometimes heat a ring gear to remove it from the flywheel, but a worn-out flexplate is usually replaced as a unit.

Remove the Camshaft and Lifters

If the engine has an overhead cam, skip this section. If the engine has overhead valves, return to Chapter 7 of this *Shop Manual*. Remove the timing gears, chain, and sprockets as explained in the section titled, **SERVICING THE CAMSHAFT DRIVE**. If you must remove the camshaft to disassemble the gears, follow the directions in that section, and skip this one. Otherwise, once the camshaft drive has been removed, turn back to this section and continue:

1. Remove the hydraulic lifters. Often you can pull them out by hand, figure 9-10. Sometimes carbonized oil will stick lifters in their bores. If so, use a valve lifter remover to pull them, figure 9-11.

   Unless you know for a fact that you will replace the lifters, keep them in order. Because they have worn to a specific cam lobe and bore, lifters must be returned to the same location if they are reused. If the lifters get mixed up, you will need to buy a new set, *and* a new camshaft as well.

2. Remove any retainer or thrust plate that holds the camshaft positioned in the block, figure 9-12.

3. Carefully pull the camshaft out through the front of the engine. Camshaft bearings wear slowly, and if you choose to overhaul, rather than rebuild the engine, you will want to reuse them. The best way to remove the camshaft is to stand the block on end and pull it straight up out of the block, figure 9-13. This prevents the cam-

Figure 9-12. Camshaft end float may be fixed by a retainer or thrust plate that must be removed before the camshaft can be taken out.

Figure 9-13. You can pull a camshaft straight up without scratching or gouging the camshaft bearings if you set the engine on end.

Figure 9-14. If the engine is on a stand, carefully draw the camshaft out with one hand as you guide it with the other.

shaft lobes from gouging the surface of the cam bearings.

If you have the engine on an engine stand, or it is otherwise inconvenient to set the engine upright, be extremely careful when removing the camshaft. You can often thread one or two long bolts into the holes on the front of the camshaft, this gives you something sturdy to hold. Draw the camshaft out with one hand as you guide it with the other, figure 9-14, remember that there is very little clearance and bearings are easily damaged.

Set the camshaft and lifters aside for inspection later. Details on thorough inspection are in Chapter 15 of this *Shop Manual*.

Remove the Oil Pan and Pump

Removing the oil pan is quite simple, but inside the pan is another place where you can find obvious clues to engine age and condition:

1. Many engines use a steel rear end plate to seal the back of the engine against the transmission or bellhousing. Remove any screws that hold the plate to the engine and detach it.
2. Turn the engine so that you can get at the underside, and remove the bolts holding the oil pan to the bottom of the engine. Tap the pan with a soft hammer to break the seal, then lift it off the engine. You may have to break the seal by working along the sides with a gasket scraper or screwdriver. Set the pan down and look inside it.

The engine sump will probably have a layer of thick oil in the bottom, but as with the lifter valley, an accumulation of sludge, charred carbon, or ash indicates abuse. Sludge accumulates because of infrequent oil changes, while charred oil or ash indicates overheating.

Disassembly, General Inspection, and Cleaning

Figure 9-15. This oil pump is bolted to the bottom of the engine and driven by a shaft off the distributor.

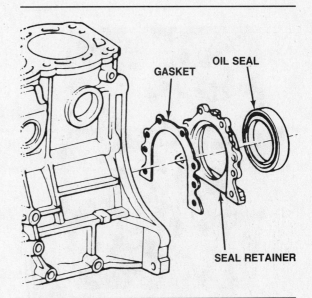

Figure 9-16. The rear seal of this engine can be removed simply by unbolting the retainer plate.

Figure 9-17. These main bearing caps have the identification numbers cast into them.

Look for signs of damage in the sump, too. You will often find pieces of valve seal, timing gear and chain, piston, and other debris in the sump. You may already know about most of these problems from the disassembly you have completed, but any extra information might be useful. The obvious clues in the oil pan can tell you which areas of the engine definitely need closer inspection.

3. Look over the oil pump pickup and inspect the filter screen for more debris.
4. You might already have removed the oil pump as part of the timing gear removal. If not, detach it now along with the pickup tube and screen, driveshaft, and any baffles that have to come off with it, figure 9-15.

Remove the Crankshaft, Rods, and Pistons

You will probably reuse many of these components, so take care during disassembly:

1. Remove any seals and seal retainers on the block that will prevent you from freeing the crankshaft. There will either be a one-piece lip seal at the rear of the crankshaft, or paired "half-moon" seals: one in the block, and another that was removed with the oil pan. Half-moon seals can be left in place during disassembly, but one-piece seals must be taken out before you can remove the crankshaft.

   One-piece rear seals may be mounted in a retainer plate that bolts to the block. Simply remove the bolts, then lift off the retainer along with the seal, figure 9-16. Some seals fit directly into a machined groove on the engine. Half of the groove is cut into the block and the other half is in the rear main bearing cap. To remove this type of seal, use a seal puller or screwdriver to pry the seal out.

2. Inspect the main bearing caps and the connecting rod caps to see if they are numbered. Main bearing caps are often numbered at the factory. The numbers may be stamped, or cast into the part itself, figure 9-17. Connecting rods are not normally numbered during production. If the engine has not been previously disassembled you will have to mark them. The best way to do this is to use a set of numbered punches, figure 9-18. If numbered punches are not available, use a series of center punch marks. Be sure to mark both the rod and the cap to avoid mismatching on reassembly, figure 9-19. *Never scribe or file identification marks*: this creates a stress riser on the part that can lead to premature engine failure.

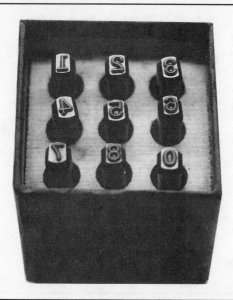

Figure 9-18. A set of numbered punches should be used to mark bearing caps and connecting rods.

Figure 9-19. Mark both the rod and the cap before disassembly, this prevents mismatching should the parts get separated.

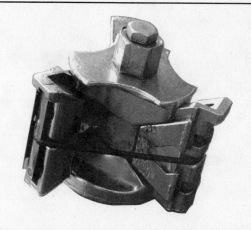

Figure 9-20. The ridge reamer cuts the ring ridge away with a cutting tool mounted in an arbor you turn with a wrench.

Mark the caps and rods before any further disassembly. Marking rods that are out of the engine and unsupported can easily distort them, and it is very easy to mismatch the rods and caps.

The standard method used by most technicians is to number all the connecting rods of an inline engine so the marks all face the same side of the engine. The connecting rods of V-type engines are traditionally marked on the sides facing the outside of the block.

3. If you find that the parts are already numbered, *check to see whether the numbering is correct.* Many engines have been destroyed by careless reassembly, and you may have one of them. Factories usually number the main bearing caps from front to rear, but distinctive caps such as those with machined-in thrust surfaces or oil seal recesses are frequently not numbered. Connecting rods are also usually marked from front to rear, with the numbers corresponding to the cylinder number of each rod. Check to be sure that the rod cap numbers are the same as the rod numbers, and that the cap and rod numbers are stamped on the same side of the assembly.

If you find any parts that are numbered differently from the traditional, front-to-rear pattern for main bearing caps and front-to-rear by cylinder number for connecting rods, then write down what you do find. Be extra careful when measuring these parts. Although unlikely, the engine might have been correctly rebuilt with the caps and rods in that position. It is much more likely that the parts are mismatched. Inconsistent numbering can also indicate a connecting rod installed backwards or a reversed bearing cap.

4. The next step is to remove the ridge at the top of each cylinder bore with a ridge reamer, figure 9-20. Begin by positioning the piston of the cylinder to be cut at BDC and placing a rag in the bore on top of the piston. Fit the ridge reamer and adjust the blade to take a light cut. When turning the reamer, keep the wrench leading the cutting blades by a few degrees. Leading the tool helps to smooth the cut and prevents undercutting or gouging, figure 9-21. Be extremely careful not to cut too deep into the bore. Removing any metal from the area of ring travel can make a rebore necessary for an otherwise good cylinder. Normally you can clean enough of the

Disassembly, General Inspection, and Cleaning

Figure 9-21. Position the wrench a little ahead of the cutter so that you avoid digging the blade into the cylinder wall.

Figure 9-22. Remove the cap from the connecting rod.

ridge away with just a few turns of the tool.

Once the ridge has been cut down, lift off the reamer, remove the rag, and clean out any remaining metal chips with a magnet or compressed air. Stray chips that lodge between the piston and the bore can damage the cylinder wall or the piston when you remove the piston. Repeat the procedure for the remaining cylinders.

5. Remove the connecting rod nuts from whichever connecting rod is at bottom dead center. A connecting rod is straight in the bore when the piston is at BDC. Jiggle the rod cap to loosen it then lift it off, figure 9-22. If the cap does not want to come off, a wooden hammer handle or a narrow block of wood can be used to free it. Tapping on the underside of the piston with the hammer handle will loosen the cap. *Never hammer on the rod or cap* to loosen it.

As soon as the cap has been removed, immediately slip some sort of protective covers over the protruding rod bolts. This prevents the bolts from damaging the crankshaft journal when the piston is removed. You can buy commercial covers, or simply use two short lengths of hose, figure 9-23. Use a wooden hammer handle to push the piston and rod assembly out through the top of the cylinder. Remove the bearing shells and set them aside. Keep them in order for later inspection. Put the cap and nuts back on the rod for the moment, and set it aside.

Rotate the engine so the next cylinder is at bottom dead center, and remove that

Figure 9-23. Two pieces of fuel line make excellent rod bolt cushions.

piston and rod. Continue for the rest of the cylinders.

Wipe the pistons and connecting rods clean with a rag, and inspect them before continuing. You can tell whether a connecting rod is bent by looking at the wear patterns on the thrust surfaces of the pistons. A normal wear pattern will extend

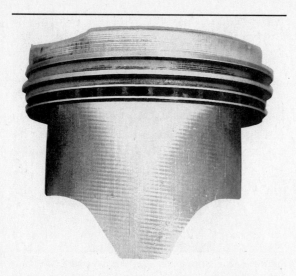

Figure 9-24. The symmetrical wear on this piston skirt shows a normal wear pattern.

Figure 9-25. Carefully lift the crankshaft out of the upper bearings.

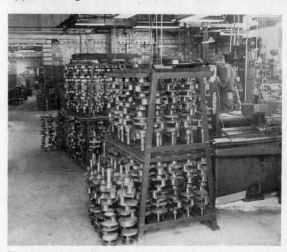

Figure 9-26. The best way to store crankshafts is in a rack designed to keep them out of the way, and safe from accidental falls.

up and down the center of the piston skirt, at right angles to the piston top, figure 9-24. Deviations from a normal pattern indicates a problem. To determine the cause see Chapter 12 of this *Shop Manual*.

6. Next remove the bolts from the main bearing caps. Main bearing caps fit snug to the cradle, they can generally be loosened by rocking them back and forth as you lift up. You may have to lightly tap the sides of the caps with a soft hammer to break them free, then lift them off and set them aside.
7. Lift the crankshaft out of the block, figure 9-25, and set it on the workbench. Bolt the main bearing caps back on their saddles to keep them organized. Wipe the crankshaft journals with a clean rag and check for obvious problems.

Check the crankshaft for sludge trap plugs. These are usually threaded plugs that have been staked in place to seal the ends of the oil passages. These plugs are removable but you may have to precisely drill out the punch mark before you can unscrew the plug.

For the time being, set the crankshaft aside. The best way to store a crankshaft is in a rack designed for that purpose, figure 9-26. If there is no way to hang up the crankshaft, stand it on end. Place it in an out-of-the-way spot and tie it to something solid to prevent it from falling. Crankshafts can bend if they fall on their sides. A good way to keep the crankshaft stable is to bolt the flywheel back on to give it a firm footing. More details on inspecting crankshafts are in Chapter 12 of this *Shop Manual*.

Inspecting the rod and main bearings
Inspect the connecting rod bearings and main bearings for damage. There is no sensible reason to reuse these parts when you reassemble the engine, even if they are in good condition, but looking them over can give you more information as to why the engine needed a rebuild. Good bearing surfaces are uniformly gray and smooth. There should not be any scratches, embedded particles, or metal flaking off the bearing shells. Wear should be greatest towards the center of the bearing, and minimal near the parting lines at the sides of the bore. Normal wear over a long period of time will eventually remove the upper Babbitt layer of a multi-layer bearing. The underlying metals, often copper-lead, will be exposed. This gives the bearing a distinct coppery color wherever the surface has worn away.

Disassembly, General Inspection, and Cleaning

BEARING WEAR PATTERNS

1. Damage caused by debris from machining left in the engine.

2. Gouging caused by cast iron particles circulated by engine oil.

3. Damage caused by dirt trapped behind the bearing shell.

4. Accelerated wear resulting from inadequate clearance.

5. Thrust bearing damage caused by improper crankshaft end play.

6. Severe damage resulting from oil starvation.

7. Surface damage caused by oil dilution, a result of excessive blow-by or a rich fuel mixture.

8. These main bearings were reversed during assembly, the solid shell was blocking the oil passage.

9. This bearing is out of a rod cap that was reversed or mismatched during assembly.

BEARING WEAR PATTERNS

10. Wear in a short arc indicates partial contact caused by excessive clearance.

11. Wear on the back of the bearing indicates that a poor fit has allowed the shells to move in the bore.

12. Even scoring across the bearing face indicates a poor finish on the crankshaft.

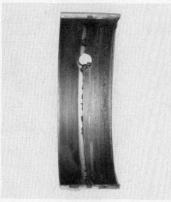

13. A tapered bore caused concentrated wear toward one edge of this bearing shell.

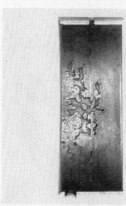

14. Distress in a V-shaped pattern is the result of a bent connecting rod.

15. Distress from an oil film unable to support the load, a result of excessive idling.

16. Distress patterns on the upper rod bearings indicates engine lugging, the lower main shells will show similar wear.

17. Distress at the parting lines is an indication of a reversed or misaligned bearing cap.

18. Corrosion can be caused by lack of maintenance, operation in extreme temperatures, or inadequate fuel or lubricants

Disassembly, General Inspection, and Cleaning

BEARING WEAR PATTERNS

19. Damage caused by lack of oil on start up, bearings farthest from the oil pump show more deterioration.

20. A bent crankshaft is indicated by severe wear on the center bearings that diminishes toward either side.

21. Distress wear on all the bearing shells is a result of the engine being operated at higher than rated speed while under load.

Look carefully for asymmetric wear patterns on the bearings. Localized smeared areas on the bearing indicate dirt particles that were trapped between the bearing shell and the saddle. Flaking or wear concentrated at the edges of the shell can indicate several problems: the shell was too wide for the journal, and was riding on the fillets, or the connecting rod or cap is bent. Connecting rod bearing wear concentrated at opposite sides and ends of the two shells also indicates a bent rod. Main bearing wear concentrated on apparently random areas across the various bearings can indicate a warped block.

Turn the bearings over and inspect their backs. If the bearings show scoring, unusual patterns, or if they appear highly polished, the saddles have not been holding the inserts tightly and they have spun in the bores. Look at the backs of the bearings to see if any of the inserts are other than standard size. If they are, the size will be engraved on the back of the bearing.

Installation mistakes can often be discovered by checking the bearings. If the bearing shells were installed reversed, then the oil holes might not have lined up, starving the bearing. If the locating tang on the bearing was not seated properly, then the bearing shell will have scraped the oil film off the journal.

Bearing wear patterns help you to pinpoint a wide variety of engine problems. Taking a good look at the bearings now will let you know what areas of the engine will need close inspection after things have been cleaned up.

Remove the Gallery Plugs, Core Plugs, and Camshaft Bearings

Once the crankshaft, pistons, and rods are out of the way, you can finish stripping the block of the rest of its removable hardware:

1. Begin by removing the dipstick tube. The tube can be tapped out from below, if it is pressed into a hole drilled into the block, or it can be carefully wiggled out. Dipstick tubes can also be bolted to the block or attached to the oil pan.
2. Remove the oil gallery plugs from the front and back of the engine. Plug design and method of removal will vary:
   - Threaded plugs may have a 1/4-inch (6 mm) square female opening, an internal hex, or a screw slot.
   - Sometimes the plugs are not threaded, but are simple cups tapped in to seal the drillways of the oil passages.

Figure 9-27. After treatment, the plugs will often come out easily.

   - Sometimes a threaded plug will be hidden behind a smooth metal cap.

Occasionally threaded plugs turn out easily, but more often they are stubborn. No matter how difficult, never leave these plugs in place. Over the life of the engine, sludge, metal shavings, and casting sand will lodge behind the plugs. The debris must come out. There is absolutely no method of cleaning the galleries with the plugs in place.

One way to remove threaded plugs is to heat them with a torch. After they cool, spray them with penetrating oil. Then with the proper tool, turn the plug out, figure 9-27. Paraffin wax can also be used as a penetrant to help loosen a plug. Heat the plug with a torch and touch a piece of paraffin to the hot exposed threads. The melting wax will flow down around the threads to provide a layer of lubrication and make removal of the plug easier. Extremely difficult plugs, or ones with rounded out heads, can be removed by center drilling the plug and using a screw extractor. Details on using screw-extractors are in Chapter 10 of this *Shop Manual*.

Non-threaded gallery plugs are easier to remove. These are simple cups, installed with their concave side out. You can remove them easily using a slide hammer with a self-tapping screw tip. To use: center-drill the plug, thread in the tool tip, and extract the plug by working the slide. An alternative method is to center-drill through the plug and screw in a self-tapping bolt. Grip the bolt with a pair of pliers and use the pliers to pry the plug out. After you remove the plug from one end of a gallery, you can often reach

Disassembly, General Inspection, and Cleaning

Figure 9-28. Use a punch to turn the core plug sideways, be careful not to damage the plug bore in the block.

Figure 9-29. Once the plug has been turned, it can be levered out with a prybar.

Figure 9-30. You can use a piece of pipe to drive the plugs out of the back end of the camshaft bore.

Figure 9-31. This camshaft bearing drift is adjustable for a variety of camshaft bearing diameters.

through the gallery with a length of quarter-inch drillrod, and tap the plug out of the other end.

3. Remove the water-jacket core plugs next. These are usually simple. Just tap around one edge with a punch to turn the plug sideways in its bore, figure 9-28. Then lever the plug out with a prybar, figure 9-29, or pull it free with a pair of pliers.

You can also tap the plugs straight through into the water jacket, and then pull them out with pliers. If you do tap them into the jacket, be sure to get them out. If you leave the old plugs inside, they will disrupt coolant flow around the cylinders and possibly cause localized overheating.

For OHV engines, there will often be a plug at the end of the camshaft bore that resembles a water jacket core plug. These are easy to get out by tapping through the bore with a long punch or a piece of pipe, figure 9-30.

4. Finally, remove the camshaft bearings. There are special tools for this purpose, and you must use them. Never try to drive them out with a punch, or you may damage the bearing bore of the block. Camshaft bearing tools are either adjustable for different size bearings, figure 9-31, or specifically-sized for only one application. In the latter case, you will need to have a different drift attachment for each size bearing. Bearing tools can use either a

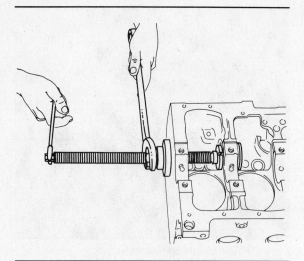

Figure 9-32. Some camshaft tools use a forcing screw to push the bearing shells out of the bore.

Figure 9-33. Insert the tool into the bearing shell, expand it to fit the bearing, and seat the lip of the driver up to the edge of the bearing.

Figure 9-34. Tap the bearing driver with a hammer to knock the bearings out.

forcing screw to pull bearings out of the block, figure 9-32, or are used with a hammer to drift the bearings out.

To use an expandable driver tool, assemble the bearing driver and insert it into the first bearing, figure 9-33. Rotate the handle until the driver expands to support the inside of the bearing. With the driving lip seated against the bearing edge, tap the tool with a hammer to drive the bearing out of the bore, figure 9-34. Keep the tool straight to prevent the lip of the driver from digging into the block. To release the bearing after it is free of the bore, turn the handle the opposite direction. Remove the remaining bearings in the same manner. Camshaft bearings are not the same size from bore to bore within the same engine, normally the largest bearing is at the front of the engine and the smallest is at the rear.

Inspecting the cylinder block

At this point, all the major assemblies should have been removed from the block. Take time now to check over the block for any remaining non-ferrous metal or plastic parts that have to be removed before the block is to be cleaned. Remove any distributor drive bushings, sending units, vacuum fittings, rubber mounts, seals, and water jacket drain plugs.

Look the block over for obvious problems, like cracks in the cylinder walls or the main bearing saddles. It is not really cost-effective to spend much time inspecting the block before you clean it. If the block is good, you will be cleaning and inspecting several times during the course of rebuilding. If you suspect the block is bad, you will probably clean it anyway to better assess the damage and determine if it can be salvaged.

ENGINE CLEANING

Now that you have the engine disassembled into its major components, they must all be thoroughly cleaned. Most of the cleaning solutions you will be using are toxic and considered to be hazardous materials, they must be handled with respect and never dumped or allowed to leach into the sewer system. Cleaning is an essential part of engine repair and rebuilding:

- Clean parts are easier to inspect.
- Clean parts are easier to work with and machine.
- Removing varnish and deposits make measurements more accurate.
- Cleaning prepares reusable parts for reassembly.

Disassembly, General Inspection, and Cleaning

Figure 9-35. These lumps of charred oil, found under the heat shield, will not be removed in the tank, but they will be softened enough to fall apart after you start the engine.

We explain further disassembly in Chapters 12, 14, and 15 of this *Shop Manual*. Because the engine block is now ready to be cleaned, we will explain all the various cleaning methods here. Some cleaning operations apply not only to the block, but to other parts of the engine as well. Skip ahead to those chapters for remaining disassembly procedures, then refer back to here for specifics on cleaning.

One point worth mentioning: Inefficient or haphazard cleaning can be worse than no cleaning at all. Over the life of an engine, a considerable amount of debris accumulates inside the parts. Water jackets develop scale and rust deposits, little piles of loose sediment lodge in nooks and crannies. Oil sludge, casting sand, and carbon deposits collect inside the engine. Intake manifold heat shields and inside the blind passages of oil galleries are places the sediment accumulates. Even crankshafts and piston ring grooves fill with various deposits. When you use brushes, solvents, or a hot tank, you must be sure to get all of the debris out of the engine. A classic mistake is to hot-tank an intake manifold without removing the heat shield from its underside. In a running engine, this area is exposed to both oil splash and the extreme heat of the exhaust passages. These conditions allow charred oil deposits to accumulate, figure 9-35. The underside of the manifold of some engines is also prone to cracking and must be inspected. Even the most thorough hot-tank session will not remove these deposits, but it will soften them enough so that they can disintegrate after the engine is started. This releases abrasive grit that circulates throughout the lubrication system during break-in. A new engine will be quickly destroyed by this seemingly small oversight. Oil galleries that were wire-brushed but not flushed thoroughly afterwards can have the same result.

Figure 9-36. Portable steam cleaners can be wheeled around or outside the shop.

Be sure to follow up glass-beading with a procedure to clear the residual beads out of cleaned parts, as explained later on. The beads are both tiny and very hard, which is why they work so well at getting deposits off metal surfaces. These same characteristics let beads destroy a rebuilt engine in a hurry if they are not all removed before assembly.

Cleaning Methods

There are a number of different techniques for cleaning engine parts. Your shop probably will not have all of them, because some are just different ways of accomplishing the same thing, some are more expensive, and some generate wastes that are more difficult to recycle or dispose of. We will explain each of them here:

- Steam cleaners
- Parts washers
- Hot tanks
- Spray booths
- Chemical dip tanks (cold tanks)
- Bead blasting (vapor honing)
- Airless shot blasting
- Tumbler and Shaker units
- Cleaning ovens

Steam cleaners

Steam cleaning is an extremely effective technique for removing the grease, oil, dirt, and general baked-on grime from the outside of an engine before disassembly. Many steam cleaners are portable, figure 9-36, and let you wash down a dirty engine outside before you bring it into the shop.

Steam cleaners do not actually clean the engine with steam. What they do is heat up soapy water under pressure to a temperature much higher than its normal boiling point. When the super-heated water reaches the end of the nozzle, it instantly boils and shoots out against whatever you are cleaning. The machine can be used without soap, but the hot water alone is not as effective.

Steam cleaners are great for removing engine grime, but you must be careful. They are also great for removing paint and decals, and for forcing high-pressure, soapy water deep inside fragile electronic components and connectors.

One of the main drawbacks to steam cleaning is that it generates a lot of grease and detergent-saturated waste water. In the past, this waste could be legally flushed into a city sewer system. This is now seldom true, and using a steam cleaner brings with it the responsibility to dispose of the sludge properly. If you work in a shop or garage that steam cleans engines, you will have to arrange for safe disposal of the waste matter. As a result, many shops today avoid steam cleaning. The work is sent out to specialists instead.

Parts washers

Parts washers are probably one of the most versatile and frequently used pieces of equipment in any shop. There are a variety of different kinds of parts washers, but most consist of a large basin with a pump that recirculates solvent through a cleaning wand or spigot, figure 9-37. You will use them repeatedly to clean pistons, connecting rods, valves, pushrods, fasteners, and crankshafts, plus your tools.

In most commercial setups, the solvent is stored underneath the basin in a replaceable drum. These designs are popular with companies that rent parts cleaners to shops, because they are easy to service. They can simply roll in a fresh drum of solvent and move the same basin and pump assembly from one drum to the other. Other designs use a large, deep basin to both wash parts in and to store the solvent. While this type will let you completely submerge parts in the solvent to soak them clean, they are more difficult to service when you need to change solvent. Both types always have a cover that closes to completely seal off the basin in case of fire. Covers are held up by a fusible link that melts at a fairly low temperature. If the solvent catches fire, the link melts and the cover swings closed, figure 9-38.

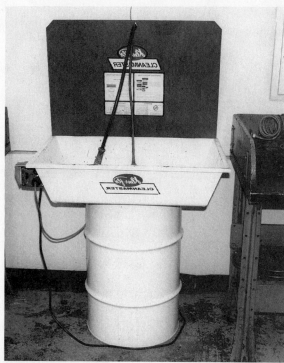
Figure 9-37. This parts washer sits on top of a drum of solvent, the pump and filter located underneath.

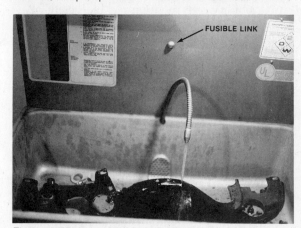
Figure 9-38. In case of fire, the fusible link will melt allowing the lid to drop and seal off the solvent.

Most parts washers use fairly gentle Stoddard solvent or mineral spirits (paint thinner). These solvents do a good job of loosening and removing small amounts of grime. Solvents are normally gentle enough that they can be used to clean plastic and other delicate parts that would be destroyed by other cleaning methods. The main advantage of a parts washer is that you can use it to hand-clean small items quickly. This advantage is lost if you fill the washer with caustics or very toxic chemicals. If your shop owns parts washers, rather than renting them, you can buy the correct solvents commercially. Never fill a parts

Disassembly, General Inspection, and Cleaning

washer with gasoline, kerosene, or any other highly volatile fluid.

Before you use a parts washer, take a moment to scrape off larger amounts of grease, tar, or dirt from the parts. The solvent will eventually remove heavy deposits, but the dissolved materials stay in the solution. As this happens, the cleaning ability lessens, and you will have to replace the solvent more frequently.

Hot tanks

Hot tanks provide an effective way of cleaning ferrous (iron or steel) parts. A hot tank will quickly clean engine blocks, cylinder heads, manifolds, and tinware, right down to the bare metal. Simple tanks are a covered tub containing a caustic solution that is heated to near boiling temperature, figure 9-39. The most common caustic solution is a mixture of water, sodium hydroxide (a type of lye, also called caustic soda), and sodium carbonate (also called soda ash). The resulting caustic soda solution is extremely alkaline, or basic — the exact opposite of acidic, and just as corrosive to certain materials. The caustic solution must be thoroughly rinsed off with clean water after the parts are removed from a hot tank.

The alkali solution dissolves grease, oil, varnish, sludge, most paints, gasket sealers, and plastics. It will also dissolve rust and remove it from the water jacket. It will *not* remove gaskets, carbon, or residual casting sand — you must clean this off by hand before you put the part into the tank. In addition, it will not remove water jacket lime or scale deposits. Scale requires an acid to remove it, not a base.

To use a hot tank, you simply lower the parts directly into the solution (or in a large parts basket), turn on the agitator if the tank has one, and leave it for a few hours. If the parts are really dirty, or if the caustic solution needs to be recharged, you might even leave the parts in the tank overnight. Some units come with automatic agitators that keep the liquid moving to speed up the cleansing action. Hot tanks may even have a built-in hoist to lift large heavy parts in and out of the solution without splashing.

The caustic soda hot tank solution is only for iron and steel parts. If you try to clean aluminum, brass, bronze, babbitt, or magnesium parts in the tank, the solution will dissolve them. Small parts will completely disappear, and large parts with machined surfaces, like aluminum heads, will be damaged, permanently. In addition, dissolving the metals depletes the solution, and makes it weaker.

Figure 9-39. A hot tank cleans by immersing parts into a vat of heated caustic soda solution.

You can buy special solutions (mainly detergents) to use in hot tanks for cleaning non-ferrous parts, such as aluminum. These detergents will also clean iron and steel, but not as quickly or effectively as caustic soda. Many shops will use two separate tanks, one for ferrous metals and the other for non-ferrous parts. A recent industry trend is to use citric acid solutions in hot tanks. The citric acid is weak enough to clean most metals without damaging them, its slower action is offset by the fact that a shop need only maintain one tank for all jobs, and waste disposal concerns are also reduced.

When you clean parts in a hot tank, be sure to separate any working assemblies. The chemical cleaning action can form deposits that can bind moving parts together. For instance, water pumps and manifold heat riser valves will be ruined in a hot tank. If you fail to remove any rubber or plastic parts, the chemical action will do it for you by dissolving them. Before hot tanking parts, you must have all the oil and water passage plugs removed. This allows the chemicals to soak the entire passage and makes rinsing off the solution much easier.

Spray booths

Spray booths (also called jet cleaning booths), use the same chemicals as a hot tank, but instead of immersing the parts, the detergents

Figure 9-40. Spray booths use a cabinet built over a tank of heated caustic solution. The parts rotate inside on a steel turntable as the solution is pressure sprayed on them from a bank of nozzles.

Figure 9-41. Be sure to hose the cleaned parts down thoroughly with water to remove all the cleaning solution.

Figure 9-42. Coat the exterior and interior of the cleaned parts with a water-dispersant oil to prevent corrosion.

are sprayed on. The parts are placed on a rotating turntable then sealed inside a cabinet. Hot, filtered caustic solution is sprayed from a bank of nozzles at the rotating parts, figure 9-40. Spray booths are quick, and can clean an engine block completely in thirty minutes or less.

Spray booths come in a variety of sizes and configurations to accommodate the specific needs of a shop. The booth itself is usually built above the caustic solution tank, the solution is heated to about 180°F (82°C) by gas, oil, or electric burners. Even the least expensive units come with a timer so that you can set the machine for a certain period of time and then leave it. In some of the best-quality spray booths, the spray nozzles oscillate back and forth while the turntable rotates so that every square inch of the part is cleaned.

When you remove parts from the spray booth, rinse thoroughly with a strong stream of clean water, figure 9-41. This neutralizes the caustic cleaning solution and flushes residue off the part and out of internal passages. You should also turn the part over several times as you rinse to let any trapped solution drain out.

Then use compressed air to clear as much water off and out of the part as you can. The parts will still be hot, so the remaining water on them will evaporate quickly. The parts will also be so clean that they will begin to rust immediately. Be ready with a can of light lubricant to spray or wipe over the parts to prevent the rust from forming, figure 9-42.

Some of the more sophisticated spray booths offer a neutralizing fresh water spray cycle to rinse off the chemical residue before you even

Disassembly, General Inspection, and Cleaning

Figure 9-43. This simple chemical dip tank lets you soak parts without heat.

open the door. Some also come with automatic timers that warm up the solution early in the morning and turn the heater off at closing time.

Chemical dip tanks
Chemical dip tanks are useful for cleaning aluminum or other small metal parts, such as pistons, rods, and carburetor bodies. The simplest kind that you can buy is a five-gallon bucket with a small-parts basket that lets you immerse parts into the cleaner, figure 9-43. When the solution is depleted, you just buy another bucket. You use these cleaners by submerging the parts basket into the container of solvent. Be sure to raise and lower the basket a few times to stir the chemicals up, let the basket rest on the bottom, and then replace the cover. Always keep the solution covered; this prevents evaporation and the escape of toxic fumes.

More sophisticated versions of this same item use a special lid with an air fitting to power an internal agitator. You clamp the lid onto these containers, attach an air line, then set the timer. The compressed air turns an agitator paddle connected to the basket to keep the solution stirred and constantly moving over the parts for more effective cleaning.

Most shops lease, rather than purchase, this type of equipment, which is often serviced by the same company that provides your parts washers.

Dip tanks operate without heat. The solvent they use is normally an extremely strong carburetor cleaner containing chlorinated hydrocarbons. These chemicals will effectively remove oil, grease, fuel deposits, and paint. Dip tanks are extremely useful for removing hard baked on deposits that are difficult to get off with a standard solvent wash. Their chief advantage is that they allow you to clean small steel and iron, as well as aluminum parts that you can not wash in the hot tank or spray booth. The dip tank solution works quickly, most parts will soak clean in less than 30 minutes.

The solvent in most commercial dip tanks is probably the most poisonous chemical you will ever have occasion to use in engine rebuilding. It is not only harmful to breathe, but instantly soaks into your bloodstream through your skin. Never reach into the bucket to retrieve a part that falls out of the basket. Use a magnet or a small parts retriever instead. Also avoid handling parts directly from the basket. Wear gloves, and neutralize the solution by rinsing the parts with water or solvent before you handle them. Most dip baskets have hooks on the side that attach them to the rim of the tank. This allows excess solution to drain off into the bucket.

Because it is so toxic and evaporates so easily, the commercial containers of solvent have a layer of water that floats on top of the solvent to prevent it from contacting the air. If you try to pour the solvent out of the bucket into another, you will likely lose most of this protective layer.

Bead blasting
Bead blasting, or vapor honing, equipment uses tiny glass beads shot from a nozzle by compressed air to scour engine parts clean, figure 9-44. It is a quick and easy cleaning process that removes any solid, dry material from most any metal part that you can fit into the beading cabinet. It is especially useful for quickly removing carbon and baked-on varnish from aluminum or smaller parts that might be damaged in a standard hot tank.

Bead blasting equipment consists of a large sealed cabinet that contains a nozzle with two hoses; one for compressed air, and one for the beads. Two tough rubber or neoprene gloves built into the front of the chamber let you direct the nozzle and hold the parts you are

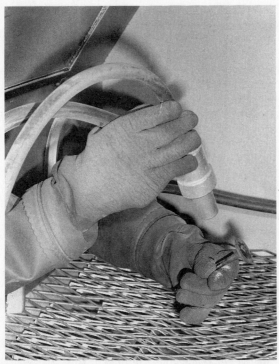

Figure 9-44. Bead blasting cleans by spraying parts with tiny glass beads to knock off caked on deposits.

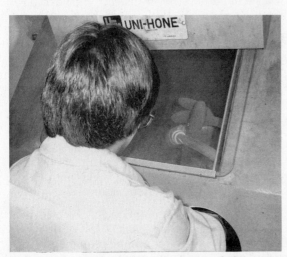

Figure 9-45. You use built-in gloves to work the parts clean while watching through a window on the front of the cabinet.

Figure 9-46. Use bore brushes to clear all the debris from inside the oil galleries after the parts have been cleaned.

cleaning. You watch through a glass window on the front of the cabinet, figure 9-45. Most units have a foot pedal that starts and stops the flow of the beads. Simply point the nozzle at the part to be cleaned and press down on the pedal. The beads, dust, and dirt from a cleaned engine part fall down through a screen and into a funnel at the bottom of the cabinet. Reusable beads are filtered out and returned to the pressure chamber.

The glass beads are available in several sizes, ranging from about 0.005 inches to 0.046 inches in diameter (0.125 to 1.12 mm). With use, the beads gradually break up into dust and fall through the filter with the rest of the debris. You must periodically remove the debris and add fresh beads.

Bead blasting is ideal for cleaning heavy deposits from delicate parts, such as the accumulated build up on the back side of valves. Cleaning with beads will not remove metal when used correctly, and it leaves behind a clean, satin finish. The parts you clean must first be degreased and dry, and free of any sticky deposits. Grease, oil, and moisture tend to catch and hold the beads clogging the filters.

There are some precautions you should take when bead blasting engine parts. You must be absolutely sure that you remove all the beads from any cavities, bolt holes, and exposed threads. The beads are extremely hard, and if they lodge in threads they will destroy them the first time you turn down a nut or bolt. Trying to save time here will create more work later on, be thorough when you clean the beads off any parts that have been blasted. When you are using a bead blaster, be careful not to direct the beads into internal casting passages, *especially* oil galleries or drillways. Blow all threaded holes, exposed threads, and accessible passages clean with compressed air, and clean threaded holes with bore brushes, figure 9-46. Use compressed air only, *Do not use water or solvent*, to clean parts after bead-blasting. Any type of liquid will cause the debris to clump together and adhere to the parts.

If you use bead blasting to clean a cylinder head with useable cam bearings in place, be

Disassembly, General Inspection, and Cleaning

Figure 9-47. Airless shot blasters use metal shot to clean parts automatically inside a sealable cabinet.

sure to protect the bearings with tape. The beads will quickly destroy the soft bearing surfaces if they are exposed. You should also be careful about where you direct the beads when cleaning pistons. Glass beads are excellent for cleaning the domes, but always protect the ring grooves with tape, and keep the beads away from the piston skirt.

You can use other cleaning compounds in bead blasting equipment besides glass beads, including sand, aluminum oxides, carborundum, aluminum beads, and crushed walnut shells. Aluminum beads and walnut shells are special-purpose abrasives that will scour very delicate parts clean without damage. Unlike glass, these beads will not damage the engine should a few particles remain behind after reassembly.

Airless shot blasters
Airless shot blasting, figure 9-47, is similar to bead blasting, but uses an entirely different method to propel the abrasive particles. Instead of using compressed air to propel the abrasive, a rapidly rotating impeller picks up steel or zinc metal shot and throws it against the part to be cleaned.

Unlike bead blasting setups, airless shot blasters are usually automatic. You load the parts to be cleaned into a rack, basket, or turntable inside the shot cabinet, turn the cabinet on, and walk away. Most setups slowly rotate the parts horizontally, but others resemble spray booths, with the parts rotating vertically on a turntable. Many cabinets have timers and automatic start and stop switches to turn the machine off after cleaning is completed. A cylinder head can be completely cleaned in about 15 minutes. A few extra minutes of tumbling the parts will clean out the residual shot.

Shot blasters are versatile tools. With the racks in an average-size machine, you can clean large parts such as engine blocks and cylinder heads. With a parts basket you can also clean smaller castings, rods, crankshafts, and other miscellaneous parts.

Like bead blasters, airless shot blasters require dry, degreased parts to operate efficiently. They are ideal as a follow up to thermal ovens. Typically, an airless shot cabinet has a tumble cycle in which the parts rotate without being blasted, giving most of the shot a chance to fall out. Even so, you must also go over the cleaned parts with a compressed air nozzle to make sure that all the shot particles are removed.

Tumbler and shaker units
Tumbler and shaker units are often used with both bead and airless shot blasters to remove and recover the residual particles left inside castings or stuck in threads and crevices. They do a thorough job of shot or bead removal in just a few minutes. Some shaker units consist of a cabinet into which you bolt the block, heads, or manifolds. The machine then tilts and shakes the parts in several directions to dislodge and recover the shot.

Tumblers operate a little differently. They tightly cradle castings, cylinder heads, and even engine blocks while the machine rolls them over and over, figure 9-48. These are quicker and easier to use than shaker units because they eliminate the need to bolt the engine parts into place. The actual tumbling takes only three or four minutes, and then the parts are ready for machining.

Cleaning ovens
Thermal cleaning ovens (also called pyrolytic ovens) can thoroughly clean engine castings without chemicals, figure 9-49. Ovens are mainly used by production shops that clean a large number of parts at once, or continuously. Instead of using a chemical bath or spray to dissolve and wash away grime and particles, cleaning ovens use high temperatures to bake the contaminants to a solid ash. Residue from baking can be easily rinsed off, shaken out, or bead blasted loose. Oven cleaning removes rust, scale, oil, grease, sludge, flammable (non-asbestos) gaskets, and all gasket cements.

Figure 9-48. Tumblers firmly hold and rotate the cleaned parts to dislodge loose debris and residual beads and shot.

Figure 9-49. A thermal cleaning oven works by baking all the oil and dirt to an ash that is easily removed by shaking, tumbling, or brushing off.

One big advantage of these ovens is that they will burn out carbon deposits from exhaust passages in intake manifolds and cylinder heads. No other automatic cleaning technique can do this as well.

There are two basic types: convection, and direct flame. Convection ovens work much like the ones in a kitchen. Electric, gas, or oil burners heat the air inside to about 400°F (200°C). Direct flame ovens work more like those in commercial incinerators; gas jets apply flames directly to the parts. Both systems have advantages or disadvantages. Convection ovens take time to heat up, and may heat parts unevenly between the top and bottom of the oven if not maintained correctly. However, a convection oven in good condition will heat individual parts evenly throughout, so localized warping and scorching is not a problem. Direct flame ovens are best for cleaning deposits out of water jackets and oil galleries, but can damage parts if not used correctly.

Parts cleaned in an oven need only to have the ash residue removed once they cool down. Ash can be removed in a hot tank or spray booth using ordinary soaps and detergents. Parts can also be bead or shot-blasted clean. A recent industry trend is toward a three-stage cleaning system that combines a thermal oven, shot blaster, and shaker or tumbler. The three machines, all similar sized, are side by side with an overhead track crane linking them together.

You can also use cleaning ovens for a variety of other machine shop tasks. Because the oven stays warm for some time after you remove the cleaned parts, you can use it to dry smaller items without turning it back on. You can heat parts evenly for press fitting or welding repairs, and you can even use it to straighten aluminum cylinder heads.

Be careful about cleaning aluminum parts, such as cylinder heads or manifolds. Aluminum is less tolerant of heat, and you must usually clean it at a lower temperature than iron or steel castings.

Cleaning ovens have two great advantages over any other cleaning method. They use heat, instead of expensive toxic chemicals, to clean and there is no liquid waste generated. Disposing of toxic cleaning sludge is a major problem. Since ovens concentrate this material into a small amount of solid ash, it is easier to dispose of properly. Cleaning in a thermal oven generally reduces the amount of waste generated by about 75 to 90 percent. Scrubbers and afterburners can eliminate the release of pollutants through the chimney.

Although the ash residue from an oven is not near as toxic or hazardous as the chemical sludge from the bottom of a hot tank, you must handle it carefully. Use a vacuum to remove it from the oven, not a broom, and a wear a dust mask or respirator.

Toxic Waste Disposal

For many years, hazardous waste disposal was never considered to be a problem by anybody. As a result, it has now become a problem for everybody, and how you plan to legally and ethically dispose of the toxic waste you generate will affect the procedures and equipment you can use in your shop.

Engine cleaning generates a large amount of poisonous waste. Liquid waste comes from

Disassembly, General Inspection, and Cleaning

used-up parts washer or cold tank solvents, hot tank or spray booth solutions, engine oil, used engine coolant, and contaminated wash water. These chemicals used to be easy to flush into the sewer systems or to pour out on the ground, but they eventually began to show up in our tap water.

Solid wastes come from the poisonous ashes and sludge from ovens and washing equipment, plus any liquids that have evaporated to solids. Even the non-toxic, non-polluting soaps and cleaning materials become hazardous once you use them and they become contaminated with oil, grease, gasoline, and engine dirt.

Your shop probably has arrangements with local waste disposal companies to collect the material periodically and haul it off for treatment and disposal. Make sure waste is being handled properly. According to law, *you are responsible*, not the company that hauls it away, for properly disposing of toxic material. Here are some things you can do to help:

- Find out exactly which types of the waste that your shop generates can be mixed together, and which types must be separated. For instance, old engine oil can be recycled fairly easily as long as it has not been contaminated with other chemicals like carburetor cleaner or mineral spirits. A very small amount of solvent can ruin a very large amount of old engine oil.
- Dried up chemicals, paints, or solid sludge should never be put into trash destined for ordinary landfills. Ground water will reactivate these chemicals the first time it rains, and will gradually transfer them into your tap water.
- Minimize the waste you produce by using up solvents, paints, and lubricants completely. Minimizing the amount of waste you generate reduces the amount you will have to dispose of.
- Use equipment that produces waste that is easier to deal with. For instance, cleaning ovens generate solid ash that can be stored in many landfills, while hot tank caustic solutions are more expensive and difficult to dispose of. Even hot tanks can be used with environmentally-safe detergents and emulsifiers to do the same job as toxic cleansers, without the waste problems.
- Lease the solvents and chemicals you use instead of buying them. Cleaning equipment companies will lease parts washers and tanks, and service the equipment regularly. Because they pick up the chemicals in bulk, they can afford the expensive treatment costs to recycle and dispose of them correctly.
- Consider farming out heavy cleaning jobs to specialists. This not only keeps your waste disposal concerns to a minimum, it also eliminates the task of heavy cleaning — the least desireable aspect of engine rebuilding.
- Be aware, whenever you perform engine repairs you generate hazardous waste, it is your responsibility to dispose of all refuse properly.

Chapter 10

Repairing Cracks and Damaged Threads

During engine repair and rebuilding, you will often need to repair damaged threads. Fasteners and threaded holes will often be broken or stripped when you receive the engine. Other threads are easily damaged during disassembly, regardless of how carefully you handle them. Using the techniques we show you in this chapter, you will be able to repair threaded holes and fasteners to their original integrity. In some instances, the threads will actually be stronger than they were originally.

We also explain the techniques you will use for detecting cracks and porosity in castings or forged parts. We show you how to repair this kind of damage, if possible, and cover the different repair techniques that are available for you to use.

REMOVING BROKEN FASTENERS

The best way to deal with broken fasteners in an engine is not to create them in the first place, but sometimes you have no choice in the matter. By the time an engine is ready for a major overhaul or rebuild, it has heated up and cooled down many thousands of times. In combination with a little moisture, this can result in fasteners that are permanently bonded to each other or to a casting with corrosion. Sometimes the exposed bolt and screw heads corrode, while the threaded parts stay in good shape. These can be just as difficult to remove because there is nothing for the tool to grip. Sometimes you break the head off a bolt or screw completely, and must remove the threaded shank from the hole.

How To Avoid Broken Fasteners

As we said earlier, try not to break, strip, or round off fasteners in the first place. There are several ways that you can minimize the number of fasteners you damage. First, never force fasteners loose during disassembly. Taking a few precautionary steps, will often prevent damage. If a bolt or nut will not come loose with normal force, try tightening it slightly and then backing it out. Sometimes turning the fastener the other way will break corrosion loose, and the fastener will then come out easily. Another method that works well is to rest a punch on the head of a stubborn bolt and strike a sharp blow with a hammer. Remember, you are not trying to drive the bolt in, you simply want to break the corrosion loose.

Left-hand threads
Although rare, left-handed fasteners are occasionally found on engine assemblies. These fasteners will loosen when you turn them

clockwise, and tighten when you turn them *counterclockwise*. Left-handed fasteners are used to fasten parts to the ends of rotating assemblies that turn counterclockwise, such as crankshafts and camshafts. Most automobile engines do not use left-hand threads, however, they will be found on many older motorcycle engines. Some left-hand fasteners are marked for easy identification, others are not.

Lubrication

Penetrating oil is a versatile tool that can prevent damage when you remove seized fasteners. A lightweight lubricant similar to kerosene, penetrating oil soaks into small crevices in the threads by capillary action. The chemical action of penetrating oils helps to break up and dissolve rust and corrosion. The oil forms a layer of boundary lubrication on the threads to reduce friction and make the fastener easier to turn. The best time to use penetrating oil is before you need to. When you get ready to disassemble the engine, apply penetrating oil to the external fasteners, especially the intake and exhaust manifold bolts, before you break them loose.

Give the oil some time to soak in before you start removing the nuts and bolts. You can increase the effectiveness of penetrating oil by tapping on the bolt head or nut with a hammer, or by alternately working the fastener back and forth with a wrench. This weakens the bond of the corrosion and lets more of the lubricant work down into the threads.

Proper tools

Another way to reduce the risk of destroying fasteners is by using the proper tools. Using good-quality, close fitting wrenches and sockets will help prevent broken fasteners. Using six-point, rather than twelve-point, sockets and end wrenches for disassembly will also help you avoid distorting and destroying nuts and bolts. These tools do not offer as many turning positions as a twelve-point, but they support the fastener better and can prevent it from rounding. Using an open-end wrench to loosen tight or damaged nuts and bolts is asking for trouble.

Stud removers can make pulling threaded fasteners out of blocks and manifolds quick and easy. There are also special pullers you can use to remove the pressed-in dowel pins from the surface of cylinder blocks or heads, figure 10-1.

Heat

Stubborn fasteners can often be easily removed with normal hand tools after they have been heated. Applying heat to the fastener, or

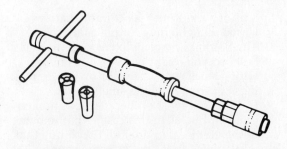

Figure 10-1. This puller is used to remove press-fit dowel pins from cylinder blocks and heads. Interchangeable collets allow you to extract different size dowels.

the casting around it, with the direct flame of a torch will cause the metal to expand and loosen the grip of the threads.

Use a minimal amount of heat to break fasteners loose, a common propane torch works nicely, if you use an Oxy-Acetylene torch keep the flame adjusted low. Never heat anything red-hot, too much heat will cause bigger problems than you already have. Always be careful when working with aluminum and alloy castings, they are extremely susceptible to damage from the concentrated heat of a torch flame. When applying heat, direct the flame either on the fastener or to the area of the casting surrounding it. Keep the flame localized, avoid heating both the fastener and casting, if you are trying to get the two metals to expand at different rates. Heating an area of the casting around the head of a bolt or stud will expand the casting away from the bolt. Directing the flame at a nut will expand the nut and loosen grip of its threads.

If heat alone fails to loosen the fastener, paraffin wax might help. Once the fastener has been heated, hold a piece of paraffin up against it. As the wax melts it will flow down the threads to lubricate them in much the same way penetrating oil will. Give the wax a moment to work its way in, then loosen the bolt or nut with a good tight-fitting wrench. *Never* use penetrating oil or any aerosol spray lubricants on a hot workpiece, most are flammable and could flare up.

Why Did the Bolt Break?

Before you start any kind of broken bolt removal, ask yourself why it broke in the first place. The reason can make a difference as to which removal method will work the best:

- Did the bolt head twist off because it was overtightened? If so, the bolt broke under

tension, and it may have immediately released all tightening torque as it broke. The bolt may be completely loose in the hole, so you might be able to turn it out with a small screwdriver or by lightly striking it off-center with a small punch.
- Did the bolt twist off because it bottomed in a blind hole? In this case, what remains of the bolt is still under all the torque that was applied to it, and getting it out will be a lot tougher.
- Did external corrosion on the bolt cause the flanks of the bolt head to distort so that a wrench cannot fit properly? Is so, the threaded portions might still be in good shape, and a gripping tool, such as a stud puller, slip-joint pliers, or locking pliers might work.
- Did the bolt break off because internal rust, corrosion, or electrolysis have seized the threads? If this has happened, the threaded portions might be permanently locked in the hole, and you will have to drill out the bolt and rethread the hole.

Answer these questions before you start, and you will save yourself much time and aggravation.

Removing Fasteners With Damaged Heads

Sometimes a bolt head or the nut on a bolt becomes so damaged that even a six-point wrench or socket spins around it without gripping. You have several options for solving this problem:

- Use a special tool to grip the damaged fastener.
- Destroy the bolt head or nut in order to remove it.

If the head or enough shank of a damaged bolt is protruding, you may be able to use a cam-action or wedge-type stud puller to remove it, figure 10-2. Stud removers fit over a fastener and are designed to grip, even on irregular shapes. The tool is designed to tighten its grip as you turn it with a wrench. Cam-action pullers use the same principle as a simple pipe wrench, but are designed for a much more specific purpose.

The simplest way to deal with a rounded nut is to chisel through it. For fasteners on iron castings, this is a quick and easy repair; however, you should never use a chisel on a nut seated against sheet metal or aluminum. Both of these metals are weaker than the nut, and you will likely cause a great deal of damage while chiseling. Seized nuts can be removed

Figure 10-2. Cam-action stud pullers can be used to remove broken bolts and studs if enough of the fastener is exposed for the tool to grip.

from most any metal by using a nut splitter, figure 10-3. Nut splitters are special tools that use pressure to deform frozen nuts to the breaking point. Once split, the nut can be removed with minimal damage to the bolt or stud. Nut splitters range in size and design from a simple hand tool for occasional use, to hydraulic powered units for production shops that experience seized nuts on a regular basis. A more time consuming alternative to using a nut splitter is to carefully cut the nut off with a hacksaw or die grinder. If you are careful, these techniques will not damage the underlying bolt or stud.

If a bolt head is too damaged to grip with any tool, you will have to do some cutting, drilling, grinding, or filing to remove it. Occasionally you may be able to re-shape the profile of a rounded-off bolt head so you can grip it with a smaller sized wrench, more often you will have to cut the bolt head down. Removing the head of a bolt often relieves tension so the shank will thread out of the hole easily.

A die grinder with a cutting wheel or stone will quickly remove a damaged bolt head, figure 10-4. To avoid injury, always wear eye protection while cutting, chiseling, drilling, and grinding. Work the grinder carefully to prevent damage to the surrounding casting or yourself. If you are using a cutting wheel, cut as much of the head off as you can with one pass, then switch to a stone to grind down what remains. Work the stone slowly from the center of the bolt toward the sides taking only a little metal off with each pass. Keeping the

Repairing Cracks and Damaged Threads

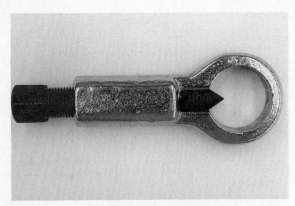

Figure 10-3. Nut splitters make removing seized nuts simple by breaking them off the bolt or stud.

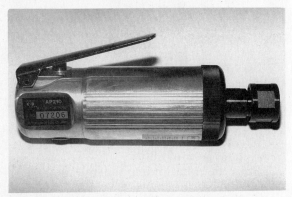

Figure 10-4. Die grinders work excellently for removing the head off a seized bolt, keeping the cut deepest in the center will make it easier to drill later.

Figure 10-5. Bolts with rounded-off heads can often be removed using a small chisel and a hammer to turn them out.

cut deepest in the center will make alignment easier should you have to center-punch and drill the bolt to remove it. Once the head has been cut down, follow one of the procedures for removing fasteners that are broken off flush or recessed.

Sometimes you will have to remove a fastener with a damaged head from one of the blind holes used to attach mounting brackets to engine blocks, cylinder heads, or manifolds. This typically occurs when the head of a bolt snaps off, leaving the shank concealed within the hole, either flush with the top, or at some depth. There are several techniques and a variety of special tools for removing these broken bolt shanks.

If the bolt is not truly fixed in the hole, but simply has no head, you might be able to cut a slot in the end of the shank with a hacksaw and use an ordinary screwdriver to remove the bolt. An alternative is to position a small sharp chisel off center on top of the bolt, angle the chisel and strike it with a hammer to drive the bolt counterclockwise, figure 10-5. Once the bolt begins to turn, you can often back it the rest of the way out with a small screwdriver.

An experienced welder may use arc welding equipment to remove large bolts that are broken off down in the hole. The flux on the welding rod will insulate it from the sides of the bolt hole as you tap the broken end of the bolt with the rod to strike an arc. Once the rod attaches itself to the bolt shank, turn off the welder and detach it from the electrode. Use locking or slip joint pliers to turn the welding rod and spin the bolt out.

Finally, you can sometimes turn a broken bolt out of a hole using a reverse, or left-handed, drill bit and a reversible drill. As the drill bites into the bolt, it will often spin it out of the hole. If not, then you are prepared for the next treatment with screw extractors.

Using Screw Extractors

Taper-bit extractors or "easy-outs" are inexpensive common tools used for removing broken bolts, studs, or screws of virtually any size, figure 10-6. Easy-outs look like a tapered punch with flutes ground into the surface. To use this type of extractor, drill a hole through the center of the bolt, each size extractor requires a specific size drill bit, and gently tap the extractor into the drill hole with a hammer. Turning the extractor counterclockwise with a wrench draws the flutes of the tool tightly into the bore, the remaining portion of the bolt turns out with the extractor. There are several taper-bit extractor designs, some have spiral shaped flutes, others use straight cut flutes. The shouldered head of some extractors prevents them from going too deep into the hole and expanding the bolt.

Figure 10-6. Screw extractors come in sizes for every bolt or stud on an engine.

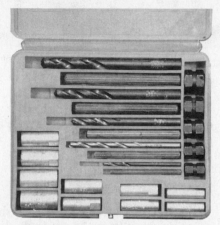

Figure 10-7. This set of straight-cut screw extractors will remove a variety of different size fasteners.

Figure 10-8. A good screw extractor will have the proper size drill bit to use stamped on the tool.

Not all extractors have a tapered shank, some straight flute bits are the same diameter from top to bottom, figure 10-7. Regardless of design particulars, all of these twist-type extractors work on the same principal, the flutes dig into the bolt and draw it out along the threads. The most important step is to get the drill hole straight and centered. Always use the proper size drill, many extractors have the correct drill to use stamped on their side, figure 10-8. If you drill the hole too small, the tool is more likely to bind and break, too large a hole will not provide enough bite to grip the bolt. To use a screw extractor most efficiently, follow this procedure:

1. Use a sharp center punch to indent the *exact center* of the broken bolt, figure 10-9. This is very important, drilling an off-center hole will make breaking the extractor much likelier.
2. Drill through the center of the bolt shank using a sharp drill bit, figure 10-10, a drill guide will help to keep the hole straight. Drill all the way through to the other side of the bolt if at all possible. Use the recommended size drill bit for each size extractor, and *choose the largest extractor that you possibly can*. If you have left-handed drill bits, use them, they will often draw the bolt out as you drill.
3. Place the extractor into the hole, and tap it lightly with a hammer several times to seat the flutes, figure 10-11.
4. Turn the extractor counterclockwise using a tap wrench or a close-fitting open-end wrench, the broken bolt will twist out with the extractor, figure 10-12. When you use a screw extractor, keep these points in mind:

- Use the extractor that will let you drill the largest hole possible through the bolt, and drill all the way through, if possible. For larger bolts, start with a small drill and work your way up to the recommended size. The large hole weakens the bolt, which loosens its grip on the hole and also lets the extractor bite into it easily.
- Use the largest extractor you can, too small a tool will not provide enough grip and might break. Extractors are made of hardened tool steel, they are very brittle, and almost impossible to drill if they break off in the hole. Never use power tools to turn an extractor, *use hand tools only*.
- Never pound the extractor into the hole. You want to tap it in just firmly enough that the sides of the extractor bite when you turn it. If you hammer it in too deeply, you will expand the sides of the broken bolt and wedge it more tightly in place.

Repairing Cracks and Damaged Threads

Figure 10-9. Center punch the exact center of the broken bolt.

Figure 10-10. Drill a hole through the center of the fastener.

Figure 10-11. Tap the extractor firmly with a hammer to seat it into the drilled hole.

Figure 10-12. Use a tap wrench to turn the screw extractor counterclockwise to remove the broken bolt.

Tool manufacturers make kits and accessories for screw extractors. Drill guides fit over the top of broken bolts, or center themselves in a bolt hole, to help you drill the hole exactly in the center of the broken bolt. You can buy kits containing the drill bits, drill guides, and correct extractors. While expensive, proper tools will ultimately pay for themselves in the time that they save you.

Because hardened screw extractors are very brittle and break easily, some professionals refuse to use them to remove broken bolts. Since removing a broken extractor is so time consuming, many machinists reason that they are not worth the trouble. Machine shops that specialize in broken bolt, tap, and drill bit removal seldom use extractors in their work.

An alternative is to use ordinary Allen wrenches to get broken bolts out. Select an Allen wrench slightly narrower than the broken bolt, and then drill a hole through the bolt that is the same diameter as the Allen wrench, measured across the flats. As with extractors, use the largest Allen wrench that you can. Once the hole is drilled, tap the Allen wrench into the hole and use it to turn the bolt out, just as if it were a screw extractor. The sharp corners of a new Allen wrench grip the hole just like the flutes of a screw extractor. Allen wrenches are softer than screw extractors, so if you reach their limits they will twist, rather than break off. This may be enough to get the broken bolt out, and eliminates the risk of breaking a hardened steel extractor in the hole.

Experienced machinists often will simply drill the broken bolt or stud out. When using this method, getting your punch mark exactly dead centered is extremely important. Start with a small drill bit and drill as deep as possi-

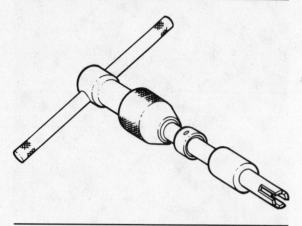

Figure 10-13. Tap extractors can sometimes remove a broken tap or straight-fluted extractor from a bolt hole.

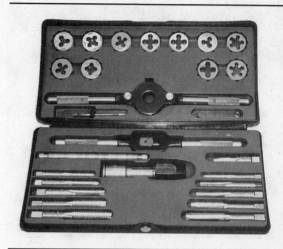

Figure 10-14. Hand taps and dies come in comprehensive sets that include taps, dies, tap wrenches, die stocks, and thread pitch gauges. Larger kits often include screw extractors, and combinations of American and metric tools.

ble, drill all the way through if you can. Change drill bits to gradually bring the bore up in size, drill up to but never exceed the minimum diameter of the bolt. Minimum diameter would be the same drill that you would use if you were going to tap the threads of the hole to fit the bolt, more information on tapping threads is covered later in this chapter. What will remain of the bolt is a thin spiral-shaped shell, if it comes out of the hole in one piece it will look like a spring. Because the wall is so thin there is barely any tension holding the shell in the hole, shells can often be walked out along the threads with a pocket screwdriver. The remains of a stubborn bolt corroded in place can be removed with a small sharp chisel. Use the chisel to break the shell into small pieces, remove the debris with a magnet and compressed air, and clean up the threads with a tap.

Broken Taps, Drill Bits, and Screw Extractors

Work carefully when removing broken fasteners, one thing you definitely want to avoid is breaking a tap, drill bit, or screw extractor off in a bolt hole. These items are made of hardened tool steel, that you cannot center punch, drill, or chisel, and they are very hard to remove. When this happens, *stop and think*. The best solution might be to farm the repair out to a specialist right away, and not to try to remove it yourself at all.

Broken taps and straight-fluted extractors can often be removed with a tap extractor, figure 10-13. The steel prongs of the tool fit into the spaces between the tap flutes and the drill hole bore. A sliding collar locks the prongs in place to grip the broken piece. You turn the tap extractor with an ordinary tap wrench, and if everything goes well, it will bring the broken tap out with it.

Never try to drill out a broken tap or extractor. The instant the cutting flutes of the drill bit catch on a projection of the hardened tap, the bit will break. You will then have to remove a broken drill bit *and* a broken tap.

You can *sometimes* make the brittleness of hardened steel work for you to extract a *short* piece of broken tap. Reach into the hole with a sharp center punch or chisel pressing the end of the punch against the broken tap. A sharp, hard blow to the punch can shatter the tap, then the broken pieces can be removed with a magnet, needle nose pliers, or compressed air. This technique is much rougher on the threads in the hole than using a tap extractor, so expect some damage to the hole. This method will not work on drill bits or screw extractors because they are stronger than taps. Trying to shatter a drill bit or extractor can wedge them in tighter and make matters worse. This technique will only work if the broken piece of tap is less than about 1/2 inch (12 mm) in length.

If you are lucky, there are machine shops in your area that specialize in non-traditional processes that enable them to remove next-to-impossible fasteners from next-to-impossible locations. One of the techniques that these companies use is electrical discharge machining (EDM). EDM uses high-amperage electrical current to produce thousands of tiny arcs between an electrode and the broken tool. Current is controlled and directed by submerging the workpiece in a non-conducting liquid during the process. Each tiny spark from the electrode vaporizes a small piece of the broken tool. The higher the amperage, the faster the removal. EDM equipment commonly uses graphite electrodes, but other metals such as brass, copper, tungsten, zinc, and various alloys can also be used.

Machinists who can remove taps, drill bits, and screw extractors advertise their skills, and most repair shops and general machinists will know who they are. Contact them *before* you get in over your head trying to remove broken tools from bolt holes. People often bring them castings containing broken bolts, taps, drill bits and screw extractors all in the same hole at the same time. The worse the situation, the more time consuming and expensive the repair will be. Their work is easier, quicker, and much cheaper for you if you bring the problem to them while it is still simple.

REPAIRING DAMAGED THREADS

Nuts, studs, or bolts with severely damaged threads should always be replaced with new ones. If the damage is minor, or the part is difficult to replace, you can repair the threads with a threading die. Damaged internal threads can be repaired by chasing the threads or by creating new ones with a tap.

Using Taps and Dies

Hand taps and dies are available in a variety of designs and sizes. Taps and dies can be purchased individually, or as a set that covers a wide range of thread sizes, figure 10-14. For most automotive work, you will be using taps and dies that cut straight threads for nuts and bolts. Fluid couplings will often have precisely-tapered threads, that require special thread-cutting tools. Specialized taps and dies are also available for cutting over or under sized threads and working with soft metals and plastics. Threaded fasteners differ in diameter, numbers of threads per inch, shape of the thread form, and how tightly the threads fit a hole or another fastener.

In automotive work, you will be using tools for both American Standard and Metric threads. American Standard threads come in two main series: National Coarse (NC) and National Fine (NF). Sometimes you will find older manuals or tools that refer to National Coarse and National Fine thread by the obsolete names U.S.S. and S.A.E., respectively. Metric threads conform to guidelines published by the International Standards Organi-

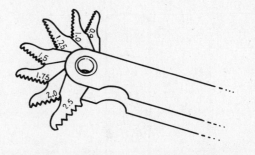

Figure 10-15. A thread pitch gauge has a series of saw-toothed blades that correspond to the thread pitch of a fastener.

zation (ISO). In addition, you are likely to encounter National Pipe Taper (NPT) threads on engines that use both American Standard and metric fasteners. NPT threads, which are tapered to seal against fluid leakage, will normally be found on oil pressure gauges, temperature senders, and oil galleries. Older British engines use Standard Whitworth fasteners that require a unique set of taps and dies, that do not interchange with any American or Metric tools.

Identifying threads

Thread identification is easy. There are seldom more than five or six different sizes on any particular engine. Once you become familiar with the engines repaired in your shop, you will be able to quickly identify thread sizes and fastener grade by sight. For additional information on fastener identification, see Chapter 14 of the *Classroom Manual*.

To measure threads you use a thread pitch gauge. These folding tools, similar to a feeler gauge, are supplied with most tap and die sets, figure 10-15. Thread gauges are precision tools, the saw-toothed patterns on the blades perfectly match standard thread pitches. To use a thread pitch gauge on a bolt, you hold the teeth of the blades against the threads of the bolt until you find one that matches. To find the diameter, measure the bolt with a pocket scale or a caliper.

The threads inside a bolt hole are harder to identify because it is difficult to see the gauge fitting into the threads. The way around this is to locate a bolt that fits the hole correctly, and then to use the gauge to determine the thread pitch of the bolt. Thread terminology is the same for both internal and external threads. There are different thread pitch gauges for use with metric and American Standard threads. Be sure that you choose the correct tool for the system used on the engine you are servicing.

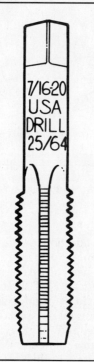

Figure 10-16. This American Standard tap will cut threads for a 7/16-inch diameter bolt, with a pitch of 20 threads per inch, you use a 25/64 drill bit to provide the correct diameter hole for the tap.

Always watch for unusual thread pitches, fluid couplings often use pipe thread, rod and main bearing caps will often use fine threads even though the other fasteners on the engine have a coarse thread pitch.

Using a tap to cut new threads

Once you identify the necessary thread pitch and bolt diameter, choose the correct tap to fit the threads. The marks on the tap list the diameter and thread pitch of the *bolt* that will screw into the threads they cut, and also the drill size that you should use if you are cutting new threads in a freshly drilled blank hole, figure 10-16. Some, but not all, American Standard taps also have an "NC" or "NF" stamped on the shank to indicate they will cut National Coarse or National Fine threads.

Metric tap markings are similar to those for American Standard taps. The marks list the diameter of the corresponding bolt in millimeters and a number corresponding to thread pitch, you may also see the letters "M" or "MM" indicating metric thread. Instead of giving the number of threads per inch, metric thread sizing gives the crest-to-crest distance between two adjacent threads. All spark plug hole threads have been metric for many years, most are either a 14 × 1.25 or 18 × 1.50 millimeter thread.

Pipe threading-taps cut threads in a tapered, conical hole, and have the mark "PIPE", "NPT", for National Pipe Taper, or "BSP", for British Standard Pipe.

Selecting the proper size tap is not the only decision you will have to make. Besides specialized taps, there are three common tap designs you can choose depending on the job:

- Taper tap
- Plug tap
- Bottoming tap.

The taper tap, often called a starter tap, is designed to make cutting threads in a straight-sided bolt hole faster and easier by gradually increasing the diameter of the tap. The taper makes starting the tap easier but can not cut full-width threads all the way to the bottom of the hole. Although taper taps are designed to align themselves when cutting new threads in an untapped hole, they can also be used to clean existing threads in through-holes or holes that are much deeper than the necessary thread depth.

Plug taps, the most common type found in sets, have only the lowest three to five threads tapered. The short taper allows the tap to start easily in a hole that has already been threaded and can cut satisfactory to very near the bottom of the hole.

Bottoming taps have only one or two of the threads tapered, just enough to get it started on existing threads, and can cut threads all the way to the bottom of a blind hole. *Never* use a bottoming tap to thread a newly drilled hole, use them only *after* cutting as much of the threads as you can with a taper or plug tap. It is too difficult to get a start and keep a bottoming tap straight, you also risk breaking the tool off in the hole.

To cut new threads in a blank hole, follow this procedure:

1. Center-punch the exact spot where you want the threaded hole to be.
2. Determine what bolt you want to thread into the hole, and choose the tap of that nominal bolt size and thread pitch.
3. Select the correct drill bit based on the information engraved into the tap. The correct drill size will be smaller than the bolt. For example, to make a hole for a 1/2-inch bolt with 20 threads per inch, use a 1/2-20 tap. This tap requires you to drill a 29/64-inch hole, figure 10-17.

You must use the correct drill bit for each tap. If you use a bit that is too large,

Repairing Cracks and Damaged Threads

195

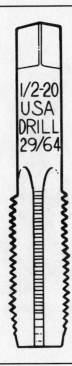

Figure 10-17. This half-inch tap requires a 29/64-inch drilled hole.

the tap will not cut deeply enough into the sides of the hole to make good threads. If you use a tap that is too small, the threads will be much harder to cut, and you run the risk of breaking the tap off in the hole.

4. Drill the hole into the workpiece, keep the drill bit straight, to the correct depth, figure 10-18. You will need extra depth at the bottom of the hole to catch the chips as you cut them. Use the correct cutting fluid for the drill bit.
5. Chamfer the top of the hole with a countersink or a small file, figure 10-19.
6. Clean all of the metal chips from the hole with a brush or compressed air.
7. Take the correct tap, (remember to start with a taper or plug tap even if you are threading to the bottom of the hole), and clamp it firmly in the tap wrench. Lubricate the tap with cutting fluid, a light mineral oil or soluble oil. *Do not use motor oil*, motor oil will dull the cutting blades of your tap. Threads can be cut into cast iron with a dry tap, the high graphite content of the casting will lubricate the tap as it cuts the new threads.
8. Gently push down and carefully turn the tap clockwise two or three turns (for normal, right-hand holes) to start the tap into the hole. Work with one hand and center the tool in your palm to get a good straight

Figure 10-18. Be sure the hole is straight. A drill press is the best tool for this job.

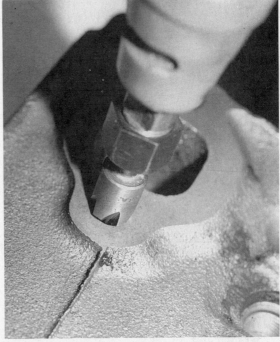

Figure 10-19. Chamfer the hole to provide a lead-in for the tap, and to prevent the threads from binding on the edge of the hole.

Chapter Ten

Figure 10-20. Get a good straight start by working the tap in with one hand.

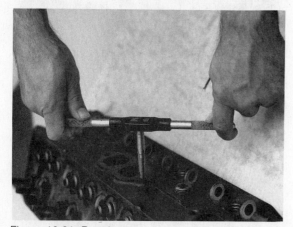

Figure 10-21. Run the tap into the workpiece with both hands, use a smooth motion and steady pressure.

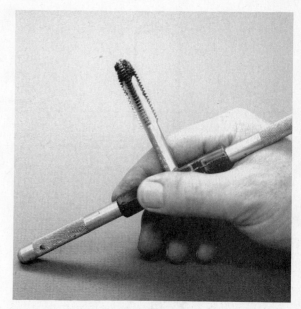

Figure 10-22. A considerable amount of debris from cleaning and machining will collect in the bolt holes, a tap can be used to remove it.

start, figure 10-20. Keep the tap exactly perpendicular to the workpiece and start slowly to establish a bite.

9. After the tap has established itself in the hole, remove the wrench from the tap. Use an angle or machinist's scale make sure that the tap is running at a right angle to the workpiece. If not, remove and restart the tap. If the angle is true, reattach the wrench and continue.
10. Continue by using both hands to turn the tap wrench slowly and evenly until you reach the necessary depth, figure 10-21. A good tap will cut cleanly with steady even pressure, if you have to force it something is not right. As the tap turns, it will lead itself into the hole cutting threads as it goes. Every so often, turn the tap backward a half-turn to break the chips, a rule of thumb is "forward two steps and back one". Be sure the tap continues straight as you work it to the bottom of the hole.
11. When you have tapped the hole to the correct depth, unscrew the tap and clean it off. Blow the tapped hole clean with compressed air. If you are tapping a blind hole, run a lubricated bottoming tap into the hole to cut threads all the way down the bore.
12. Clean away all chips and test fit the new threads with a good bolt, the bolt should thread easily to the bottom.

Using a tap to chase old or damaged threads
The procedure to chase damaged or worn threads in the bolt holes of engine castings is similar to that of tapping threads onto a blank hole. Even threads that are not damaged need to be chased as a normal part of a rebuild. Debris from wear, teardown, and cleaning will collect in the bolt holes, figure 10-22. If not cleaned out, this debris can prevent the bolt from torquing properly and may result in premature engine failure. Plug and bottoming taps work well to chase threads, there is no need to use a taper tap because threads already exist. Keep these guidelines in mind when you chase or tap to repair threads:

Repairing Cracks and Damaged Threads

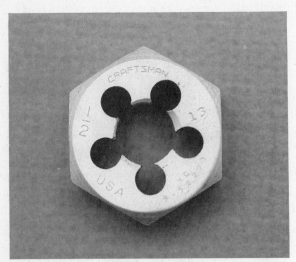

Figure 10-23. Dies come in as many sizes and styles as taps, the diameter and pitch of the threads they cut will be stamped on the face.

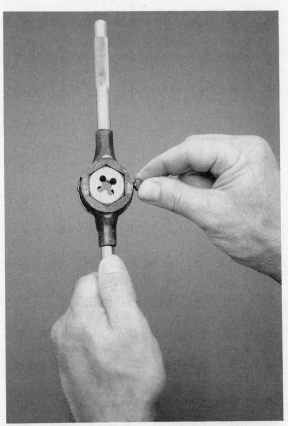

Figure 10-24. Use a die stock to support the cutting die, the handles on either end help to equalize pressure and keep the die running straight, a set screw locks the die into the tool.

- Make sure that the hole is clean, and free of any loose metal shavings, dirt, grease, or glass beads. Use a rifle brush and solvent to clean out dirty bolt holes, then blow them out with compressed air.
- Lubricate the tap, then carefully turn it into the existing threads to get it started. The tap will lead itself down the threads and into the hole as you continue turning. Keep the tap perpendicular as you run it down with steady even pressure.
- Use a bottoming tap, if at all possible. Plug taps will clean out through-holes, but only a bottoming tap will get down to the bottom of threaded holes in blocks, manifolds, and cylinder heads.
- Follow the thread chasing by chamfering the rim of the bolt hole to smooth off any burrs and rough edges with a countersink, small file, or stone.

Remember that you are merely cleaning the threads, not cutting new ones. You should never feel more than just a slight resistance, the tap should glide along the threads as it removes burrs and debris. Try not to use new taps for chasing threads. A worn tap will slip around the existing threads a little more smoothly, and will be less likely to cross-thread. However, brand-new taps that have thread classes less than 6 (metric) or either 1A or 1B (American Standard) will work well chasing threads. Taps and dies with these designations fit threaded holes and fasteners a little more loosely than tools graded at the normal 6 or 2A and 2B (metric or American Standard, respectively).

Using a threading die to cut new threads

As with taps, you can use a threading die to both chase worn and damaged external threads, and to cut new threads on blank stock. Most blank rod is actually made 0.005 to 0.010 (0.125 to 0.250) *under* its nominal size, which provides the correct clearance for normal threading. The only preparation this rod requires is for you to chamfer the end to let the die lead in correctly. To cut new threads on blank stock follow these steps.

1. Choose the correct die to cut the required threads, similar to a tap, diameter and pitch will usually be stamped on the face of the tool, figure 10-23.
2. Place the die into a die stock and tighten the set screw to hold it in position, figure 10-24. Lubricate both the die and the workpiece lightly with cutting fluid.
3. Place the open end of the die squarely over the chamfered end of the bolt stock. Hold the die stock with the die centered in the palm of your hand. Gently push down and carefully turn the die stock clockwise (for normal, right-hand holes) to start the

die over the rod, run it down two or three turns. Keep the die exactly perpendicular to the workpiece.
4. Once a firm bite has been established, use both hands to work the stock and keep the die cutting straight. Continue working the die down the rod until you reach the necessary length of thread. As it turns, the die will lead itself around the rod cutting threads as it goes. Just as with tapping a hole, every so often, turn the tap backward a half-turn to break the chips. Be sure the die continues straight as you tap.
5. After you have cut the threads, back the die and die stock off of the workpiece, then buff the threads with a wire brush.
6. Test fit your threads by running a nut down the entire length of the threads.

Using a threading die to chase worn or damaged threads

You will rarely have the need to cut new threads on a piece of blank stock, however, you will be using dies to clean existing threads on every engine you rebuild. To chase damaged or worn threads on bolts or engine fittings use the same procedure as that outlined to cut new threads. Be sure not to cross-thread the die as you start and keep the cutting surfaces lubricated. Dies are designed to cut threads into blank stock, and will happily cut new threads across the old ones if you are not careful. Hold the die stock perpendicular to the bolt or stud, and keep it straight, you should not have to force it onto the threads.

Thread Chasers, Restorers, and Thread-cutting Files

You can also use a variety of hand tools, besides taps and dies, to clean up threads. Universal thread chasers are adjustable tools that you clamp around a bolt and turn to clean threads. The interchangeable blades will accommodate a variety of different thread pitches. These tools can be used on bolts of any diameter, as long as the pitch is the same. You can also buy adjustable thread chasers to clean up internal threads.

Thread restorers are tools that look similar to taps and dies, however, they are designed to roll the threads back into shape rather than cut new ones. Internal thread restorers are used to clean up topped holes, figure 10-25. You can dress the threads of bolts and studs with external restorers, figure 10-26. Like taps and dies, a restorer will fit only one diameter and thread pitch, and are usually sold in sets.

Figure 10-25. Internal thread restorers are used to recondition the threads of bolt holes.

Figure 10-26. External thread restorers will quickly repair minor damage on the threads of a bolt or stud.

Thread files are even simpler. These tools are usually four-sided files that you draw across the damaged threads to clean off burrs, figure 10-27. Thread files commonly fit as many as eight different thread pitches.

INTERNAL THREAD REPLACEMENT

Sometimes the threads inside a bolt hole are too damaged to clean up with a tap, such as when a bolt was overtightened and stripped the hole. If so, you have several options:

- Drilling and tapping to oversize
- Helical inserts
- Self-tapping inserts
- Key-locking inserts.

Each of these methods are discussed below.

Drilling Oversize

In the past, the only way to repair a stripped hole in a casting was to drill the hole to a larger diameter, and then tap it to fit a larger bolt. There is still nothing mechanically wrong with this approach for most fasteners, simply follow the directions on tapping explained in the previous sections. However, it makes assembly

Repairing Cracks and Damaged Threads

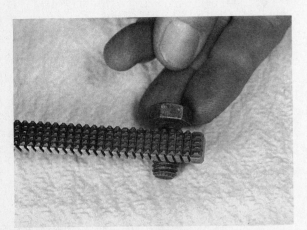

Figure 10-27. Thread files are simple inexpensive tools that will clean minor damage to the external threads of bolts, studs, and fittings.

Figure 10-28. Helical inserts look like small, coiled springs. The outside is a thread to hold the coil in the hole, and the inside is threaded to fit a standard fastener.

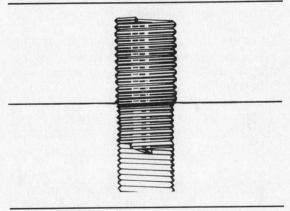

Figure 10-29. The insert provides new stock-size threads inside an oversize hole so that you can reuse the original fastener.

more involved because you must use several different size fasteners, and keep track of which fasteners fit the oversized holes.

This method also causes problems when used for critical fasteners, such as head bolts. If you attempt to use an oversized bolt for a precision clamping job, like cylinder head bolts, then you will have a problem. If you tighten the larger bolt to the specified torque value, you may not stretch the bolt far enough into its elastic deformation region. As a result, you may have problems with the bolt loosening repeatedly. On the other hand, if you tighten the bolt to the greater torque value normally specified for bolts of that diameter, then you may distort one or both of the castings you are bolting together.

Finally, there is no way to use anything but the factory-specified fastener in situations where torque-to-yield bolts, or other specialized fasteners are used. For these situations, you must use thread inserts to bring the hole back to the original factory dimensions.

Helical Inserts

A helical insert looks like a small, stainless-steel spring, figure 10-28. To install a helical insert you must, drill the hole to a specified oversize, tap it with a special tap designed for the thread inserts, then screw the insert into the hole, figure 10-29. The insert stays in the casting as a permanent repair, bolts can be removed and replaced without disturbing the insert. One advantage of a helical insert is that you can use the original bolt because the internal threads are the same size. When correctly installed, an insert is often stronger than the original threads, especially in aluminum castings. In fact, many high-performance engine rebuilders install inserts in blocks, manifolds, and cylinder heads as a precaution.

One of the best known of the helical fasteners are Helicoils®, manufactured by Helicoil Products. To install Helicoil inserts you will need to have a thread repair kit. The kit includes a drill bit, tap, installation mandrel, and inserts. Repair kits are available for a wide variety of diameters and pitch to fit both American Standard and Metric threads. A simple kit contains the tooling for one specific thread size, master kits that cover a range of sizes are also available. Once you have the tooling, you can purchase the inserts separately as you need them. Installing an insert is similar to tapping new threads, follow this procedure:

Figure 10-30. Helicoil® kits, available in a wide variety of sizes, contain everything you will need to repair a damaged hole back to standard size.

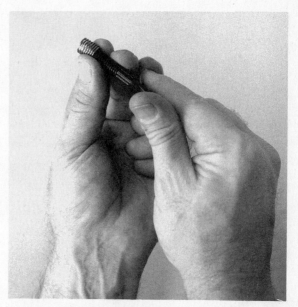

Figure 10-31. Thread the insert onto the mandrel and apply a thin coat of thread sealing compound.

1. Select the Helicoil kit designed for the specific diameter and thread pitch of the hole you need to repair, figure 10-30.
2. Use the drill bit supplied with the kit, drill size is also specified on the Helicoil tap, to open up the hole to the necessary diameter and depth.
3. Tap the hole with the Helicoil tap, remember to lubricate the tap and turn it in slowly. Follow the previously detailed tapping procedures to break the chips off and *keep the tap straight* as you go. The Helicoil tap is unusual, the large diameter and course pitch are not the standard thread pattern for the fastener you will be using. Never try to substitute, the correct tap is included with the Helicoil kit.
4. Thread an insert onto the installation mandrel until it seats firmly, figure 10-31. Apply a light coating of the recommended thread locking compound to the external threads of the insert.
5. Use the mandrel to screw the insert into the tapped hole, figure 10-32. Once started spring tension prevents the insert from unscrewing, so you cannot back the insert out once it has established itself in the hole. Stop when the top of the insert is 1/4 to 1/2 turn below the surface of the workpiece:
   - If the hole is deeper than the standard insert, continue to thread it down far enough to make room for a second Helicoil on top.

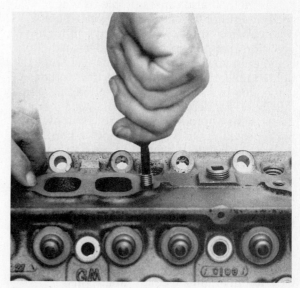

Figure 10-32. Turn the Helicoil® into the threaded hole with the mandrel.

   - If the hole is shallower than the insert, trim the insert to the required length with cutting dikes before you thread it on the mandrel, figure 10-33.
6. Remove the mandrel by unscrewing it from the insert, then use a small punch to break off the tang at the base of the insert, figure 10-34. Never leave the tang in the bore. If you have room, you can use needle-nose pliers to gently wiggle the tang back and forth until it breaks off. The finished thread is ready for use immediately.

Repairing Cracks and Damaged Threads

201

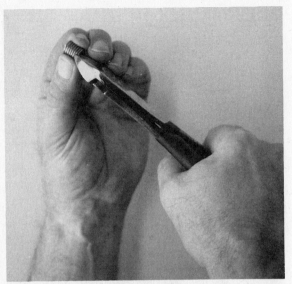

Figure 10-33. To repair a shallow hole, cut the insert to size before threading it on the mandrel.

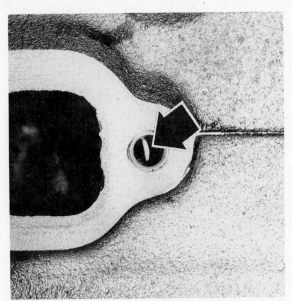

Figure 10-34. Use a punch, chisel, or needle-nosed pliers to snap the tang off the base of the insert.

Threaded Inserts

Threaded inserts are tubular, case-hardened, solid steel wall pieces that are threaded inside and outside. The inner thread of the insert is sized to fit the original fastener of the hole to be repaired. The outer thread design will vary, these may be self-tapping threads that are installed in a blank hole, or machine threads that require the hole to be tapped. Threaded inserts return a damaged hole to original size by replacing part of the surrounding casting so drilling is required. Most inserts fit into three categories:

- self-taping
- solid-bushing
- or key-locking.

Self-tapping inserts

The external threads of a self-tapping insert are designed to cut their own way into a casting. This eliminates the need of running a tap down the hole. To install a typical self-tapping insert, follow this procedure:

1. Drill out the damaged threads to open the hole to the proper size. You must use the specified drill bit.
2. Select the proper insert and mandrel. As with Helicoils, the drill bit, inserts, and mandrel are usually available as a kit.
3. Thread the insert onto the mandrel, use a tap handle or wrench to drive the insert into the hole. Be sure to keep the tool straight as you run it in. Since the insert will cut its own path into the hole, it may require a considerable amount of force to drive the insert in.
4. Thread the insert in until the nut or flange at the bottom of the mandrel touches the surface of the workpiece. This is the depth stop to indicate the insert is seated.
5. Hold the nut or flange with a wrench, and turn the mandrel out of the insert. The threads are ready for immediate use.

Solid-bushing inserts

The external threads of solid bushing inserts are ground to a specific thread pitch, so you will have to run a tap into the hole. Some inserts use machine thread so a standard tap can be used, others have a unique thread and you have to use a special tap. To install solid-bushing inserts follow this procedure:

1. Drill out the damaged threads to open the hole to the proper size, figure 10-35. You must use the drill bit supplied with the kit, or one that is properly sized to the tap.
2. Chamfer the top of the hole with a countersink or a small file, then clean the hole with a brush or compressed air.
3. Use the previously detailed tapping procedures to thread the hole, figure 10-36. Be sure to tap deep enough, the top of the insert must be flush with the casting surface.
4. Thread the insert onto the installation driver, use the driver to screw the insert into the hole, figure 10-37. Some inserts require that a thread locking compound be applied, others go in dry.
5. Remove the installation driver, and the new threads are ready for service with the original fastener.

Figure 10-35. Drill out the damaged thread with the correct bit.

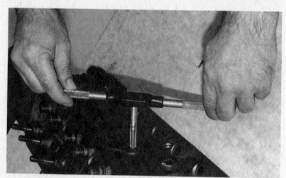

Figure 10-36. Use a special tap for the insert.

Figure 10-37. Thread the insert into the hole using the driver.

Key-locking inserts

Key-locking inserts are similar to solid-bushing inserts, but are held in place by small keys. After the insert has been installed, the keys are driven into place — perpendicular to the threads — to keep the insert from turning out. To install a key-locking insert, follow this procedure:

1. Drill out the damaged thread with the correct drill size, figure 10-35.
2. Tap threads into the drilled hole with the special tap, figure 10-36.
3. Use a mandrel to screw the insert into the tapped hole until it is slightly below the surface, figure 10-38. The keys act as a depth stop and prevent the insert from turning, figure 10-39.
4. Drive the keys down using the driver supplied with the insert kit. You can also do this with a small punch and hammer if you are careful. Be sure to drive the keys down so they are flush with the top of the insert, figure 10-40.

Reamer-cut spark plug hole inserts

You may encounter stripped spark plug holes that have been repaired using a self tapping reamer and a threaded insert. This type of repair is done using a simple inexpensive kit available from auto supply stores. The self-tapping reamer is used to enlarge the hole, the insert is threaded onto the spark plug, then the plug and insert are threaded into the hole together. These kits, aimed primarily at the do-it-yourselfer who needs to repair a damaged spark plug hole on a running engine, are not a recommended repair for a professional facility. When you encounter an engine that has been repaired in this manner, you should rework the plug hole threads using one of the previously described methods.

CRACK DETECTION

We explained in the last chapter that you should inspect all of the major parts of the engine for damage and cracks — the block, head, crankshaft, and rods — before spending any money or time on them. On the other hand, major engine parts are expensive, and some types of damage you can either repair or safely ignore. Cracks due to overheating are common in cylinder heads, figure 10-41, and you can often repair them easily. Cracks in engine blocks are more serious, but can still sometimes be fixed. On the other hand, cracks in a crank-

Repairing Cracks and Damaged Threads

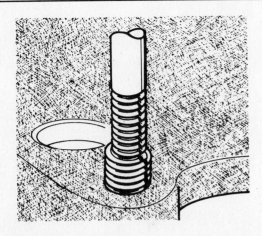

Figure 10-38. Use the mandrel to screw the insert into the threaded hole.

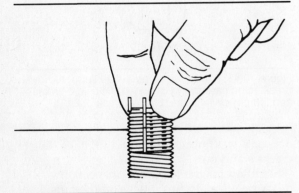

Figure 10-39. The keys lock the insert in the threads to prevent it from turning out.

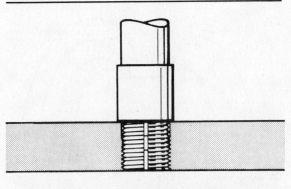

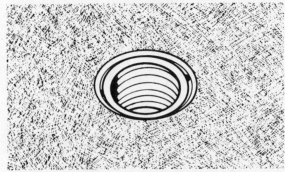

Figure 10-40. Use the driver to drive the keys down flush with the surface of the work.

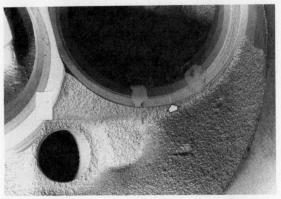

Figure 10-41. Cylinder heads frequently crack from overheating across or between the valve seats.

shaft camshaft, or connecting rod are terminal. If you find damage in one of these highly-stressed parts, throw it in the scrap bin and replace it with a new one.

Almost all cracks start out as fairly small and inconspicuous defects. Before attempting repairs, you need determine where a crack starts and stops. Often, a small crack in a cold, unstressed casting sitting on the bench will become a major compression, oil, or coolant leak once the engine is warmed up and running. There are several techniques machinists use to inspect for cracks. We will explain each of these methods here:

- Magnetic particle testing
- Dye penetrant testing
- Pressure testing.

Before you perform any crack testing, parts must be thoroughly cleaned. Use the cleaning procedures detailed in chapter 9 of this *Shop Manual*. All crack detection techniques require a clean, grease-free surface to be effective.

Magnetic Particle Inspection

Magnetic particle inspection is a detection technique that lets you discover cracks so small and narrow that they are invisible to the naked eye. Magnetic inspection is often called *magnafluxing®*. The Magnaflux Corporation is the company that first offered magnetic particle inspection service. Most machine shops have a magnetic particle inspection machine.

A draw back to magnetic particle inspection is that it works only on ferrous metal parts that can be magnetized. Be aware that some engine parts — such as valves and valve seats

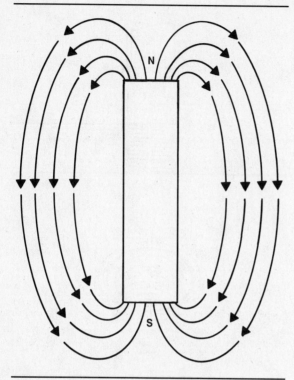

Figure 10-42. Magnetizing the casting, polarizes it and forms a magnetic field that will attract the iron powder.

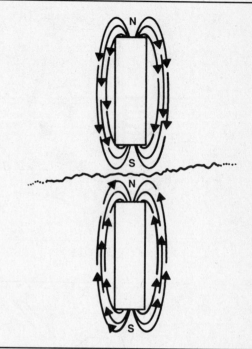

Figure 10-43. If the lines of force are interrupted by a break in the casting, two magnetic fields are created and the powder will lodge at the crack.

— can be made of nonmagnetic steel alloys. To tell whether magnetic particle inspection will work for a particular part, just touch it with a magnet. If the magnet can attract it, you can use magnetic methods.

Prime areas to inspect with magnetic particle inspection equipment are the external surfaces of cylinder heads and engine blocks:

- Core plug holes
- Main bearing saddles, webs, and caps
- Cylinder walls
- Lifter bores
- Deck surfaces
- Combustion chambers
- Ports
- Valve seats
- Water jacket passages.

Magnetic particle inspection works because cracks interrupt the magnetic field in a iron or steel part, figure 10-42. All magnets have a North and South pole, which you can see if you dust iron filings onto a piece of paper with a magnet under it. If you break the magnet in half, you interrupt the magnetic field, and you end up with two smaller magnets — each with its own North and South poles. A crack in a temporarily magnetized engine part acts the same as breaking a magnet in half, figure 10-43. You can use either dry magnetic powder, or a wet fluorescent liquid to make the cracks stand out.

Standard magnetic particle inspection is a quick easy way to find cracks on the outside of a casting or forging, where you will be able to see the crack once the operation has been performed. This is sufficient for most of the pieces you will be inspecting, but will not detect internal cracks in the oil passages or water jackets. However, there are magnetic particle testing methods that will locate cracks that are under the surface of the metal. This procedure, which is routine in the aerospace industry, requires specialized equipment. The technique is simple: you can detect underlying cracks before they reach the surface by increasing the amperage you use to magnetize the metal. In fact, equipment manufacturers use this high amperage technique to calibrate their magnetic particle machines.

When you use a very high amperage, internal cracks near the surface will cause exactly the same patterns as surface cracks. If you use too much current, though, the magnetic field becomes so strong that every tiny imperfection in the surface stands out like a major flaw — even polishing marks will look just like cracks. You must therefore "sneak up" on the level of current that shows enough of the internal structure without being overwhelmed by the surface texture.

Repairing Cracks and Damaged Threads

Figure 10-44. This small magnetic particle inspection machine is useful for hand inspection of heads and blocks.

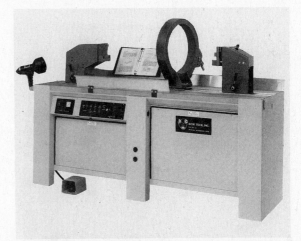

Figure 10-45. This larger magnetic particle inspection machine magnetizes crankshafts or cylinder heads using induction from the surrounding ring.

Dry magnetic particle inspection

Dry magnetic particle inspection can be done either on a workbench with small tools, figure 10-44, or in special piece of equipment that holds the part in a fixture, figure 10-45. When you dust the magnetized engine part with the iron powder, it collects where the magnetic field is interrupted, and makes cracks visible.

To perform a magnetic particle inspection test, first clean the part thoroughly so that all traces of dirt and grease are removed. Place the electromagnet on the part and turn it on to set up the necessary magnetic field, figure 10-46. Spray the iron powder onto the surface with an atomizer, magnetic action will cause the powder to collect along any crack that crosses the magnetic field.

Once you have inspected the area between the ends of the electromagnet, turn it off, rotate the electromagnet 90°, and repeat the test on the same area. You must test the part in both directions because the dust only collects on cracks that *cross* the magnetic field. Once you have checked in both directions, turn the magnet off, move it to the next area of the casting to be checked, and repeat the procedure.

Production shops use larger magnetic particle inspection machines to work over an entire cylinder block or head at once. These tools suspend the part inside a cage of magnetizing cables, so you can dust and inspect the entire outer surface.

Wet magnetic particle inspection

Wet magnetic particle inspection uses the same principle as dry techniques, but is more sensitive because it uses much higher amperage. Instead of using a dry powder to pinpoint

Figure 10-46. Place the magnet on the part you want to test and turn it on.

surface cracks, wet techniques use a fluorescent iron-oil liquid. The liquid is flushed over the engine part and collects along breaks in the magnetic field, figure 10-47. A black light is used to make the liquid that accumulates in the cracks glow brightly.

Dye Penetrant Testing

Dye penetrant testing is a chemical technique that will work for both ferrous and non-ferrous metals. That means that you can use it for parts made of aluminum, magnesium, and zinc, as well as those made of iron and steel — essentially any metal engine part. Different manufacturers have somewhat different procedures, but most use a dye penetrant, a dye remover, and a developing agent. Follow these general steps:

1. Clean the surface thoroughly to remove grease and dirt. Use a non-residual solvent like lacquer thinner or brake-cleaning solvent. Follow this cleaning with any special chemical cleaners included in the test kit.

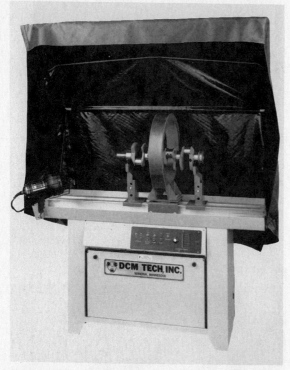

Figure 10-47. Wet magnetic particle inspection uses a fluorescent liquid that flows over the entire part.

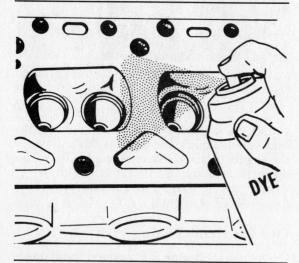

Figure 10-48. The first step is to apply the dye penetrant to the area you want to test.

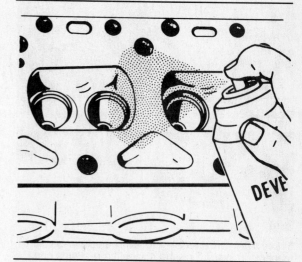

Figure 10-49. Apply the developer to the same area you treated with penetrant.

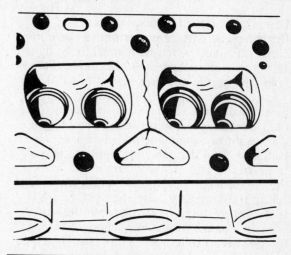

Figure 10-50. The developer causes a color change that makes the crack visible.

2. Apply the dye penetrant to the surface of the part, figure 10-48. After a few minutes, the penetrant will have soaked into the cracks, even those too narrow to see.
3. Apply the remover, if any, and wipe or blot both the remover and the excess penetrant from the surface. Some techniques require you rinse the surface with water. Cleaning the surface removes all the dye penetrant except what remains down inside of the cracks.
4. Apply the developer, and wait for it to dry, figure 10-49. The developer draws the dye penetrant out of the cracks and reacts with it to make it visible, figure 10-50.

Some procedures produce a fluorescent chemical that stands out under black light, while others produce a stain that you can easily see under normal light.

Pressure Test Inspection and Equipment

Visual inspection for cracks would be effective if all areas of the part were visible. But some parts are difficult to inspect completely, espe-

Repairing Cracks and Damaged Threads

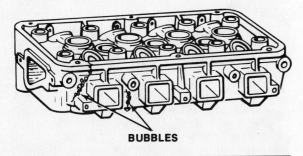

Figure 10-51. Air pressure escaping from the casting causes bubbles in the soap solution.

Figure 10-52. This pressure tester holds a cylinder head on the fixture, sealing pads on adjustable clamps plug the coolant passages to make them air tight.

Figure 10-53. Position the sealing pads over all the accessible water jacket openings in the casting and tighten them down.

cially cylinder heads. Because of these limitations, visual inspection methods cannot guarantee that a part will not leak when under pressure. Cylinder heads typically operate with a coolant pressure of about 15 psi (103 kPa). The best way to test them is to seal off and apply pressure to the coolant passages. There are several methods that can be used to pressurize the passages, and a wide variety of equipment available to do the job.

Simple systems operate by pressurizing the passages with compressed air and a coating of liquid soap is applied to the suspect areas. If there is a leak, the escaping air will cause bubbles to form in the soap solution, figure 10-51. More elaborate testers submerged the cylinder head in a water bath and then pressurize it with compressed air. Leaks are revealed as a stream of bubbles rising off the part to the surface of the water. The water in the dip tank is often heated to simulate engine running temperature, this allows the technician to spot even small cracks that might not leak on a cold engine. Another type of equipment uses heated water instead of air to pressurize the test piece. Dye is usually added to the water so that it can be easily spotted as it works to the surface of the workpiece.

The increased use of non-magnetic aluminum and alloy engine castings has created a big demand for pressure testing. Pressure testing, the only effective way to test non-magnetic castings, will also locate hard to find cracks in cast iron and steel parts. One drawback to pressure testing is the elaborate equipment that is required. Unless you have a universal test fixture, you need individual closure plates for each different engine you will be checking as well as an assortment of plugs to seal heater, thermostat, and other outlets. All of the passages must be blocked so that the internal chambers of the workpiece are air tight. As a result, most repair shops will sub-contract this work to a machine shop.

Universal testers, although not as effective as the submersible type, are comparatively inexpensive and relatively easy to use. These units support the workpiece in a fixture, adjustable clamps are used to position and hold sealing pads over the coolant passage openings, figure 10-52. Take the following steps to pressure test a cylinder head with a universal rig:

1. Place the cylinder head on the fixture, with the combustion chamber surface facing up.
2. The machine has adjustable arms with rubber-faced pads to seal the coolant openings. Arrange the pads over the openings and tighten them until they are sealed, figure 10-53.

Figure 10-54. A rubber pad on the back of this plate seals off the thermostat and temperature sender outlets.

3. Seal the larger holes, such as heater and thermostat outlets with the proper plugs, plates, and gaskets, figure 10-54.
4. Attach a compressed air supply to the head. One of the pads, usually at the end of the head, will have a quick-disconnect hose fitting.
5. Check to be sure all passages are blocked off, then open the air valve to pressurize the head.
6. Apply liquid soap solution to the combustion chambers and other suspect areas. Any escaping air will cause the solution to bubble along any cracks.

With some universal testers, you can pump liquid containing a dye penetrant, rather than compressed air, into the sealed water jacket. The dye in the liquid will leave a permanent mark after the casting dries. Again, this technique will only help you find cracks in the visible parts of the casting.

Invisible cracks deep in the passageways and seepage from metal fatigue can often be detected using a submersible pressure tester, figure 10-55. The cylinder head is sealed, pressurized with compressed air, and placed in a tank of water. A trail of bubbles will rise off the part anywhere air can escape. Engine running conditions can be closely simulated by heating the water in the tank. This type of testing will reveal even marginal cylinder head defects and can keep a potential disaster out of service. Once a cylinder head passes a through pressure test, it is ready to be reconditioned and put back into service.

CRACK REPAIR

Many times repairing a crack is a time-consuming process which can be cost-prohibitive even if the crack is repairable.

Figure 10-55. Submersible pressure testers will cause a trail of bubbles to rise to the surface of the water bath from air leaks in the pressurized part. Even small cracks deep in a casting will be revealed.

Often it is difficult to determine which cracks are repairable and which are not. Before you begin any crack repair procedure, you should also consider all additional work that is required to restore the service life of the casting. It may be more expedient to simply replace the part. Crack repair may involve;

- welding
- stop-drilling
- pinning
- or chemically sealing.

Each one of these techniques is almost an art form in itself, and there are specialists in the field. The customer's engine should not be your first attempt, practice the repair technique on a discarded part out the scrap bin first. Every crack is unique and will react differently to your repair efforts. Learning the "feel" of crack repair methods takes practice.

Welding Cracks

Welding cracks in cast parts usually requires that the workpiece be preheated before repairs can be made. Attempting to weld on a cold casting can cause the crack to travel further along the workpiece.

Thermal ovens, as used in parts cleaning, can be used to bring the casting up to temperature. Temperatures will vary for the type of metal to be welded and the welding equipment being used. Most welds also require some type of surface preparation to guarantee adequate penetration for a good weld.

Aluminum and alloy castings can be repaired with TIG (tungsten inert gas) or MIG (metal inert gas) welding equipment. Prepare the surface and preheat the casting to between 200°

to 300°F (100° to 150°C), then run a bead along the crack. After the crack has been welded, you can grind the bead down to finish the surface. Holes caused by corrosion in the water passages, combustion chamber damage, and mating surfaces etched by electrolysis can often be repaired by inert-gas welding methods.

You must *pressure test a head*, as detailed earlier in this chapter, after welding. Additional machining may also be necessary depending on the location of the crack, you may have to resurface the gasket face. Heat distortion from welds you make in the bearing support areas of overhead cam engines and some engine cases, can make it necessary to machine the journals by align boring or honing. These machining operations are detailed in chapters 11 and 14 of this *Shop Manual*.

Until recently it was not easy to do a good job welding cast iron because of the many impurities in the iron as a result of sand casting. However, an expert welder may be able to use standard arc welding equipment to seal cracks in cast iron. Welding cast iron is not easy, and requires skills that even many professional welders do not have. Many of the welders that do cast iron repairs are very protective of their skills, and are unwilling to divulge their techniques. In general, proper cast iron welding requires that the entire casting be slowly heated to a constant temperature before welding, kept hot during the repair, and slowly cooled to room temperature afterwards.

Stop-Drilling Cracks

Minor cracks often have no effect on the strength and integrity of an engine part and will have no bearing on engine performance. If you can prevent the crack from spreading, then it will require no further action. Blind cracks radiating from water jacket holes, core plug bores, or other non-stressed, non-critical areas can often be treated by stop-drilling. Stop-drilling will not repair the crack, but it will stop a crack from spreading any further along the casting. Drilling a 1/8-inch (3-mm) hole into the casting at the end of the crack, will effectively stop the crack at the drill hole. The round hole spreads the stress that caused the crack, and prevents it from progressing across the casting. Be sure to drill at the end of and all the way to the bottom of the crack. Be careful to avoid drilling into oil galleries or through water jacket walls, this can create bigger problems.

Remember, stop-drilling is only effective in stress free non-critical areas, *never* try to use

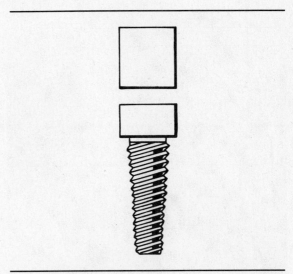

Figure 10-56. Crack pins come in various lengths and sizes, but all have a tapered portion and a head that you can turn with a wrench.

this technique on cracks that cross oil or water jacket passages, or that leave a bolt hole.

Pinning Cracks

Pinning cracks is a time-honored technique that works to make permanent repairs on the exterior surfaces, ports, and combustion chambers of cylinder heads and blocks. Essentially, you fill the crack by screwing in a series of tapered iron pins into overlapping holes along its length. You can only use this technique in any area where the casting is thick enough to support the threaded pins. Because you install the pins cold, the problem of warping or further cracking, that can occur with welding, is greatly reduced.

Crack pinning requires a drill, a selection of tapered pins, taps, and a few hand tools. Pins come in a variety of lengths and diameters for different shapes of cracks, figure 10-56. This is the general procedure:

1. Locate the exact beginning and end of the crack with a dye penetrant or magnetic particle inspection technique. Mark the crack all along its length with a series of small center-punch marks.
2. Select a drill bit to match the tap for the size pins to be installed, then stop-drill both ends of the crack, figure 10-57. The drill hole should project beyond the end of the crack by about half the diameter of the pin you will install.
3. Inspect the drilled hole. You must drill all the way to the bottom of the crack.

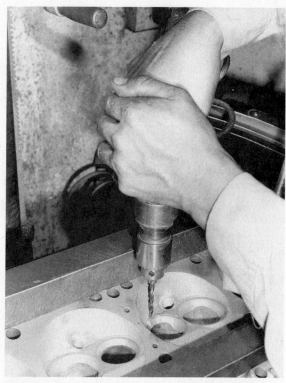

Figure 10-57. Drill both ends of the crack to stop it from spreading while you work.

Figure 10-58. Use a tapered tap to thread the hole for the pin.

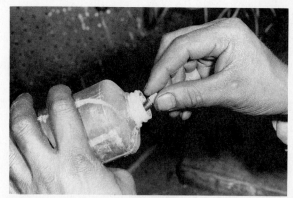

Figure 10-59. Use a sealer on the pin's threads to lock it in place and prevent leaks.

Figure 10-60. Screw the pin into the hole and tighten it.

4. Ream the hole with a tapered reamer, and then thread it with the tapered tap, figure 10-58. Do not tap the hole too deep or the pin will go in too far before the threads clamp it in place. If you do, you will have to drill the hole oversize and fit a larger pin.
5. Coat the pin with a thread-locking compound or ceramic sealer, figure 10-59. Screw the pin into the hole and tighten it securely, but not so much that you spread the hole out, figure 10-60. After the pin is in place, lightly tap the end of it with a hammer to set it in the hole.
6. Cut the pin off just above the surface of the workpiece. There are several ways of removing the extra length of pin:
    - Cut the pin off with a key-hole hacksaw, figure 10-61. If you cannot cut all the way through, cut as much as possible, then tap the head of the pin with a hammer to break it off, figure 10-62.
    - Drill through the pin to weaken it, and then cut it off with a sharp chisel.
    - File the pin off.
    - Use a die-grinder or a drill motor with a grinding stone or cut-off wheel to trim the pin down.

    Choose whichever method lets you remove the pin most easily. The pins will often be in awkward locations.
7. Peen the stub of the pin to lock it into the hole, figure 10-63. and grind it off flush with the surface.
8. Drill the next hole overlapping the first pin and partway along the crack. Install the next pin in this hole, tightening the pin to the same torque setting as the first. It will both seal the next segment of the crack, and lock the first pin in place.

Repairing Cracks and Damaged Threads

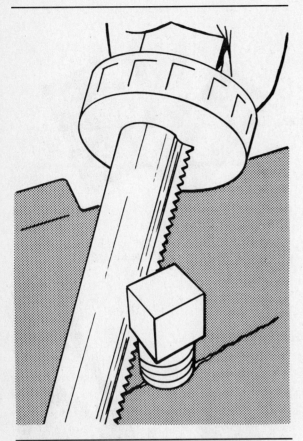

Figure 10-61. Cut the pin off just above the surface of the work.

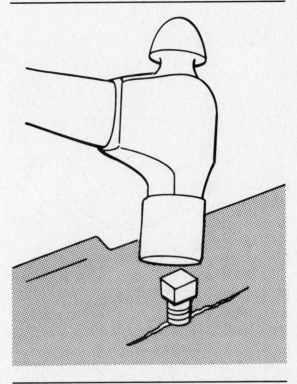

Figure 10-62. If you are unable to cut all the way through the pin, cut as much as you can, and tap it with a hammer to break it off.

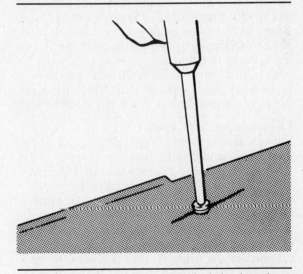

Figure 10-63. Peen the pin stub to lock the threads just below into the hole.

9. Continue installing tapered pins along the length of the crack until you reach the end, figure 10-64.
10. Finish the surface with a grinder to blend the tops of the pins with the existing contours of the surface.
11. Once again, perform a dye, magnetic particle, or pressure test to make sure that you have sealed the entire length of the crack.

Sealing Cracks with Ceramic Sealer

On engine castings with inaccessible cracks in the water jacket, you can sometimes run a ceramic sealer through the passages to seal them. This metal-coating technique requires a piece of equipment similar to a pressure tester. A heated ceramic solution is circulated through the sealed jackets, the passages are pressurized driving the fluid into the cracks. After treatment, the casting is drained and the sealer allowed to cure.

Metal coating automotive parts is a relatively new process and a field where the technology is changing rapidly. There are a variety of coatings and processes being used, however, there are very few companies that can perform these services. Coating processes are used not only for repairs, but can also be used to help reduce wear. Coatings can be applied to pistons, rockers, and valves to reduce friction and extend service life. Metal coated parts are expensive and used primarily for heavy-duty and high-performance applications.

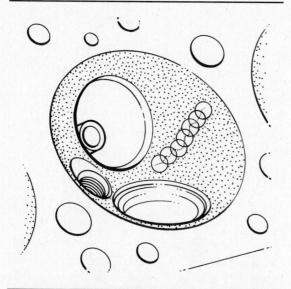

Figure 10-64. Install a line of overlapping pains all along the crack.

Figure 10-65. Several manufacturers make epoxies or plastic-metals for casting or thread repair.

Epoxies or Metallic Plastics

Casting and thread repair with space-age epoxies and metallic plastics has only been possible in recent years. There are many different substances available, some of which are very strong, and can be used on iron, steel, and aluminum, figure 10-65. Although not as strong as the original casting, epoxies can often be machined, drilled and tapped. You can use epoxies to repair engine blocks, cylinder heads, intake manifolds, front and rear covers, and valve covers, typically, anywhere the temperature remains below about 500°F (260°C).

Casting repair with epoxies

Epoxies are excellent for repairing engine parts with small sand-casting flaws or porosity, 1/4 inch (6 mm) or less. Epoxies are not designed to repair cracks. A crack indicates a tensional shear or torsional load that the original metal could not withstand, and the epoxy will not do any better. Some engine manufacturers specifically recommend epoxy repairs, but even they caution against using it in areas where the epoxy will have to hold back oil, coolant, or extreme pressure. For most epoxies, you can follow this general procedure:

1. Clean the surface thoroughly with a file or grinder, until you have a clean, rough metal surface. The rough surface gives the epoxy a "tooth" to bite into, and hold more firmly.
2. If you are sealing an exposed hole, try to undercut the hole with the grinder or a drill to give the epoxy a lip to hang onto.

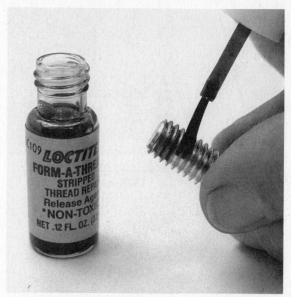

Figure 10-66. Lightly coat the threads of the fastener with the releasing agent.

3. Clean the area with a non-residual solvent such as lacquer thinner or brake and clutch parts cleaner.
4. Mix the epoxy resin and hardener thoroughly on a clean surface using a clean tool. Follow the manufacturer's instructions for mixing the epoxy.
5. Spread the epoxy into the hole with a putty knife or wooden stick. Make sure that you force the filler into the hole.
6. Let the epoxy harden, typical epoxies take 10 to 12 hours to cure at room temperature. A heat lamp placed about a foot away from the epoxy will speed the curing. If you use heat, let the epoxy set for several hours first, heat will cause freshly applied epoxy to run.

Repairing Cracks and Damaged Threads

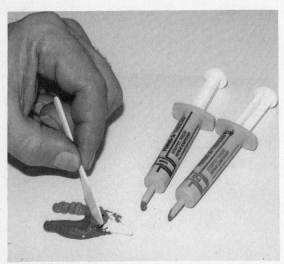

Figure 10-67. Mix the resin and hardener together and spread the mixture evenly around the threads inside the hole.

Figure 10-68. Screw the coated fastener into the hole by hand, do not tighten it until the mixture has set up.

Thread repair with epoxies

You can also use epoxies to repair damaged threads, as long as they are in a location that does not require a great deal of fastener torque. Do *not* use epoxies to repair threads in holes for critical fasteners, such as internal engine parts or cylinder head bolts. You can use epoxy to repair threads for distributor clamp bolts, oil drain plugs, oil pan bolts, front cover bolts, and even water pump bolts.

Thread repair epoxy kits usually contain a two-part epoxy and a releasing agent. Follow this general procedure to use them:

1. Clean the threads of the hole and the fastener you will be using. Clean the hole with a rifle brush and compressed air. Follow up with a non-residual cleaner like lacquer thinner or brake-cleaning solvent.
2. Spread a *thin* coat of the releasing agent on the fastener threads. Be sure to spread it onto the entire threaded surface, but avoid using too much — the release agent neutralizes the epoxy so you want to use it sparingly, figure 10-66.
3. Mix enough resin and hardener to fill the threads in the hole, and spread it as evenly as possible inside the hole, figure 10-67. For open-end holes, tape the open end shut to keep the epoxy from running out.
4. Screw the releasing-agent coated fastener completely into the threaded hole, but *do not tighten* it, figure 10-68. For open-end holes, turn the fastener in until it just contacts the tape.
5. Let the epoxy cure. Once the mixture sets, you can tighten the bolt and return the part to service.

Some thread repair epoxies set-up in less than an hour, and fasteners can be tightened completely in about two hours. The epoxy threads provide a substantial amount of clamping force, fasteners can normally be tightened to their recommended torque specification.

PART THREE

Engine Inspection, Machining, and Rebuilding

Chapter Eleven
Servicing Blocks

Chapter Twelve
Servicing Crankshafts, Flywheels, Pistons and Rods

Chapter Thirteen
Engine Balancing

Chapter Fourteen
Servicing Cylinder Heads and Manifolds

Chapter Fifteen
Servicing Camshafts, Lifters, Pushrods, and Rocker Arms

Chapter 11
Servicing Blocks

The cylinder block is the central casting of the engine. It supports and aligns the crankshaft, cylinder head, and parts of the valve train. It also contains the cylinder bores, or holds the liners that the bores are machined into. Finally, it locates and supports the internal and external engine accessories and brackets, and is usually drilled and tapped to locate the mounts that hold the engine in the car. Short block reconditioning requires attention to all of these areas, together called the bottom end of the engine. In an engine overhaul, you will:

- Resize the main bearing bores to correct misalignments, and to compensate for stretch or damage.
- Resurface the block to cylinder head mating surface.
- Deglaze the cylinders, without altering bore size, to prepare them for new rings.

In a complete rebuild, you will still attend to the deck surface and bearing bores, but rather than deglazing the cylinders to prepare for new rings, you will:

- Bore the cylinders to oversize
- Hone the cylinders to fit new pistons

In this chapter, we explain the fundamental measurements and machining operations you will perform to the cylinder block for either a major engine overhaul or for a complete rebuild, including boring, honing, sleeving, and deck resurfacing.

TOLERANCES AND OIL CLEARANCES

During most stages of engine rebuilding, you will be working with specifications given as tolerance values. You must understand and be able to work with these specifications during the rebuild process. You will be required to use tolerance values to calculate oil clearances, as well as, to determine how much metal is to be removed in machining operations. Virtually all cylinder block reconditioning operations require you to calculate and work within specific tolerance ranges.

Tolerances

A tolerance is the range within which you can consider a part to be "ready for service." The lower limit of the tolerance is the minimum below which the part or diameter is undersize, and the upper limit indicates the value above which it is oversize. The more critical the dimensions, the narrower the tolerance will be.

Servicing Blocks

These limits apply to both new parts that you install, and to used parts that you are machining to use again.

Oil Clearances

Oil clearance is the precise space purposely left between moving parts in an engine to accommodate a layer of lubricating oil. These clearances are critical, especially for rotating shafts such as the crankshaft and camshaft, and must be accurately provided for during machining operations. In the *Classroom Manual*, we explained how the oil pumped into the clearance space between a bearing insert and a journal becomes temporarily trapped. As the journal rotates, the oil forms a wedge that lifts the journal off the insert and lets it spin, encased in a protective, low friction oil film.

Over time, the bearing inserts and journals wear away, the main bearing and connecting rod caps stretch, and the clearance increases. If this space becomes too great, the oil runs out too easily, and the oil wedge becomes harder to build up and maintain. Eventually, the oil runs out so rapidly that the journals hammer against the bearing inserts, causing knocking sounds and rapid wear.

In cylinder block reconditioning, you will restore these clearances to within the factory-recommended tolerances by machining the various bores and journals, and installing new bearing inserts. To accomplish this, you must be able to calculate *what oil clearance will result*, given the existing diameter of the journal, the bearing bore, and the precise thickness of the bearing inserts. You must also be able to calculate what diameter journal, housing bore, and inserts will be necessary to end up with the oil clearance specified by the engine manufacturer.

Calculating Oil Clearances from Specifications

To calculate oil clearance from specifications, you subtract the journal diameter from the bearing inside diameter. Remember that clearances are the spaces between two engine parts. The manuals give oil clearances as total values for a given journal and bearing inside diameter, which means that you must include the space on *both* sides of the journal to get the clearance. The specifications you will work with are, figure 11-1:

- Housing bore diameter (minimum/maximum)
- Insert thickness
- Bearing inside diameter (minimum/maximum)

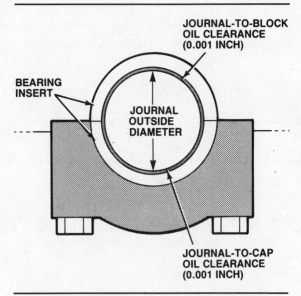

Figure 11-1. Oil clearance specifications include the gaps on *both* sides of the journal. A 0.001-inch gap gives an oil clearance 0.002-inch across.

- Journal outside diameter (minimum/maximum)
- Oil clearance (minimum/maximum).

Housing bore diameter
Housing bore diameter is the diameter across a bearing bore without the bearing inserts installed, figure 11-2. Housing bore diameter is a tolerance value, there will be a maximum and minimum specified for this dimension.

Insert thickness
Insert thickness is the true thickness of the bearing insert or shell, it is *not* the same as the nominal oversize or undersize, figure 11-3. Bearings are precision-made, so you can usually ignore the extremely minor variation in their dimensions. However, you will have to measure the inserts to determine clearance. Insert thickness is an exact dimension, not a tolerance value.

Bearing inside diameter
Bearing inside diameter is the diameter across the hole for the journal, with the bearing inserts in place, figure 11-4. This is a tolerance value because the housing bore diameter has a minimum and maximum, even though the inserts do not. To calculate bearing inside diameter you need to know the minimum and maximum housing bore and the actual thickness of the bearing shell.

To calculate the maximum bearing inside diameter, multiply the insert thickness by 2, and subtract your answer from the maximum housing bore. To determine minimum inside

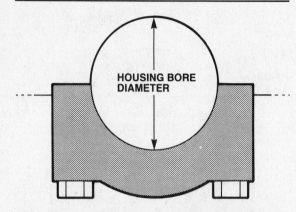

Figure 11-2. Housing bore diameter is the diameter across a connecting rod or main bearing bore *without* the inserts in place.

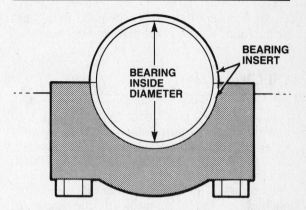

Figure 11-4. Bearing inside diameter is the diameter across the bore with the bearing inserts in place.

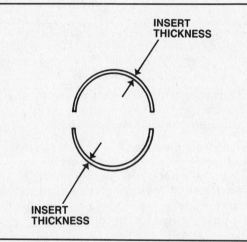

Figure 11-3. Insert thickness is the true thickness of the bearing insert or shell. You can safely ignore the tolerance values on these precision-made parts.

bearing diameter, subtract double the insert thickness from the minimum housing bore. For example, the main bearing bore has a minimum diameter of 2.4981 inch (63.452 mm) and a maximum diameter of 2.4986 inch (63.464 mm). The bearing insert thickness measured 0.094 inch (2.388 mm). Calculate bearing inside diameter as follows:

Maximum bearing inside diameter
 2.4986 − 0.1880 (2 × 0.094) = 2.3106 inch
 63.464 − 4.776 (2 × 2.388) = 58.688 mm

Minimum bearing inside diameter
 2.4981 − 0.1880 (2 × 0.094) = 2.3101 inch
 63.452 − 4.776 (2 × 2.405) = 58.676

Journal outside diameter
Journal outside diameter is the diameter of the shaft at the point where it fits into the bearing, figure 11-5. Journal outside diameter is a tolerance value, there is a maximum and minimum acceptable diameter. Typical journal diameters for the example shown above would have a minimum diameter of 2.3076 inch (58.613 mm) and a maximum diameter of 2.3081 inch (58.626 mm).

Calculating oil clearance
Finally, oil clearance is the difference between journal outside diameter and bearing inside diameter. Oil clearance is a tolerance value. You calculate the *maximum* oil clearance by assuming you have the narrowest shaft in the widest housing bore. You calculate the *minimum* oil clearance by assuming you have the widest shaft in the narrowest housing bore. Using the dimensions from the above example, calculate as follows:

Maximum oil clearance
 2.3106 − 2.3076 = 0.003 inch
 58.688 − 58.613 = 0.075 mm

Minimum oil clearance
 2.3101 − 2.3081 = 0.002
 58.676 − 58.626 = 0.050

Any oil clearance you measure that falls within the tolerance range is acceptable.

You may not have to calculate these values very often. For most of your work you will simply be measuring and machining parts to the tolerance limits specified in the manufacturer's service manual. Manufacturers often do all the calculating for you, and list the numbers you need to see when you measure the parts.

Measuring Oil Clearances

The tolerances specified by the manufacturers are what you arrive at after you have machined the bottom end. Before machining, you

Servicing Blocks

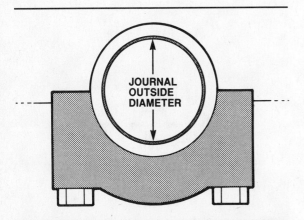

Figure 11-5. Journal outside diameter is the shaft diameter where it must fit into the bearing.

Figure 11-6. The most accurate way to measure the housing bore diameter is with a dial bore gauge.

determine *existing* oil clearance for the bearings in the engine by measuring the journal, bore, and bearings and then computing it. You will use a variety of precision tools to take these critical measurements.

Housing bore diameter
There are several ways to measure the housing bore diameter. The most accurate, and quickest, way is to use a dial bore gauge, figure 11-6. Dial bore gauges are not only fast and accurate, they are also expensive. Unless you work in a shop that does a lot of engine rebuilding, you probably wont have access to a dial bore gauge. Bore diameter can also be accurately measured with an inside micrometer. A telescoping, or snap, gauge and an outside micrometer can also be used, figure 11-7.

Insert thickness
An outside micrometer with a ball-shaped anvil is the proper tool to use when measuring the thickness of a bearing insert, figure 11-8. Take your reading in the center of the bearing and 90 degrees from the parting line. A standard micrometer will give a false reading because the arch of the bearing prevents the anvil from seating against the shell.

If you do not have a micrometer with a ball-shaped anvil, you can get a fairly accurate reading using a standard micrometer and a small ball bearing. First measure the ball bearing with the micrometer and record the reading. Next place the ball bearing in the center of the insert shell and take a total measurement of both the ball and bearing shell with your micrometer. Subtract the diameter of the ball from the total reading and you have the thickness of the shell.

Remember that there are two shells so you have to double the thickness of the bearing insert to calculate clearances.

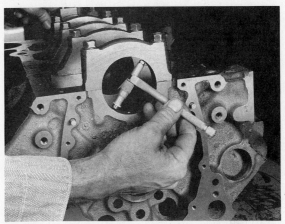

Figure 11-7. Housing bore diameter can also be measured using an inside micrometer.

Bearing inside diameter
Bearing inside diameter can be measured using the same tools and techniques used to measure housing bore diameter. Be extremely careful when measuring new bearings, the surface can easily be damaged.

Crankshaft journal outside diameter
An outside micrometer is the only tool that will give you an accurate measurement of the crankshaft journals, figure 11-9.

BLOCK INSPECTION

At this point your engine block, which has previously been cleaned and inspected, is ready for machining. Before you begin the reconditioning process, spend some time looking over the casting very carefully for problems that you might have overlooked previously.

On the underside of the block, look for hairline cracks in the main bearing saddles. Overtorqued main bearing caps can crack these critical parts. Make sure the bearing caps fit

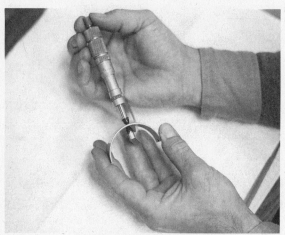

Figure 11-8. Measure the thickness of the bearing insert using a micrometer with a ball-shaped anvil.

tightly into the notches machined into the webs of the block, there should be no play. Check the bottoms of the cylinder bores for cracks and breaks caused by metal fatigue or poor disassembly.

On the top of the block inspect the deck surfaces. Look around the cylinder bores for grooves etched into the surface from compression leaks. Also check for pitting near the coolant passages caused by electrolysis. If any of the grooves or pits are too deep to clean up by resurfacing, you will have to weld them or scrap the block.

Finally, check all the threaded bolt holes for stripped or pulled threads, or for broken bolts. Also look for damaged locating dowels or dowel holes in the deck surface and in the bellhousing area.

If any thing at all looks questionable, recheck using magnetic particle inspection, dye penetrant inspection, or pressure testing, as explained in Chapter 10 of this *Shop Manual*. Repair any problems before continuing, make sure that they are completely fixed.

Once the block has passed a visual inspection, you will determine the actual condition of the casting by measuring with precision instruments. Taking these critical measurements now, before you begin the machining operations, will help you determine what steps will be necessary to put the block back into service. To evaluate the block you will:

- Inspect the main bearing bores
- Inspect the deck surface
- Inspect the cylinder bores.

It is important to check all these aspects of the block now. You do not want to find out that the block cannot be salvaged after you have begun your machining.

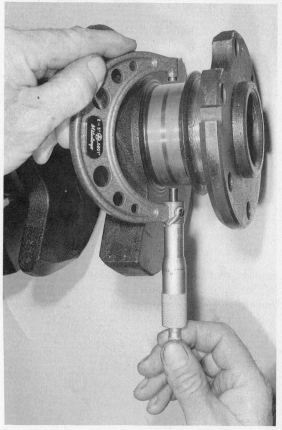

Figure 11-9. Measure the crankshaft journal outside diameter with an outside micrometer.

Inspecting the Main Bearing Bores

There are three main causes for main bearing bore problems:

- Bore misalignment
- Bearing cap stretch
- Damage.

Main bearing bore *misalignment* results from block or cap warping and distortion. As we explained in Chapter 5 of this *Shop Manual*, the factory machines an engine from an unseasoned, green casting. Over time, normal engine operation causes the casting to slowly relax, and the block finally settles down to a final, stable, shape. The main bearing bores in a seasoned casting, which may have been machined perfectly straight on the factory floor, are now out of alignment, figure 11-10.

Because the warping takes place very slowly during the service life of the original engine, the bearing inserts can take up the alignment shift by wearing unevenly. However, if you install new inserts without realigning the bores, the crankshaft may be hard to turn, or it will bind at some point. If you run the engine that way, the new bearings will overheat and prematurely fail.

Servicing Blocks

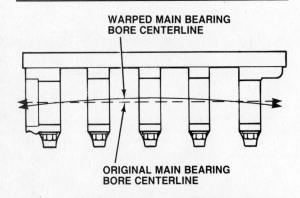

Figure 11-10. The main bearing bores of a warped block usually bend into a bowed shape, the greatest distortion is in the center bores.

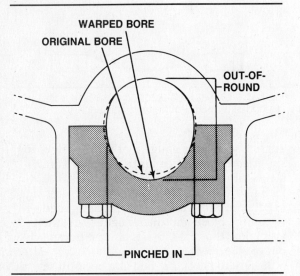

Figure 11-11. When the main bearing caps bow away from the block, they also pinch in at the parting line.

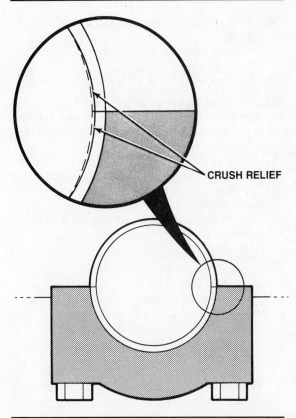

Figure 11-12. Crush relief refers to the way the manufacturers thin the very ends of the bearings near the parting lines.

Main bearing bore *stretch* is also a condition that develops slowly during the life cycle of an engine. Stretch occurs because the main caps must withstand all the combustion pressure from a running engine. The bearing bores gradually bow away from the cylinder head, elongating vertically, and pinching in at the parting line, figure 11-11. As with block warping, stretch usually happens slowly enough so that the main bearing inserts can wear unevenly to compensate.

Both rod and main bearing inserts come from the factory just a little thinner near the parting lines than at the middle of the insert, figure 11-12. This *crush relief* allows the bearing shell to deform a precise amount as the cap is brought to torque. The deformation helps to retain the bearing in the bore, and promotes oil pooling for improved lubrication. In addition, crush relief prevents a stretching bore from closing off the oil clearance space, and also prevents inserts in a cap machined on the tight side of the specifications from pinching inward as you tighten the bolts.

Installing new inserts in a stretched bearing bore will result in excessive vertical crankshaft clearance, and minimal horizontal clearance. If the cap is seriously pinched, the new bearings can completely close off the oil clearance space at the parting lines. New bearings installed in a stretched bore are destined to fail.

Main bearing bore *damage* is generally found in engines that have experienced a major internal failure. The most common type of damage you will encounter are housing bores that have been scored by spun bearings. If the engine has spun a bearing, the only way to repair the problem is to refinish the bores.

There are several ways to evaluate the condition of the main bearing bores in order to determine what process to use to bring them back to within specification:

- Inspect the old main bearing insert wear patterns.

- Use a straight edge to gauge the housing bores.
- Measure with an inside micrometer or telescoping gauge.
- Measure with a dial bore gauge.

Main bearing insert wear patterns
Bearing wear patterns reveal a variety of alignment problems, as detailed in Chapter 9 of this *Shop Manual*. However, they will not tell you the extent of the damage. Remember that normal bearing wear is fairly even over the entire surface of the bearing, although the center is normally the first section to finally wear through to the underlying metals. Watch out for these exceptions:

- A warped crankcase will cause the bearings to wear more on one side of the engine than the other. This direction might be up, down, left, or right, or anywhere in between. The effects might or might not be concentrated on one of the inserts in a pair, and may differ between the center, front, and rear bearings.
- A bent crankshaft usually causes main bearing wear on both inserts at a journal. Because crankshafts usually bend in a bow-shape, the wear is normally greatest at the middle bearings, and decreases toward the front and rear of the engine.
- Bad main bearing housing bores cause asymmetric main bearing wear. Warped bores can cause diagonal wear patterns across the face of a bearing insert. Bores can also become tapered, with different diameters front to rear, look for excessive wear along the front or the back of the insert. Heat and distortion can actually cause the housing bores to become *undersize*.

Straightedge in the housing bores
Severe block warping can be detected using an ordinary machinist's straight edge and a feeler gauge. Reinstall the main bearing caps, without bearing inserts, and tighten to the correct torque. The stresses exerted on a block by the main caps being bolted in position are considerable, and it makes no sense to check the block unless the caps are torqued in place.

Set the block upside down and place a straight edge across the housing bores, align the straight edge as carefully as you can along the saddles of the block. Shine a flashlight at one side of the straight edge and look for light passing through along the surface of the block saddle on the opposite side. Any light leaks indicate saddle warpage or misalignment.

To get an idea as to the extent of the warp, try to pass a feeler gauge under the straight edge. Use a feeler gauge that is one-half of the specified oil clearance for the engine. For instance, if the specified oil clearance is 0.002 inch (0.050 mm), use a 0.001-inch feeler gauge (0.020-mm). If you can fit the feeler gauge between the block and the straight edge, then you know that the block is at least that much out of alignment.

Be sure to check the housing bores in several positions, including along the inner surface of the main bearing caps.

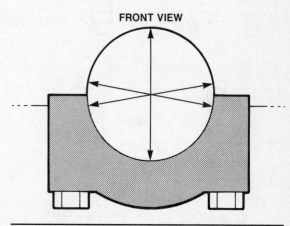

Figure 11-13. Measure across the bores in both directions, and also at the front and back.

Measure with an inside micrometer or telescoping gauge
You can use an inside micrometer to measure the housing bores directly, figure 11-7. The inside micrometer gives you an exact measurement, rather than an estimate. Measure each individual bore from top to bottom, and from side to side, figure 11-13. Because the parting line between the block saddle and the cap tends to deflect the micrometer, measure at a slight angle just above the parting line on one side and just below the parting line on the other side. Be sure to check the diameters at the front and back of the bore to make sure there is no taper.

Inside micrometers are expensive, and you will need several different sizes to accommodate the various bore sizes. A cheaper alternative is to use a telescoping, or snap gauge, figure 11-14. Retract the spring-loaded plung-

Servicing Blocks

Figure 11-14. Telescoping, or snap, gauges give you indirect inside measurements.

ers of the snap gauge and gently tighten the ferrule to hold them in place. Position the gauge in the bore, and loosen the ferrule. The plungers will snap out against the inner surfaces of the bearing bore, center the gauge in the bore, then tighten the ferrule to lock the plungers in position. Remove the gauge from the bore, and measure the width across the plungers with an outside micrometer, figure 11-15. As with inside micrometers, measure across the bore at several locations, and at the front and the back to check for taper.

Write all your measurements down for future reference. There is no way you will be able to remember all these numbers and keep them straight.

Measure with a dial bore gauge
Without question, the quickest and most precise tool for measuring the main bearing bores is a dial bore gauge, figure 11-16. This tool is expensive, but the time it will save you makes it well worth the money.

The dial bore gauge has small, spring-loaded plungers at the working end that roll against the walls of the bore being measured. An anvil between the plungers rotates the needle on the dial face as it moves in and out. Opposite the anvil and rollers, the indicator has a threaded hole to accept extensions of various lengths, these adjust the tool to the diameter of the bore being measured.

As explained in Chapter 3 of this *Shop Manual*, dial bore gauges do not read the actual size of the bore — you set the gauge to a particular dimension and the dial tells you how much over or under that dimension the bore actually is. The dial indicator face of a typical bore gauge, figure 11-17, is graduated in increments of 0.0005-inch, so every other tickmark is a whole thousandth. The needle travels from zero in either direction to indicate how far over or under the setting of the tool the bore actually measures.

Adjusting the setting fixture
You can use an ordinary micrometer to set the dial bore gauge, or you can use a fixture de-

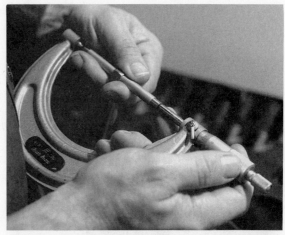

Figure 11-15. You measure the snap gauge with an outside micrometer to get the dimension you need.

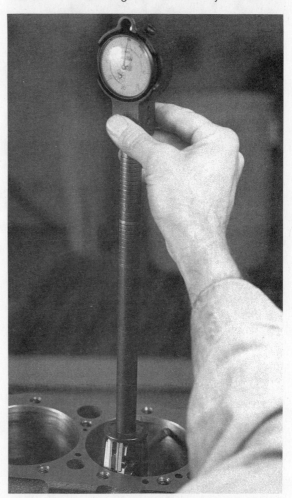

Figure 11-16. The dial bore gauge is quicker and easier to use than any other inside measuring device.

signed for setting the gauge, figure 11-18. This setting fixture is usually more expensive than the gauge itself, but it makes the dial bore gauge much more convenient to use.

Figure 11-17. The dial face has small tickmarks that read in increments of 0.0005-inch each — "5 tenths."

Figure 11-18. The setting fixture is expensive, but worth the money because of the time it will save you.

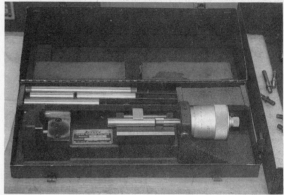

Figure 11-19. The various spacers set the tool for the whole-inch portion of the measurement.

Look up the manufacturer's specified main bearing bore diameter in the factory shop manual, a bearing manual or another source. For instance, suppose the specifications called for a bearing bore diameter of "2.9419/2.9429 inch". The exact middle of the range is 2.9424, the total tolerance is 0.001 inch. or plus-or-minus 0.0005 inch.

Adjust the dial bore gauge setting fixture to the finish bore size, in this case 2.9424 inch. You will set the fixture by dropping in a precision-machined spacer bar for the inch increments, and then using the micrometer head to set the decimals:

1. Select the spacer marked "2.000-inch", wipe both ends with a clean rag, and place it on the fixture, resting against the micrometer head, figure 11-19.
2. The thimble moves the spindle 0.025 inch for each revolution, just like a normal micrometer. However, the micrometer head is oversize, and the smallest tickmarks read directly to 0.0001 inch. This allows you to read directly to four decimal places without a vernier.
   Rotate the micrometer head on the fixture to zero. At this point, the fixture is set at exactly 2.0000 inch.
3. To add in the 0.9, rotate the thimble counterclockwise until the zero mark lines up with the "9" on the sleeve. You are now at 2.900 inch.
4. Add in the 0.042, by rotating the thimble counterclockwise one complete turn, this gives you an additional 0.025, continue turning counterclockwise to line up the "17" mark on the sleeve. Twenty-five plus seventeen give the oh-four-two necessary to set the tool to 2.9420 inch.
5. To add in the last 0.0004, simply turn the micrometer head counterclockwise to the fourth tickmark past "17".

Now the gap between the end of the spacer and the anvil of the setting tool is exactly 2.9424 inch, figure 11-20. You will use the gap to adjust the dial bore gauge so it will read over and under for that diameter.

Setting the dial bore gauge

Once the setting fixture is correct, use it to set the dial bore gauge as follows:

1. Place the dial bore gauge in the fixture, and attach an extension that will fill the gap between the base of the gauge and the end of the spacer. The extension must push the plungers of the gauge back into the base slightly, to give the gauge enough

Servicing Blocks

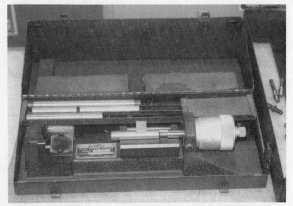

Figure 11-20. Once adjusted, the gap between the end of the spacer and the anvil will be exactly the same as the manufacturer's specification for the bore you will be measuring.

Figure 11-22. Zero the indicator face by turning the extension in and out, then tighten the lock nut.

Figure 11-21. Place the dial bore gauge into the setting tool with the extension loosely screwed in.

Figure 11-23. Measure the bearing bore.

reserve travel to read larger diameters than the size you set it to, figure 11-21.
2. Approximately zero the gauge needle by turning the extension in or out, then *Gently* tighten the locknut at the base of the extension, figure 11-22.
3. *Perfectly* zero the gauge by rotating the faceplate using the serrated rim.

At this point, the gauge is set to read zero at a bore diameter exactly 2.9424 inch across. The needle will point to the negative numbers at the right of the zero if the bore is undersize, and to the positive numbers at the left of the zero if the bore is oversize. If the bore is exactly 2.9424 inch across (our target), then the gauge will read exactly zero.

Measuring a bearing bore with the dial bore gauge

Before measuring the bore, remove and inspect the bearing caps. Look for burrs and raised areas on the bearing seat surface. Pay particular attention to the area around the bearing retaining notches machined into the saddles and caps. Clean up any surface irregularities with a fine file, also run the file across the parting faces to remove any burrs. Reinstall and *torque* the bearing caps to specification, then use the following procedure to measure the bores:

1. Insert the gauge head into the bearing bore, figure 11-23. Gently rotate the gauge until the small extension and the plunger rest as horizontally across the bore as you can get them. Avoid resting the tool on the parting line. Watch the needle as you slowly rock the dial end of the gauge back and forth to move the plungers up and down in the bore. The position where the needle is *as negative as possible*, is the reading you want to record. The dial indicator needle can indicate the correct bore diameter, and any diameter above that, within its range of travel. The only *correct* reading is therefore the smallest, or the most negative. It is also the reading where the needle swings the farthest *clockwise*.

For example, in figure 11-24, the needle reads 0.0005 inch undersize, but might read as high as 0.010 inch oversize (or any value in between) as you swing the tool back and forth. The farthest clockwise reading of the dial is at 0.0005 inch, so that is the difference between the setting of the tool and the actual bore. Because you set the tool at 2.9424 inch, the true diameter of the bore is 0.0005 inch under that, or 2.9219 inch. This reading is within the 0.0005 inch tolerance, so although the bore is slightly pinched in at the parting line, it is still within specifications.

2. Next, check the diameter of the bearing bore at 45 degrees to the parting line, figure 11-25. Swing the tool to find the most negative reading. In our example, the reading is 0.001 inch. This indicates the bore is stretched oversize in this direction, and is out of tolerance by 0.0005 inch.
3. Finally, measure the bore at 90 degrees to the parting line, figure 11-26. Here, the bore face is 0.002 inch oversize, stretched well out of specification. If you measured the bore in this direction with an inside micrometer, you would find that the actual diameter was 2.9444 inch, well beyond the acceptable tolerance range.
4. Measure all the other main bearing bores the same way. If any of them are out of specification, you will align hone or line bore the main bearing bores.

These numbers are typical of the values you will see when actually working on engines. A stretched main bearing cap will pinch in at the sides, but will be egg-shaped and oversize in line with the cylinders. To check for distortion and taper, measure each bearing bore in all three positions at the front, and also at the back. Be especially suspicious if you see asymmetric wear patterns on the bearing inserts. The main bearing bores of an engine with readings like those in our example can be reconditioned by align honing or line boring. This is usually the first machining operation you will perform after you have inspected and measured the rest of the block.

Inspecting the Deck Surface

The first inspection you make of the cylinder deck is with your eyes and hands. Look for excessive scoring, corrosion, erosion, threads pulling up around bolt holes, cracks, dents, and scratches. Run your fingernail across any irregularities you find. If a scratch can catch your fingernail, then it probably is deep enough to need correcting.

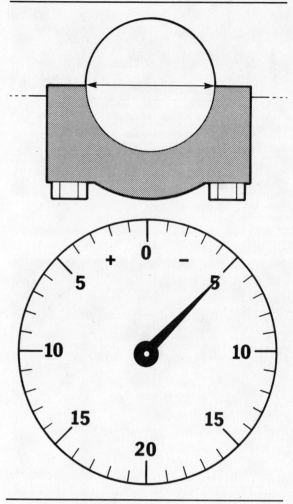

Figure 11-24. This needle position indicates a diameter 0.0005 inch undersize.

Engine manufacturers require a specific finish, generally in the 90 to 100 microinch range, on the deck surface. Some modern cleaning methods, such as thermal ovens, shot blasters, tumblers, and shakers, can have a detrimental effect on the surface finish. Even if the block deck is perfectly flat, you may have to take a light cut simply to get the necessary finish for proper gasket sealing.

Evaluate surface condition first, then check the deck for flatness.

Check for flatness

Start with a completely clean surface — any gasket material left on the deck will make your results meaningless. Carefully lay a straightedge along the deck surface, and try to insert a feeler gauge between the straightedge and the deck. You can gauge how distorted the surface is by the thickness of the feeler gauge that you can slip underneath.

Servicing Blocks

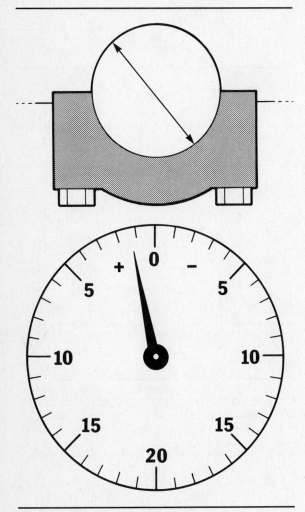

Figure 11-25. This needle position indicates a diameter 0.001 inch oversize.

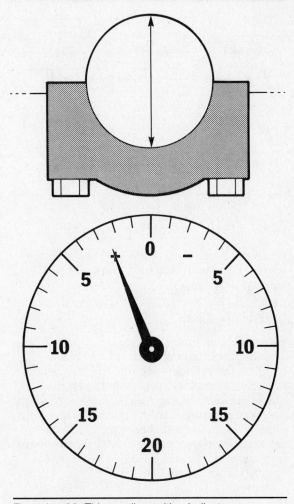

Figure 11-26. This needle position indicates a diameter 0.002 inch oversize.

Check the deck surface at the cylinder centerline parallel to the crankshaft, along all four sides, and diagonally from corner to corner, figure 11-27. If any of the surfaces have too much clearance, the deck needs resurfacing.

Limits for distortion

The amount of distortion that is allowable varies, and manufacturers specify the tolerance of the deck surface in different ways. Some give two numbers: a normal range, and a maximum limit. Others simply give a maximum value, before resurfacing is necessary.

A few manufacturers do not even specify a value for deck distortion at all. Instead, they assume you will resurface the deck, and provide only a maximum depth for the surface cut. If the surface does not "clean up" within the specified depth, replace the block.

If you do not have any specifications at all, you can use these general guidelines:

- If the deck surface is distorted by 0.004 inch (0.100 mm) or more, then you should resurface it.
- If any warping approaches 0.003 to 0.004 inch (0.075 to 0.100 mm) within about ten inches (25 cm) of casting length, then the surface *definitely* needs attention.

For engines with specified surfacing limits, you must follow the manufacturer's advice. For most older domestic engines, you can usually cut up to 0.015 inch (0.38 mm) without trouble. If you are not sure, find out before you cut the block.

Inspect the Cylinder Bores

During your teardown inspection you already checked the cylinder bores for major damage such as cracks, gouges, scoring, and broken cylinder walls. Now you will use precision measuring equipment to find out exactly how

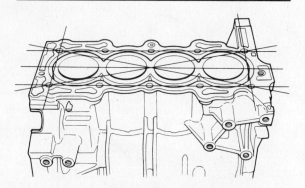

Figure 11-27. Check for warp both at right angles and diagonally across the block surface.

much the bores have worn, and to determine what repair method to use. You will check for:

- Previous overbore
- Taper
- Out-of-round.

Check for previous overbore
The first check you should make is to see whether the cylinders are the original diameter, or whether they were bored oversize during a previous rebuild. To check for a previous overbore measure near the bottom of the cylinders, you can use a dial bore gauge, inside micrometer, or telescoping gauge. Compare your findings with the specifications listed by the engine manufacturer.

Domestic engines built to U.S. Customary dimensions formerly used 0.030 inch (0.75 mm) increments for overbore. For example, an engine with a factory bore size of 4.000 inch would have been bored thirty-over to 4.030 inch, or sixty-over to 4.060 inch. To accommodate today's thin-walled castings, you can also routinely purchase pistons in the intermediate 0.020 and 0.040 diameters. These additional sizes often let you get an extra rebore out of the block before you must scrap it.

Manufacturers of engines built to metric specifications commonly use 0.50 mm (0.020 inch) increments for overbore. For example, if the standard bore size for a given engine is 81.000 mm, then bores measuring 81.500 or 82.000 mm are already one or two piston sizes over, respectively.

Check for taper and out-of-round
Cylinders never wear out evenly. Wear is always greatest in the central area of the cylinders that is swept by the piston rings, figure 11-28. Because the rings do not travel to the absolute top of the cylinder, there is no wear in the extreme top portion of the cylinder. As

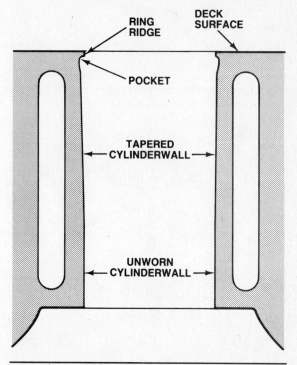

Figure 11-28. Cylinders wear in a taper, with most of the enlargement taking place at the top, under the ring ridge.

a result, a ridge develops in the cylinder wall at the upper limit of ring travel. By the same token, the lower half of the cylinder merely supports the piston skirts at bottom dead center, and seldom wears very much. It is very common to still see the original factory honing marks on the cylinders below the ring travel, even on engines with fairly high mileage. Because the thrust forces are greatest at top dead center, this is the area that will show the most wear. As the piston travels toward bottom dead center the thrust forces decrease and the amount of wear diminishes. This uneven wear results in a tapered cylinder that is wider at the top than at the bottom.

The cylinders also wear unevenly around the circumference of the bore. As explained in Chapter 12 of the *Classroom Manual*, each cylinder has a major and a minor thrust surface. Because the forces exerted on the cylinder wall varies from side to side, so does the amount of wear. The cylinder bores in any reciprocating will wear out-of-round, or oval shaped, due to the difference in the thrust forces.

Checking for taper and out-of-round is very simple. You can use an inside micrometer, figure 11-29, but the best tool to use is the same dial bore gauge you used earlier to check the main bearing bores, figure 11-30. Go back to that section for information on how to set the

Servicing Blocks

Figure 11-29. You can use an inside micrometer to measure the cylinder bores.

Figure 11-31. This saddle bracket lets you use a standard dial indicator to measure cylinders for taper and out-of-round.

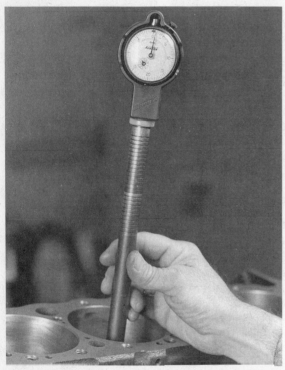

Figure 11-30. The dial bore gauge is the quickest tool to use for measuring cylinder bores.

gauge up. Some shops may have a special saddle bracket, figure 11-31, that lets you use a standard dial indicator and extensions to measure cylinder diameters. Once set up, the dial indicator and straddle bracket is used basically the same way as a dial bore gauge.

Set the dial bore gauge so that it will read zero in the original, factory bore. You can get this value from the service manual, or from an aftermarket piston or ring manual. If the engine you are working on has been bored previously, set the gauge to zero on the proper overbore. Insert the gauge into a cylinder, and measure the diameter across the bore at the top and bottom of the cylinder, both in line with the crank and at right angles to it, figure 11-32. Taper is the difference between the readings at the top and bottom of the cylinder. Out-of-round is the difference between the bore diameter measured in line with, and across the crankshaft.

A realistic limit for cylinder taper on automotive engines is about 0.005 inch (0.125 mm). The limits for diesel engines are much lower, usually about 0.003 inch (0.075 mm).

A realistic limit for cylinder out-of-round on automotive engines, both diesel and gasoline, is about 0.001 inch (0.025 mm).

If the cylinders are out of specifications for either taper or out-of-round, the correct procedure to follow is to rebore all the cylinders to a standard oversize. After boring, you will hone the cylinder surface to prepare them for new rings. If the cylinders are not worn enough to justify a rebore, you can deglaze the cylinder walls and install new rings on the old pistons.

MAIN BEARING BORE RECONDITIONING

Main bearing line boring or align honing should be your first step in machining the block. Aside from the fact that the housing tolerances are quite small — typically 0.0005 to 0.001-inch (0.0125- to 0.025-mm), many of the machines designed for deck surfacing and cylinder boring and honing use the main bearing bores to line up the tool. It stands to reason that you will want fresh, round, bearing bores for these tools to index from. In addition, resizing the main bearing bores changes their location and corrects the clearances — how

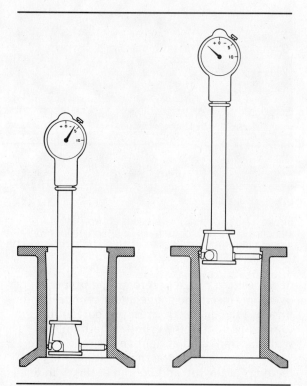

Figure 11-32. Measure the bore diameter at the top, just below the ridge, and at the bottom, below ring travel. Take measurements in line with and at right angles to the crankshaft.

Figure 11-33. The align honing machine can quickly realign, resize, and surface main bearing bores.

much depends on how stretched they were to start with, and how careful you can be in selectively removing metal from the block or from the caps. When all of the machining operations have been performed, you want cylinder bores and deck surfaces that are correct in reference to the *new* bearing bores.

There are two methods you can use to recondition the bearing bores, *line boring* and *align honing*. Line boring can remove a substantial amount of metal from the bearing saddle and cap, on the order of 0.020-inch (0.500-mm), and will correct major damage or misalignments. Align honing removes smaller amounts of metal, such as 0.005-inch (0.125-mm), and is a much quicker way to correct the smaller problems that you will encounter during a typical engine rebuild. We will explain both of these procedures here.

Align Honing

An align honing machine is a large stationary power tool, figure 11-33. The machine uses grinding stones mounted on a long mandrel to resurface all the housing bores simultaneously. Since the cutting stones are attached to a rigid mandrel, the bores automatically line up with each other. Align honing brings the bores back to their original size, so you can use standard-sized main bearing inserts when you reassemble the engine.

Because the stones in the mandrel are under tension and cut equally in all directions, align honing always removes the same amount of metal from both the cap and the block. This means that align honing will always raise the crankshaft in the block by several thousandths of an inch. This is negligible for almost all the engines you will routinely work on, but keep it in mind if piston-to-head or piston-to-valve clearance is marginal. You may also have to check, and possibly correct, the valve timing or timing gear mesh during reassembly.

In order to return the bores to their standard size, first you must grind metal from the bases of the main bearing caps. This makes the bores slightly undersize when you bolt the caps onto the block. After the caps have been ground and bolted back on the block, the undersize bores are returned to standard size by opening up the internal diameter with the hone.

The machine we show in the illustrations is a Sunnen CH-100 Horizontal Hone. This is the only align honing machine commonly available in the United States at this time.

Grinding the main caps

The bearing caps can be ground using a rod grinding machine. We detail how to use this machine in Chapter 12 of this *Shop Manual*. Remove no more than about 0.002 inch (0.050 mm) of metal from the main bearing caps at a time. Keep in mind, it is hard to remove metal from the cap evenly in all directions, and easy to end up with a cap that may be flat but does not sit squarely in the main bearing saddles. Even though you can hone the bore to the correct diameter, you can not correct the slightly

Servicing Blocks

misaligned bolts or thrust bearing surfaces. The problem is compounded because the thrust bearing main cap is often much wider than the others, and therefore harder to center in the grinding machine.

You can avoid alignment problems by using a special bolt-on surface plate in the rod grinding machine. These surface plates are designed especially for grinding main bearing caps. Use a dial indicator to set the surface plate up square with the grinder.

Many machinists prefer to remove unequal amounts of metal from the bearing caps because of an uneven wear pattern, figure 11-10. Since the greatest amount of wear is toward the center of the engine, these caps are reduced more than those at the ends of the crankshaft. This practice allows all the bores to "come in" at about the same time when you hone them. For an engine with five main bearings, an experienced machinist will typically grind 0.002 inch (0.050 mm) of metal from the two end bearing caps, and reduce the three center caps by 0.004 inch (0.10 mm).

Honing the bores

When you hone, remember that the machine removes metal very quickly. Hone for only 10 or 15 seconds, about three or four complete forward-reverse strokes of the mandrel, then measure the bores. If you removed 0.002 inch (0.050 mm) from each of the caps, your total honing time will be less than a minute.

When you turn the drill motor off, be sure to keep stroking the honing mandrel until it comes to a complete stop. This prevents grooves forming on the stones and extends their service life.

Measuring your progress

After you hone the bores for 10 or 15 seconds, remove the mandrel and measure each of the bores with the dial bore gauge. This will allow you to see how much metal you have removed, and calculate how much more honing will be required to bring the bores out to size. If all the bores are still undersize, insert the hone mandrel back into the bores and hone for another 10 or 15 seconds.

It is likely that some bores will reach the finish size before others. Whenever a bore diameter reaches specifications, loosen its main bearing cap bolts. Then continue honing — the hone will not take any more metal from the bores which you have loosened, but will continue to cut the rest.

If you happen to hone some bores too far, and they measure oversize, remove the caps and grind away a *small* amount of metal in the rod and cap grinding machine — perhaps 0.001 inch (0.025 mm). Then reinstall them on the block and hone a little more.

The following sequence details how to align hone the main bearing bores of a V-type engine using the Sunnen CH-100. The procedure also applies to inline engines, however; the set up will vary slightly. This procedure is to be used when removing up to 0.050 inch from the bores. If more metal needs to be removed, the hone must be adjusted with a dial indicator, rather than the centering pins, according to the Sunnen operator's manual.

ALIGN HONING MAIN BEARING BORES

1. Remove the main bearing caps, inspect the parting faces and clean off any burrs. Grind about 0.002 inch (0.050 mm) of metal off the bearing cap parting surfaces.

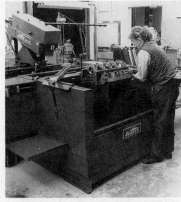

2. Lift the block into the align honing machine, and set it on the riser blocks. For inline engines, turn the riser blocks on their sides.

3. Slide the clamping bars through the camshaft bores and fix the yokes in place with the drawbolts. Protect the cam bearings with shop rags if they are still installed.

ALIGN HONING MAIN BEARING BORES

4. Select the proper size honing mandrel for the engine block, and inspect the cutting stones to make sure they are in good condition.

5. Retract the shoes all the way by loosening the adjuster screw on the end of the honing mandrel.

6. Lay the honing mandrel into the main bearing saddles of the block, with the stones facing up. Fit the centering pins so they rest on the center bearing saddle.

7. Rock the mandrel gently on the pin as you extend the shoes with the adjuster screw. When rocking stops, turn an additional 1/4 turn, then remove the pins.

8. Bolt the main caps back on the block and torque them down. You do not need to remove the mandrel from the bores.

9. Adjust the height of the block in the machine by turning the hand crank, line up the main bearing bore with the drive motor.

10. Use the chuck wrench to pry the wedge in the mandrel back as far as you can. This retracts the stones without changing the centering adjustment you made earlier.

11. The coupling is a universal joint that connects the mandrel to the drive motor, it also contains the cutting pressure and stone feed controls.

12. Slip the coupling onto the honing mandrel, and turn it 1/4 turn clockwise to engage the linkage. Flip the quick release lever closed, and tighten the set screw.

Servicing Blocks

ALIGN HONING MAIN BEARING BORES

13. Turn the feed-up and cutting pressure controls counterclockwise as far as they will go to retract the stones and release the spring tension.

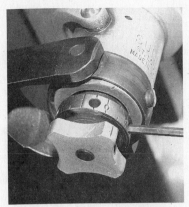

14. Set the cutting pressure by turning the control clockwise with the chuck wrench to align the "0" on the dial with the witness mark.

15. Attach the coupling to the drive motor.

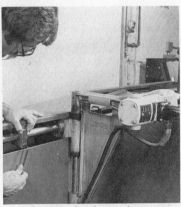

16. Position both stroke stops so you center the mandrel with equal overstroke on both ends of the block. Be sure the stones can not slide out of the bores.

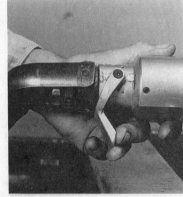

17. Turn the stone feed-up control all the way in, then back-off two turns. Check the quick release lever, it should not be able to move past the position shown.

18. Swing the oil feed arm down, and set the clips to block any fluid holes that do not flow directly between main bearing webs. Raise the front splash shield.

19. Turn on the power and immediately start stroking. Stroke the mandrel back and forth from stop to stop for about 10 to 15 seconds.

20. Turn the machine off, stroke until the mandrel stops rotating. Swing the oil feed up, uncouple the drive motor, then slip the hone mandrel out of the bores.

21. Measure all the bores with a dial bore gauge. If more honing is needed, refit the mandrel and repeat steps 18 to 21 until you bring the bores up to size.

Line Boring

The alternative to align honing is to line bore the bearing housings to straighten and resize them. Line boring machines use a long, rigid mandrel and a cutting bit to remove metal, figure 11-34. Like align honing, line boring the housing bores ensures that they are all straight, in line with each other, have the correct diameter for adequate bearing crush, and provide good heat transfer from the bearing shells to the caps and block.

Unlike align honing, line boring can rebore the correct-sized hole slightly off center. This means that after you grind the main bearing caps, you can position the cutting bit to barely clean the uppermost surface of the block saddles, while removing most of the metal from the caps. This reduces the risk of camshaft timing problems and piston to valve interference that can result if metal is removed equally from both the caps and the block saddles. Because the cutting bit does not follow the preexisting hole contours the way a hone does, line boring will accurately locate the housing bores and make them perfectly round.

As with all machining, the care and skill of the machinist counts for as much as the type of equipment he or she uses. For routine engine rebuilding, differences between line boring and align honing are fairly unimportant. Align honing is definitely the quicker method.

We have already explained main housing bore reconditioning on an align honing machine, so we will not detail using the line boring tool. However, the preparation for line boring is basically the same as for align honing. Clean and deburr the block saddles and the caps, and then cut about 0.002 to 0.005 inch (0.050 to 0.125 mm) of metal from the cap surfaces, torque the caps back into place and bolt the block into the line boring machine. Line boring machines can tolerate greater depth of cuts — as much as 0.020 inch (0.050 mm) — so professionals routinely remove as much as 0.010 inch (0.025 mm) from the caps to speed production.

Follow the instructions for the machine to set the cutting bit to open the hole to the correct diameter. Once you have centered the boring bar, check the position within each of the housing bores and adjust the bar so that the minimum material will be removed from the block side of the bore.

DECK RECONDITIONING

Reconditioning the cylinder block deck surface will be the next major machining operation, af-

Figure 11-34. The line boring machine cuts the housing bores to size, rather than grinding them.

ter align honing or line boring. If the cylinder boring equipment uses the cylinder deck surface to locate the boring bar, the deck surface *must* be reconditioned before boring the cylinders. Obviously, if the old deck is not square with the newly machined main bearing bores, then any cylinder bores you cut using it as a reference will not be correct either. On the other hand, if your boring equipment uses the pan rails and the main bearing bores to line up the block, then it does not matter whether you bore the cylinders or resurface the deck first. If you are installing, or replacing, cylinder sleeves they must be fitted before any other machining is performed. Install the sleeves as detailed later in this chapter, then return here and continue servicing the block.

Surface and Alignment Considerations

Cylinder deck condition is very important to engine life and performance. In routine engine rebuilding, the main purpose of deck reconditioning is to give the new cylinder head gasket a flat, freshly machined surface to seal against. If the deck is not flat, then some parts of the gasket will not be clamped securely, and the engine will have a tough time holding back combustion pressure. If the surface does not have a good "tooth", then it will be unable to hold the head gasket in place, and internal leaks are likely.

Because the head gasket has a top and a bottom surface, reconditioning both the block deck and the base of the cylinder head is equally important. A poor gasket surface on one of the castings can leak combustion gases no matter how well you finish the other casting, so resurfacing both castings is often neces-

Servicing Blocks

sary. We discuss the specifics of cylinder head reconditioning in Chapter 14 of this *Shop Manual*, but remember that machining techniques detailed here also applies to heads.

For routine work, you will simply take a surface cut from the deck by indexing a grinding or milling machine to the old surface. This gives a freshly machined deck with a good tooth for the head gasket, and is sufficiently accurate for most engines.

Remember, that setting a machine to cut a surface parallel to the old one will not correct any indexing problems that already existed. The new surface may not be quite parallel to the crankshaft, and it may be tipped slightly from the cylinder centerlines. These inaccuracies are not a problem for most engines, but correcting them is important for rebuilding diesels and for blueprinting gasoline engines.

Surface Finishes

As explained in the *Classroom Manual*, every surface has minute grooves or imperfections, which can be measured with a surface roughness indicator. The indicator moves a diamond stylus over the surface of the metal. As the stylus rises and falls over the uneven surface, its movement is magnified and measured by the machine. The result is a measurement in millionths of an inch (microinches: fin), or millionths of a meter (micro-meters or microns: f).

As an automotive machinist, you may never measure the actual surface roughness of a cylinder head or block. However, you will often need to produce a manufacturer's recommended surface roughness during machining. Normally, the machine tool manufacturer will provide tables or other information to let you know what type of surface finish you can produce by using various stones or cutters, traversing speeds, and depth of cut. Older single-speed surfacing machines may not be able to give you the various surfaces you need to work on different types of engines. You must use a variable-speed grinding or cutting tool, or a variable-speed traverse.

How Much Should You Cut?

In general, keep the total amount of metal you remove down to the minimum necessary to clean up the surface. Be aware some manufacturers list maximums beyond which you should not cut the block, and this amount may be quite small. Removing more metal than this aggravates all the problems we discussed.

Consider that you might not be the first machinist to take a cut off the deck surface (or the cylinder head, for that matter). The engine might have run successfully with the reduced clearances from the previous overhaul, but have trouble when your cuts are added to those already taken. If the customer tells you that the engine has been overhauled before, inspect it very carefully. Remember, it is very easy to remove metal from a surface, but impossible to put it back on. Before you take a cut, you must consider the:

- Amount of cylinder head surfacing
- Piston-to-head and piston-to-valve clearance
- Compression ratio
- Intake manifold to cylinder head alignment
- Pushrod length
- Valve timing changes.

Amount of cylinder head surfacing

Keep the total amount of metal you remove from both the cylinder head and the block deck surfaces in mind. Generally the cylinder head will warp before the block will, as a result, you will have to remove more metal from the head than from the block. Keep this in mind, and only remove a minimal amount of metal from the deck surface of the block. The combined cuts from both castings must never exceed the amount allowed by the design of the engine.

Piston clearances

The problems that excess resurfacing will cause with respect to piston clearance are obvious: every cut will bring the piston that much closer to contact with the cylinder head and valves as it reaches top dead center. Different engines have different degrees of safety with respect to piston clearance — diesel engines are especially vulnerable. All other things being equal, piston clearance problems are more likely if you cut metal from an engine that has:

- High compression
- Radical valve timing and higher lift
- Larger valves
- Interference design.

Compression ratio

Cutting metal from the deck or head also increases the compression ratio, and the size of the increase depends on the amount of the cut. You can usually neglect this effect for narrow bore, low compression engines. You can absolutely *not* neglect it for diesel engines. Diesels have extremely high compression ratios, often as high as 22 to 1, and even a very small surface cut may cause interference or detonation problems. For gasoline engines with computer-controlled ignition timing and a knock sensor, the increased compression ratio can *decrease* performance if the resulting spark knock forces the computer to constantly retard the ignition timing.

Intake manifold to cylinder head alignment
For V-six or V-eight engines, removing too much metal from the deck surface can cause alignment problems with the intake manifold, figure 11-35. You may get misaligned bolt holes, poorly matched ports, and faulty sealing at the top of the block. Solving this problem requires careful compensation by machining the manifold, the cylinder heads, or both. We discuss these processes in detail in Chapter 14 of this *Shop Manual*.

Another area of consideration when machining V-type engines is the manifold seating area on top of the block. Although this operation is standard procedure for high-performance machine work, it is often overlooked during standard engine rebuilding because of the large amount of crush allowed by the manifold gasket. Surfacing equipment manufacturers provide charts to let you know how much metal to remove from the top of the block based upon how much was removed from the deck surface, figure 11-36.

For engines that have a cast front timing chain cover, you must also resurface the cover. Machine the top surface of the cover to match the amount of metal removed from both the deck and the head.

Pushrod length
In OHV engines, removing metal from the deck surface can cause problems with pushrod length, especially in engines that use over- and undersize pushrods to bring the valve clearance into the range of the hydraulic lifter travel. Removing excessive metal might require undersize pushrods to correct the problem. The manufacturers usually anticipate this, and you can often buy pushrods in at least one standard undersize for many engines.

Valve timing changes
For overhead-cam engines, decreasing the distance between the cylinder head and the crankshaft can cause slight problems with valve timing. Manufacturers carefully match the length of the timing chain or belt with the sprockets or pulleys to ensure that the valve timing is correct when the timing marks line up. Whenever you remove metal from the

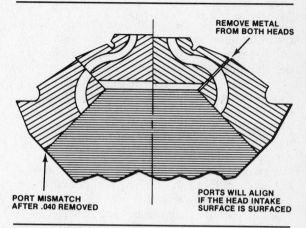

Figure 11-35. Cutting too much metal from the decks of a V-type engine will cause manifold misalignment.

deck or the cylinder head, the camshaft position will lag a little behind the crankshaft position, when the engine runs the valve timing is slightly retarded.

In addition to driveability problems, removing too much metal may cause interference between the pistons and the exhaust valves, if the situation was marginal before. The exhaust valve is more vulnerable than the intake, because during the exhaust stroke the rising piston is "chasing" the exhaust valve as it closes. Because intake valves close much earlier during the compression stroke, they are less likely to contact the pistons.

Sometimes you can correct camshaft timing problems by advancing the camshaft timing slightly. Slotting the bolt holes for the camshaft sprocket and using aftermarket adjusting hardware are two methods you can use. Anytime you change the camshaft timing, you will need to reset it using a degree wheel.

Resurfacing the Deck Surface

There are two machining methods, grinding and milling, that can be used to restore the cylinder block deck surfaces. Surface grinders true the deck by removing metal with a rotating grinding stone, figure 11-37. Milling machines use high-speed cutting bits to take the surface down, figure 11-38. In some machines, the block is stationary and the grinding stones or milling cutters pass over the surface. In others, the cutters are stationary and the casting moves across them. Cylinder blocks, manifolds, or cylinder heads can generally all be resurfaced using the same machine — the limits are most often the shear size of the casting that will fit into or onto the machine.

The machine we use to illustrate cylinder block resurfacing is a Kwik-Way Model 860

Servicing Blocks

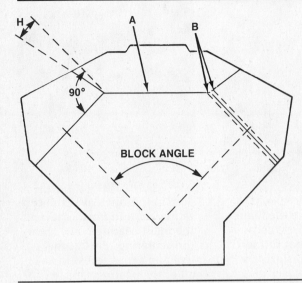

H	M	
CYLINDER HEAD ANGLE (Degrees)	MULTIPLIER CONSTANT	
	BLOCK ANGLE (Degrees)	
	60	90
5	0.8	1.5
10	1.3	1.7
15	1.4	1.9
20	1.5	2.2
25	1.6	2.6
30	1.7	3.3
35	1.9	4.7
40	2.2	9.0

EXAMPLE: E = 90°, H = 15°, B = 0.02 inch
B × M = A
0.02 × 1.9 = 0.038 inch

Figure 11-36. This chart is used to determine how much metal should be removed from the top of the block (A). Multiply the total cuts taken from the deck and head surfaces (B) by the multiplier constant listed on the table. You must know the engine block (E) and cylinder head (H) angles.

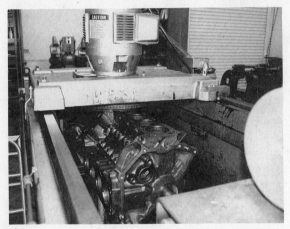

Figure 11-37. This surface grinder passes a rotating stone across a stationary block to restore the deck surface.

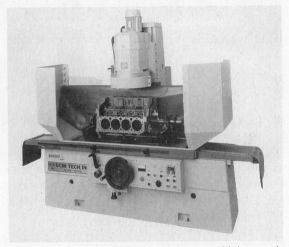

Figure 11-38. This milling machine uses a high speed cutting bit to remove metal from the deck of the block.

Surface Grinder. This machine tool holds the workpiece stationary as a rotating grinding stone moves across it to finish the surface. You make precise cuts that remove about 0.002 inch (0.050 mm) of metal from the deck surface with each pass of stone.

A word of caution: always double-check your measurements and set-up when you cut or grind the cylinder block deck surface — this is the one block machining operation that does *not* forgive mistakes. If you take too much metal off when you recondition the main bearing bores, you can always cut the caps again and re-hone. If you make a mistake and go oversize with the boring bar or cylinder hone, you can always fit the next oversize piston or even sleeve the bore. You cannot restore metal removed from the deck surface, and there is no machining technique to compensate for an inaccurate cut. If the engine block is not properly set up and adjusted in the surface grinder, you can quickly and permanently transform it into scrap metal. Be sure the block is properly set-up before you take a cut.

The following sequence details how to resurface the block decks of V-type engine using the Kwik-Way 860 Surface Grinder, although many other surfacing machines operate in a similar fashion. The procedure also applies to in-line engines; however, the set up will vary slightly. Keep in mind, what we present here is a general procedure, always follow the specific instructions provided by the equipment manufacturer.

RESURFACING THE CYLINDER BLOCK DECK SURFACE

1. Remove any pins or sleeves that protrude above the deck. If the VIN is stamped on an area that will be ground, write it down so you can restamp it later.

2. Turn the block upside down and remove the main bearing caps. Select the two centering rings that fit the main bearing bores exactly.

3. Slip the rings over the centering bar and lay the bar in the main bearing bore, with the centering rings in the front and rear bores. Install all of the bearing caps.

4. Move the grinding head all the way left to get it out of the way. Raise the table using the coarse height adjustment lever to make loading the engine easy.

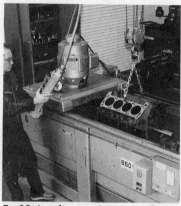

5. Hoist the engine into the machine, resting the bar in the towers.

6. Roll the engine over to rest the side of the block on one of the leveling posts.

7. Place the tower caps over the bar and tighten the bolts to hold the bar to the towers.

8. Place a level on the deck surface. Turn the leveling post until the block is approximately leveled from left to right.

9. Turn the level 90 degrees, loosen the large lock-bolt, and turn the large nut until the block is approximately leveled from front to rear.

Servicing Blocks

RESURFACING THE CYLINDER BLOCK DECK SURFACE

10. Place the straddle bar across the machine and set the dial indicator in the slot of the bar. Position the dial indicator at one side of the block, note the reading.

11. Move the dial indicator to the other side. Use the leveling post to tilt the block so the indicator reads the same on both sides. The block is level, left to right.

12. Now use the straddle bar at the ends of the block. Adjust the large nut up or down until both ends of the block read the same value, and the block is level, front to back.

13. Tighten the lock-bolt, and traverse the grinding head over the block. Raise the block until the deck surface just touches the grinding stone shield.

14. Turn the motor and coolant on. Use the handwheel to raise the block, 0.002 inch at a time, until you can hear the stone touch the deck surface.

15. Loosen the handwheel lock-screw, set the pointer to zero, and tighten the screw. Turn the handwheel to lower the block about 0.010 inch.

16. Traverse the grinding head all the way left, raise the block so the micrometer head again reads zero. You are now ready to take a zero cut on the deck.

17. Push the traverse rod left, and let the stone make a pass. When it reaches the stop, reverse the traverse rod to return the stone.

18. Turn the handwheel to raise the block 0.002 inch, then engage the traverse rod and the grinding head will take a 0.002 inch cut.

Reverse the traverse then repeat the last step, continue until the deck surface is even and level. Remember that each pass of the stone removes an additional 0.002 inch, do not cut more than the manufacturer recommends.

CYLINDER RECONDITIONING

Cylinder reconditioning is necessary whether you rebuild or overhaul the engine. As we explained in Chapter 6 of this *Shop Manual*, piston ring sealing affects performance directly by containing the combustion pressure, and indirectly by limiting oil consumption. In an overhaul, you will remove the ring ridge and deglaze the cylinder wall surface to allow new rings to be installed on the old pistons. In a rebuild, you will bore and hone the cylinders oversize to accommodate new pistons as well as new rings.

Deglazing Cylinders

Cylinder deglazing is a repair procedure that you use only when both the cylinder walls and the pistons themselves are in good condition. Deglazing roughens the surface of the cylinders without significantly changing their overall diameter, figure 11-39. A typical deglazing tool might remove only 30 to 40 *millionths* of an inch from the bore (less than half of one ten-thousandth). The rough wall surface is necessary to break in the new rings, and allow them to conform exactly to the cylinders and form a gas-tight seal. If you install new rings into the old, polished surface of the cylinders, they may never bed-in correctly. The engine will have poor compression, excessive blow-by, high oil consumption, and perform poorly.

Machinists commonly use a Flex Hone®, sometimes called a "ball hone", as a glaze breaker, figure 11-40. Another tool you can use resembles a hand-held hone body or honing head, figure 11-41, and some technicians even call them hones. It has three or four stones, held against the cylinder wall under spring tension. It differs from a true honing head in that the stones are free to pivot on the arms and can not be adjusted to an exact size.

A variable speed drill is used to power the glaze breaker. Operate the drill at low speed, between 200 and 500 rpm, as you stroke the tool up and down in the cylinder bore. Stroking creates the required cross hatch pattern on the cylinder wall. When deglazing cylinders, lubricate the stones with a light weight oil, never use solvents.

Remember that although the glaze breaker will remove cylinder wall glaze, it will not re-

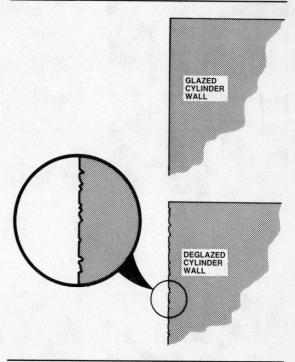

Figure 11-39. Deglazing provides a fresh crosshatched surface, the amount of metal removed is insignificant.

Figure 11-40. The Flex-Hone comes in sizes large enough to handle cylinder bores and small enough to fit into a hydraulic lifter bore.

store roundness or reduce taper in the cylinder. The deglazer simply follows the old tapered, out-of-round surface and gives it a better finish. You can use only cast iron rings in deglazed cylinders, hard chrome rings will not seat readily and the surface finish is too rough for molybdenum rings.

Cylinder Boring

If either cylinder taper or out-of-round are excessive, then refinishing with a glaze breaker will not bring the cylinders back up to standards. The engine may run well for a short time, but the rings will quickly wear out as they squeeze in and out to accommodate the tapered or out-of-round cylinder bore.

The correct fix is to bore the cylinders to oversize and install new pistons. As men-

Servicing Blocks

Figure 11-41. This glaze breaker uses stones that are mounted to spring loaded fingers.

Figure 11-42. Portable boring bars do an acceptable job if they are used very carefully.

tioned before, domestic engines typically use 0.030 and 0.060 inch (0.075 and 1.500 mm) as the standard overbore sizes, while metric engines often use 0.500 and 0.100 mm (0.020 and 0.040 inch) oversize. Of course, you can custom order any size pistons you are willing to pay for, but the only people who do that are race engine rebuilders who must keep engines under specified displacement limits.

Normally boring to oversize does not require any special compensation when you reassemble and tune the new engine. The existing fuel and ignition systems can handle the very slight increase in displacement and compression ratio that results, and the extra piston weight is negligible.

Boring machines

There are two main types of boring machines, the difference is in how they index to the cylinder block. One type is portable, and many machinists refer to it as a boring bar, figure 11-42. To use this machine, you bolt the block to a rigid surface, such as a bench or engine stand, the boring bar is then solidly attached to the block. This type of machine is totally dependent on the condition of the deck surface for its alignment. If the deck is warped, then the bar might cut the hole off-center, or it might lead the hole off slightly rather than straight down towards the crankshaft.

A more precise way to bore cylinders is to use a machine that mounts the block on a cradle and positions the boring head according to the pan rails and main bearing bores. This is the way cylinders are bored during the manufacturing process, and there are a variety of machines available that let you overbore in a similar manner, figure 11-43.

Regardless of how the block mounts in the machine, all boring equipment operates on the same principal. A single-point cutting bit clamps into a boring head at the bottom of a heavy steel sleeve. The boring head is centered in the bore using extendable fingers, figure 11-44. The machine rotates the cutting bit as it simultaneously feeds the boring head down through the cylinder, cutting an oversize hole.

How much should you bore?

When you bore a cylinder oversize, you will not use the boring machine to take the diameter out to the final size. The surface in a freshly bored cylinder is far too rough and irregular to seat piston rings correctly. Instead, you will set the tool bit to cut a diameter not quite as large as you need, and you remove the last material by honing. This way the cylinder is

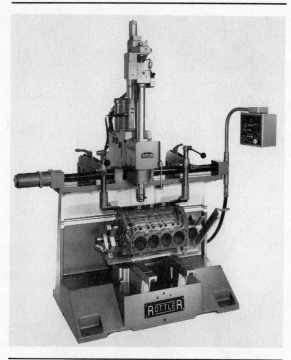

Figure 11-43. This computerized boring machine indexes on the pan rails to keep the bores straight and parallel. Once programed, the machine automatically bores all the cylinders to the same size.

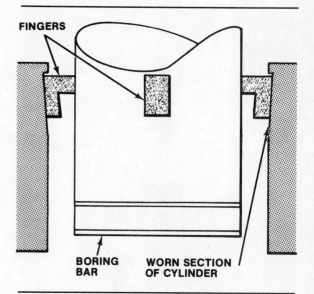

Figure 11-44. The extendable fingers center the boring head in the cylinder bore.

brought out to the final size and at the same time the surface is prepared for the new rings.

To an experienced, professional machinist, time is money, and the less time spent in each operation, the quicker the job is completed. In order to bore a cylinder 0.030 inch (0.075 mm) oversize, a machinist might set the boring head to cut all but the last 0.003 inch (0.075 mm). Then remove the last few thousandths from the bores with a honing machine.

For students with more time than experience, it is more important to perform the honing a little more slowly. You should therefore set the boring machine to a diameter about 0.005 in (0.125 mm) *less than* the final bore size required for the pistons. If the pistons are 0.030 inch (0.075 mm) oversize, set the boring machine to remove 0.025 inch (0.640 mm). This way you spend a few more minutes honing each cylinder, but you decrease the risk of removing too much metal.

Boring cylinders to oversize

As with all machining processes, you must be familiar with the particular equipment you will be using. Because of the wide variety of machines in the field, we will not detail step-by-step procedures for boring the cylinders. We will highlight some important considerations that you can apply to all boring jobs. Centering the cutting bit to the crankshaft centerline is your top priority. The boring tool must be square to the block or the cylinders you bore will not be parallel.

Another consideration is at what point of the cylinder you will center the boring head. Keep in mind that boring opens up a perfectly round cylinder of a precise diameter around the center of the boring head. The boring equipment indexes to the block in a worn out-of-round cylinder. When you adjust the tooling to the top of the cylinder, where most taper occurs, the bores you cut will not align to each other. This is because wear is never even on all of the cylinders of an engine. If you index to the bottom of the cylinder, where there is the least amount of wear, all of the cylinders may not clean up. This is because cylinders wear unevenly with most wear on the major thrust side, if you take a minimal cut the tool bit may not contact all the way around at the top of the bore just below the ridge. High-performance machinists set up the boring equipment using a special indexing fixture that bolts to the deck surface of the block. The fixture, aligns the cutting bit to bore the cylinders concentric to each other based on the head bolt position, figure 11-45.

There is not a right or wrong method for indexing the bore, the choice is more a matter of personal preference. The important point is to be accurate in your set up. You should have the new pistons on hand so that any minor inconsistencies in size can be taken into consideration before you make a cut. Bore each of

Servicing Blocks

Figure 11-45. An indexing plate, used in high-performance blueprinting operations, positions the boring head to the exact center of the cylinder.

the cylinders for a specific piston, mark the pistons so they are installed into the proper cylinder on final assembly.

Once the block is positioned, you can prepare the cutting bit. Keep in mind that the tool bit operates under extreme loads as it lifts chips of metal off of the cylinder wall. If the bit is not properly dressed, it cannot perform its task. Throw-away bits, the type with replaceable cutter inserts, must be in good condition, if not replace the insert. Standard cutting bits should be sharpened and lapped before each use. A precision tool sharpener, figure 11-46, can be used to dress the cutting and relief angles of the bit. Some boring machines have a built-in lapping fixture, an adaptor holds the cutting bit at the correct angle while you dress it, figure 11-47.

After you have dressed the bit, you place it in the tool holder, properly position it, then clamp it in place with a set screw. This special fixture uses a micrometer spindle to precisely set the cutting bit into the tool holder, figure 11-48. The tool holder is then fit into the boring head and adjusted to take the proper cut. A special gauging micrometer positions the tool holder with the correct overhang, a set screw locks it in position, figure 11-49.

Most boring machines have a feed stop that must be adjusted before you begin cutting. The feed stop automatically stops the machine when the cutter reaches the bottom of the cylinder. If the feed stop is not properly set, the tool can cut past the bottom of the cylinder and into the bearing webs, destroying the block.

The last set-up operation is to adjust the spindle speed and feed rate of the machine. Follow the equipment manufacturer's recommendations. The block material, as well as the type of cutting bit you are using, will determine what speeds to use.

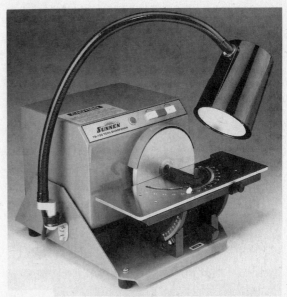

Figure 11-46. This precision tool sharpener is used to dress cutting bits, the unit is adjustable to grind various angles.

Figure 11-47. This lapping unit is part of the boring machine, the adaptor holds the bit at the correct angle to the lapping wheel.

Double check you adjustments before you begin. Switch the power on to start the boring head rotating, then, engage the feed control lever to begin feeding the boring head into the cylinder. As the bit cuts through the cylinder, the sound should be even and regular. You can tell by listening when the bit has cut all the way through the cylinder. After you have cut to the bottom of the bore, reposition the cutting bit so you can back the head out without scratching the bore. Repeat the procedure for the remaining bores.

Cylinder Honing

Cylinder honing is the final major machining step you will perform to the cylinder block.

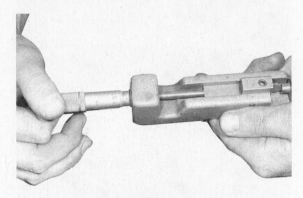

Figure 11-48. The cutting bit is installed in the tool holder in a special micrometer fixture, once positioned, the set screw locks the bit in place.

Figure 11-49. A micrometer gauge is used to install the tool holder in the boring head. Set the bit to cut the proper diameter, then tighten the set screw to hold it in place.

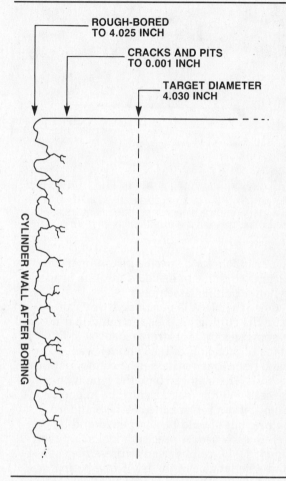

Figure 11-50. After boring, the cylinder surface is rough, pitted, and fractured to a depth of about 0.001 inch.

The main bearing bores will have been line bored or align honed, the block surface will have been decked, and each of the cylinders will have been bored.

Boring produces a straight, round cylinder, but the surface left by the cutting bit is not finished enough to seat the piston rings. As the cutting bit works its way into the cylinder it leaves behind a surface grooved with tiny jagged peaks and valleys that can be as deep as 0.001 inch (0.025 mm), figure 11-50. The surface also has a network of tiny fractures. These fractures result because cutting bits work by lifting and tearing away metal particles to open up the bore.

The surface left by boring is weak. If the engine were reassembled at this point: the rings and piston skirts will ride against the sharp peaks on the surface, without an oil film, and will quickly overheat and wear. Scuffing and local surface welding will result. The rings will also bend the sharp peaks over, trapping oil and abrasive metal particles behind them. The horizontal washboard grooves left by the cutting bit will fracture the ring's edges, interrupt any oil film on the cylinder wall, and hold oil away from the rings, letting it burn during the power stroke. The engine will suffer from poor compression, poor performance, excessive blowby, high oil consumption, and a short life. You will also have a very dissatisfied customer back at your shop in a short time.

Unless you correct the cylinder walls by honing, the engine will never perform satisfactorily. The cylinders must be honed down a minimum of 0.002 inch (0.050 mm) after boring to cut below the rough surface and provide an adequate finish.

Honing leaves a plateau surface that can support the oil film for the rings and piston skirts, figure 11-51. Honing also leaves the necessary cross-hatched pattern, figure 11-52. The pattern retains oil long enough to lap the piston rings to the cylinder walls forming a gas-tight seal.

Servicing Blocks

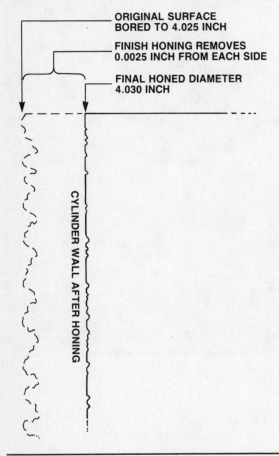

Figure 11-51. Honing opens the cylinder to final size, and leaves a plateaued surface that retains oil and laps the rings and cylinder walls together.

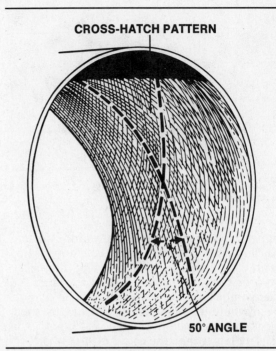

Figure 11-52. The cross-hatched pattern holds oil to keep the rings from wearing excessively, and also laps the rings against the cylinder wall for a gas-tight fit.

Actual piston size

Most cast pistons require about 0.002 inch (0.050 mm) of cylinder wall clearance, and most oversize pistons are actually not quite as large as their nominal size. Manufacturers often size their pistons slightly smaller than nominal size as a convenience to machinists. For instance, a piston for a four-inch bore labeled "0.030" on the top will usually not be exactly 0.030 inch oversize, but instead will probably measure about 4.028 inch across. To prepare an engine for oversize 4.030-inch pistons, you simply hone the bores to 4.030 inch, and the piston will automatically have the necessary 0.002 inch clearance. Not all manufacturers mark their pistons this way, so you must measure the pistons before you bore. Also keep in mind, the amount of wall clearance will vary depending upon piston design, the alloy used, and the intended use of the engine. Always provide the amount of clearance recommended by the manufacturer of the pistons you are using.

In fact, careful machinists routinely measure each piston individually, and then hone a particular bore for that piston. For instance, if the oversize piston actually measures 4.0275 inch, they will hone one bore to exactly 4.0295 inch, and use that piston in that bore. The next piston might measure 4.0285 inch, and they will hone a cylinder to 4.0305 inch for that piston, and so on. For routine rebuilding, this high a level of precision is unnecessary as either of these pistons would be within tolerance for a 4.030 inch bore. However, the extra precision provides an engine with well balanced combustion pressures and is a sign of quality craftsmanship.

Torque plates

Torque plates, also called deck or honing plates, are thick slabs of carefully machined metal that bolts to the deck surface of a cylinder block during honing, figure 11-53. Torque plates simulate the stress exerted on the engine block by the cylinder head being in position. Since correct gas sealing depends on round, straight cylinder bores for the piston rings to contact, any problems with that seal translate directly into poor performance, blow-by, and oil consumption. Race engine builders discovered that tightening a cylinder head onto a correctly honed block actually pulled the

Figure 11-53. Torque plates duplicate the distortions of an assembled engine during honing.

upper parts of the bores out-of-round, causing gas leaks and poor ring wear. To compensate for the distortion during machining, they developed the torque plate. Bolting a torque plate on distorts the block just like the cylinder head would if it were installed. The tensioned block can be honed through access holes, slightly larger than the actual cylinder bore, cut into the torque plate. When you remove the torque plate the block relaxes and the cylinder bores are no longer straight and round. However, the bores return to their correct shape when you bolt the cylinder head on during final assembly.

In the past, using torque plates was exclusively a blueprinting operation and machinists did not consider them necessary for routine operations. With the introduction of lightweight, thin-walled castings, all that has changed. Thin castings are much less rigid than the older heavy pieces, and the cylinder head bolts can easily pull them out of shape. Many modern engines are honed with torque plates at the factory, and you *must* do the same for routine rebuilding.

When you use a torque plate you also use a head gasket under it. The gasket must be just like the one you plan to install in the completed engine. In addition, be sure that the bolts you use penetrate to the same depth in the block as the stock head bolts would with a cylinder head installed. Some torque plates come with short towers above the bolt holes so that you can actually use stock head bolts to attach the torque plate.

Normally, the distortion caused by the head bolts is less than the 0.005 inch (0.125 mm) you leave after boring, so a torque plate is unnecessary until you begin the honing operation. However, if you are evaluating condition in a marginally tapered or out-of-round bore, make your measurements with the plate on the block. It might make enough of a difference to matter.

Figure 11-54. The Sunnen CK-10 is a top-quality honing machine widely used in the industry.

Automatic honing

The quickest and most consistent way to hone cylinder bores is to use an automatic honing machine, such as the Sunnen CK-10, figure 11-54, or the Rottler HP3A, figure 11-55. Automatic honing machines are quick and easy to use, and give you consistent results, time after time. They also automatically produce a mechanically-perfect cross-hatch pattern on the cylinder walls, and let you vary the exact angle of the cross-hatching according to the type of piston rings you will be using.

Some manufacturers provide instructions for using their honing machines to open up a worn and tapered cylinder all the way to the next oversize, without using a boring machine at all. To do this, you start with coarse, roughing stones, in the 80- to 150-grit range. These stones remove metal rapidly as you take the bore out to about 0.002 to 0.003 inch (0.050 to 0.075 mm) under the finish size. Once the rough cut is completed, then you switch to finer finishing stones, in the 220- to 280-grit range. These finish stones remove the last few thousandths, and provide a surface finish that is ready for service. If you choose to use a honing machine instead of a boring machine, be sure to switch to the finishing stones for this final operation.

The machine we illustrate is the Sunnen CK-10. The CK-10 has a number of features that

Servicing Blocks

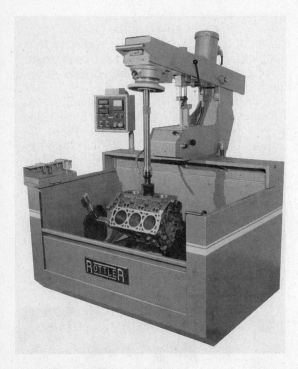

Figure 11-55. The Rottler HP3A is electronically-controlled honing machine that you can program for bore size and feed rate.

Figure 11-56. The CK-10 controls let you modify honing action to correct cylinder taper without stopping the machine to measure.

make it easy to use both for traditional finish honing after cylinder boring, and for complete cylinder resizing. We have already covered cylinder resizing with the boring machine, so all of the features of the CK-10 will not be covered here. We only explain the procedure you follow to correctly finish hone a cylinder after rough sizing with the boring machine.

On the CK-10, you can set the controls so that the machine automatically stops after it removes the amount of stock you specify with the indicating ring. The control panel has an emergency stop button, load meter, dwell control, and an indicator light, figure 11-56:

- **Load meter.** The load meter indicates the force that the stones exert on the sides of the cylinder wall. During normal honing, the gauge should read between 30 and 60 percent, but it will fluctuate if the cylinder is tapered. Less than a 10 percent variation means a cylinder is acceptably straight, within 0.001 inch (0.025 mm).
- **Indicator light.** The indicator light flashes at the instant the honing head reaches the bottom of the cylinder. Watch where the load needle points when the light flashes:
  — If the needle reads *high*, the *bottom* of the cylinder is narrow.
  — If the load reads *low*, the *top* of the cylinder is narrow.

- **Dwell control.** To correct taper, the dwell control allows the honing head to remain at the bottom of the cylinder for several revolutions. The load will begin to balance out after a few dwell cycles. You can either turn the pointer on the control panel for multiple dwells, or just press the green button for a single dwell. Dwell control is used only when honing tapered cylinders that were not previously bored.

Always hone with the main bearing caps in position, and use a torque plate if required. When you hone cylinders, the area around the cylinder heats up because of friction. Never hone adjacent cylinders right after each other. Instead, hone every other cylinder. This allows the metal to return to normal temperature.

Be sure to select the correct stones for honing the cylinders. The manufacturers provide extensive instructions for choosing the stones for their machines, and for using them properly. Take advantage of their knowledge and read the instructions. In general, for cast iron or chromium-faced rings use 220 grit stones, molybdenum-faced rings require a smoother surface finish and 280 grit stones.

Stones also vary in length and the length of the stones being used must be considered when setting up the machine. Be sure the correct scale for the stones you will be using is mounted to the machine. The amount of overstroke, how far the stones protrude from the bore at the top of the stroke, also varies for the length of stone being used.

Stone feed rate is another adjustment you will need to make. This is done by rotating the selector cover to the proper numbered setting, the higher the number the faster the feed. Most cast iron automotive blocks can be rough honed at the #6 setting, and finish honed at the #4 setting. For faster stock removal select a higher number, a lower number will give you longer stone life and a smoother finish.

Most blocks can be honed using standard stones, guides, and shims. However, if after the overstroke has been set the main guide contacts the bearing web of the block at the bottom of the stroke, the guide will have to be modified. This can be done by removing part of the guide shoe backing plate, bronze strip, that protrudes below the holder.

An engine block is fitted into the honing machine, leveled and adjusted in much the same manner as for other machining operations. Because of the similarity, these steps will not be detailed here. The following procedure details the finish honing operations performed on a freshly bored cylinder. We do not cover the steps required to remove taper from the cylinders. Once the block has been bored and properly mounted into the honing machine, proceed as follows:

FINISH-HONING THE CYLINDER BORES

1. Fit the setting gauge into a cylinder, center it, tighten the knob snugly, and move the graduated slide to zero.

2. Select the correct length and grit stones. Inspect the stones, they must be in good condition. Do not use stones that are chipped or cracked.

3. Place a stone on the gauge and add shims to take up slack. A properly shimmed stone will slip in and out of the gauge easily. Shim both stones to the same thickness.

4. Place a guide in the gauge, move the slide until the pin contacts the guide. Install the correct shims as indicated by the guide. Shim the other guide the same way.

5. Place an alignment guide in the gauge and zero the gauge. Loosen the set screw, adjust the bars to rest on the gauge shoulder, and tighten. Then set the other guide.

6. Install the stones into the honing head. The stone with the greatest amount of wear goes in position #1, the other stone fits position #2 on the opposite side.

Servicing Blocks

FINISH-HONING THE CYLINDER BORES

7. Place the main guide in position #3 and the centering guide in position #4 of the honing head.

8. Complete assembling the honing head by inserting the two small alignment guides in the upper part of the fixture in line with the stones.

9. Measure the length from top to bottom of the cylinder to set the honing stroke. Reach into a cylinder with the hooked ruler and record your reading.

10. Rotate the feed wheel all the way to the right, squeeze the stones together until they are fully retracted.

11. Position the hone head over the cylinder, push the centering guide against the spring loaded pin, and lower the head into the bore using the lift lever.

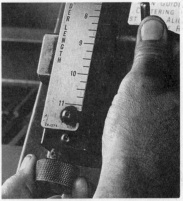
12. Open the right side guard to expose the stroke adjuster. Loosen the adjustment bolt, and turn the knob until the index marks indicate the right length, tighten the knob.

13. With the honing head at the top of its stroke, adjust the cradle with the elevating crank so the stones protrude about 3/8 inch (10 mm) from the top of the cylinder.

14. Check that the hone clears the bearing webs at the bottom of the stroke. Then, select the honing speed and stroking rate using the chart on the machine.

SPINDLE SPEED & STROKING RATE						
	CYLINDER DIAMETER					
	UP TO 3½"		3½ TO 4½		4½ & OVER	
CYL. LENGTH	UPPER BELT	LOWER BELT	UPPER BELT	LOWER BELT	UPPER BELT	LOWER BELT
UP TO 5"	C	D	B	D	A	D
5" TO 7"	C	E	B	D	A	D
7" TO 9"	C	E	B	E	A	D
OVER 9"	C	E	B	E	A	D

15. If necessary, loosen the tensioner, and adjust the belts. Use the upper belt to set the honing speed, and the lower belt to set the stroking rate.

FINISH-HONING THE CYLINDER BORES

16. Turn the feed wheel to the left until you feel a slight resistance — you want the stones just snug in the bore. Shake the drive tube to seat the stones.

17. Raise the front splash panel, start the motors, and position the oil spout to direct a good flow of cutting oil into the cylinder to be honed.

18. Now slowly pull the clutch control lever forward to engage the drive belts, it will click into position and start the machine stroking.

19. Watch the load meter, adjust the feed wheel to keep readings at 30 to 60 percent. Hone for about 30 seconds, push the clutch control lever to stop the machine.

20. Remove the hone and check your progress with the dial bore gauge. Continue honing, checking regularly, until the cylinder is opened out to the finish size.

21. After the cylinders have all been honed, use a chamfering cone to finish the top edge of the bores.

Honing hints
Experience is the only way to develop a feel or technique for any machining operation. The more familiar you are with a precess, the easier it becomes. Listed here are some helpful tips to develop your honing technique:

- **Compensating for stone wear.** When setting the feed dial you must compensate for stone wear unless you are using diamond bonded stones. A starting point is to figure the stone will wear at about half the rate of the stock being removed during rough honing. For finish honing, the stones will wear faster. Expect a stone wear to stock removal ratio of about 1 to 1 when finish honing. The stones will wear slower in the first cylinder you hone because they true themselves to conform to the cylinder.

- **Finishing with fine grit stones.** Stones that are 280 grit and finer require precise stone pressure to operate. Low pressure will cause glazing, high pressure will cause the stones to crumble. Advance the feed handwheel smoothly and at a moderate speed to regulate pressure.

- **Consistent high load readings.** If the load meter consistently reads over 90, reduce the feed rate by selecting a lower number with the selector cover, or switch to a softer stone.

- **Stones break down too fast.** Start honing at a lower load rating, always apply load after rotation has started. If stone wear is still excessive, set the selector cover to a lower number or use harder stones. If you get poor stone life at low loads, less than

Servicing Blocks

60 while roughing or less than 20 while finishing, increase the spindle speed.
- **Stock removal rate is too slow.** Increase the spindle speed — you must keep load rating within range. Set the selector cover to a higher number, or switch to softer stones to maintain load rating.
- **One stone wears faster than the other.** Switch the stone positions in the honing head to equalize wear.

Hand honing

Some very experienced technicians still hone cylinders by hand, but this is a dying art in both production and general machine shops. Hand honing equipment consists of a honing head, similar in appearance to the ones in an automatic honing machine, that fits into the chuck of a drill motor, figure 11-57.

Using a hand-operated honing tool requires a great deal of artistry. With a hand-held hone you must vary the stone's position while stroking to correct taper. You must also alter the stroke rate or the drill speed to obtain the correct cross-hatch pattern. As with an automatic honing machine, you can use roughing stones to open the cylinders up to the next oversize, and then prepare the final surface with finishing stones.

Professionals who use hand-held hones rely on the sound of the drill motor as a guide to determine cylinder condition. The motor slows when the hone is in the narrow part of the cylinder, and speeds up as it moves into the wider bore at the top of the cylinder. As taper is removed the drill speed evens out, and as the bore approaches finish size the scraping sound of the stones diminishes.

Reconditioning Aluminum Cylinder Bores

Most of the engine blocks you will work on will be either cast-iron or aluminum with cast-in or pressed-in iron sleeves or liners. However, some light-weight engine blocks use no liners at all — the pistons bear against a specially prepared aluminum cylinder wall. These are often referred to as "linerless" engines.

Most linerless alloy engine blocks use a special hypereutectic casting process that distributes small particles of very hard, pure silicon throughout the aluminum. During production machining, the manufacturers use an acid etching process to dissolve away the soft aluminum from the surface layer of the cylinder bores. This leaves only the exposed silicon particles to support the piston and to keep it from wearing the aluminum away, figure 11-58. The

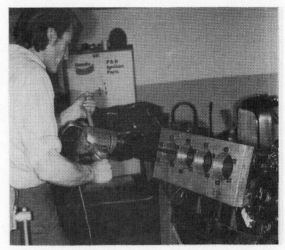

Figure 11-57. A drill motor provides the power for hand honing the cylinder bores, note the torque plate bolted into position.

piston skirts are generally plated with a thin layer of iron to protect them from the abrasive silicon particles.

Machining limitations

Aluminum cylinder blocks can not be machined in the same way that cast iron can. Hypereutectic castings have to be reconditioned so that the silicon particles stand away from the cylinder walls to prevent the aluminum from contacting the piston skirts and rings. After you open up the bore, you perform a lapping process to scrub the aluminum away from between the silicon particles.

An alternative method is to fit the block with iron sleeves rather than trying to duplicate a factory bore. If you choose to recondition to stock specifications, be sure to check with both the engine and the machine equipment manufacturers for specific recommendations.

Although you can open up the bores using a boring bar, most manufacturers recommend you perform the entire resizing job using an automatic honing machine. If you use a boring bar, it is *essential* that the cutting bit is extremely sharp and freshly lapped. If the bit is dull, it will gouge chunks of silicon from the bores instead of clipping them off cleanly. If this happens, the cylinder wall is destroyed and the block must be sleeved or scrapped.

Following is the procedure Porsche recommends for reconditioning cylinders in their hypereutectic engine blocks. The entire operation is performed using a Sunnen CK-10 honing machine. Keep in mind that this is a brief summary of a detailed procedure. Our intent is to acquaint you with the process, *not* to substitute for factory information.

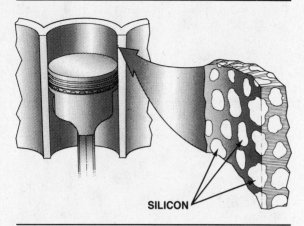

Figure 11-58. In a hypereutectic cylinder bore, the iron-plated pistons ride only on the protruding silicon particles.

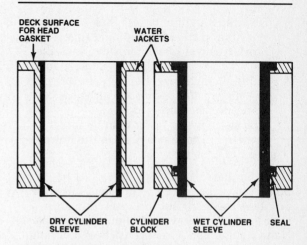

Figure 11-59. Cylinder sleeves, both the wet and dry types, are replaceable liners. Dry sleeves can sometimes be installed to repair a iron block that you might otherwise have to scrap.

Refinishing Procedures

The refinishing procedure is a four step process. The bores are enlarged in three stages with different grit stones, the final step is the lapping process that removes aluminum and exposes the silicon. Before mounting the block in the honing machine, bolt on the main bearing cradle and torque to specification.

Your first step is to open up the bores with a course-grit roughing stone. Be careful, these roughing stones cut quickly. Your rough cut should enlarge the bore to within 0.004 inch (0.10 mm) of the finish diameter.

After completing the rough cut, replace the course stones in the cutting head with fine-grit finishing stones. Honing with the finishing stones will bring the bore diameter to within 0.001 inch (0.02 mm) of final size.

The last 0.001 inch (0.02 mm) of metal is removed by honing with extremely fine-grit polishing stones. These stones are specifically designed for working on aluminum and are much finer than what you use in a cast iron bore. If the shop you work in does not normally machine aluminum blocks, the correct stones will not be readily available. Do not make substitutions — you must follow the factory recommendation.

At this point, the cylinders are the correct diameter, but the surface layer is a mixture of aluminum and silicon. You must perform a final lapping procedure to scrub the aluminum down below the level of the silicon. The lapping process removes aluminum, but not silicon — no further bore enlargement occurs.

Before you begin lapping, wash the cylinder walls with filtered honing oil to remove all trace of abrasive, then wipe the walls dry with a clean rag. Replace the grinding stones and guide shoes in the honing head with the special felt lapping pads. Use the setting gauge to adjust the pads to the final diameter of the cylinder bore. Soak the pads and coat the cylinder wall with silicone grinding compound. If the silicon mixture is too thick, it can be diluted by thoroughly mixing with fresh honing oil. Fit the honing head into the cylinder, and adjust the machine to operate at 185 rpm with a load rating of 20 to 30 percent, switch the automatic oiler off. Turn the power on and allow the pads to lap the cylinder for 80 seconds. Be sure the oiler is off — if not, it only washes the grinding compound off the pads and cylinder walls. Coat the pads with fresh silicon compound before lapping the next cylinder.

The abrasive-covered pads wear down the aluminum from the surface without affecting the silicon. The final surface will have a dull, gray finish that will not look anything at all like the standard cross-hatch you need for traditional cast iron cylinders. Finish off by cutting a 30 degree chamfer into the top of the cylinder bore.

Cylinder Sleeving

Some engines use replaceable cast iron liners in their cylinders called sleeves. The pistons ride against the sleeves, rather than against the block casting itself, figure 11-59. As discussed in the *Classroom manual*, cylinder sleeves can be wet or dry. Both types look similar when out of an engine.

Servicing Blocks

Figure 11-60. Leave a small step at the bottom of the cylinder to support the base of the sleeve.

Figure 11-61. If the sleeve does not already have a chamfer on the lower edge, cut one yourself.

A dry sleeve is a thin-walled liner that is pressed into the cylinder with a 0.001 to 0.002 inch (0.025 to 0.050 mm) interference fit. The metal in the cylinder wall casting of the block supports the sides of the sleeve, and the interference fit ensures that heat passes quickly into the cylinder block. Many aluminum-block engines use iron liners to get the advantage of both long-wearing cast iron cylinders and a lightweight aluminum block. Worn dry sleeves are removed by machining them down with a boring bar, the sleeve is bored out until only a thin shell remains. The thin wall of the liner can then be carefully pried away from the block and crushed as you lift it out. If prying results in damage to the block, you bore the block to accommodate an oversize sleeve.

A wet sleeve is a sturdier tube that is supported by the block only at the top or the bottom. Engine coolant flows around the outside of the liner and is kept out of the crankcase by O-rings or gaskets. To repair an engine with wet sleeves, you bore out the supporting flanges in the block that hold the liner in place, then remove the sleeve. Wet sleeves are often removed with a special tool that fits down the bore and is tightened to firmly grip the liner, the tool is then attached to a slide hammer, and the sleeve is removed by working the slide. Once the old sleeve is removed, fit a replacement with new O-rings or gaskets.

Aside from repairing an engine block by removing a worn or damaged sleeve and installing a new one, sleeving a block with one or two iron liners is an economical fix for repairing partial damage, such as a single cylinder with cracks or serious gouges.

When you do choose to re-sleeve a cylinder bore, make sure you do the necessary boring and pressing before working on any of the other cylinders. The sleeve tends to distort the block slightly, and if you finish the other cylinders first, they will no longer be round after you install the sleeve.

Recent manufacturing and material developments have led to the use of carbon fiber and ceramic compounds, rather than cast iron, cylinder sleeves. Although rare at the present time, these composite liners are bound to become more common in the future as the technology develops. Working with these materials requires special considerations. They can often be bored to oversize or replaced, but you must follow the manufacturer's specific procedure. The following repair methods apply only to cast iron cylinder liners.

Sleeving

The following procedure is a typical method used to repair an engine block by sleeving:

1. Carefully measure the sleeve diameter with an outside micrometer. Check the diameter in several locations.
2. Bore the cylinder with the roughing bit to a diameter 0.002 (0.050) inch less than the diameter of the sleeve. This will give the 0.0015 to 0.0025 interference fit you need to support the sleeve and prevent the engine from overheating.

Figure 11-62. A sleeve can be driven into the block by hand using a broad flat driver and a large hammer.

Instead of boring all the way through the cylinder, leave a 1/4-inch step at the bottom, figure 11-60. This will prevent the sleeve from coming out the bottom when the engine is running. The cylinder head will hold it in at the top.

3. Cut a chamfer on the lower edge of the sleeve to prevent it from galling as you press it in, figure 11-61.
4. Coat the sleeve with high pressure lubricant and install it in the block in one of two ways:

   - On V-style engines, you can hammer in the sleeve with a broad, flat driver, and a sledgehammer, figure 11-62.
   - On inline engines, you can set the block in a press so that it rests on the pan rails, figure 11-63. Press in the sleeve from the top, and seat it against the stop at the bottom with about two tons of pressure.

You can make the process easier if you heat the block in a thermal oven, and refrigerate the liner before installation.

5. Cut off the excess sleeve at the top. You can do this with a hacksaw, or with a special cutoff bit in the boring machine.
6. Resurface the cylinder block in the usual way to blend the top of the sleeve with the block's deck surface.
7. Bore and hone the cylinder to its final size.

LIFTER BORE RECONDITIONING

Lifter bore reconditioning is the last of the block machining operations. In a rebuilt OHV engine, it is critical that non-roller hydraulic lifters rotate freely in their bores during the first few minutes of operation. This rotation allows the lifters to seat properly onto the new

Figure 11-63. A large hydraulic press is a quick and easy way to install sleeves into an inline engine block.

camshaft. As explained in the *Classroom Manual*, the bottoms of the lifters have a slight crown perhaps as much as 0.002 inch (0.050 mm). The nose of the individual camshaft lobes are also ground to a slight taper. As a result, new lifters do not sit squarely on the lobes of a new camshaft.

Crown and taper encourage the lifters to spin in their bores when the engine is running. This helps to equalize wear symmetrically on the bottom of the lifter and the surface of the camshaft lobe. Without the spinning, the camshaft and lifters will be ruined in just a few minutes of running.

When rebuilding an engine you need to dress the lifter bores so that the lifters will spin freely when the engine is started. This can be done by honing the bores using a drill motor and a small glaze breaker. Remove any rust, burrs, or glazing that might prevent the lifter from rotating when the engine is started.

You do not need to resize the hole — making it any larger than stock will let oil leak past the lifter, resulting in noise, poor valve action, and wear. Besides, oversize lifters are seldom available, and re-sleeving a lifter bore is an unnecessary expense.

Chapter 12

Servicing Crankshafts, Flywheels, Pistons and Rods

In the last chapter, we covered the steps necessary to recondition the cylinder block and return it to service. We now turn to the crankshaft and reciprocating parts attached to it. In this chapter we detail inspection and reconditioning procedures for the crankshaft, pistons, connecting rods, and related parts.

CRANKSHAFT SERVICE

The crankshaft is the heart of the engine, and reinstalling a crankshaft in poor shape without proper reconditioning can quickly and thoroughly destroy a new engine. Reconditioning includes disassembly, cleaning and inspection, crack detection, journal grinding, straightening, and balancing.

Crankshaft Disassembly

Disassembly of the crankshaft is fairly simple, most are a single casting or forging. If the crankshaft has threaded sludge trap plugs sealing the ends of the oil drillways, you must remove them. Remove them in the same way that you removed the plugs from the engine block. This will allow the cleaning solvent to flow freely through the oil passages.

Remove the pilot bearing from the rear of the crankshaft, figure 12-1. This bearing fits tightly into the back of the crankshaft to center the input shaft of the transmission and to distribute the side loads caused by disengaging the clutch. Pilot bearings are generally one of two types, either a solid bushing or a sealed ball bearing. Ball bearing type pilots are usually held in place with a snap-ring. Once you remove the snap-ring, the bearing will come out of the crankshaft easily.

Either type of pilot can be removed using a slide hammer fitted with a blind bearing puller adaptor, figure 12-2. The adaptor is slipped through the inner bore of the bearing and expanded to provide a firm grip. Attach the slide hammer and work the slide to pull the bearing out. All pilot bearings are installed with an interference fit and can require a considerable amount of force to remove.

An alternative time-honored removal technique is to tightly pack the space inside and behind the bearing with very heavy grease, and then insert a close-fitting bolt or rod through the hole. Strike the end of the bolt with a hammer and the hydraulic lock from the grease behind the bearing will lift it out.

You can also tap threads into the inside of a bushing, screw in a bolt, and then pull the bolt and bushing together with a slide hammer. Whatever method you choose, be careful. If

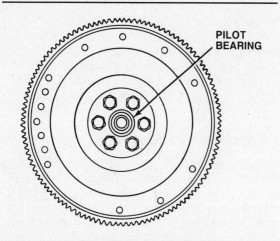

Figure 12-1. The pilot bearing, or bushing, must be removed from the rear of the crankshaft.

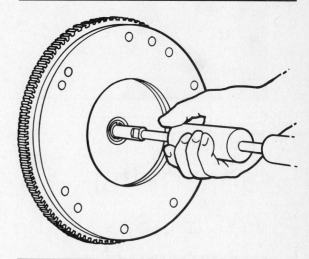

Figure 12-2. A slide hammer with a blind bearing puller adaptor makes pilot bearing removal easy.

you damage the pilot bearing bore, you cannot always repair it easily.

Clean the crankshaft using one of the methods detailed in Chapter 9 of this *Shop Manual*. After cleaning, spray the crankshaft with a light oil to prevent corrosion.

Inspecting Main Bearings

We discussed bearing inspection in Chapter 9 of this *Shop Manual*, and will not go into detail here. Evaluating the main bearing insert wear will help to determine what areas of the crankshaft will require close examination. In general, most of the same wear patterns can occur on both connecting rod and main bearing inserts. The primary difference is that main bearings receive their worst shock loads on the *lower* bearing shell, while connecting rod bearings receive their worst shock loads on the *upper* bearing shell.

Crankshaft Inspection

The areas that you must pay attention to when considering a crankshaft for reconditioning and reuse include:

- Threads, keys, keyways, and pilot bearings
- Cracks
- Journal and thrust bearing condition
- Straightness, including snout and flange runout.

Threads, keys, keyways, and pilot bearings

Look over the crankshaft for damage that might have occurred in the past or during disassembly. You should replace damaged keys, and dress the edges of burred keyways with an oilstone. Check the threaded holes for stripped threads and broken fasteners. Finally, look into the rear of the crankshaft and inspect the pilot bearing bore. If you plan on installing the crankshaft into an engine with a manual transmission, this bore must be round and undamaged. Pilot bearing bore is not as critical on engines with automatic transmissions because the torque converter bolts to the flexplate and there is no load on the pilot.

Cracks

We explained crack detection in Chapter 10 of this *Shop Manual*, and you can use most of those techniques as easily for a crankshaft as for a cylinder head or block. Cast-iron crankshafts normally announce the appearance of a serious crack by vigorous self-destruction, so you will seldom find a cracked cast-iron crankshaft in a good-running engine. Forged-steel cranks are much stronger, and can occasionally run for quite some time before a developing crack finally gets bad enough to stop the engine. Whether the crankshaft is cast or forged, a serious crack — such as in the fillet area — is a fatal flaw. Toss the shaft in the scrap bin and locate a replacement.

As a point of interest, you can sometimes detect cracks in a one-piece forged crankshaft by performing a sound check. Once you remove the woodruff keys, balancer, flexplate, and bolts, tap on the crankshaft with a ball-peen hammer. A solid forged crankshaft will ring like a bell, whereas cast cranks and forged cranks with a serious crack will not. This technique should not be used as a substitute for a thorough inspection, but it can be used to cue you to serious damage so you can avoid wasting further inspection time on a hopelessly damaged part.

Servicing Crankshafts, Flywheels, Pistons and Rods

Figure 12-3. Measure each crankshaft journal with an outside micrometer, take three readings, one at either end and one at the midpoint.

Figure 12-4. Take a second set of measurements 90 degrees around the circumference of the journal from the first.

Journal and thrust condition
Your next step is to inspect the journal areas of the crankshaft. A slight discoloration is normal, excessive discoloration, scoring or pitting indicate a problem. Small localized nicks, such as caused by improperly removing the pistons, are not critical and can be treated with a fine hand-held stone. Use the stone to smooth out any burrs and ridges, you will not be able to clean up the entire depression, you simply want to true the surface.

If the entire journal is scored the crankshaft will have to be polished or ground to undersize. A quick way to check the crankshaft journal condition is with a copper penny. Take the penny and scrape it across the journal several times, if the shaft picks up copper from the penny, then it is too rough and must be reground. Be sure to use a fairly new penny, the layer of copper is very thin, and you can easily scrape through it to the zinc underneath.

The next step is to measure each main and connecting rod journal with an outside micrometer. You will be checking for diameter, taper, and out-of-round. To get an accurate picture of wear, you will have to measure each journal at three locations: near the front and rear fillet radii, and at the midpoint of the journal, figure 12-3. In addition, take two readings at each location. The second measurement should be 90 degrees around the circumference of the journal from the first, figure 12-4.

Crankshaft journals that have been worn out-of-shape cannot provide adequate oil clearance for the bearings. Putting a worn shaft back into service will lead to premature failure. It is important to measure at all three locations because different conditions will result in different wear characteristics. Three common bearing journal wear patterns are:

- End-to-end taper, wear is greatest at one end of the journal, figure 12-5.
- Hourglass taper, wear is greatest in the center of the journal, figure 12-6.
- Barrel taper, wear is concentrated near both of the fillets and minimal at the center, figure 12-7.

Measuring diameter will also allow you to determine whether or not the crankshaft was reground to an undersize in a previous rebuild. Typical crankshafts are ground to 0.010, 0.020, or 0.030 inch (0.025, 0.050, and 0.075 mm) below the standard value.

Compare the measurements for taper and out-of-round to the manufacturer's tolerance for crankshaft journals. These values vary between manufacturers and must be followed,

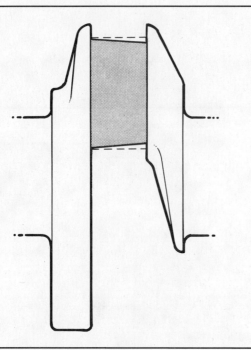

Figure 12-5. Crankshaft journal wear with end-to-end taper.

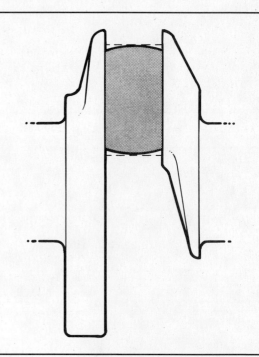

Figure 12-7. Crankshaft journal wear with barrel taper.

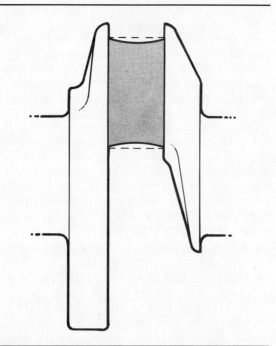

Figure 12-6. Crankshaft journal wear with hourglass taper.

you should not use rules-of-thumb. In general, older domestic engines tolerate up to 0.001 inch taper and out-of-round in either the main or the connecting rod journals. Modern limits are much tighter, especially for high-output engines: usually only 0.0002 to 0.0005 inch (0.0051 to 0.0125 mm), and sometimes even less. Even if the crankshaft journal has no taper or out-of-round, any wear that reduces the diameter to 0.001 inch (0.025 mm) below standard will condemn the journal.

If a journal is out of specifications for diameter, taper, or out-of-round, then all the journals should be reground to the next undersize. If only a single journal is bad, you can sometimes save the customer money by building up that journal by submerged-arc welding and then grinding it to standard size. However, the usual technique is to resize all the journals (both main and rod) to the next undersize. By doing this the machinist can be sure that all journals index to the same centerline, all the strokes are equal, and the bearings will be a matched set.

You can also save a severely damaged crankshaft by building up *all* the journals by welding and then grinding them to standard size. However, this is extremely time consuming, and often more expensive than buying a usable crankshaft core. Save this technique for the customer with an extremely rare engine or a custom one-of-a-kind stroker crankshaft.

Look over the thrust bearing surface of the crankshaft, figure 12-8. It should be smooth and free from scoring or galling. If you find that crankshaft endplay is insufficient during assembly, it is possible to sand the thrust bearing insert slightly to increase it. If, however,

Servicing Crankshafts, Flywheels, Pistons and Rods

Figure 12-8. Inspect the thrust bearing surface, it must be smooth and free from scoring or galling.

Figure 12-9. Checking crankshaft straightness using V-blocks and a dial indicator.

the endplay is excessive, and no undersize thrust bearings are available, the only alternatives may be to scrap the crankshaft or to build up the surface and then grind it to size.

Straightness, snout, and flange runout
Crankshafts turn at phenomenal rates, at 6000 rpm the crankshaft rotates 100 times each second. If the crankshaft is not straight, the resulting pulses will wear the main bearings asymmetrically. Just as installing a straight crank and new bearings into a warped block can cause rapid failure, installing a bent crankshaft into a reconditioned block can give the same results. Crankshafts usually bend in a bow-shape, with the greatest amount of deflection in the middle.

You should check crankshaft straightness using a dial indicator, with the crankshaft supported on V-blocks, figure 12-9. You can also install the upper end main bearing inserts in a straight block, and use that setup to substitute for V-blocks. Set the dial indicator so that the plunger contacts the center main bearing journal, zero the dial indicator, and rotate the crankshaft one complete turn. The total indicated runout (T.I.R.) is the sum of the greatest reading above zero, and the greatest reading below zero. It should not be greater than specifications. Like taper and out-of-round, the value varies from engine to engine, typically between 0.001 and 0.002 inch (0.025 and 0.050 mm). Remember that a 0.001 inch (0.025 mm) bend will produce a T.I.R. of 0.002 inch (0.050 mm) on the dial indicator.

When you check for T.I.R. on a crankshaft by using V-blocks, the rise and fall of the dial indicator plunger indicates *both* crankshaft runout and out-of-round on the journal. For instance, a journal worn 0.002 inch (0.050 mm) out-of-round and bent 0.002 inch (0.050 mm) in the same direction will have a T.I.R. of 0.004 inch (0.10 mm). Sometimes, but not always, the manufacturer's specifications allow for this. In order to cut the journals along the same centerline when reconditioning, you must correct the bend first, before correcting the journal shape.

If you have no runout specifications, a rule of thumb is that the total indicated runout of a crankshaft should not exceed half the specified main bearing clearance. For instance, if the oil clearance is 0.002 inch (0.050 mm), then total runout should be less than 0.001 inch (0.025 mm). This assumes that the bearing surfaces are straight and within specifications.

Once you have checked the bearing journal runout, you should check snout and flange runout. These are checked with a dial indicator in the same manner as the bearing journals. Rotate the crankshaft one complete turn, and compare the T.I.R. against specifications. If either reading is out of the tolerance range the shaft will have to be straightened.

Crankshaft Straightening

If your inspection reveals that the crankshaft is bent beyond specifications, then you must either find a replacement or straighten the bent shaft before any machining operations are performed. Crankshaft straightening used to be limited only to forged cranks, but there is no reason not to straighten a modern high-quality cast crank. The risk of breaking a crankshaft is always there, whether cast or forged.

It is important that you determine the T.I.R. of all the journals before you begin the straightening operation. To straighten a bent crankshaft you always begin at the rod journal that is farthest out of specification, then work

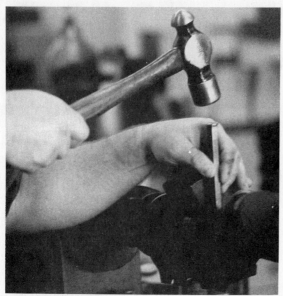

Figure 12-10. Bent crankshafts can often be straightened by striking them with a hammer and a rounded bronze-tipped chisel.

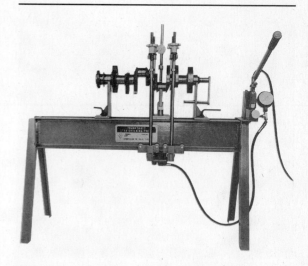

Figure 12-11. A crankshaft straightening press uses hydraulic pressure to remove a bend.

your way down the shaft in either direction until the entire bend has been removed. Crankshaft bend is always corrected at the rod journals, never at the main journals.

There are two methods for straightening a crankshaft, both require that the shaft be supported by the end journals in V-blocks. Additional V-blocks, or adjustable clamps are positioned to support the main journals adjacent to the rod journal being straightened. Adequate support is important to avoid overcorrecting the bend.

Once the crankshaft has been properly supported, a bend can often be removed by peening. This is accomplished by striking the crankshaft a sharp blow with a hammer and a rounded bronze-tipped chisel, figure 12-10. Dress the end of the chisel so that it fits the fillet radius, this will prevent unwanted marring on the journal. Position the crankshaft so that the highest point of deformation is facing down. Be sure the adjacent main journals are well supported. Place the chisel into the fillet radius of the journal, then strike the chisel a hard sharp blow with your hammer. This will straighten the crankshaft slightly, succeeding blows will bring it back into alignment. Check your progress frequently, one too many hammer blows will bend the shaft in the opposite direction. Obviously, crankshaft straightening is a *highly-skilled task*, requiring a certain touch that can only be developed through experience. Practice on some scrap crankshafts before you attempt to straighten one from a customer's engine.

An alternative method is to mount the shaft in a straightening press and bend it with a hydraulic ram, figure 12-11. The press has V-blocks to support the end main journals and adjustable clamps that support the main journals adjacent to the bend, position the ram, and mount a dial indicator. Press straightening will only work with a forged crankshaft, the extreme pressure can break a cast shaft.

To use this machine, you lay the crankshaft in the V-blocks and use the dial indicator to measure the T.I.R. for each journal separately. When you know which journal is the farthest out of line, you slide the clamps in to support the main bearing journals on either side. Rotate the crank so that the maximum bend is down, position the clamps to hold the crank down as you apply hydraulic pressure from below. Once the crankshaft is set up and pressurized, use your blunt chisel to strike the fillet radius of the bent journal in the direction of the bend to relieve stress. After you remove the bend, check the other journals to see whether those bends were decreased, and if additional straightening is required.

Crankshafts normally bend in a bow shape, with all the deflections pointing more or less in the same direction. Once you correct the largest bend, the others will usually be decreased as well, and perhaps removed completely. If the crank still has other bends between other journals, you can simply slide the clamps over and remove those bends independently.

Runout on the ends of the crankshaft can be corrected by careful use of the round-nosed bronze chisel and hammer.

Servicing Crankshafts, Flywheels, Pistons and Rods

NOTE: When you straighten a crankshaft, either by peening or pressing, you do not want to bend the shaft back into shape, you simply want to relieve the stress tension built up in the metal. Once stress has been relieved, the shaft will have a tendency to return to its original shape. Always strike a bent crankshaft in the direction of the bend. Although it seems this would make the bend worse, it actually relieves stress allowing the journals to rebound back into position.

Building Up Crankshaft Journals

Once you straighten the crankshaft, any severely damaged or undersized journals must be brought back to size before they can be finish ground or polished. The journals of most crankshafts, both forged and cast, can be brought back to size using the submerged-arc welding process. A specialized piece of equipment is required. The crankshaft welder is the wire feed type that uses flux rather than inert gas to displace oxygen. Although modern crankshaft welders are automated and electronically controlled, crankshaft repair still requires an expertise that can only be developed through experience. Following is a general description of how journals are repaired by submerged-arc welding.

Before welding prepare the crankshaft by straightening to remove any bend and by pre-grinding any damaged journals. Pre-grinding removes any embedded bearing material, pinholes, and hair-line fractures from the journal surface that might cause an adhesion problem for the weld. Carbon plugs that fit the journal oil passages are commercially available. Install the plugs by tapping them in with a hammer. Since the weld will not adhere to carbon, the plugs are easily removed and the shape and position of the oil passage is retained.

Position the crankshaft between the chucks of the machine and adjust the welding head and flux tube so they are in line with the journal to be welded. Switch on the power, then strike an arc and run a bead along the fillet radius at one side of the journal. Move the welding head to the opposite side of the journal and run a bead along the fillet radius. Be sure to get a good solid build up of metal in the radius areas, these are the most likely places to develop pinholes or other defects in your weld. Welding a bead at either side preheats the journal for good adhesion.

The crankshaft welder builds up the journal by rotational welding. The shaft is rotated while a bead is spiraled around the journal as the welding tip works from one side of the journal to the other. Most machines are fully automated. Once set up properly, all the operator is required to do is chip off the slag that forms on the weld.

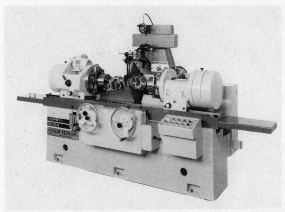

Figure 12-12. Crankshaft grinders restore journals by removing metal with a grinding stone.

After welding, recheck the crankshaft for straightness, the extreme heat of welding can distort the shaft. Straighten the shaft as necessary, remove the carbon plugs from the oil passages, then finish the journals by grinding and polishing. If a journal has been welded correctly, using the proper wire and technique, the weld will often be stronger and harder than the original surface.

Crankshaft Grinding

Grinding crankshaft journals is a complicated process that requires a high degree of skill, as well as a crankshaft grinder, an expensive and delicate piece of machinery, figure 12-12. The crankshaft grinder rotates the shaft as a large-diameter grinding stone removes precise amounts of metal from the journals. We will not go into exact procedures for grinding a crankshaft, but we will cover the basics so that you understand how and why it is done. You grind a crankshaft for any of these reasons:

- To remove out-of-round and taper in the journals
- To restore the proper oil clearance
- To give the proper surface finish to the journals
- To provide a fresh bearing surface on a built-up journal
- To repair thrust surfaces, seal surfaces, and the flywheel and damper mounting surfaces.

A crankshaft is difficult to grind because it is flexible. During grinding, the crank must be supported from the opposite side of the grinding wheel with a steady rest, figure 12-13, to

Figure 12-13. A steady rest is used to prevent the grinding wheel from deflecting the shaft during the machining operation.

Figure 12-14. Indexing offsets the main bearing journals so rod journals can be centered in the machine and ground.

Figure 12-15. Inline engines have two rod journals on the same axis, both throws can be ground from the same crankshaft index.

Figure 12-16. Each crank throw on a V-type engine has a separate axis, you must index each journal before grinding.

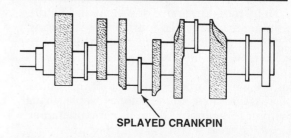

Figure 12-17. Even firing V-6 engines with splayed crankpins requires you to index twice for each throw.

prevent grinding wheel pressure from deflecting the shaft. Using this steady rest, knowing how to set it up, as well as how much grinding pressure to use, requires a skill you can only gain through experience.

As mentioned before, you can sometimes save the customer money by building up a single journal and then grinding it to standard size. The usual technique, however, is to resize all the journals (both main and rod) to the next undersize, usually 0.010, 0.020, or 0.030 inch (0.025, 0.050, or 0.075 mm) less than the standard diameter. The exact amount of metal you remove depends upon crankshaft condition and the availability of undersize bearings to fit the engine.

When both the main and rod bearing journals are to be ground it is important to machine the rod journals first. The reason for this is that grinding the rod journals will often relieve stress in the crankshaft, this can cause the main journals to go out of alignment. Although this is not always the case, you should

Servicing Crankshafts, Flywheels, Pistons and Rods 263

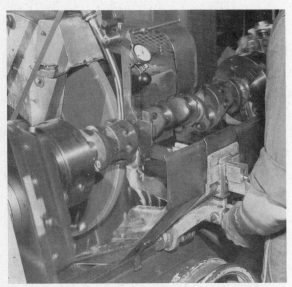

Figure 12-18. With this grinding set up adjustable chucks support the crankshaft ends, drive is provided directly by the spindle chuck.

Figure 12-19. Another method is to position the crankshaft between centers.

avoid taking chances. *Grind the rod journals first*, then finish off the operation by grinding the main journals.

Setting up rod journals

You must grind the rod journals around their own centers, not the crankshaft centerline. There are two methods to accomplish this, depending upon the type of equipment you use. The most common approach is to offset the entire crankshaft in the machine so that it will rotate on the rod journal centerline, this is called indexing. When a crankshaft is indexed it is offset from the main bearing journals the same distance as from the center of the main journal to the center of the crank throw, figure 12-14.

On inline engines, two of the crank throws always rotate on the same axis, so you can grind these two throws without re-indexing the crankshaft, figure 12-15. On V-type engines, every crank throw is on a different axis, figure 12-16, so you must re-index the crankshaft for each journal. Even firing V-6 engines with splayed, or staggered, crankpins requires you to re-index for each cylinder, figure 12-17.

The other grinding method uses more elaborate equipment that you program to follow the crankshaft profile. The crankshaft is rotated along the main journal centerline and the grinding head moves in and out, according to a preset program, to follow the contour of the rod journal.

Setting up the machine is the most important step. If not indexed properly the grinder will quickly turn your serviceable crankshaft into scrap metal. Set up procedures vary depending upon the equipment you are using, so you must follow the manufacturer's instructions for the particular machine you are using. In addition to proper indexing, you must adjust the machine for the correct spindle and cutting speeds, and the stone must be properly dressed.

Setting up main journals

You grind the main journals concentric with the centerline of the crankshaft. To get the journals perfectly round and true, you center the crankshaft in the machine between the spindle and a tail stock. You adjust the position of the shaft in the machine so that it will spin true around the axis of the main journals.

Depending on the type of equipment, and operator preference, the crankshaft will either be held in place with adjustable chucks at both ends, figure 12-18, or positioned between centers, figure 12-19. The spindle and tail stock of most machines are designed to accept either chucks or centers.

When turning with chucks, the shaft is supported at the ends and driven directly by the spindle. You use a dial indicator to adjust the chucks so that the shaft will run true. Since the shaft is supported at the extreme ends you can take indicator readings on the snout and at the rear seal area. This eliminates the possibility of wasting your time grinding a crankshaft with a bent snout.

If you are grinding between centers, a drive dog is used to rotate the shaft. The crankshaft is supported by the centers, the drive dog is clamped around the snout and coupled to the spindle, figure 12-20. A dial indicator is used to adjust the shaft to run true. Because the dog is clamped to the snout, you will have to take

Figure 12-20. Power to drive a crankshaft when turning between centers is provided by clamping a drive dog to the snout of the shaft.

Figure 12-21. The grinding wheel must be in good condition, true, and balanced for a successful job.

your readings off the front and rear main journals. The dial indicator will not detect any bend forward of the front main journal.

The choice of which method to use is a matter of preference. Both methods will provide an adequate finish, but only if the crankshaft is straight and properly adjusted.

Stone preparation

Grinding wheel condition is critically important to a successful job, figure 12-21. The wheel has to be in perfect balance; otherwise the heavier side of the wheel will swing out further than the lighter side. This will cause the wheel to remove too much metal as the heavy side passes the crankshaft journal.

The cutting surface finish of the grinding stone is also critical, it is impossible to get a true grind and good finish unless the stone is properly prepared. The face of the stone must be true, clean and square.

In addition, the corners of the stone must be properly chamfered to conform to the fillet radius of the journals, figure 12-22. Too large a radius can contact the edge of the bearing inserts, too small a radius can create a weak point in the crankshaft. Crankshaft interchange manuals specify the correct radius for every given crankshaft, and you should adhere to those recommendations when grinding.

A diamond dressing tool prepares the cutting surface of the stone. Most machines have an automatic diamond stone dresser, figure 12-23, mounted to the machine that is fed across the stone. When dressing a stone keep in mind the following:

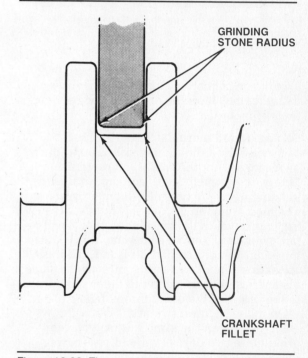

Figure 12-22. The corners of the stone must be properly chamfered to conform to the fillet radius of the journals.

- The diamond must be at a slight angle, between 10 and 15 degrees, to the stone. Too small an angle will dull the diamond, too large an angle can remove metal from the tool and loosen the diamond.
- Keep the diamond as close as possible to the tool holder. The farther out the diamond is the more likely it is to vibrate. Vibration causes a poor finish on the stone.
- Keep the diamond cool during dressing. Either turn on the coolant pump and direct the flow at the diamond, or wait between passes to allow it to air cool.

Servicing Crankshafts, Flywheels, Pistons and Rods

Figure 12-23. A diamond dressing tool is used to prepare the stone, some machines have an automatic dresser.

- Run the diamond across the face of the stone smoothly with a slow steady sweep, keep the rate of travel even.
- Only take a small cut on the stone with each pass. Begin dressing at the highest point on the stone, stones never wear evenly.
- Occasionally rotate the diamond in the tool holder. This will equalize wear, keep the diamond sharp, and extend its life.

Grinding

There are two grinding styles: plunge grinding, and sweep grinding. Almost all automotive machine shops use the plunge grinding method. In sweep grinding, you pass the stone back and forth across the journal until you bring it down to the target size. In plunge grinding, you dress the stone so that it is the exact shape of the new journal, fillet-to-fillet. Then feed the stone straight into the crankshaft journal.

Sweep grinding requires several passes before the journal is ground, but the same stone can be used for journals of different widths. The major disadvantage to a sweep grind is that the stone wears unevenly. Since the stone is moving across the journal the leading edge is doing the majority of the work and will wear faster than the trailing edge. In addition, sweeping is more time consuming because you must make more than one pass.

When plunge grinding, stone wear is equal across the grinding face because the stone is fed straight into the work. The plunge method is also quicker than sweeping because you can finish the journal with one motion rather than with multiple passes. The draw back to plunge grinding is that it requires a different stone for each journal width. A double plunge is used when journal width is greater than your largest stone. To double plunge, select a stone that is slightly larger than half of the journal width, grind one half of the journal, then reposition the stone and grind the other side so the cuts overlap at the center of the journal. If done properly, a double plunged journal will be perfectly flat from fillet to fillet, and show absolutely no sign of the stone overlap.

With either method the secret to a successful grind is to use light pressure with both the stone and the steady rest. Too much pressure on the stone will result in an unacceptable finish. Too much pressure on the steady rest forces the journal into the stone. When the rest is released, the shaft springs away from the stone and the grind is no longer true. If the shaft does not clean up under light pressure, dress the stone for a smoother finish.

Be sure to follow the recommendations of the equipment manufacturer with respect to grinding the thrust surface and flywheel flange. Briefly, you can grind these surfaces by gently pressing them against the flat sides of the grinding wheel.

The final step to grinding a crankshaft is to engrave the size of the rod and main journals on the face of the front counterweight. This last step is done as a courtesy. It eliminates guesswork the next time the engine is torn down for rebuilding.

Crankshaft Polishing

After the crankshaft has been ground, the journal surface is much too coarse and irregular for the bearing to ride on, so the journals must be polished. Standard industry practice is to grind the crankshaft while rotating it in one direction and polish it while rotating it in the opposite direction. The theory is that grinding pushes the grain of the metal up to form tiny peaks; polishing in the reverse direction is a quick and effective way to knock down the high spots. Polishing can be done either by hand lapping or with a motor driven micro polishing machine.

To hand lap you wrap a length of 320 grit emery cloth around the journal, then work the emery cloth around the journal by alternately pulling on either end. Wrapping a leather thong over the emery paper can give you a little extra leverage to make the job go quicker.

Most shops that perform crankshaft grinding will also have a micro polisher. Micro polishers finish the journals by belt-sanding. This piece

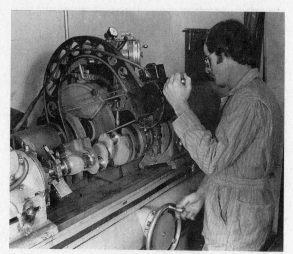

Figure 12-24. This polishing belt is an accessory that attaches to the crankshaft grinder.

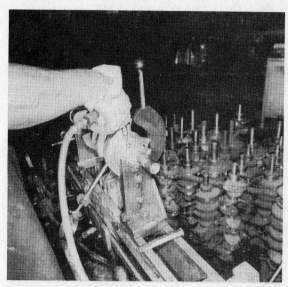

Figure 12-25. After polishing, deburr and chamfer the oil passage openings on the journals with a grinding stone.

of equipment may be an accessory that attaches to the crank grinder, figure 12-24, or a separate free standing machine. With either machine, you rotate the crankshaft, and run the belt back and forth across the journal to provide a smooth and shiny surface finish.

The purpose of micro polishing is to improve the surface finish, it is not a final sizing operation. The amount of metal you remove, never more than 0.0002 inch (0.005 mm), is negligible. Micro polishing can also remove any light score marks from a crankshaft that does not require grinding.

After polishing, you must deburr and round the edges of any oil holes that run through the journals so the bearing will not be scored or marked when the crankshaft rotates. You can do this quickly with a die grinder and a small stone, figure 12-25.

Once you grind and polish the crankshaft, you must clean it thoroughly. Grinding material, a combination of metal chips, stone dust, and coolant, is especially likely to collect in the oil passages. Clean the oil holes with a rifle brush, rinse them with solvent, and blow them dry with compressed air. If debris is left in the oil holes it can ruin the crankshaft within a few hundred miles. After cleaning, spray the journals with light oil to prevent rust.

PISTON AND CONNECTING ROD SERVICE

Piston and rod service consists of disassembly, inspection, cleaning, and reconditioning. We will detail the inspection procedures as well as machining operations. We show you how to: straighten bent connecting rods, resize the wrist pin and crankshaft journal housing bores, and knurl piston skirts.

When you disassembled the engine, you measured the wear in the cylinder bore in order to determine whether a rebore was necessary. If so, you can dispense with any further care or cleaning of the pistons because they will have to be replaced with oversized parts. However, if the cylinder bores were within specification, and the pistons were in good shape, then you can install a new set of rings on the old pistons.

Separating Pistons and Connecting Rods

If you are going to reuse the pistons, you do *not* need to separate the pistons from the rods. You can replace the rings, clean the pistons, and check the straightness of the rods with the pistons attached to the rods. If the pistons are going to be replaced, the connecting rods can generally be reconditioned and reused so you will have to remove the wrist pins. As explained in the *Classroom Manual* there are a variety of designs for attaching pistons to the connecting rods, the two most common are "full-floating" and the "press-fit" pins.

Removing full-floating wrist pins

Full-floating wrist pins fit through the piston bosses with 0.0001 to 0.0005 inch (0.003 to 0.013 mm) of clearance. There is also about 0.0003 to 0.0007 inch (0.008 to 0.018 mm) of clearance between the pin and the small-end bore of the connecting rod. The pin is held in place by snap-rings that fit a machined slot on

Servicing Crankshafts, Flywheels, Pistons and Rods

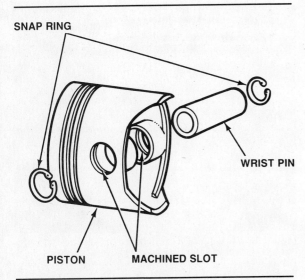

Figure 12-26. Snap-rings seat into a machined slot on the piston to hold free-floating wrist pins in place.

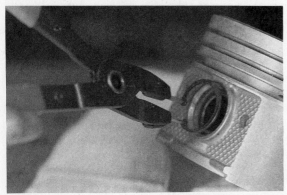

Figure 12-27. Carefully remove the snap-rings, twisting or closing the ring too far can damage it.

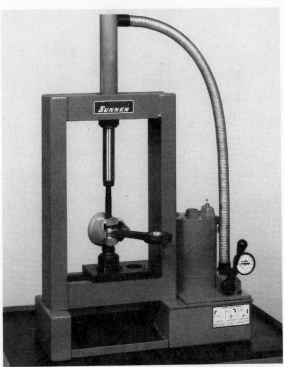

Figure 12-28. Press-fit wrist pins are removed with a hydraulic press.

the inside bore of the piston boss, figure 12-26. Because of the clearance, the wrist pins are easy to remove, simply release the snap-rings and slide the pin out.

There are several snap-ring designs in use, most can be removed with snap-ring or needle-nosed pliers, figure 12-27. Be careful when removing the lock-rings, never compress them more than the minimum amount necessary to free them from the groove. Compressing a ring too much can stress the metal, cause a loose fit and may lead to premature failure.

In theory, once the snap-rings are removed the pin will slide freely out of both the piston and rod. In reality, the pins have a tendency to stick in the bores as a result of temperature changes and the build-up of sludge and varnish. You can generally remove stubborn wrist pins with a hammer and a drift. The drift must be wide enough to get a good seat on the end of the pin, yet narrow enough to easily slip through the bore. Strike the drift with light hammer blows to drive the pin out, be careful not to gouge the bore with the drift. Gently heating the piston bosses with a propane torch or hot water will often help to loosen the most difficult wrist pins.

Removing press-fit wrist pins

Press-fit wrist pins normally fit the connecting rod with an interference of about 0.0008 to 0.0012 inch (0.02 to 0.03 mm), and have 0.0003 to 0.0005 inch (0.008 to 0.013 mm) of clearance in the piston bosses. Because of the interference fit, you must use a press to remove the pins. A variety of adapters are used to hold the rod and piston in position as the press forces the pin out, figure 12-28. Take care because aluminum pistons crack easily if they are not properly supported.

Selecting the proper press adapters is important. The lower adaptor must be able to firmly support the piston and have a large enough inside diameter so that the pin will fit through as it is pressed off the rod. The pressing adaptor must be large enough to get a good even footing on the pin, yet narrow enough so that it can pass through the piston and rod without damaging the bores.

Be careful whenever you are handling connecting rods, they can be damaged very easily. Always keep the connecting rods from an engine together as a matched set. Once removed

Figure 12-29. Broken piston ring lands are often the result of detonation.

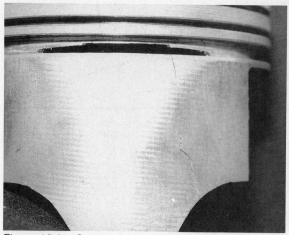

Figure 12-31. Carefully examine piston skirts, excessive wall clearance can crack and eventually break a skirt.

Figure 12-30. Preignition can burn through or melt the top of a piston.

from the pistons, store the connecting rods in a safe place, preferably hanging, until you are ready to recondition them.

Reconditioning Pistons

The following operations detail the steps necessary when you can reuse the pistons. If the engine requires new pistons, skip these procedures and continue with the section titled "Reconditioning Connecting Rods".

Removing piston rings

Whenever the pistons are removed from an engine the rings must be replaced, even if the pistons are to be refitted. Piston rings are brittle and fragile and have a tendency to break as you remove them. Ring breakage should be of no concern, however, you do need to be extremely careful not to damage the ring lands of the pistons as you remove the rings. If there is a lot of built up sludge and varnish on the pistons, do a preliminary cleaning before you attempt to remove the rings. Soaking the pistons in carburetor cleaner for a few minutes, then rinsing with solvent or water, will dissolve enough sludge to make ring removal easier.

Once the rings are loose in the ring lands you can often remove them by hand. Hold the piston with both hands, place your thumbs on the ends of the ring, push out with your thumbs to expand the ring, and slip the ring up and over the top of the piston. Keep in mind that the ring land on the piston is both delicate and critical, be careful not to gouge or nick the piston as you remove the ring. Avoid screwing the ring off of the piston, this stresses the ring land and can cause damage. Remove stubborn rings by spreading them open with a ring expander or a pair of snap-ring pliers. It is fairly common for rings to break as you remove them.

Your next step is to thoroughly clean the pistons. Once clean, you inspect and measure each piston to make sure it is within specifications and can be safely returned to service. Replace pistons with obvious problems. Problems you might encounter are: broken ring lands caused by detonation, figure 12-29, burned through or melted tops caused by preignition, figure 12-30, cracked or broken skirts caused by excessive wall clearance, figure 12-31, extensive scuffing or scoring caused by inadequate lubrication, figure 12-32, or damage from a foreign object passing through the combustion chamber, figure 12-33. Discard any damaged pistons and replace them with new ones. To avoid doing the job over again in the near future, you must correct any operating conditions that may have contributed to piston failure.

Servicing Crankshafts, Flywheels, Pistons and Rods

Figure 12-32. Inadequate lubrication leads to extensive scuffing or scoring of the pistons.

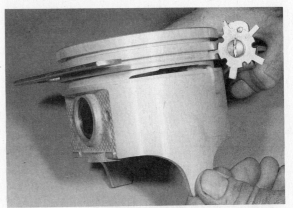

Figure 12-34. Using this tool makes cleaning piston ring grooves a quick and easy task.

Figure 12-33. A carelessly handled bolt found its way through the intake and into the combustion chamber to totally destroy this piston.

Figure 12-35. You can clean ring grooves with a piece of broken ring, be careful not to remove any metal.

Cleaning the pistons

Begin by carefully removing any heavy deposits on the tops of the pistons with a gasket scraper. Keep in mind that pistons are forged or cast from light-weight soft alloy and are easily damaged. You do not want to scrape down to the metal, you simply want to remove the heavy carbon build up. Avoid using stiff wire brushes, these can remove metal and damage the pistons. Remaining deposits on the tops, as well as the rest of the piston, can be removed by soaking in carburetor cleaner or solvent washing with a soft brush. You can also clean pistons by bead blasting, however bead blasting can damage the ring lands. Protect the ring lands by wrapping them with tape before bead blasting.

Next, clean the piston ring grooves. The best way is to use a ring groove cleaner, figure 12-34. Select the proper blade to fit the ring groove, position it in the tool, and tighten the locking nut. Slip the tool over the piston, fit the cutting blade into the ring groove, and adjust the tool so the blade will cut to the exact depth of the groove. Setting the tool too tight will cut the groove too deep and destroy the piston. Work the tool around the circumference of the piston with slow, steady, even pressure. It is important that you remove all traces of dirt and debris, any small particle that remains in the ring groove can prevent the new rings from properly seating. If a ring groove cleaner is not available, clean the grooves with a broken piece of the old ring. Carefully work your way around the piston with the sharp broken edge of the ring, figure 12-35. Remember that the groove must be perfectly clean and the piston can be easily damaged. Never scrape hard enough to remove metal from the piston.

Once the ring grooves are clean you will have to clear the oil bleed passages inside the oil control ring groove. Whether these passages are drill holes or machined slots, they must be open to allow oil to flow freely through them. Clean the oil passages by hand,

Figure 12-36. You must clear the piston oil bleed passages to ensure adequate oil control.

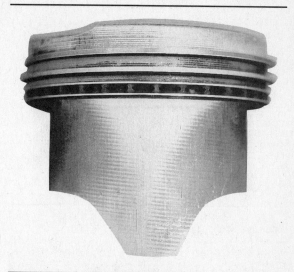

Figure 12-37. Good piston wear is indicated by skirts with a smoothly polished straight up and down pattern.

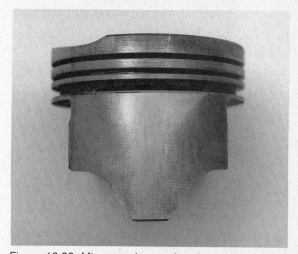

Figure 12-38. Minor scoring on the piston skirt is a sign of abrasive particles in the oil or intake charge.

use a drill bit, torch tip cleaner, small pick, or any other suitable tool that will fit through the passage without removing metal, figure 12-36. After clearing the oil bleeds, give the pistons a final solvent wash to remove any residual build-up and dislodged particles. The pistons are now ready for inspection.

Inspecting and measuring the pistons

Now is the time to look for piston damage that was not so obvious before cleaning. Examine the entire piston, keeping a close eye out for unusual wear patterns on the skirts, ring lands, and wrist pin bosses.

In a running engine, the major and minor thrust forces push the piston against the cylinder wall in a line perpendicular to the wrist pin. Pistons from a good running engine will wear a symmetrical pattern along the skirts where thrust reduces cylinder wall clearance to a minimum. Ideally, this area of the piston should appear smoothly polished and straight, figure 12-37. Small abrasive particles that have passed through the engine's air or oil filters can cause minor scoring on the piston skirts, figure 12-38. This type of wear can often be ignored and is no reason to scrap the piston. Also examine the skirt for cracks, especially near the pin bosses. Engine overloading, high mileage, and metal fatigue, can cause hairline stress fractures.

Every time a piston crosses through top and bottom dead center it changes direction, as this happens the rings are hammered from side-to-side in their grooves. Over time this hammering can cause the ring grooves to wear to a point that even new rings cannot seal properly, figure 12-39. The top compression ring groove is the most susceptible to this type of wear, particularly in the area closest to the exhaust valve. Inspect the ring grooves and ring lands. Good pistons will show even wear and have crisp sharp edges. Replace the piston if you find any scuffing, hairline cracks, or chips on the ring lands.

Examine the wrist pin bosses of the piston because this area is under extreme loads when the piston changes direction and must be in sound shape. Check for signs of cracking and metal fatigue on the inside of the piston. If the wrist pins are the free floating type, held in place by a snap-ring at either end, pay close attention to the condition of the snap-ring grooves. A snap-ring that works loose in a running engine can cause considerable damage to the piston and cylinder walls.

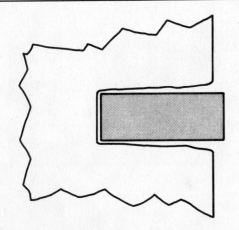

Figure 12-39. Piston ring grooves can be worn out of shape from the punishment they take under normal conditions.

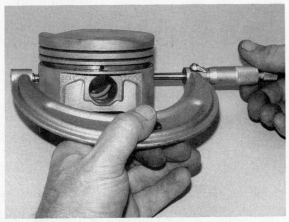

Figure 12-40. Piston diameter is generally measured at the wrist pin centerline, perpendicular to the pin.

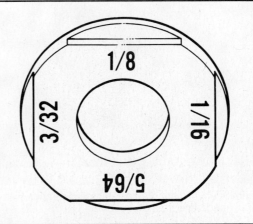

Figure 12-41. A ring wear gauge is used to measure the width of a ring groove.

After the pistons pass a visual inspection, they must be accurately measured for diameter and ring groove wear. Diameter is checked with an outside micrometer, ring groove wear is evaluated using either a ring groove wear gauge or a new set of rings and a feeler gauge.

Piston diameter is generally measured on the thrust surface at the centerline of the pin bore and perpendicular, at a 90 degree angle, to the wrist pin, figure 12-40. Most automotive pistons will measure 0.001 to 0.003 inch (0.025 to 0.076 mm) less than the cylinder bore diameter. Some engine and piston manufacturers specify the measurement be taken at a different point, such as just below the bottom ring groove or at the bottom of the skirt; always check the manufacturer's recommendation. If aftermarket pistons are being used, follow the piston manufacturer's guidelines. Pistons that measure to minimum diameter, or are slightly out of tolerance, can often be reconditioned by knurling. The knurling process is detailed later in this chapter.

Check for collapsed skirts by measuring across the thrust surfaces at the bottom of the skirt. Compare this reading to your previously measured diameter reading. Because this is the coolest running area of the piston, and is less susceptible to heat distortion, expect your reading to be about 0.0015 inch (0.038 mm) wider at the bottom of the skirt. If the reading is less, the skirt has collapsed and the piston must be replaced. Collapsed skirts allow the piston to rock in the bore and result in noise, increased blow-by, poor oil control, and excessive oil consumption.

The final step in evaluating a piston is to measure the piston ring side clearance. Side clearance is a tolerance value that can be found in the manufacturer's specifications. Typical side clearance for automotive engines is in the 0.001 to 0.003 inch (0.025 to 0.076 mm) range.

Ring wear gauges work like a go-no-go gauge, figure 12-41. The widest side of the gauge that will fit into the groove equals the groove width. Subtract the width of the new ring from the width of the groove to determine side clearance.

You can also measure side clearance with a feeler gauge and a new piston ring. Fit the ring backward into the groove, then slip a feeler gauge blade between the ring and the side of the groove, figure 12-42. The largest blade that will fit easily to the bottom of the ring groove will equal the side clearance.

Whenever you measure for side clearance keep in mind that the ring grooves do not wear evenly around the piston. It is important to take measurements at several points on the piston circumference. Top compression ring

Figure 12-42. You can measure side clearance with a feeler gauge and the new piston ring.

grooves that are unevenly worn or out of tolerance can sometimes be repaired by cutting the groove oversize and installing a shim. We detail ring groove repair later in this chapter.

Machining pistons
Reconditioning pistons by knurling the skirts and cutting the ring grooves, once common overhaul procedures, is rarely done anymore. Longer engine life, tighter clearance tolerances, the availability of inexpensive replacement pistons, and the high cost of labor have all contributed to make these operations virtually a thing of the past. On occasion, especially when working on antique or rare engines, you may be able to salvage a marginal piston by reconditioning it. A brief description of knurling and shimming ring grooves follows.

Knurling piston skirts
Piston knurling increases the diameter of the piston to reduce the piston-to-cylinder wall clearance in an engine. Knurling can reliably expand pistons by about 0.002 to 0.004 inch (0.051 to 0.102 mm).

The piston knurling machine has two bits that displace metal on the thrust surface of each skirt of the piston, figure 12-43. Knurling leaves a pattern on the piston where the bits push metal out to expand the skirt, figure 12-44. Because the knurled area will help retain oil between the piston and the cylinder wall, knurled pistons should fit the cylinder bores with about half of the recommended clearance. The metal displacement can be controlled to give the amount of piston-to-cylinder wall clearance desired.

Repairing ring grooves
A piston ring groove lathe, a specialized piece of equipment no longer readily available, and a 1/16 inch shim are required to repair piston ring grooves. The piston lathe is rotated around the piston to undercut the top compression ring groove.

Figure 12-43. The piston knurling tool uses two special bits to displace metal on the piston skirts.

To cut a ring groove, the piston is clamped in a vise with the lathe straddling the piston. The lathe rides on guides that fit into the second ring groove, while the cutting bit opens up the top groove. The bit undercuts the top of the ring groove to accommodate the shim. Once you install the shim, the ring has a fresh surface to seal against, figure 12-45.

Evaluating Connecting Rod Condition
Unless an engine has suffered a catastrophic breakdown, you can usually recondition the connecting rods and return them to service. Now that they have been separated from the pistons they can be closely inspected. Under normal engine operating conditions, the big end bore will stretch to an out of round condition. Most of the stretch will occur in the rod cap, the weakest point, and in a vertical direction, figure 12-46. In addition, rods may be bent, twisted, cracked, or otherwise damaged. You must evaluate rod condition and correct any problems before you assemble the engine. The place to start is by inspecting the rod bearings, remove the bearing shells and then reinstall and torque the rod caps.

Inspecting rod bearings
Many of the same conditions that affect main bearings, including contamination, metal fatigue, excessive oil clearance, lugging, inade-

Servicing Crankshafts, Flywheels, Pistons and Rods

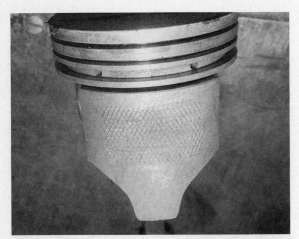

Figure 12-44. Knurling leaves a pattern where metal was pushed out to expand the skirt.

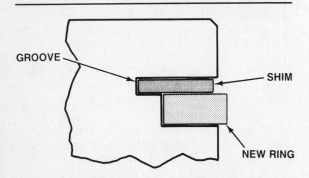

Figure 12-45. A shim installed in the groove provides a flat new surface for the ring to seat against.

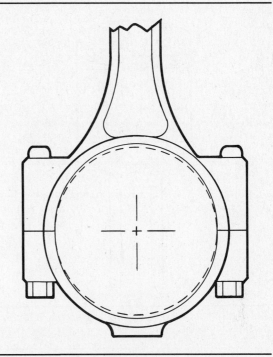

Figure 12-46. Rod bearing bores normally stretch from top to bottom, with most of the wear concentrated on the cap.

quate oil supply, insufficient clearance, and poor seating, also affect connecting rod bearings. Refer to Chapter 9 of this *Shop Manual* for information on inspecting bearing inserts and recognizing wear patterns. Bearing wear patterns can also be used to detect bent connecting rods, improper assembly, and radius ride.

Bent connecting rod
Bent connecting rods place a twisting load on the bearing insert. This load will wear the bearing insert in a characteristic V-shape, concentrated at one side of the bearing crown. Also check the piston skirts for scuffing if you find this pattern.

Improperly assembled offset rods
Manufacturers sometimes install offset connecting rods, in which the big-end bore of the rod is not centered exactly under the piston. If the connecting rod is installed reversed during a repair, the base of the rod — and the bearing insert — will be forced against the crankshaft fillet. Side pressure causes the bearing to develop a wear pattern along the edge on one side of the inserts. Only those rods installed backwards will show the wear.

Radius ride
Radius ride occurs with reground crankshafts. If a machinist grinds the fillet radius too large, the bearing rides on the fillet. This causes wear along both edges of the insert, all the way around the circumference of the bearing. In engines with one rod per throw, the contact can reduce or even cut off oil flow through the bearing, causing a lubrication failure. If radius ride is the problem, it is likely that all the bearings will be affected, rather than just one. Crankshaft journals worn into an hourglass-shape can cause a similar pattern and result in the same problems.

Inspecting rods, caps, bolts
Thoroughly clean the connecting rods so that they can be properly examined. Rods can be cleaned by hot or cold tanking, solvent washing, bead or shot blasting, or in a thermal oven. After cleaning the rods, you will check them for:

- Pin bore condition and diameter
- Bend and twist
- Big end bore out-of-round
- Cracks, stress risers, and irregularities.

In addition, you will have to take a close look at the bolts and nuts that attach the cap

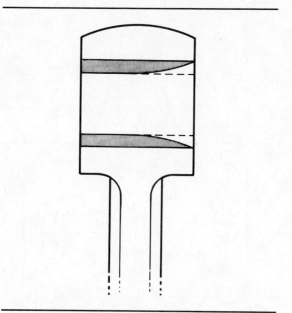

Figure 12-47. Rod bushings that are worn to a taper must be replaced.

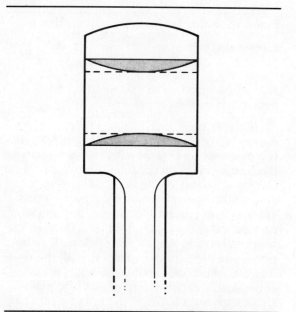

Figure 12-48. Bell-mouth bushing wear is the result of too much pin clearance.

Figure 12-49. A precision pin bore gauge is the best way to determine the wrist pin bore diameter.

to the rod. The bolts must be in good sound condition; discard them if you find any pulled, chipped, or stripped threads. For maximum performance and reliability the rod nuts should be automatically replaced with new ones whenever you remove them. Keep in mind, the nuts are a critical link that holds the engine together. Every time the piston changes direction (twice per crankshaft revolution) the force of the reciprocating mass is transferred to the nuts holding the cap to the rod. The first time rod nuts are reinstalled about 20 percent of their tension is lost, tension loss increases each time the nuts are reused. Trying to save the insignificant price of new rod nuts can cost a considerable amount of money if the old ones fail. Always replace the rod nuts.

You also must take into consideration whether or not the rods have been previously reconditioned. If the big end bores were previously reworked, machining them again may not be possible. Resizing the bearing bore reduces the overall length of the connecting rod, this alters the compression ratio and can effect engine performance. Manufacturers usually list rod length, often as center-to-center between the big and small end bores, in their service specifications as a tolerance value.

Pin bore condition and diameter
The wrist pin bore of a press-fit type connecting rod is machined to provide the precise interference fit. The bore must be round and smooth; any nicks, scratches, or gouges in the surface can cause problems with fit, as well as performance. A pin that does not fit properly cannot transfer heat effectively and may result in premature failure. Full-floating wrist pins ride on a bronze bushing that is pressed into the small-end bore of the connecting rod. The bushing must be straight, provide the specified clearance, and have a surface finish that will support an oil film.

The soft bronze bushings of a free-floating design tend to show more wear than press-fit pins. Pin bushings have a tendency to wear to a taper, figure 12-47, or develop a bell-mouthed shape, figure 12-48. If either of these

Servicing Crankshafts, Flywheels, Pistons and Rods

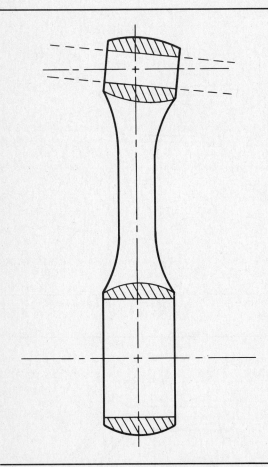

Figure 12-50. The small and big end bores will not line up vertically on a bent connecting rod.

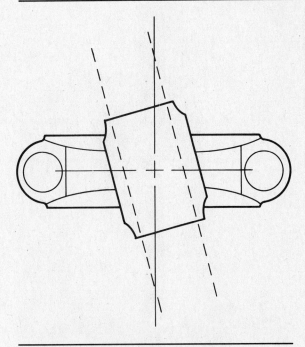

Figure 12-51. If the end bores do not align horizontally, the connecting rod is twisted.

conditions exist, you will have to replace the bushings. Bushing replacement is detailed later in this chapter.

The most accurate way to measure pin bore wear is with a pin hole gauge, figure 12-49. The gauge is calibrated using either two good wrist pins or an outside micrometer. Once calibrated, the dial indicator will zero to the precise diameter of the wrist pins. Place the pin bore of a connecting rod on the expandable fingers of the gauge and note the reading on the dial indicator. The indicator will display the interference or clearance between the pin and the bore. You can repair full-floating connecting rods that are out of specification by replacing the bushing. Sometimes you can salvage press-fit rods that are out of specification by boring the small-end to accept an oversize wrist pin.

Bend and twist

Connecting rods must be accurately checked for both bend — vertical misalignment between the two end bores, figure 12-50, and twist — horizontal misalignment between the end bores, figure 12-51. As a rule of thumb, bend or twist must not exceed 0.001 inch (0.025 mm) per six inches (152 mm) of connecting rod length. Connecting rods with a slight amount of bend or twist can often be straightened and returned to service, this procedure is explained later in this chapter.

Bend and twist are checked using a rod alignment fixture, figure 12-52. Most rod alignment fixtures can be used either with or without the piston installed on the rod. To gauge a rod without the piston the wrist pin must be fitted to the connecting rod. To check for bend and twist without the piston attached, install the pin and proceed as follows:

Adjust the mandrel so that the clamp collars will grip the center of the big end bore of the rod, be sure the contact bars are fully seated in the mandrel. Place a connecting rod on the mandrel with the major thrust (right hand) side facing out, rock the rod slightly to seat it. Hold the bend indicator fixture flat against the surface plate so that the two parallel contact bar surfaces straddle the end of the rod and rest against the wrist pin, figure 12-53. If the bars fully contact the pin on both sides, the rod is straight. If light shows between the rod and one of the contact bars the rod is bent and will have to be straightened or replaced.

Checking for twist is similar, however you hold the twist indicator fixture against the

Chapter Twelve

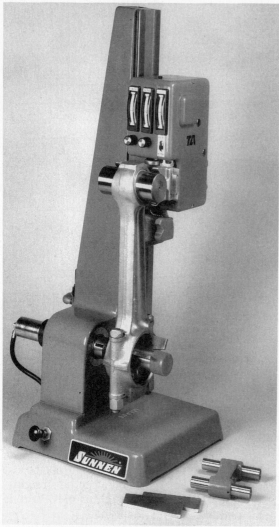

Figure 12-52. This rod alignment fixture electronically checks connecting rods for bend and twist.

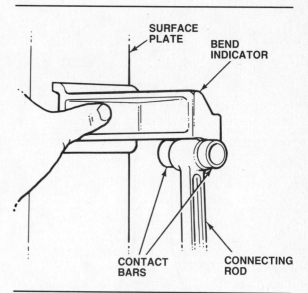

Figure 12-53. This accessory is used to check rod bend.

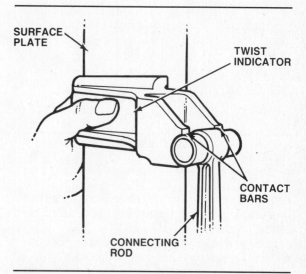

Figure 12-54. Rod twist is checked using this attachment.

surface plate. The twist fixture straddles the side of the rod to contact the wrist pin, figure 12-54. If light shows between the rod and one of the fixture contacts the rod is twisted and must be repaired or replaced.

Big end bore out-of-round

The reciprocating action of the engine will stretch the big-end bore of a connecting rod out-of-round under normal operating conditions. The out-of-round bore can be brought back to standard size much the same way the main bearing bores of the engine block were. The machining procedures are detailed later in this chapter, now we will simply measure the bores to determine wear and condition.

The quickest and easiest way to measure the bore is with a special dial bore gauge designed for connecting rods. Rod honing machines often have an accessory bore gauge fixture attached to them, figure 12-55, or you can use a hand held model. To use the gauge, adjust the floating pin to fit the bore and zero the dial indicator. Now slowly rotate the rod and watch the dial, the indicator will display how far over or under the set figure the bore actually is.

You can also measure big-end bore with an inside micrometer, a telescoping gauge and an outside micrometer, or the same dial bore gauge you used to check the cylinder and main bearing bores. Chapter 11 of this *Shop Manual* details the use of these tools, so we will not repeat the procedures here. Always take several measurements at different points of the bore, figure 12-56. Keep in mind, the

Servicing Crankshafts, Flywheels, Pistons and Rods

Figure 12-55. Both big- and small-end bores are measured quickly and accurately with a special bore gauge.

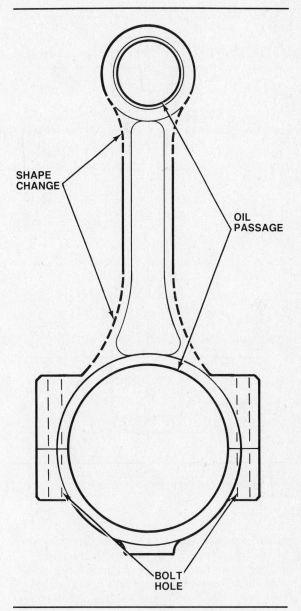

Figure 12-57. Inspect all the stress areas of the rod. Pay close attention to bolt holes, oil passages, and places where the rod changes shape.

rod bore stretches at the top and bottom, most of the stretch normally occurs in the cap.

Cracks, irregularities, and stress risers

You will have to separate the connecting rod and rod cap in order to inspect them. Before you separate the two pieces check the thrust faces on the sides of the big-end bore. Any chips, nicks, or gouges should be dressed flat with an oilstone, they can prevent the cap or rod from seating properly in the grinder.

Carefully look the rod and cap over for any sign of cracking or deformation. Closely scrutinize oil drill passages, bolt holes, and other

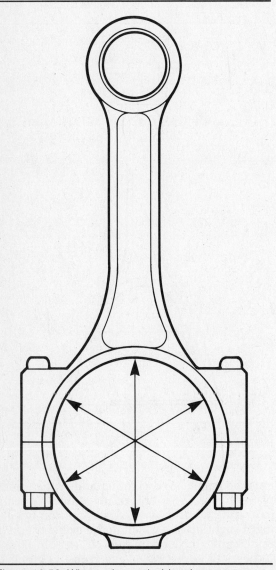

Figure 12-56. When using an inside micrometer or telescoping gauge, measure the rod bore at several different positions.

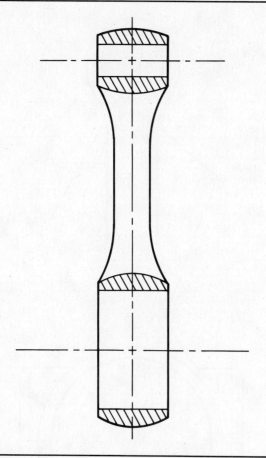

Figure 12-58. The piston pin bore and the crankpin bore must be parallel to each other and in the same plane.

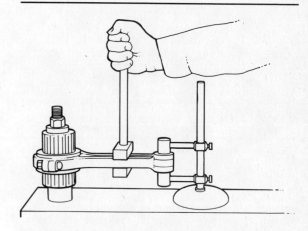

Figure 12-59. This fixture is used to cold-bend connecting rods back to their original shape.

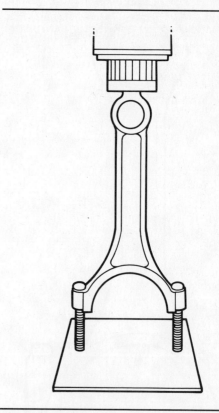

Figure 12-60. Rod bolts are quickly removed in a hydraulic press.

stress zones (wherever the metal changes shape), figure 12-57. If you find a crack, no matter how small, replace the rod. Keep an eye out for any misshaped bolt holes or surface irregularities. Should anything look suspicious, check the rod using magnetic particle test equipment, see Chapter 11 of this *Shop Manual*. Keep in mind that any chip, nick, or gouge on either the rod or cap is a stress riser that could lead to premature failure.

Reconditioning Connecting Rods

Once a connecting rod has passed inspection, you can recondition it and return it to service. Reconditioning might require you to:

- Correct bend and twist
- Grind the rod and cap to undersize
- Machine the big-end back to standard size
- Bore or hone the small-end bore
- Replace the pin bushing.

Correcting rod bend and twist

The piston pin bore and the crankpin bore must be parallel to each other and in the same plane, figure 12-58. This relationship is critical to bearing life and also keeps the piston traveling straight up and down in the cylinder bore. Rod aligning is done by carefully bending the rod back to its original shape. You straighten connecting rods with the rod caps installed and torqued to specification.

Connecting rods commonly distort near the top, just below the small-end bore. You can

Servicing Crankshafts, Flywheels, Pistons and Rods

Figure 12-61. Clamp the rod in the grinder so the parting face is square to the grinding stone.

Figure 12-62. Swinging the workpiece across the grinding wheel removes metal from the parting face.

correct slight amounts of bend and twist by carefully bending the rod back to its original shape on a rod straightening fixture. The fixture supports the rod by the big-end bore, a bending bar fits on the beam of the rod giving you enough leverage to pry or twist the rod back into position, figure 12-59. Remember to handle rods carefully; any nick, scrape, or scratch you make on the surface must be considered a stress riser, and could be a potential disaster. The rod must be firmly supported in the fixture, and the bending bar must closely fit onto the rod.

Straightening connecting rods, like many other engine rebuilding techniques, is something you develop a "feel for" through experience. Practice on some scrap rods before you try to bend one for a customer's engine. Bend and twist the rod in small increments, use the rod alignment fixture to check the results of each adjustment as you go. To make the job of checking press-fit rods easier and quicker you can grind down an old wrist pin so that it will slip-fit the small-end bore.

Reconditioning the big-end bore

Prepare the rods for machining by removing the bolts, this can be done easily with a small press, figure 12-60. Replacing the rod bolts is recommended practice, although it is not absolutely necessary unless they are stripped or stretched. Before you can bore or hone the rod, the diameter of the hole must be reduced. Do this by grinding material from the parting surfaces of the rod and cap to reduce the overall diameter, then open the bore to its original size by boring or honing.

Grinding the parting faces

To grind the parting faces of the connecting rod and cap proceed as follows:

1. Place the rod or cap in the grinder so that it is resting on the squaring rod, tighten the vise to hold the cap securely, figure 12-61. Back off the feed wheel to be sure that the rod will not contact the stone yet.
2. Position the cap over the grinding stone and adjust the feed wheel until the cap just barely touches the stone. Then swing the cap out of the way.
3. Start the motor and run the cap over the grinding wheel to make a zero cut, then adjust the feed to take a cut of about 0.002 inch (0.050 mm) and make another pass, figure 12-62. Inspect the surface and take another cut if necessary, keep the amount of metal you remove to a minimum.

NOTE: Many experienced machinists prefer to remove most of the metal from the cap rather than from the rod. Any metal machined off the rod will lower the height of the assembled piston when it is at TDC. This alters the compression ratio, and can cause timing and performance problems. Removing metal from the cap does not change piston height, so compression remains the same.

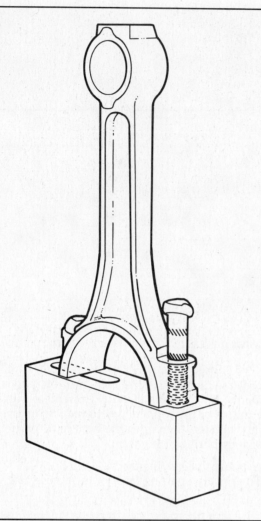

Figure 12-63. Using a fixture to install rod bolts protects the parting face.

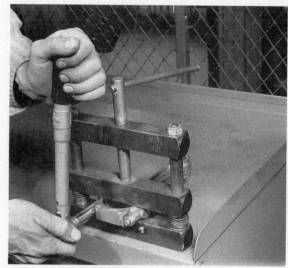

Figure 12-64. This special vise keeps the rod and cap in perfect alignment as you torque them together.

Figure 12-65. A rod honing machine is used to recondition both the big- and small-end bores.

Honing the bearing bore

After you grind the rods and caps, reassemble them in order to resize the bearing bore. Replace or reinstall the rod bolts; protect the parting faces by using a fixture, figure 12-63, and seat the bolts with a hammer. Fit the rod cap, replace the nuts, and tighten them to the specified torque. Be sure the socket wrench you use fits squarely over the rod nut, if the socket contacts the cap it cannot torque properly. There are special vises designed to accurately align the rod and cap during assembly, figure 12-64; if you have one available, use it.

There are a variety of machines available for resizing connecting rod bores, some use a cutting bit on a boring bar to remove metal, others hone the metal away with abrasive stones, figure 12-65. Because of surface finish considerations, and the small amount of metal removed, honing is more popular. Most machines can be set up to refinish both the big and small ends of the rods, oversize the pin bosses of the pistons, and remove and install small-end bushings. Some equipment automatically strokes the stones through the rod, others have to be stroked by hand.

The operating procedures are different for all machines, you must follow the equipment manufacturer's recommendations. The following steps provide a general guideline for honing the big-end bore back to standard size:

Servicing Crankshafts, Flywheels, Pistons and Rods

Figure 12-66. Select the proper honing unit to fit the bore, then install the hone into the machine.

Figure 12-68. Rotate the spindle so the mandrel shoes face up, then loosen the clamp screws holding the shoes.

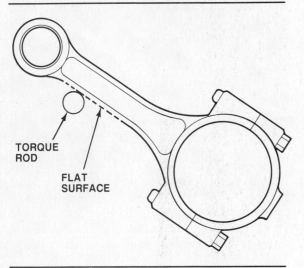

Figure 12-67. Adjust the torque rod, it should support the rod by contacting a flat area close to the small-end bore.

Figure 12-69. Place a rod on the honing unit, depress the foot pedal and turn the feed dial until the rod is held firmly in place by the hone.

1. Select a honing unit that will fit the diameter of the bearing bore and has the proper grit stones. Be sure the power to the machine is off, then install the honing unit into the machine, figure 12-66. Check your setup by turning the feed wheel in both directions while watching the cutting stones. The stones should move in and out as you adjust feed. If not, the hone is not seated, remove and reinstall the honing unit.
2. Check the spindle speed, adjust if necessary by repositioning the belt on the pulleys. In general, any diameter over 2 inches (50 mm) should be honed at 200 rpm, the smaller the bore, the faster the spindle speed.
3. Adjust the torque rod and clamp it into place. The torque rod should support a flat surface on the connecting rod as close to the pin bore as possible, figure 12-67.
4. Set the proper cutting pressure by turning the dial. Most rods can be rough honed at a cutting pressure between 2 and 2½, reduce pressure about 1 to 1½ when finish honing.
5. Position the oil lines so that you will get a good flow of coolant over the workpiece.
6. Rotate the spindle until the mandrel shoes are at the top and loosen the clamp screws that hold the shoes in place, figure 12-68.
7. Place a rod on the honing unit so that it is straight up, slowly depress the foot pedal while turning the feed dial until the hone expands to firmly hold the rod in place,

Figure 12-70. Back off the feed control to zero the honing dial.

Figure 12-71. Reposition the rod so it rests on the torque rod, depress the foot pedal and begin honing. Stroke the rod over the stone as you adjust the feed.

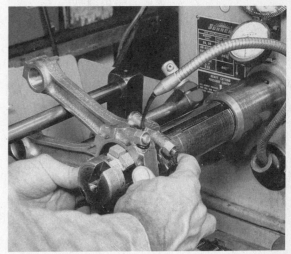

Figure 12-72. You can hone two narrow rods at once, this helps keep the bores square and also saves time.

figure 12-69. Then, firmly tighten the shoe clamp screws.

8. Back off the feed until the honing dial reads zero, figure 12-70. Once zeroed, the dial will read how far the stones expand as you hone the bore.
9. Check the rod diameter on the bore gauge to determine how much material has to be removed from the bearing bore.
10. Reposition the rod on the mandrel so that it rests on the torque rod, depress the foot pedal slowly to begin the honing operation. Stroke the rod over the entire length of the rotating stones, figure 12-71. Increase the feed dial to maintain honing pressure on the rod.
11. Only take off small amounts of metal at a time, measure the rod frequently to determine when you reach the proper size.
12. You may want to hone narrow rods two at a time. This helps keep them square on the hone and saves time, figure 12-72.

NOTE: As you stroke the rod be sure to run it along the entire length of the stones, let the rod travel past the end of the stones by about half the width of the bore. This will help keep wear on the stones uniform, and prevent a ridge from building up on the ends. Flipping the rod over end-to-end between cuts will help keep the shoes and stones, as well as the bore, running true.

Another thing to keep in mind is that the friction of the stones on the rod will generate heat that is held in the rod. This heat expands the housing bore, making the diameter larger during honing than it will be after the rod cools down. Keep this in mind as you work, hone off a little at a time. Alternating between rods will allow them to cool between cuts.

Reconditioning the small-end bore

The steps you take to recondition the small-end bore of the connecting rod will depend upon whether the wrist pins are the full-floating or press-fit type.

Press-fit pins require a precise amount of interference between the pin and the rod bore, honing off any metal can reduce the interference and allow the pin to spin in the rod bore. If the small-end bore of a press-fit rod is out of tolerance, you can correct it by boring or honing to accept an oversize wrist pin. When you use oversize pins, the piston boss bore will also have to be opened up to a larger diameter.

Small-end service for full-floating rods will generally involve removing and replacing the bronze bushing the pin rides on. Since you

Servicing Crankshafts, Flywheels, Pistons and Rods

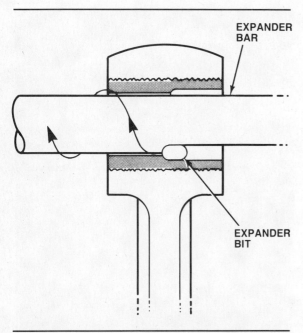

Figure 12-73. A new bushing must be expanded into place for good heat transfer and to control pin clearance.

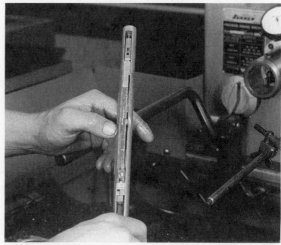

Figure 12-74. Select the proper mandrel for the bushing size.

finish the new bushing to the original inside diameter, stock wrist pins can be used.

You can usually use the same equipment you used to open up the big-end bore of the rod to machine the small-end. Different honing units, attachments, and spindle speeds are used for the various pin fitting operations.

Pin fitting - stationary type

The stationary piston pin is the most common type of wrist pin in use today. Most pins press into the rod with about a 0.001 inch (0.025 mm) interference fit. A clearance of about 0.005 inch (0.013 mm) between the piston bosses and the pin allows the piston to oscillate on the pin. If the clearance of the pin in the piston pin bore exceeds 0.0005 inch (0.013 mm), or the small end bore of the rod is damaged, machine the rod and piston to accept a larger or oversized pin.

Fitting oversize wrist pins is time consuming and requires a considerable amount of precise machine work. In addition, removing metal from the rod and piston, plus the extra weight of the new pin, will upset engine balance. Because of these factors, it is generally more cost effective to simply replace the pistons or defective rods.

Since installing oversize pins is no longer a common practice, we will not detail the operations here. Should you decide it is necessary, follow the equipment manufacturer's instructions for the machine you are using.

Pin fitting - full-floating type

A full-floating pin rotates in both the piston pin and the rod pin bores, both bores have some clearance. To prevent seizing and provide adequate lubrication, the small-end of the rod is fitted with a soft-metal bushing for the pin to ride on. Problems with pin-to-rod clearance, bore taper, and out-of-round can be corrected by simply replacing the bushing.

The old bushing is pressed out in an arbor press using an adaptor that will fit the bushing without damaging the rod bore. A new bushing is installed with the press, be sure to position the bushing properly before pressing it in. If the bushing is drilled for oil passages, the drill holes must line up with the oil passages in the rod. Some bushings have a parting line, split, or slot to direct oil flow. These oil passages must be positioned to the minor thrust side, away from the area of the rod stressed by the power stroke.

After installation, you expand the bushing and finish it to size on the rod honing machine. Expanding the new bushing seats it to the rod bore, provides good heat transfer, and controls pin to bushing clearance, figure 12-73. The tolerance for oil clearance between the pin and the bushing is so small that if the bushing were simply pressed into place the bushing would pound itself into the pin bore when the engine was run, causing excessive clearance. Follow these steps to expand the bushing:

1. Select the proper mandrel for the bushing size, figure 12-74.
2. Adjust the spindle speed to 200 rpm and turn the cutting pressure all the way to the right, figure 12-75.

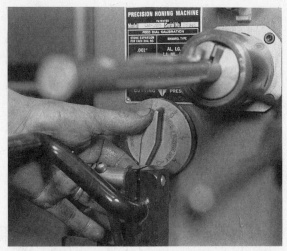

Figure 12-75. Adjust the spindle speed then turn the cutting pressure dial all the way to the right.

Figure 12-76. Fit the pin bore over the mandrel and direct coolant flow onto the hone.

Figure 12-77. Increase pressure with the feed dial until you can feel the expander tip contacting the bushing.

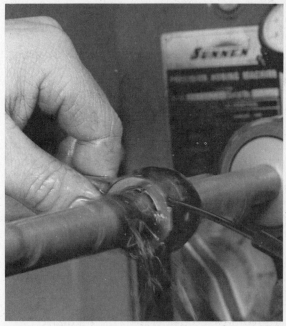

Figure 12-78. Slowly stroke the rod across the hone as you advance the feed dial an additional two or three numbers.

3. Fit the pin bore over the expanding mandrel tip and direct a light stream of oil onto the expanding mandrel, figure 12-76.
4. Begin stroking the rod over the mandrel while slowly increasing pressure with the feed dial until you can feel the expander tip contacting the bushing, figure 12-77.

NOTE: If the bushing has a large oil drill hole or groove make sure that the expander bit does not fall down into it.

5. Continue slowly stroking as you advance the feed dial an additional two or three numbers, figure 12-78.
6. Remove the rod from the mandrel and examine the inside of the bushing. It should have a burnished appearance all the way across, figure 12-79. If not, expand the feed dial several thousandths and repeat the operation.
7. Use the facing cutter on the honing machine to trim off any bushing material that may have been squeezed out past the end of the rod bore.
8. Finish hone the bushing to the proper clearance and test fit the wrist pin. The pin should slide in easily by hand.

Servicing Crankshafts, Flywheels, Pistons and Rods

Figure 12-79. A properly expanded bushing will have an even burnished appearance all the way across.

Figure 12-80. Fit the ring to be checked into the cylinder.

Figure 12-81. Invert a piston and push the ring down until it is below the ring travel area of the bore.

Piston and Connecting Rod Reassembly

Unless you are going to balance the engine, you are now ready to assemble the pistons, connecting rods, and rings. The procedures for balancing can be found in Chapter 14 of this *Shop Manual*. Balancing, although not always required is highly recommended, and must be done before the parts are assembled. If you are going to balance the engine, skip ahead to the next chapter, perform the balancing operations, and then return here for reassembly.

Final pre-assembly inspection
Your first assembly step should be to double check clearances and measurements. Your inspection will include: piston and cylinder bore diameters, piston ring side clearance and end gap, wrist pin clearance and interference fit, rod bearing bore diameter, and rod bearing oil clearance. Most of these operations were detailed previously in this chapter, and will not be repeated here. The technique for determining rod bearing clearance is similar to that used when reconditioning the main bearing bores, follow the steps detailed in Chapter 11 of this *Shop Manual*.

Checking bearing clearance
A quick and easy way to check the rod bearing oil clearance is to use the bore gauge on the honing machine. Fit the bearing shell inserts to the rod and cap, install the cap and tighten the nuts to proper torque. Now, measure the crankshaft rod journals and set the bore gauge to that dimension. Carefully slip the big-end of the rod, with the bearing installed, onto the bore gauge. Oil clearance, the difference between the set figure (journal diameter) and the bearing inside diameter, will be displayed by the dial indicator.

Checking piston ring end gap
The only operation not previously explained is checking the piston ring end gap. Do this by fitting the rings into the cylinder bores and measuring the space between the ends. Fit a ring into the cylinder, figure 12-80, invert a piston and use it to squarely push the ring down into the cylinder below the bottom of ring travel, figure 12-81, then use a feeler gauge to measure the gap between the ends of the ring, figure 12-82. End gap is usually listed in the manufacturer's specifications as a tolerance value that corresponds to piston size. You can expect to see about three thousandths of gap for each inch of piston diameter.

If the gap is too small, you can open it up by filing the ends of the ring with a special ring gap cutter, figure 12-83. Ring gap can also be opened up using a small file, be careful not to damage the ring. Always file from the outer face of the ring towards the inside diameter, this prevents chipping any surface coating or leaving any burrs on the ring face. After you

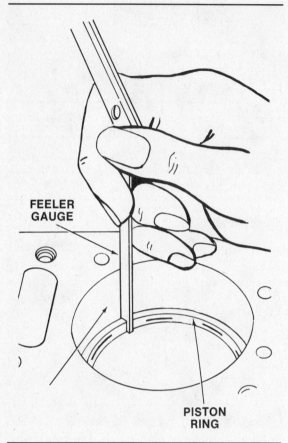

Figure 12-82. Use a feeler gauge to measure the end gap.

Figure 12-83. This ring gap cutter will quickly remove metal from ends of the ring to open up the end gap.

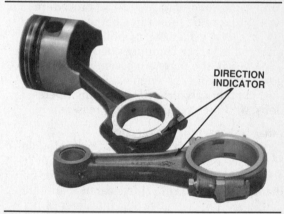

Figure 12-84. Two different methods manufacturers use to indicate connecting rod direction, at the top is a Honda rod engine, the lower rod is from a Volkswagen engine.

cut down the ring ends, dress the area with a soap stone to remove any rough edges.

Most piston ring sets are packaged in units, each unit contains the rings for one piston, keep the rings together as units when you check them. Mark the rings so that on final assembly, you can install them in the bore they were fitted to. If all of your measurements are within specification, you can fit the pistons to the rods, then install the rings on the pistons.

Installing Pistons On Rods

Connecting rod installation methods depend on whether the wrist pins are full-floating or press-fit. With either type, your first step should be to organize the parts so that they are ready to assemble in the correct position and order. As mentioned in the *Classroom Manual*, the pistons, rods, or both may be offset from their centerlines. Since the amount of offset is usually slight, and not visually apparent, many manufacturer's mark their parts for ease of assembly. Pistons often have an arrow or a small notch stamped or machined on the top as a directional indicator. When properly assembled the arrow or notch generally points towards the front of the engine. Connecting rod markings are usually more subtle, you may have to orientate them in relation to the part or casting number stamped on the face, a raised indicator on the side of the rod, an oil bleed at the parting line, or the position of the tangs on the bearing shells, figure 12-84. If you disassembled the engine and are reusing the rods, proper alignment should be no problem, the numbers or punch marks you stamped on the rods and caps will line up the same way they were before teardown. Putting a small

Servicing Crankshafts, Flywheels, Pistons and Rods

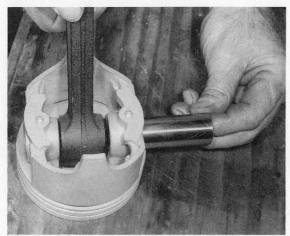

Figure 12-85. Full-floating wrist pins should slip through the piston and connecting rod easily by hand.

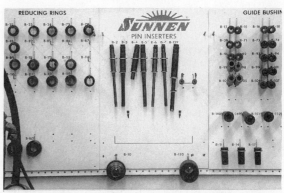

Figure 12-87. An assortment of arbors and adapters are used to press-fit wrist pins.

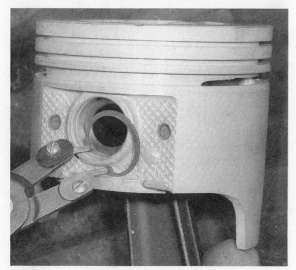

Figure 12-86. The open end of the snap-ring should always face down, toward the crankshaft.

dab of brightly colored paint on the outside of the rod will help you position the parts accurately and quickly during assembly.

Installing rods with full-floating wrist pins
Installing the pins in full-floating pistons is relatively easy because the pin will slide through the piston and rod by hand. Start by fitting the pin into the boss on one side of the piston so that it extends slightly past the boss inside the piston. Lightly lubricate the connecting rod bushing, slip the rod into the piston and hook the small-end bore over the end of the rod; check to be sure the parts are positioned correctly to each other, then push the pin through the rod bushing and into the boss on the other side of the piston. The parts should fit together easily by hand, figure 12-85, and the pin should be free to rotate in both the piston and the rod. If not, disassemble and recheck the clearances. The final step is to install the snap-rings into the grooves in the piston at either end of the pin.

Remember to handle the snap-rings carefully, do not compress them any more than is absolutely necessary to fit them into the piston. Also, make sure that the snap-ring firmly seats into the groove, they must fit tightly and bottom in the groove around the entire circumference. The open end of the snap-ring should always face the bottom of the piston, toward the crankshaft, figure 12-86. The greatest stress a running engine puts on the snap-ring occurs at the top of the ring when the piston changes direction crossing over top dead center. Positioning the strongest part of the snap-ring at the top of the pin bore can help insure against failure. Snap-rings that are a one-ear design are installed so that the ear points out, away from the wrist pin and toward the cylinder wall.

Installing rods with press-fit wrist pins
Press-fit, or stationary, wrist pins are more difficult to install because you have to overcome the interference between the pin and rod. Stationary pins can be installed in one of two ways: press them in using a hydraulic press, or heat the small-end bore of the rod to expand it so that the pin can be slip fitted. Both methods require the use of special tooling, the procedures for both are described below.

Press fitting wrist pins requires not only a hydraulic press, but also an assortment of arbors and adapters to support the piston and drive the pin, figure 12-87. This is basically the reverse of the operation you used during disassembly. Cutting a slight chamfer on the leading edge of the rod bore will help the pin lead itself in, and reduce the chance of galling. Special chamfering tools are available for dressing the rod bore.

Figure 12-88. The press uses hydraulic force to drive the pin through the piston and center it on the rod.

Figure 12-89. Rod heater heaters expand the small-end bore so the wrist pin can be fitted by hand.

Carefully select your press attachments, keep in mind that using the wrong adaptor will destroy the piston. The bottom adaptor must squarely fit a reinforced area of the piston, properly support the piston, and keep the piston boss parallel to the press ram. The arbor must be slightly smaller than the pin outside diameter, yet large enough to get a firm footing on the end of the pin to provide adequate push. Lubricate the pin with a light-weight pressing oil, then use the press to draw up the slack between the ram, arbor, wrist pin, piston, and support fixture, figure 12-88. *Recheck your alignment*, push the pin through the piston and rod by applying steady even pressure with the press. Be sure the rod is properly centered on the pin.

Installing wrist pins by heating is generally the preferred method because it does not require extreme pressure to overcome the interference-fit. Pressing in pins places a high level of stress on a small area of the piston, this tends to distort the assembly. If not done properly the piston can break during the press operation. There is also the possibility of creating a stress riser that could lead to premature failure once the engine is up and running. Heating the small-end bore of the rod expands its diameter and allows you to fit the pin with hand pressure, this controls distortion and ensures a quality assembly.

Rod heaters may be either electric or gas powered, a quality rod heater will have a special vise to hold the piston and a fixture to guide the pin as you press it in, figure 12-89. Heaters work by raising the small-end temperature of the rod, usually to about 425°F (204°C), this expands the bore just enough for the wrist pin to slip through. *Avoid overheating* the rod, if the metal begins to turn blue it is too hot, remove it immediately. Handle the big-end of the rod with care, wear a good heat resistant glove. Keep in mind, rods are designed to transfer heat, even the very end of the cap can get hot enough to cause severe burns. The piston is held firm in the vise, the fixture aligns the piston boss and guides the pin as you insert it through the hot rod bore. The fixture also limits the travel of the pin to prevent it from over-centering on the rod.

Like all shop equipment, the variety of rod heaters in the field is vast. Once you are familiar with the set up, controls, and the feel of the machine, the procedure is fairly simple.

Place the rod in the heater and turn the power on. As the rod is heating up, clamp the piston in the vise, adjust the limiting fixture, and fit the pin onto the insertion tool. Once the rod is up to temperature remove it from the heater, place the small-end bore inside the piston, be sure you have the piston and rod correctly positioned to each other. *You must work quickly*, the rod will cool rapidly once you remove it from the heat chamber. Use the insertion tool to push the pin through the piston and the rod bores, the pin must be inserted all the way until it contacts the limiting fixture.

Servicing Crankshafts, Flywheels, Pistons and Rods

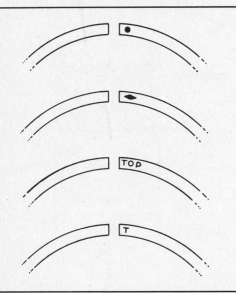

Figure 12-90. Piston ring manufacturers use a variety of different markings to show proper installation position.

Piston ring installation

After properly hanging the pistons on the rods you are ready to install the piston rings. This is the final step before the assemblies can be fitted into the block.

Before you begin, look the rings over. Keep in mind that most compression rings have a top and bottom, rings installed on the piston upside down cannot effectively hold back compression and will cause increased oil consumption. Look for markings on the top face of the rings; they may be stamped with "TOP", "T", a dot, a notch, or some other symbol, figure 12-90. Some rings are cut with perfectly square sides and can be installed either side up; however, most are beveled, grooved, or tapered, and will only seal in one direction. If there are no marks on the rings, follow these general guidelines: install counterbored rings (with a groove on the inside facing the piston) with the groove facing up toward the piston top, install rings with a groove on the outside (facing the cylinder wall) with the groove facing down toward the piston skirt, figure 12-91. Install rings with a tapered face with the taper up, the widest side of the ring towards the skirt and the narrowest side towards the piston top, figure 12-92.

Oil control rings are available in a wide assortment of designs, see Chapter 12 of the *Classroom Manual*. Most designs are a multiple piece assembly, there are so many variations, we could not possibly cover the installation of every type. We detail assembling and install-

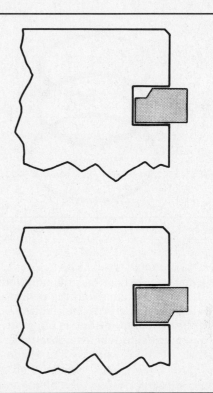

Figure 12-91. An undercut on the inside of the ring should face up, a groove on the outside faces down.

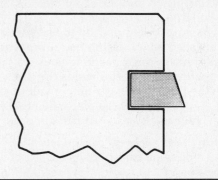

Figure 12-92. Piston rings with a tapered face are installed with the taper toward the top of the cylinder.

ing two of the most popular design configurations, two and three piece oil ring sets.

Checking ring side gap

The first step in ringing the pistons is to check the ring side gap. Side gap is the difference between the width of the ring and the side-to-side distance of the ring groove. As a general rule, the side gap of a compression ring should be greater than 0.0015 inch (0.038 mm) and less than 0.006 inch (0.15 mm). Too much side gap can allow the ring to twist past its limit, leak compression and possibly break. Too little gap can cause the ring to expand wider than the groove, seizing it to the piston.

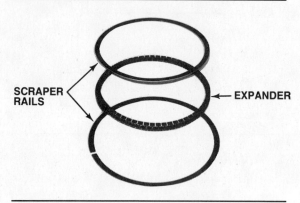

Figure 12-93. Three-piece oil control rings use a flexible expander and two narrow sealing rails.

Side gap is easy to check, the only tool you need is a feeler gauge. Hold the piston and slip a ring into the groove, so that the outer face of the ring bottoms out in the groove. Side gap is equal to whatever size feeler gauge will easily fit into the groove along side the ring, figure 12-42. Check the gap of all the compression rings at several points around the ring and on the piston. Keep your parts well organized as you work. Rings should be gapped and fitted to a specific groove on a specific piston and the assembled piston is installed into a specific bore.

The lowest ring, usually the oil control ring, is installed on the piston first. Then you fit the compression rings, working your way up the piston from the skirt towards the top. Assemble the pistons one at a time, fit all the rings on one piston, then go on to the next. Organize your work and keep parts in order.

Fitting three piece oil control rings

Three piece oil control rings consist of a spring steel spacer and two narrow sealing rails, figure 12-93. They can be assembled by hand, first fit the expandable spacer, then the top rail, and finally the lower rail. Once you assemble the pieces of the ring, the spacer provides tension to hold the sealing rails against the cylinder wall and direct oil flow. The two open ends of the spacer must butt together without an end gap, *never allow the ends to overlap*. When first fitted to the piston the spacer may appear too large in diameter. *Do not trim* the ends of the spacer to make it fit in the groove. Keep in mind the spacer is a spring that pushes the rails away from the piston against the block to scrape oil from the cylinder wall. Cutting the ends of the expander will reduce spring tension and lead to early failure. Once the bottom and top rails are fitted, the entire assembly will be properly drawn into the ring groove.

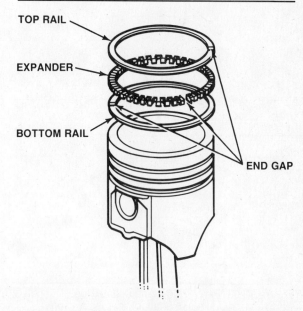

Figure 12-94. Assemble three-piece oil control rings as shown. Position the scraper rail end gaps at 45° to either side of the expander gap, the expander gap is at a 90° angle to the pin bore.

Begin by placing the spacer into the ring groove, butt the ends together at a point perpendicular to (90° from) the wrist pin. Positioning the ring ends is important. When assembly is complete, the end gaps of the two rails should be about 45° to opposite sides of the spacer ring parting line, figure 12-94. Hold the butted ends of the spacer firmly in place with your thumb while you fit the top rail. Start the open end of a sealing rail into the ring groove on top of the spacer, now carefully wind the ring around the piston and into the groove. Sealing rails are for the most part extremely flexible and easy to fit, on the other hand, they are fragile and easily damaged. *Avoid sharply bending* the ring, any scratch, kink, or heat riser you create on the surface can lead to early failure. If the top rail is seated correctly it can be easily rotated to position the end gap. Most rails will also hold the butted spacer ends together, so you can relax your thumb. The bottom rail installs similar to the top. Start a rail end in the ring groove below the spacer, make sure both rail and spacer are seated in the groove. Hold the end in place with your thumb, work your way around the piston carefully feeding the rail over the top rail and spacer then into the groove. Oil control rings will rotate freely around the piston as an assembly if they are properly installed, you can also easily reposition the end gaps of the individual rings.

Servicing Crankshafts, Flywheels, Pistons and Rods

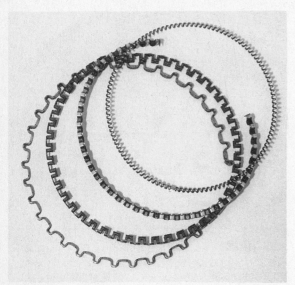

Figure 12-95. There are a variety of flexible expander designs used to tension oil control rings.

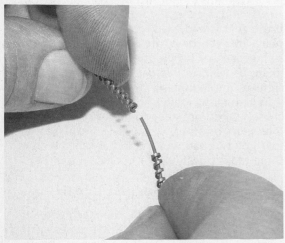

Figure 12-96. Coil wire expanders have a stake pin that holds the ends of the spring together during assembly.

Figure 12-97. A piston ring expander spreads the ring evenly during installation.

Fitting two piece oil control rings

Two piece oil control rings generally consist of a cast iron ring and a flexible expander. Treat the cast iron ring like a compression ring, check both the end gap and side gap, and always handle with care. The expander fits into the ring groove behind the ring, expanders often fit into a slot machined in the back of the ring. An expander is a spring, many designs are used, figure 12-95, the ends butt together to form a ring that loosely fits around the piston ring groove. The cast iron ring fits over and compresses the expander. The tension created by the compressed spring pushes the assembly into the cylinder to scrape the wall and direct oil flow.

First fit the expander around the piston and into the ring groove with the ends butted together. Coil expanders usually have a wire "stake pin" on one end, figure 12-96. Insert the pin all the way into the opposite end of the coil so that the two ends of the coil firmly butt together. The ends of hump-type expanders are also butt fitted to each other, *never overlap the ends* — damage and premature failure can result. Hump-type expanders are normally split on the crest of a hump so the butted ends face out of the groove toward the ring and away from the piston. All expanders should be positioned on the piston with their parting line at a 90° angle to the wrist pin. The ring is installed over the expander with the open end on the opposite side of the piston, 180° degrees from the expander parting line.

Handle the ring very carefully, cast iron is fairly brittle and breaks easily. Avoid twisting and screwing the ring down onto the piston. Use a piston ring expander to spread the ring, figure 12-97, open the ring just far enough to fit it over the expander. The installed ring assembly should rotate freely in the groove.

Fitting compression rings

The majority of compression rings used in automotive applications are made of cast iron, so they are brittle and you must handle them with care. Although cast iron is a reliable long-wearing material, the amount of twist a ring can tolerate is minimal. Install compression rings using a ring expander, figure 12-97. This easy to use tool equalizes stress around the ring as it is stretched open.

Fit the lower compression ring into the ring expander so that the top of the ring is facing up. The two jaws of the tool fit into the ring end gap, squeezing the handle moves the jaws away from each other and opens the ring by spreading the end gap. Open the ring just far enough to slip the ring over the piston top and down to the groove, slowly release the handle to fit the ring into the groove, then remove the tool. Check the ring-to-groove fit, the ring should spin freely. Repeat the process for the top compression ring.

After the top ring is in position, set the piston assembly aside, it is ready for installation in the block. Repeat the process with the remaining pistons, fit the oil control ring first, then the compression rings from bottom to top. Store the assembled pistons in a well protected place until you are ready to install them into the engine block

FLYWHEEL AND FLEXPLATE SERVICE

Flywheels and flexplates are a part of the reciprocating mass of the engine and perform several important functions. They couple the engine and transmission, provide the gearing for the starter to turn the engine, and also assist in balancing the engine. Do not overlook their importance, give them a thorough inspection. You must replace a flexplate that is not up to standard, you can often repair a flywheel and return it to service.

Flexplate Inspection

Flexplates are prone to wear and should be inspected. A visual check is normally all you need to determine if the flexplate is in usable condition. If something looks questionable, you can follow up with magnetic particle inspection. Pay close attention to bolt holes, both where the plate bolts to the crankshaft and where the torque converter bolts to the plate. Also take a good look at the flanges or any reinforced area surrounding the bolt holes. Closely inspect the starter ring gear, it is a prime wear zone.

Bolt holes should be round and concentric, look for any signs of distortion. Damage from bolts that were not tightened properly is fairly common, movement of the flexplate will cause the bolt holes to wear in an oval shape. Also look over the flanges for hairline cracks, stress risers, or any other signs of metal fatigue. Cracks in the plate, also fairly common, can be the result of improper installation, careless handling, or simply age. If any stress indicators are found replace the flexplate now, you do not want to have to do it once the engine is installed in the car.

Most flexplates are made of stamped steel, a hardened ring gear attached to the outer edge provides a coupling for the starter to turn the engine. The ring gear may be fitted to the

Figure 12-98. If the flywheel surface has hard spots, heat checks, or cracks it will have to be resurfaced.

plate with an interference fit, welds, bolts, or rivets, and is generally not a serviceable item. If the gear is bad, you replace the entire flexplate. Inspect the entire circumference of the ring gear, look for chipped, missing, or broken teeth, as well as unusual wear patterns and signs of misalignment.

In general flexplates are easy to deal with because you do not have to be concerned with tolerances or other variables. A flexplate is either good or bad with no in between. If anything looks questionable, replace the unit.

Flywheel Inspection and Repair

Flywheels are a wear item, the machined surface that the clutch friction disc presses against takes a lot of abuse. After long service the flywheel surface will develop hard spots, heat checks, and possibly cracks, figure 12-98. If the flywheel is returned to service with a new clutch disc, the clutch is likely to grab and chatter when it is engaged. This is because the surface is not smooth and even, and the new disc has a hard time seating properly on this irregular surface. A damaged flywheel can often be salvaged by machining the surface to ensure good clutch contact.

Like flexplates, flywheels also have a ring gear that engages the starter motor to turn the engine over. Unlike flexplates, flywheel ring gears are usually replaceable so gear damage alone is no reason to scrap a flywheel.

Ring gear replacement

The starter motor ring gear is frequently overlooked on flywheels. In many cases the ring gear teeth are worn or missing because of poor

Servicing Crankshafts, Flywheels, Pistons and Rods

Figure 12-99. A flexplate with starter ring gear damage should be replaced.

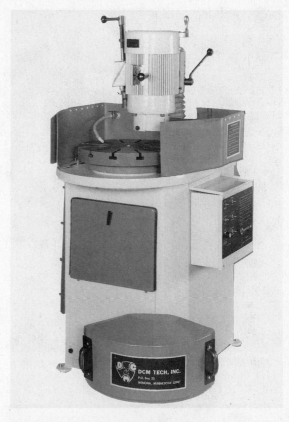

Figure 12-100. The flywheel grinder rotates the flywheel as the surface is reconditioned by a grinding stone.

starter motor engagement, figure 12-99. The ring gear is a shrink-fit on many flywheels and can easily be replaced. This is done by heating the gear to expand the metal, an oxy-acetylene torch is the only tool you need. To replace a ring gear proceed as follows.

Begin by removing the old ring gear. Most flywheels are undercut to fit the gear, so the gear will only come off and go on in one direction, usually toward the engine side. In addition, the gear teeth are often cut at an angle or beveled on one side for positive starter engagement. Place the flywheel on a workbench and support it with a block of wood, the flywheel should be flat so that the gear can drop off as you heat it. Fit the torch with a large tip, preferably the rose-bud type, that will produce a wide flame and heat as much of the gear as possible. Light and adjust the torch, then begin applying the flame to the ring gear. It is important to concentrate the flame on the gear, the gear will not loosen if the flywheel itself is heated. Heat the entire gear evenly, work the torch in slow circular sweeps along the circumference of the gear. Be patient, bringing the temperature up high enough will take time. Eventually, the heat expands the gear to the point where it will drop free of the flywheel, a few raps with a punch and hammer will help it along. *Handle with care*, both the ring gear and the flywheel can be hot enough to cause severe burns. Allow the flywheel to completely cool before you attempt to install the new gear.

To install the new ring gear lay the flywheel flat so the gear can be fitted from the top. Place the ring gear over the flywheel so it is in position to fall into place. Some flywheels have a slight chamfer leading into the ring gear seat, this makes it easy to position the gear for heating. Light your torch and apply even heat to the gear, just as you did for removal. The heat will slowly expand the gear to the point where it will slip into place. Be sure it bottoms out in the seat of the flywheel, a few careful raps with a hammer will guarantee a good seat. As the ring gear cools it shrinks and holds tightly to the flywheel. Let the flywheel air cool, trying to speed things up with water, oil, or compressed air can cause the part to cool unevenly, lose its temper, and possibly crack or warp.

Flywheel machining

If a flywheel has any hard spots, heat checks, or surface cracks it is possible to restore the surface. You can resurface flywheels by turning them on a lathe or by grinding them with a surface grinder. If there are any hard spots on the surface of the flywheel it cannot be reconditioned using a lathe. Surface grinding will remove hard spots, as well as heat checks and cracks, so grinding is a more common practice. The flywheel grinder, figure 12-100,

Figure 12-101. Position the flywheel in the center of the turntable and bolt it down using the proper adapters.

Figure 12-103. Fasten a dial indicator to the machine with the probe of the indicator contacting a smooth flat surface on the side of the flywheel.

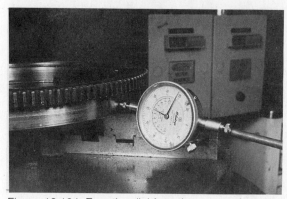

Figure 12-104. Zero the dial face then rotate the table as you watch the needle, the indicator displays how far off center the workpiece is.

Figure 12-102. Flywheels that have a stepped surface require special care, the flywheel must be exactly centered on the rotating table.

has a turn table that rotates the flywheel and a grinding stone that produces a smooth surface on the flywheel face as it rotates.

Of course there are many different flywheel grinders available, even some milling machines have attachments that allow you to grind flywheels. You must be familiar with the equipment, operating techniques vary from machine to machine. The following is a general procedure for restoring the surface of a flywheel by grinding:

1. Select an adaptor ring that will correctly fit and support the flywheel flange. Place the ring in the center of the turn table, then lay the flywheel face up on top of the adaptor. Attach the flywheel to the bed using the proper size centering cone and through bolt, figure 12-101.
2. If the flywheel has a step in it where the pressure plate is mounted, figure 12-102, you must center the flywheel exactly on the bed. This ensures that the grinding stone will be concentric to the flywheel so that when the stone gets near the step it will hit evenly on both sides. Center the flywheel as follows:

- Fasten a dial indicator to the machine with the probe resting on the edge of the flywheel, figure 12-103.
- Then rotate the bed slowly as you watch the dial indicator. The amount that the flywheel is off center will be indicated by the needle, figure 12-104.
- Correct runout by shifting the position of the flywheel. This is done by adjusting the chucks that support the flywheel in their ways.

Servicing Crankshafts, Flywheels, Pistons and Rods

Figure 12-105. Before taking a cut, measure the distance from the step to the clutch surface with a depth micrometer.

Figure 12-106. True the grinding stone with a diamond stone dresser to ensure a good surface finish.

Figure 12-107. Move the grinding head into position, start the coolant pump and direct the flow over the flywheel.

- Recheck the concentricity by rotating the bed and taking another reading with the dial indicator.

3. When grinding stepped flywheels, the face of the step must be ground the same amount as the clutch surface. This is because as the clutch surface is ground, the surface is moved away from the pressure plate. The increased distance reduces tension on the clutch disc and will cause it to slip in service. Measure the distance from the step to the clutch surface with a depth micrometer before machining, figure 12-105. Write the figure down, you will use it later to determine how much metal to remove from the step after the clutch surface has been ground. Grind the clutch surface, then measure the step to surface distance again. The difference in the two readings is the amount of material you must remove from the step.

4. Before you begin grinding the flywheel, true the stone with a diamond stone dresser, figure 12-106. Turn on the machine and run the diamond tip slowly across the face of the stone.

5. Position the stone over, but not touching the flywheel, direct the flow of cutting fluid onto the face of the flywheel, figure 12-107, and turn the machine on.

6. Slowly adjust the feed control to bring the grinding stone into contact with the flywheel surface. Continue to gradually feed the stone in until the flywheel surface is smooth and uniform, figure 12-108.

If the flywheel is flat, the job is now completed. If the flywheel is stepped, you will have to remove an equal amount of metal from the face of the step as you did from the surface. Leave the flywheel on the grinder, then use your depth micrometer to measure the dis-

Figure 12-108. Gradually feed the stone into the flywheel, continue grinding until you have a smooth uniform surface.

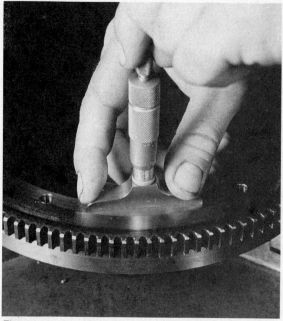

Figure 12-109. After you grind the clutch surface, take another reading with your depth micrometer to determine how much metal to remove from the top of the step.

tance from the step to the surface, figure 12-109. Subtract the reading you got before grinding from the new reading, the result is the amount of metal that must be removed from the step. Reposition the grinding stone so that it will contact the step, adjust the coolant flow, turn the power on, and slowly feed the stone in. Keep an eye on the feed micrometer, note the reading when you first hear the stone contacting the flywheel. Continue to make feed adjustments until the required amount of metal has been removed. Double check your cut with a depth micrometer before you remove the flywheel from the machine, do not rely entirely on the machine's micrometer dial.

Surfacing flywheels, like all other machining operations, is something you gain a feel for only through experience. Practicing on scrap flywheels will help you develop your technique, and familiarize you with the equipment. Keep the stone in good condition. A stone that is not sharp and true causes heat build up, cuts slowly, and cannot provide an adequate finish. A properly dressed stone will create a large even shower of sparks from the workpiece as it cuts the new surface. Dress the stone before you use it. Stones are available in different grit and hardness, be sure to select the right one for the job at hand.

Chapter 13
Engine Balancing

As we learned in Chapter 11 of the *Classroom Manual*, balancing an engine consists of balancing both its reciprocating and rotating parts. Balancing the reciprocating parts such as pistons, wrist pins, and connecting rods involves weighing each part and making each part conform to the lightest one. The smaller the difference in weight between parts, the better balanced the engine.

Balancing the rotating parts of an engine includes the crankshaft, flywheel or flexplate, pressure plate, and damper. This is done by spinning them in a balancing machine, figure 13-1 and removing or adding weight in the right places. To balance any engine, V-type or inline, you need the following parts:

- Crankshaft
- Vibration damper or harmonic balancer and any pulley that bolts to it
- Flywheel or flexplate
- Pressure plate
- All bolts, lock washers, keys, and spacers required to assemble the above parts on the crank
- Connecting rods
- Pistons
- Wrist pins.

To balance a V-type engine, in addition to the parts above you need *one* set of each:

- Rod bearings
- Piston rings
- Wrist pin locks (if the engine has full-floating pins)

It might seem that you would need each set of these last three parts to weigh with the individual piston or rod to get an accurate figure. However, variations in the manufacture of these parts is so slight that you can assume that the weight of one set is interchangeable with any of the others.

BALANCING PRECAUTIONS

Before you begin the balancing procedure, all parts must be clean and all new parts selected. The rods and pistons must be separated and the rod caps installed with the bolts snugged, but there is no need to torque them. It's safer and easier to work on clean parts, and the rods and pistons will need minor but essential grinding and machining as part of the balancing procedure, which is not possible if they are pressed together.

Also, all the other machining should be complete. A ground and indexed crank is a good example of why this is important. Besides changing crank weight and balance, this machining operation alters the relative position of

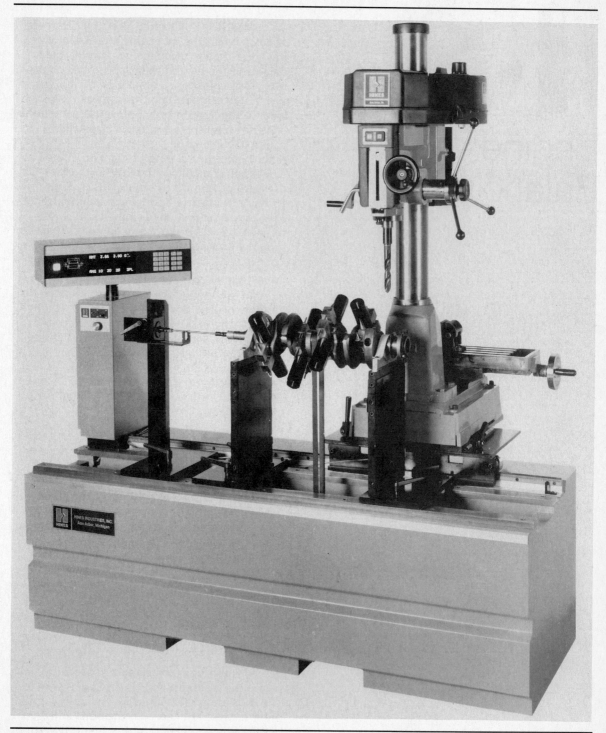

Figure 13-1. Computerized balancing machines, like this Hines HC500, greatly simplify engine balancing.

the rods, bearings, and pistons, which further changes engine balance.

The rods are another good example. They must have been resized and, if necessary, new rod bolts installed because each of these things can change the weight of the rod and the balance of the engine. Any post-balancing operation should be limited to micro polishing the crank and giving all the parts a final cleaning.

Custom Balancing

When an engine is rebuilt, many times new parts are mixed with resized or refinished orig-

Engine Balancing

```
Customer: _____        V8 Engine
Address:  _____        Job # _____
Phone #:  _____        Date  _____
Engine:   _____

Rod (big end)    _____    Complete Engine     _____
Rod (big end)    _____    Less Flywheel       _____
Rod Bearing      _____    Less Pressure Plate _____
Rod Bearing      _____    Less Damper         _____
Oil Allowance    _____
Piston           _____
Locks (one set)  _____
Rings (one set)  _____
Rod (small end)  _____
Bob Weight Total [_____]
```

Figure 13-2. A sheet like this will help you keep track of engine part weights and make balancing calculations.

Figure 13-3. Digital scales can give accuracy down to one tenth of a gram.

inal parts. Sometimes only one rod or one piston is replaced. Or maybe all the pistons are replaced but none of the rods. Since aftermarket parts often weigh more or less than original parts, when the engine is assembled it can be badly out of balance. An out-of-balance engine increases bearing wear and also places more strain on all moving parts, greatly reducing the life of the engine. The customer may notice that the engine doesn't run as smoothly as it did before it was rebuilt and may be dissatisfied with your work for that reason alone.

Balancing is important for both engine life and customer satisfaction. However, balancing the rotating parts requires a special balancing machine and adds expense that a customer may not want. For this reason, some rebuilders balance only the reciprocating parts, which is acceptable only as a minimum effort. Even the original carmakers balance both the reciprocating and rotating parts, but they do not balance them as closely as you can in your shop when you perform a custom balance job. Balancing both the rotating as well as the reciprocating parts is an essential part of a *quality* engine rebuild.

Balancing Records

The balancing procedure requires you to take many weight measurements and to make calculations. You must record these on a special sheet to keep the many numbers straight and to refer back to during the procedure, figure 13-2.

Also, keeping these records on file can help your customer in the future and be a good reason for repeat business. If he or she rebuilds the engine in a few years and changes a piston or rod, you already have a record of what that part should weigh. It would be an easy job to grind that part to the proper weight, keeping the engine in balance without having to go through the entire balancing procedure again.

BALANCING RECIPROCATING WEIGHT

The first step in the balancing procedure is to equalize the reciprocating mass, which includes the pistons and rods. Weigh each piston and find the lightest one, figure 13-3. Some machinists weigh the pistons by themselves and ignore the wrist pin during this part of the balancing procedure. They can get away with this because weight variations between wrist pins are very slight. However, it is best to weigh the piston and wrist pin together as a set to produce the finest balance job.

Trim each piston to make it equal to the others within $1/2$ of a gram. This is where experience is very important. Remove too much metal from a piston and you have to start over again, balancing to your new, lightest piston. If you weighed the piston and pin together as a set, you are equalizing each piston and pin set, although you remove metal from the piston *only*. It is best not to remove metal from the wrist pin because it is such a highly stressed part.

Where to remove the metal from the piston depends on the part. You can use a lathe to remove weight from balance pads under the

Figure 13-4. Lighten each piston to within 1/2 gram of the lightest piston.

Figure 13-5. An end mill can remove metal from some forged pistons.

Figure 13-6. Some cast pistons have balance pads on either side of the pin boss.

Figure 13-7. A grinder is an acceptable method for removing metal from a connecting rod.

piston skirt, figure 13-4. If the piston has no pads, some forged pistons are strong enough to allow metal removal from underneath the pin boss with an end mill, figure 13-5. Some cast pistons have balance pads on either side of the pin bosses where metal can be removed without weakening the piston, figure 13-6. If you are unsure about a safe location for removing metal, consult the piston manufacturer.

Next, weigh each connecting rod and find the lightest rod. Use a belt sander or a grinder to remove the necessary metal from the other rods to make them weigh within 1/2 of a gram of the lightest one. Connecting rods often have a balance pad at the pin end and another pad on the cap, figures 13-7 and 13-8. If the rods have no pads, it is acceptable to carefully remove metal from the top of the small end and from the bottom of the cap. Remove metal first from the big end, then from the small end. If you remove more than 6 or 7 grams from the small end, the balance of the big end may change, so you must double check them.

Ordinary balancing is done by simply equalizing overall connecting rod weight. A better job is done with a double scale, which measures the weight of each end of the rod, figure 13-9. You then grind the pads so that each end of the rod weighs the same as the others while still keeping total rod weight equal to the other rods within 1/2 of a gram.

With the reciprocating weights equalized, the crankshafts of inline engines are ready to be balanced. If you are working on an inline engine, skip down to the section on balancing rotating weight. If you are balancing a V-type engine, go to the next section.

THE BOB WEIGHT

V-type engines require you to calculate a bob weight measurement and add one bob weight to each rod journal on the crankshaft, figure 13-10. When you spin the crank to balance it,

Engine Balancing

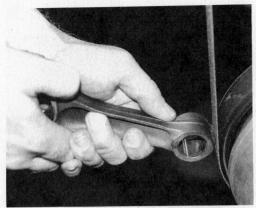

Figure 13-8. A belt sander removes weight more quickly than a grinder.

Figure 13-10. The bob weights simulate the reciprocating weight when you spin the crank on a balancer. Bob weights are essential for proper balancing of V-type engines.

Figure 13-9. A fixture that hangs the rod allows you to weigh each end individually.

Buick

3.2-, 3.8-, and 4.1-liter 90-degree V6 .... 36.6%

Chevrolet

3.3-liter V6 .... 46%
3.8-liter V6 .... 46%
4.3-liter V6 .... 35.2%

Oldsmobile

4.3-liter diesel (even fire) 35.2%

Ford

3.8-liter 90-degree V6 .... 39.4%

Figure 13-11. This is a partial list of engines that do not use a 50-percent balance factor for calculating the bob weight.

the bob weights simulate the weight of the rods and pistons. You must install a bob weight on each journal. The bob weight must equal the total of the following for each journal:

- The rotating mass
- A percentage of the reciprocating mass
- An oil allowance.

In this case, the rotating mass includes the big end of the connecting rod and the rod bearings. On engines, such as V8s, where two rods share a crank pin, you add *both* rod big ends into the equation.

The reciprocating mass includes the small end of the rod (or rods) and the piston assembly(s) (including rings, pin, and locks). To determine the percentage, called the balance factor, engineers use a formula which takes in-

to account a common rpm range for the engine, cylinder block angles, and other factors. This percentage is most often 50 percent, but there are exceptions. For example, most Buick 90-degree V6 engines use a 36.6 percent balance factor, figure 13-11.

The oil allowance is the estimated amount of oil flowing through the crank pin. Experienced machinists add four grams for regular cranks and 15 grams for cross-drilled cranks.

Calculating the Bob Weight

Let's do two examples, a 3.8-liter Buick V6 and a 5.7-liter Chevrolet V8. The Buick engine has a 90-degree block and a split-pin crank. None of its rods share a crank pin, so we must pre-

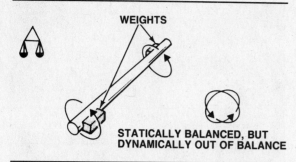

Figure 13-12. The shaft will wobble if it is not dynamically balanced.

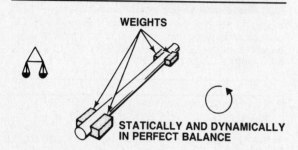

Figure 13-13. This shaft is both statically and dynamically balanced. The weight is distributed evenly around the center of the part from end to end.

pare six bob weights. The following are the Buick's weights:

- Big end of the rod..............386 g
- Rod bearings (one set)...........80 g
- Oil allowance....................4 g
  - Rotating weight...........470 g
- Piston........................450 g
- Pin..........................110 g
- Locks (one set)..................0 g
- Rings (one set)..................45 g
- Small end of the rod...........150 g
  - Reciprocating weight.......755 g

Next, take 36.6 percent (the balance factor) of the reciprocating weight of 755 grams.

.366 × 755 g = 276.33 g

Then, add the rotating weight to the adjusted reciprocating weight.

470 g + 276.33 g = 746.33 g

So, each of your six bob weights for the Buick engine must weight 746.33 g.

The Chevrolet 5.7-liter V8 is a typical 90-degree V8. Each pair of rods shares a common crank pin, so you need to prepare only four bob weights. Balancing this type of engine requires a special step and allows one short cut. The special step is that the rod big end and bearing weight must be doubled because there

Figure 13-14. The crank rides on bearings attached to sensors.

are two per crank pin. The short cut, made for the same design reason, involves the reciprocating weight. Normally, each of the reciprocating weights, piston, rings, rod small end, etc., would need to be doubled as well, then, multiplied by the balance factor. However, the balance factor is 50 percent. Since multiplying by 50 percent would halve the number, you simply obtain a single weight measure for each of the reciprocating parts to add to the equation.

- Big end of the rod x 2 ....798 g
- Rod bearings x 2.................84 g
- Oil allowance....................4 g
  - Rotating weight...........886 g
- Piston........................580 g
- Pin..........................198 g
- Locks.........................8 g
- Rings (one set)..................65 g
- Small end of the rod...........186 g
  - Reciprocating weight......1037 g

Since you don't have to multiply by a balance factor, simply add the rotating weight to the reciprocating weight.

886 g + 1037 g = 1923 g

Therefore, each of your four bob weights for the Chevrolet engine must weight 1923 g.

BALANCING ROTATING WEIGHT

You must balance rotating weight in two planes. All of the parts that rotate in line with the crankshaft are balanced so that the weight of the part is distributed equally around its center of rotation. This is called static balance.

Engine Balancing

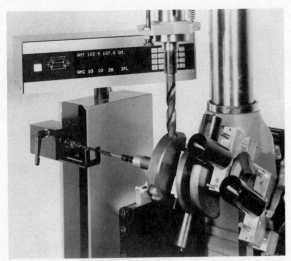

Figure 13-15. You usually remove unwanted weight on a crankshaft by drilling a hole.

Figure 13-16. On the flywheel, a hole is drilled to reduce weight.

Figure 13-17. Drill holes in the pressure plate to reduce weight and bring the assembly into balance.

Since the crankshaft is quite long, you must also check to see that it is balanced from end to end, which is termed dynamic balance.

Although the crankshaft may be balanced so that its weight is distributed evenly around its center of rotation, it's possible for the weight at one end of the crankshaft on its top side to be balanced by weight at the other end of the crankshaft on its bottom side, figure 13-12. This condition of static balance but dynamic imbalance will set up a rocking couple and cause the crankshaft to wobble from end to end as it rotates. A balancing machine eliminates this condition because it balances the rotating parts of the engine both statically *and* dynamically, figure 13-13.

Internally and Externally Balanced Crankshafts

Internally balanced crankshafts rely completely on the counterweights for their balance. Therefore, the crankshaft, flywheel or flexplate, and pressure plate of internally balanced engines are balanced independently. Externally balanced cranks, on the other hand have some of their final balance in external parts. So, they require some of the rotating parts — usually the flywheel or flexplate and vibration damper — to be assembled on the crankshaft and the entire unit spun on the balancer.

Crankshaft Balancing Procedure

You spin the crankshaft in a balancing machine that detects any imbalance and its location. If any counterweight is heavy, you drill it to remove metal. If light, you add weight. Flywheels, flexplates, pressure plates, and dampers are also balanced this way.

To ready the crank for balancing, it must have the bob weights attached if it is from a V-type engine. If it is an externally balanced type of crank it needs the damper and/or the flywheel attached. Lists of externally balanced engines and what must be attached to them is usually included with the operational instructions of any balancing machine.

The crankshaft is set into the machine and rotated at a relatively low rpm. The two end main bearing journals rest on bearings that are attached to sensors, figure 13-14. As the crankshaft rotates, the sensors pick up vibrations that the computer uses to determine the

Figure 13-18. Supply companies offer flange weights that can be cut to size and welded on flexplates or flywheels to balance them.

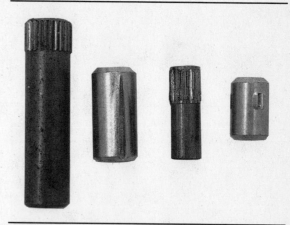

Figure 13-19. Installing plugs is a popular method for balancing dampers. Dampers often have a selection of holes drilled at the factory for the purpose of installing plugs to obtain a final balance.

amount and location of the imbalance. Computer balancing machines are the only type currently available for sale, although many non-computer balancers are still in service.

On older balancing machines, a strobe light mounted at the end of the crankshaft indicates the location of the heavy part of the shaft by flashing once each revolution when the weight is in a specific position. By putting a reference mark on the end of the crankshaft and observing the angle of this mark when the crankshaft is rotating, you can put the mark in the same position after the crankshaft has stopped, and then remove weight at the proper location. To correct dynamic imbalance on the older machines, each half of the crankshaft is balanced separately. This is done by locking the sensors and bearings at one end of the crankshaft so that only one half of the crankshaft is checked.

No matter which kind of balancing machine you use, a drill press is located on the machine and you remove the unwanted weight by drilling a hole, figure 13-15.

Once you have balanced an internally balanced crankshaft, you must then balance the flywheel or flexplate, pressure plate, and damper. Externally balanced crankshafts are a little simpler because you balance the whole assembly as a unit. However, you drill *only* the crank counterweights to bring the whole assembly into balance.

Balancing the Additional Rotating Parts

Install the flywheel or flexplate on the crankshaft and balance it in the same manner as the crankshaft. Drill a hole to reduce weight in the heavy area, figure 13-16, or add weight 180 degrees from the heavy spot. Next, install the damper on the crankshaft followed by the clutch pressure plate. Each time, you spin the whole assembly on the machine to locate the imbalance. Drill holes in each part to balance it, figure 13-17.

Mandrels are available that allow you to balance these additional items on the balancing machine without using the crankshaft. However, it does not usually provide as good a final balance as when you mount each item on the crankshaft itself.

Adding weight for balance

It is rare for a crank counterweight to be too light and need metal added to balance it. However, it can happen if you install high-performance, heavy-duty rods and pistons. In these cases, you can drill a hole in the counterweight and press in a plug of "heavy metal", usually a tungsten alloy that has twice the mass of steel. Since this alloy is very expensive, adding metal is not a common method for balancing cranks.

It is more common to add metal to flywheels, flexplates, pressure plates, and dampers than to cranks because they do not need a heavy metal. Many supply companies offer steel plugs and weld-on flange weights for these parts, figures 13-18 and 13-19.

Chapter 14
Servicing Cylinder Heads and Manifolds

HEAD RECONDITIONING

The cylinder head performs several important functions. When bolted to the block, the head seals off the top of the cylinders to form the combustion chambers. The intake charge is channeled through the head and into the cylinders, and exhaust gases exit through the exhaust ports. The valves and valve train components regulate the flow of the fuel mixture and the exhaust gases. The spark plugs, also located in the head, ignite the fuel charge. Because of the critical nature and multiple functions of the cylinder head, you must take extreme care during the rebuild process.

There are two main purposes of reconditioning the head. The first is to restore the sealing action of the valves. To achieve this you must accomplish three things:

1. The valve guide and stem must have the proper clearance, and the valve must be concentric in the port relative to the seat. This ensures that the valve will always seat correctly to seal the port.
2. The valve spring tension must be sufficient to enable the valve to seat tightly, but it must not be so tight as to damage the valve or seat or to cause excessive camshaft or follower wear.
3. The valve and seat must be ground or machined to achieve total sealing around the entire face of the valve.

Second, you must restore the mating surface between the head and cylinder block. Overheating can blow a head gasket and may warp the cylinder head. You must check the head's gasket surface and machine it flat, if necessary. When you resurface the cylinder heads of a V-type engine, you must also machine the intake manifold to maintain the proper angle for head to manifold sealing.

You can use many different processes to achieve these results. We will cover those processes and other head rebuilding procedures in this chapter.

Head Disassembly

There are a wide variety of cylinder head designs in current automotive use. One thing they all have in common is that before you can take accurate measurements, perform any machine work, or even thoroughly clean the head, it must be stripped down to a bare casting. You must remove all of the valve actuation components, including the camshaft on an OHC engine. Then, you compress the valve springs to remove the keepers, rotators,

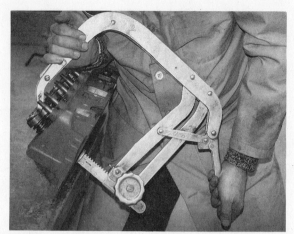

Figure 14-1. A valve spring compressor relieves valve spring tension to release the keepers.

Figure 14-3. With the spring compressed, lift off the keepers. A small magnet works well when access is limited by the retainer or the compressor.

Figure 14-2. Lightly rapping the tops of the springs helps to loosen varnish and sludge making the parts easier to disassemble.

springs, shims, seals, and valves. You must also remove any sending units or electronic sensors that install on the head, as well as some, or all of the manifold mounting studs. We detail all of the required procedures here.

A good place to start, whether you are doing the complete rebuild yourself, or work in a machine shop where you receive the heads already removed from the engine, is to give the head a good steam cleaning. Cleaning removes caked-on deposits in the valve spring area, which makes disassembly much neater and easier. Clear enough space on your work bench so you have room to spread the parts out. You must keep them in order as you remove them.

Disassembling overhead valve cylinder heads
First remove the rocker arm retaining nuts, then lift off the rocker arm and pivot. Lay the parts out in order so you can inspect them later. Next, you will remove the keepers, springs, seals, and valves. You must use a valve spring compressor to relieve valve spring tension before you can remove the keepers. Valve spring compressors use lever action to depress the springs, figure 14-1. A solid pad on one jaw of the tool seats on the face of the valve, the tool clamps around the cylinder head, and split fingers on the other jaw straddle the valve stem as they seat on the spring retainer. Valve spring compressors may be either manually or pneumatically operated. The amount of compression is adjustable. The tool locks in place when the lever reaches the end of its throw.

WARNING: An accumulation of sludge, varnish, and other deposits forms between the valve stem, keepers, and spring retainer as a normal result of engine operation. This accumulation, plus the force of spring tension and combustion, has a tendency to stick the parts together and make them difficult to remove. A quick easy way to loosen the parts is to lightly rap the top of the valve spring with a hammer before you attempt to compress it, figure 14-2. Use *soft hammer blows*. Striking the spring too hard can release the keepers and send parts flying across the shop. Always *wear eye protection* when working with valve springs.

Once the spring is compressed, remove the keepers from the top of the valve stem, figure 14-3. A pencil magnet works well to lift off keepers when access is limited. Slowly release, then, remove the valve spring compressor. Lift off the valve spring, retainer, and shims, figure 14-4. Set the parts aside, keeping them in order, then continue until all of the springs have been removed. A good way to keep

Servicing Cylinder Heads and Manifolds

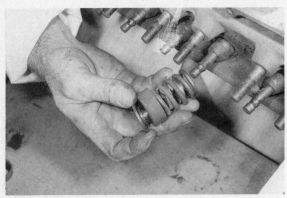

Figure 14-4. Release and remove the compressor, then lift off the valve spring and retainer.

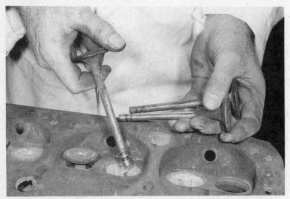

Figure 14-5. Slip the valves out of their guides. Remember, valves must be kept in their original order.

Figure 14-6. You can remove press-fit rocker studs with a socket, large washer, and nut. Tightening the nut draws the stud out of the head.

springs and keepers in order is to string them on a length of wire, twisting the wire ends together to form a loop. Parts can be cleaned by soaking or solvent washing while wired together in order.

Next, remove the valve stem seals. Umbrella and O-ring seals will generally lift off easily. Positive-lock seals can be removed with a special tool, small screwdriver, or pliers. Be careful not to damage the seal seat on the top of the valve guide.

Look over the ends of the valve stems. Excessive lash clearance and high mileage operation can mushroom the top of the valve stem. Forcing the mushroomed end through the guide will damage the guide. Remove the mushroom edges with a file so the valve will slip out of the guide easily. If varnish buildup on the valve stem is preventing the valve from coming out, remove it with a solvent such as lacquer thinner or acetone. As a last resort, you can remove the valve by driving it through the guide with a punch, but this will damage the guide and most likely the valve as well. Remove all of the valves and set them aside keeping them in the original order they fit to the head, figure 14-5.

Remove any manifold mounting studs that will interfere with attaching the head to your resurfacing equipment. Also remove any rocker arm mounting studs that show damage and must be replaced. Rocker studs are either thread-in or press-fit to the head. Remove threaded studs following procedures in Chapter 10 of this *Shop Manual*.

Special tools are available for removing press-fit studs. An alternative is to fit a socket and large washer over the stud, then install a nut and draw it up tight, figure 14-6. As you continue to tighten the nut, the stud is pulled free of the bore.

Double-check to make sure everything has been removed from the cylinder head. Remove all core plugs following the procedures in Chapter 9 of this *Shop Manual*. Then, after a thorough cleaning, the head will be ready for inspection and machining.

Disassembling overhead camshaft cylinder heads

You can apply many of the OHV disassembly procedures to an OHC engine. First, you must remove the camshaft from the cylinder head. There are several methods of camshaft attachment: Supporting the shaft with towers, fastening the cam to the head with bearing caps, or running the camshaft directly in the cylinder head on machined bearing bores, figure 14-7. Any of these methods may be used with or without bearing inserts. Before you begin, you must have factory specifications. Be aware, the camshaft is under pressure from the valve springs. Improper loosening can

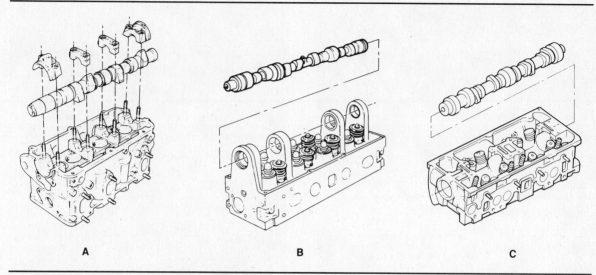

Figure 14-7. Overhead camshafts may be held in place with bearing caps (A), supported by towers (B), or fitted into bearing bores machined directly into the head (C).

cause damage. Be sure to make index marks on all bearing caps and towers. They must be kept in their original order.

Tower-mounted camshafts usually actuate the valves through a rocker arm assembly with an adjustable pivot or a hydraulic lifter. If possible, relieve spring pressure on the cam by disconnecting and removing the rocker assemblies first, figure 14-8. Then, follow the manufacturer's instructions to remove the camshaft.

Camshaft bearing caps may simply hold the camshaft, or also support the rocker shaft assembly. Keep in mind that the camshaft is under full valve spring tension. You must remove the retaining bolts in proper sequence, figure 14-9. Remove the caps, then lift off the camshaft, rockers, lifters, and other components. Reinstall the caps, unless they will interfere with valve removal.

Once you remove the camshaft, set all related parts aside for now. Service procedures for these items are in Chapter 15 of this *Shop Manual*. Continue disassembly by removing the valves from the head.

Many overhead-cam engines allow you to use a standard valve spring compressor to relieve tension, release the keeper, and remove the spring. Some require a special compressor because the assembled spring fits a recessed bore in the cylinder head. Remove the valves from the cylinder head using the techniques previously described in this Chapter for overhead valve disassembly. Then you can strip and clean the cylinder head to prepare it for inspection and machining.

Cleaning

When the head is disassembled, clean the head casting, valves, and other valvetrain parts with one of the approved methods we detailed in Chapter 9 of this *Shop Manual*. While most of these methods do a good job removing grease, oil, dirt, and varnish, some carbon will probably still cling to the cylinder head and back sides of the valves. Use an electric drill with a rotary wire brush to finish cleaning the ports and combustion chambers, figure 14-10. Brushes of various sizes and shapes are available. Avoid contacting the gasket surfaces with the wire brush. Take extra care with aluminum heads, a steel wire brush can easily damage the aluminum. Brass wire brushes are available and should be used when cleaning aluminum.

Most technicians clean the valves using a bench grinder with a wire wheel, figure 14-11. Grip the valve firmly because the spinning wheel can quickly grab it out of your hand. The valve may be damaged from striking the grinder frame, or propelled across the shop. Hold on tightly, and always wear eye protection when cleaning valves.

On cast-iron heads, use a scraper to remove traces of the head gasket not eliminated by your original cleaning method. Sometimes the gasket will etch into the metal and leave its impression. Use a block of wood or a file, wrapped with emery cloth to remove this.

You must take more care with the gasket surfaces of aluminum heads. A scraper can easily gouge the aluminum. Instead, use only the block and emery cloth. You can also safely

Servicing Cylinder Heads and Manifolds 309

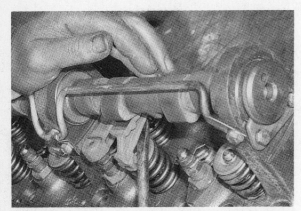

Figure 14-8. Always loosen or remove finger-type cam followers to relieve spring tension before you remove the camshaft.

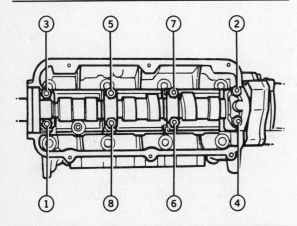

Figure 14-9. Loosen the bearing cap retaining bolts according to the proper sequence to prevent valve spring tension from bending the camshaft.

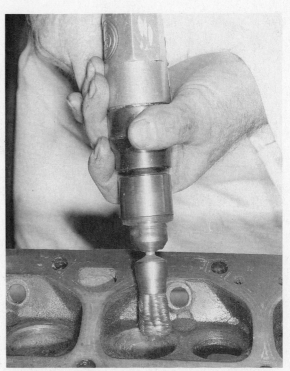

Figure 14-10. Clean any residual deposits from the combustion chambers and ports using a wire brush and a drill motor.

Figure 14-11. Use a bench grinder with a wire wheel to clean the valves. Be sure to have a firm grip on the valve, and protect your eyes.

clean an aluminum gasket surface with nylon-mesh pads or Scotch-Brite® pads that chuck into a drill motor.

Valve guides rarely need extra cleaning after being baked or hot-tanked. However, if you see carbon or gum left in the guide, clean it out with brushes and scrapers designed for that purpose. Although some shops clean cylinder heads and valves by bead blasting, we do not recommend this procedure. The tiny glass beads can get stuck in the water jackets and oil passages of a cylinder head and are very difficult to remove. The blasting can also remove the finely polished surface of valve stems. You must be extremely careful should you decide to clean these parts by bead blasting. Keep the high pressure stream of abrasive away from the polished stem when cleaning valves. After bead blasting cylinder heads, be absolutely certain that all trace of residual shot is cleared out of the oil galleries, water jackets, ports, and valve guides.

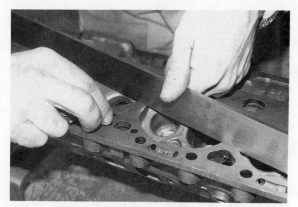

Figure 14-12. You can measure cylinder head warpage using a straightedge and a feeler gauge.

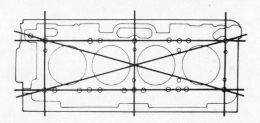

Figure 14-13. Take straightedge readings horizontally, vertically, and in an "X" pattern to get an accurate picture of surface condition.

Figure 14-14. Valve failure from burning and guttering is a result of high valve temperatures.

Inspection

With the head clean, you can then inspect the casting properly for cracks, as detailed in Chapter 10 of this *Shop Manual*. If the casting is sound, inspect the head and other valve train parts for damage as we describe below. Some manufacturers provide a minimum head thickness specification. Use a vernier or dial caliper to measure from the deck surface to the valve cover surface. Check head thickness now to avoid wasting your time rebuilding a head that is unusable.

Head gasket surface

The cylinder head gasket surface must be a relatively smooth, flat area on which the gasket can seal. Some technicians damage the gasket surface with a pry bar when they remove the head, or they damage the surface with a scraper when removing the gasket. Inspect the surface for gouge or scrape damage. This kind of damage can keep the gasket from sealing properly. Any but the most minor damage is cause to resurface the head.

Engines that overheat and blow the head gasket often warp the head, also. Head warpage is especially common on aluminum heads that have overheated. To check the surface of the head for flatness, place a straightedge across it, figure 14-12. Try to slip a feeler gauge under it to determine if the head is warped. Repeat this check horizontally, vertically, and in an "X" pattern, figure 14-13, record your findings. The warpage is equal to the largest feeler gauge that will slip under the straightedge without lifting it.

How much warpage a manufacturer allows varies. As a rule of thumb, you should not be able to measure more than 0.003 inch (0.08 mm) over a 6 inch (15 mm) span in any direction, or more than 0.006 inch (0.15 mm) total.

In most cases, if the head is warped or damaged you can restore the surface by machining. Give special consideration to OHC cylinder heads because camshaft timing can be affected when machining reduces the overall height of the head. Check the manufacturer's specifications. There is a maximum amount of metal that can safely be removed. You can use a special procedure to heat and straighten some warped aluminum heads prior to resurfacing them. Straightening reduces the amount of metal that you must remove to obtain a flat surface for the head gasket.

The sequence in which you perform straightening is very important. On most heads with press-in valve guides, if the guides need replacing, you should remove them before straightening. The heating process involved in straightening can seize the guides in the head, making the guides very difficult to remove. However, you must straighten the aluminum head before you grind the valve seats. If you perform the straightening afterwards, you can distort the freshly machined seats. Therefore, perform head resurfacing after completing all other machine work, before assembling the head. The flow of cutting fluid from the resurfacing machine also helps to clean the head by washing away grinding residue. We detail cylinder head straightening and resurfacing procedures later in this chapter.

Servicing Cylinder Heads and Manifolds

Figure 14-15. Small pieces of hard carbon that land on the valve face creates hot spots that can eventually lead to guttering.

Figure 14-16. Coolant, or other chemicals, in the combustion chamber can cause erosion.

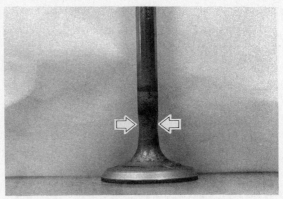

Figure 14-17. Necking is when the valve stem stretches just above the valve head.

Valve Inspection

Valve problems range from the obvious to the nearly invisible. Most of the obvious problems involve some kind of valve damage that requires you to throw the offending valve away. Identifying these problems can help you understand what went wrong in the engine and alert you to what else needs to be fixed. Some of the problems are usually taken care of routinely in the course of the valve job or complete engine rebuild. Others, such as a blocked cooling system or a lean air/fuel mixture, are not. Stay alert to the signs this damage is telling you and be sure you fix the problem in the course of the rebuild. Beginning with a visual inspection, check for the following:

- Burning
- Guttering or channeling
- Necked valve stem
- Cracks
- Valve face wear
- Valve stem wear

Visual inspection

Burning and guttering, figure 14-14, are caused by excessive valve temperatures. Local temperatures on the valve can reach the melting point of the metal as a result of preignition, poor seating, deposit accumulation, or metal erosion. Remember, you must repair the cause of the failure to prevent it from happening again.

Valve face temperatures are elevated and heat dissipation is interrupted by preignition. The valve face develops hot spots that can rapidly melt away metal. Combustion pressure works the hot metal like a cutting torch as it melts through the valve to cut a gutter for the gases to escape.

Poor seating can allow hot gases to escape at various points around the valve face. Gas flow through the leak creates a temperature channel across the valve head. The valve face in the hot gas channel will progressively burn away.

Hard combustion deposits that fall onto the valve seats can cause a leak and form a gas channel. A small piece of carbon can weld to the valve face and burn an escape gutter for the hot gases, figure 14-15.

Metal erosion is the result of coolant, or some other chemical, in the combustion chamber. Chemical reaction will etch away metal on the valve face, and eventually allow gases to escape, figure 14-16.

Valves do not always burn through from overheating. Sometimes the heat softened valve will deform as it is forced into its seat by the springs and combustion pressure. The head of the valve will pull down in the center to form a cup.

Necking of the valve, figure 14-17, is the condition in which the head of the valve pulls away from the valve stem, stretching the metal just above the head and causing it to thin. Necking is caused by overheating or too much valve spring pressure. If left unattended, the valve head will eventually break off of the stem causing severe damage.

Figure 14-18. Uneven valve cooling causes "hoop" stressing that can crack the valve.

Figure 14-19. Inspect the valve stems. Too much valve lash, excessive guide clearance, worn rocker arms, and weak springs can stress the tip end.

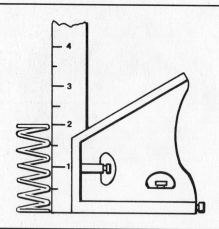

Figure 14-20. Check the valve springs. They must be square and equal in height.

Cracks are often a result of "hoop" stressing, a common overheating failure. Valves that are exposed to temperatures higher than they are designed to withstand cool unevenly. The outer edge of the valve cools quickly and contracts around the hot center. The temperature differential can stress the valve to the breaking point, figure 14-18.

Look the valve face over for signs of wear. Normal conditions will wear a nice even groove around the middle of the valve face where it makes seat contact. Check for signs of hot spots, peening from debris, and recession caused by high seating forces.

Inspect the valve stems. Look for galling, scoring and stress risers, figure 14-19. High seating forces and uneven seat pressure can cause a stress crack in the stem just above the valve head. Excessive valve lash and guide clearance, worn rocker arms, and weak springs can cause the valve to cock and stress the tip of the stem near the keeper seats. Check the keeper grooves and valve tip for mushrooming and other deformities.

Measuring valves

If the valves pass the visual inspection, use precision measuring tools to check them more closely. First, measure the valve stem with an outside micrometer to check for wear and taper. Take three micrometer readings along the stem. The overall diameter of the stem should not be smaller than factory specifications. Taper should not exceed 0.001 inch (0.025 mm). Replace valves that exceed these specifications.

Make sure the valve margin is adequate to allow you to grind the face. Use a vernier caliper to verify that the margin is greater than $1/32$ inch (0.8 mm). You must verify valve margin again after grinding the valves. Any valve put into service with less than $1/32$ inch (0.8 mm) margin will quickly burn and fail.

Not all bent valves are obvious, and even a slight bend makes the valve unusable. A valve grinding machine can help you identify mildly bent valves. When you are ready to grind the valves, chuck each valve in the grinding machine and allow it to spin close to the grinding wheel. Watch to make sure the head spins true in relation to the wheel. If the head wobbles, discard the valve without bothering to grind it. Even if the valve does not obviously wobble, it may still be bent. An experienced operator can tell a bent valve by the sound and the feel fed back through the machine while grinding it.

Valve spring inspection

Valve springs have a difficult environment inside the engine. Heat, poor lubrication, over-

Servicing Cylinder Heads and Manifolds

Figure 14-21. A valve spring tension gauge is used to test the integrity of the valve springs.

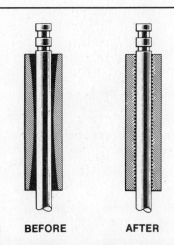

Figure 14-22. Valve guides wear to a bell-mouth shape at both ends due to the unequal forces created by the valve train components.

revving, or simply many miles of regular use can cause the valve springs to collapse, lose tension, or crack. First give the springs a good visual inspection. Obviously, you must replace broken springs. Also look for nicks, pitting, corrosion, or cracks that would cause the spring to break while in service.

Check all valve springs for proper squareness and free length. To test the spring for squareness, place the spring against the side of a square and rotate it, figure 14-20. It should not vary in length at any location more than 1/32 inch (0.4 mm). Without compressing the spring, measure the free length using a good straight edge, such as a steel ruler. Free length should be within 1/16 inch (0.8 mm) of specifications.

Test spring tension with a special gauge. The tool looks like a small arbor press with a scale, figure 14-21. Manufactures usually give tension specifications for the two extremes at which the spring operates, called open pressure and seat pressure. Open pressure indicates spring tension with the spring compressed and the valve fully open. Seat pressure indicates spring tension with the spring at its installed height and the valve resting on its seat. Place the spring on the table and pull down on the lever to the specified height. Observe the scale below the table that indicates spring length and the dial that shows the pounds of pressure on the spring. If the tension is not within 90 percent of the manufacturer's specifications at each length, discard the spring and install a new one.

Valve guide inspection

Valve guides must have a minimal amount of clearance, generally 0.001 to 0.002 inch (0.025 to 0.050 mm), for the valve to seat properly and for the guide to provide adequate oil control. Most valves are set into the cylinder head at a slight angle to the combustion chamber. This angle causes uneven guide wear as the valve stem rubs the guide on one side at the top, and the opposite side at the bottom, figure 14-22. Because bell-mouth wear at the extreme ends of the guide is normal, you must measure inside diameter at the top, bottom, and center of the guide to get an accurate picture of guide condition.

There are three ways to measure valve guide clearance. The best way is to use a special dial bore gauge, figure 14-23. A vise-like fixture is used to zero the gauge to the valve stem diameter. Place the gauge in the fixture between two identical valves, then zero the gauge, figure 14-24. Once zeroed, insert the gauge into the valve guide. The gauge will indicate valve guide clearance, the difference between valve stem diameter and the internal bore of the guide, figure 14-25.

A small hole, or split-ball, gauge also works well to accurately measure valve guide clearance. Insert the gauge into the valve guide, adjust the thumb screw so the fingers just contact the sides of the guide, lock the tool to hold the setting, then remove it. Use an outside micrometer to measure across the gauge,

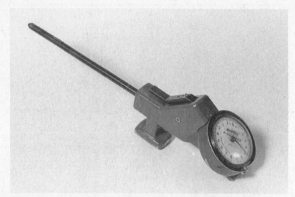

Figure 14-23. This special dial bore gauge is used to measure valve guide inside diameter.

Figure 14-24. A Setting fixture is used to adjust the gauge to read zero at the exact diameter of the new valve stems.

Figure 14-25. After setting the gauge, insert it into the valve guide. The gauge will display guide clearance.

then measure the valve stem diameter. The difference in the readings is the amount of valve guide clearance.

The least accurate method of checking guide clearance is the valve rock method. Use this method only if no other is available. Place a new valve in the guide and attach a dial indicator to the head so that it touches the edge of the valve. Rock the valve back and forth to get a reading on the dial indicator. You must use a new valve so you will read only the wear on the guide, and not the wear on the stem of the valve as well.

Valve seat inspection

There are two types of valve seats, the integral type and the insert type. Integral valve seats are used only on cast-iron cylinder heads. Aluminum and other alloy heads use insert type seats to provide a hard, solid surface for the valve to close against.

Integral valve seats are machined directly into the cylinder head casting, figure 14-26. The seat area is induction-hardened to provide longer service life and prevent recession. A draw back to induction-hardening process is that it creates stress in the casting. The built-in stress makes the valve seats prone to cracking. Inspect integral seats for cracks, burned areas, recession, and other damage. A cylinder head with extensive seat damage can often be salvaged by boring out the old seat and installing an insert-type seat. This procedure is covered later in this chapter.

Insert valve seats are made of a different material than the cylinder head, usually high-grade iron, chrome-steel alloy, or stellite. The seat is pressed into the head with an interference fit, figure 14-27. The amount of interference will vary according to the head and seat materials. Always check the manufacturer's specifications. Inspect the seat for cracks, burns, and other damage. Also look for erosion on the cylinder head around the outside circumference of the seat. Check for looseness by gently prying up on the seat. The seat must be replaced if you detect any movement. Oversize seats are generally available to correct erosion and loose fit problems. Seat replacement procedures are detailed later in this chapter.

Camshaft bore and saddle inspection

Overhead camshaft bearings operate under relatively light loads and pressure, so failure due to normal wear is unusual. Most camshaft bearing failures are the result of misalignment due to head warpage, lack of lubrication, abrasion from contaminated oil, improper assembly, or other problems elsewhere in the engine. You must inspect the bearing surfaces and repair any irregularities before the head can be returned to service.

Begin your inspection with a visual examination. Look for signs of scoring, galling, and seizing on the bearing surfaces. If the engine uses bearing inserts, check for indications of bearing shell movement in the bore and debris trapped behind the shell. Many of the inspection procedures for main and connecting rod

Servicing Cylinder Heads and Manifolds 315

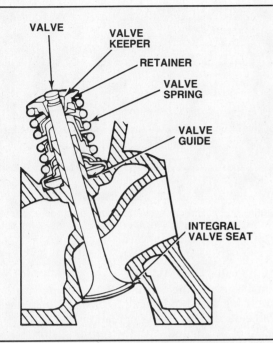

Figure 14-26. Integral valve seats are machined directly into the cylinder head casting. Late-model engines have induction-hardened seats to prevent recession into the head.

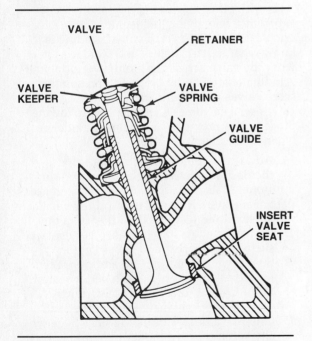

Figure 14-27. Insert valve seats are a separate part that is interference fitted to a counterbore on the cylinder head casting.

bearings, detailed in Chapters 11 and 12 of this *Shop Manual*, can be applied to camshaft bearings as well.

If bearing caps are used to retain the camshaft, install the caps and torque them to spec-

Figure 14-28. A valve grinding machine is used to restore the valve face and correct stem length.

ifications. Then measure the internal diameters with a dial bore gauge, inside micrometer, or telescoping gauge. Take several measurements in different positions, as instructed in Chapter 11 of this *Shop Manual*, to determine if there is any taper or out-of-round in the bores. Record your findings. Measure the camshaft journal diameters with an outside micrometer. Then subtract journal diameter from bore diameter to determine oil clearance. Record your findings in your notebook. Clearances must be within the tolerance range specified by the engine manufacturer.

Do not attempt to correct camshaft clearance and alignment problems at this point. Head resurfacing, straightening, and other machining will affect the position of the camshaft in the cylinder head. You must perform all other cylinder head machining operations before you correct the camshaft bore.

Valve Repair

In general, valves that are straight and not excessively worn can be reconditioned and returned to service. Reconditioning is done on a valve grinding machine, figure 14-28. The valve grinder uses a rotating stone to restore the face surface and reduce the stem length to maintain proper assembled height.

Before you grind a valve, check for oversize stems. An oversize stem can indicate that the

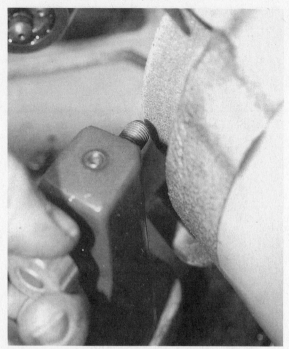

Figure 14-29. Prepare the grinding face of the stone using the diamond dressing tool.

Figure 14-30. Clamp the valve into the spindle. The chuck should grip the valve on the flat of the stem just below the head.

valve is sodium-filled to promote better heat transfer and requires special handling. Most manufacturers advise against grinding Sodium-filled valves. Some also have special disposal procedures. Sodium will explode when it comes into contact with water and it can cause a serious burn if it comes in contact with your skin. If the hollow stem of the valve is cracked or broken, you have a potentially dangerous situation. Be very careful when handling sodium valves. Not all sodium valves are marked, but they generally have oversized stems.

Oversize valve stems

It is possible, in some cases, to buy valves that have stems that are larger in diameter than the original stems. Then all you do is ream the guide to fit the larger stem. This is an especially good repair when you are replacing the valves anyway and guide wear is at or beyond the specification limit.

Valve facing

A valve grinder is a precision instrument that must be operated according to the specific instructions for the machine. Although valve grinders all perform the same function, the order of the steps you take to recondition the valves can vary. With some equipment you must grind the stem first, before you reface the valve. Machining the valve tip corrects for "mushrooming", a normal wear characteristic of the valve stem. When the stem is mushroomed, the chuck of the machine you are using may not be able to properly grip the valve during the facing operation, the valve will not be aligned with the grinding stone, and the face will not be ground concentric. If this is the case, skip ahead to the section titled *Grinding the valve stem*, then return here to reface the valve.

As with other machining operations, only experience will give you a feel for the machinery. An important part of valve grinding is to know what to look, listen, and feel for as you work the valve across the stone. Develop your grinding technique by practicing on some scrap valves before you attempt to grind the valves on a customer's engine. The following procedure is provided as a general guideline when reconditioning valves:

1. Adjust the spindle head so the valve is at the desired relative angle to the grinding stone. Machinists that use an interference angle generally prefer to grind the smaller angle on the valve face. In theory, an interference angle forms a narrower contact surface on the seat and keeps carbon buildup to a minimum. In addition, the interference angle promotes a better seal until the engine runs long enough for the valve to properly mate itself to the seat. One-half a degree of interference is generally sufficient on a modern engine. For most engines you will grind the valve face to 44.5 degrees and the seat to 45 degrees.
2. Dress the grinding stone to a smooth and square cutting surface by slowly passing the diamond dressing tool across the face of the stone, figure 14-29. Turn the coolant pump on and take light cuts, never more than 0.0005 inch (0.013 mm) per pass. Occasionally rotate the diamond tip to equalize wear and extend service life.

Servicing Cylinder Heads and Manifolds

Figure 14-31. Direct coolant flow onto the valve face and stroke the valve smoothly across the stone as you take a light cut.

Figure 14-32. Keep metal removal to a minimum, take the valve down just enough to get a clean, smooth, and concentric face.

3. Insert the valve stem into the spindle and clamp it down with the chuck, figure 14-30. The chuck should grip the valve just above the wear area on the stem, and as close as possible to the fillet to prevent wobble. Switch on the spindle motor and visually check valve runout as you watch the valve head spin.
4. Turn on the grinding stone motor, then adjust the coolant stream so it will keep the valve cool and flush away grinding dust without splashing. Position the spindle head so the entire width of the valve face will contact the cutting surface of the stone and the edge at the margin is at the edge of the smallest diameter of the stone.
5. Slowly advance the stone feed wheel until there is light contact on the valve face, figure 14-31. Immediately begin stroking the valve face across the dressed stone face using the spindle advance lever. Never allow any part of the valve face to pass beyond either edge of the stone while you stroke it. Stroking equalizes wear on the stone and also provides a fine surface finish on the valve.
6. *Watch and listen* as the stone cuts a new face on the valve. You will clearly see the new face develop. Width should be even around the entire circumference of the valve. The seat wear area in the center of the face will gradually diminish as you continue grinding. You should hear a chirping sound as the stone removes metal from the valve, the pitch will change rhythmically as you stroke the spindle. If the face develops unevenly, or the sound is irregular, the cut is not concentric and you most likely have a bent valve.
7. A shower of sparks will be emitted when the valve first contacts the stone. Allow the sparks to dissipate, then advance the stone to take another cut. Keep your cuts to a minimum. Never feed the stone in more than about 0.001 to 0.002 inch (0.025 to 0.050 mm) at a time as you constantly stroke the spindle.
8. Remove only enough metal to get a clean, smooth, concentric face on the valve, figure 14-32. Once the surface is properly ground, wait till the sparks stop, then back off the stone feed to remove the valve. Always *reverse the stone feed*. Backing the valve away from the stone distorts the outer edge of the valve face.
9. After resurfacing the valve, make certain there is at least a $1/32$ inch (0.8 mm) of margin on the edge of the valve, figure 14-33. This margin is very important to prevent overheating and burning of the valve.
10. Repeat the grinding procedure until all of the valve faces have been reconditioned. Frequently check the stone dressing as you work. If the stone is properly prepared and accurately feed into the valves, you should be able to surface eight valves between dressings. Watch for any change to the surface finish on the valve faces that indicate the stone is getting dull.

Figure 14-33. You must have at least 1/32 inch (0.8 mm) of margin on the edge of the valve after grinding.

After refacing all of the valves, give them a thorough inspection. The face must be perfectly smooth and even. There must never be more than 0.002 inch (0.05 mm) of face runout. Look for unground areas, chatter marks, and hair-line cracks. Unground areas can indicate the valve was not seated squarely in the chuck, or the valve stem is bent. Chatter marks and cracks result from excessive stone pressure or heat buildup from a lack of coolant. Discard or regrind any valves that do not have a perfect face finish.

Grinding the valve stem
Resurfacing, or "tipping", the valve stem end gives the rocker arm a smooth, flat, square surface to ride on, this reduces friction and extends service life. In addition, a precise amount of metal is removed from the valve stem to compensate for the valve sinking into the head as a result of valve seat reconditioning and face grinding. Most valve grinders have a separate stone driven by the tail shaft of the main motor for grinding stems. The valve is clamped into a fixture and feed into the stone. Grind the valve stems now to repair any surface damage. You may have to repeat the procedure to correct the length after you recondition the valve seats. Valve seat reconditioning procedures appear later in this chapter. Once you service the seats, return here and grind the valve stems. The following steps provide a general guideline for "tipping" the valve stems:

1. Fit the valve into the cylinder head and hold it tight to its seat. Measure how far the valve stem protrudes from the top of the head casting. Compare this finding to the reading you took during tear-down. The difference is the amount of metal you will remove from the end of the stem.

Figure 14-34. The valve stem is ground to compensate for metal removed from the face and seat, and to restore the surface finish.

Keep in mind that valves are surface hardened. If the stem requires more than 0.020 inch (0.50 mm) of grinding, check to make sure the seat is not too low in the head. If the seat position is good, you will have to replace the valve.

2. Use the diamond dressing tool to prepare a 90 degree angle on the grinding stone.
3. Clamp a valve into the holding fixture according to the equipment manufacturer's instructions. You must be familiar with the equipment you are using.
4. Rotate the feed wheel until the end of the stem just contacts the stone. Hold the valve in postion as you zero the micrometer dial, then back the feed off slightly.
5. Start the motor, adjust the coolant flow, then *slowly feed* the valve stem into the grinding stone, figure 14-34. Continue grinding until the micrometer dial indicates the required cut has been taken. Many holding fixtures allow you to rotate the valve by hand as you grind it. This promotes contact and will produce a nice swirl finish on the end of the stem.
6. Finish off by repositioning the holding fixture, or installing an adaptor to grind a 45 degree chamfer on the stem, figure 14-35.

Wash the valves in clean solvent, dry them, then give them a final inspection. Look over all the freshly machined surfaces, test fit the valves to check seating and clearance, and measure the valve stem height. You can dress any minor irregularities with a small soapstone. Once the valves pass inspection, they are ready for assembly. Keep the valves in order and store them in a safe place until you are ready to fit them in the cylinder head.

Servicing Cylinder Heads and Manifolds

Figure 14-35. Complete your valve by grinding a 45 degree chamfer on the edge of the stem.

Valve Guide Repair

There are many different methods of restoring valve guides to their original clearance specifications. When choosing the right method for the vehicle you are working on, remember that the best (longest lasting) methods are the most expensive. Also consider how many of the guides in the engine need repair and how excessive the clearance is.

Knurling

Knurling is the least expensive type of valve guide repair, but it also gives the least amount of service life. Knurling involves expanding the metal in the worn valve guide with a special arbor. The arbor threads the inside diameter of the guide. As the thread peaks of the arbor push their way into the guide, the metal around it is forced out into the bore, creating an inside diameter that is smaller than the valve stem, figure 14-36. To resize the guide, you run a reamer through the guide.

The size reamer you select should provide *one-half* the minimum specified clearance. The small clearance is necessary to provide as much service life as possible. Since the guide is threaded, not all of the surface area of the guide bore is actually used. Do not worry about the valve stem sticking in the guide. The knurling grooves in the guide tend to retain oil and provide an extra measure of lubrication. Knurled guides require extra oil control. Use effective valve guide seals, such as umbrella seals, on knurled guides.

Use knurling only if the excess valve guide clearance is less than 0.005 inch (0.13 mm). Knurling is more satisfactory on some engine designs than others. Do not use it on engines that have a history of poor valve guide life.

Figure 14-36. The knurling arbor displaces metal as it is run through the valve guide.

To perform knurling:
1. Make sure the guides are thoroughly clean before you begin.
2. Select a knurling arbor that is the nominal size of the guide you want to knurl. The size of the arbor is marked on the shank. For example, for a $11/32$-inch valve stem, use a 0.343-inch arbor.
3. Lubricate the guide and the arbor.
4. Use an electric drill with a speed reducer to drive the arbor at the correct rpm. Be sure to keep the arbor in visual alignment with the guide.
5. Clean the guide with a brush and solvent, then blow it dry with compressed air.
6. Run the reamer through the guide to provide the proper clearance.

Replacing valve guide inserts

The valves on all aluminum cylinder heads ride in valve guide inserts that are press-fit to the head casting. The hard cast iron or bronze inserts maintain valve train geometry and minimize wear. Valve guide inserts are a serviceable item, and should always be replaced whenever you do a valve job. The valve is free to rotate, so wear is equalized around the circumference of the stem. On the other hand, the guide is stationary in the head so it will wear in a pattern. Under normal operation, both ends of a valve guide will wear out-of-round. The wear pattern will be perpendicular to the camshaft, figure 14-22.

Figure 14-37. Valve guide inserts that have a tapered shank, positive stop, or snap-ring seat can only be removed and installed from one direction.

Figure 14-38. Measure how far the old guide extends beyond the cylinder head casting.

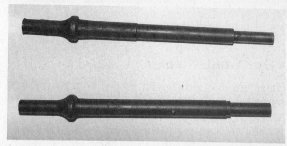

Figure 14-39. The driver must fit well into the guide to prevent it from collapsing, the shoulder must be slightly smaller than the outside diameter of the guide.

Valve guide inserts are replaceable. You repair these guides by pressing out the old guide and installing a new one. Guide removal procedures vary depending on engine design, so you must follow the manufacturer's instructions for the engine you are working on. With some engines, the old guides can be removed and installed either from the combustion chamber side or the valve spring side of the head. Other guides can only be fitted from one side because they have a slight taper, a positive stop, or a snap-ring groove to locate them in the casting, figure 14-37. Follow the specific requirements of your engine.

Valve guide inserts must have an interference fit, usually about 0.001 to 0.002 inch (0.025 to 0.050 mm), to stay in place and provide proper heat transfer from the head. Many machinists prefer to preheat the head to make removal and installation easier. You can heat the metal around the guide by carefully using a propane torch, or warm the head in a thermal oven. Simply steam cleaning the head will often raise the temperature enough to make overcoming the interference easier. However, you must work quickly.

To replace a valve guide insert, follow these steps:

1. Measure and record how far the old guide protrudes from the head casting, figure 14-38. New guides must be installed to the *exact* same height.
2. Select the proper driver to fit the guide. Guide drivers have a long shank that fits into the guide and a shoulder that butts to the end of guide, figure 14-39. The shoulder must be slightly narrower than the outside diameter of the valve guide.
3. Remove the guide using an arbor press, hammer, or air hammer. Never allow the driver to bounce on the end of the guide. This can crack or break the insert.
4. Make sure the guide bore in the head is clean and free of any burrs, nicks, or gouges. Measure the bore so you can determine the proper size for the new guide. This is important. Some standard replacement guides are as much as 0.008 inch (0.20 mm) oversize. You may have to open up the bore with a reamer, or turn the guide down in a lathe.
5. Most replacement guides have a chamfered end to lead them into the head and prevent galling. If not, grind a chamfer into the guide.
6. Coat the sides of the guide with a high-pressure lubricant designed for press-fit installations.
7. Install the guide into the head using the same driver you used to remove the old one. Remember, the insert can be damaged if you let the driver bounce. Use steady even hammer blows or press pressure as you work the guide in. Measure the installed guide. Be sure you have the exact amount of protrusion above the spring seat as the original.

Servicing Cylinder Heads and Manifolds

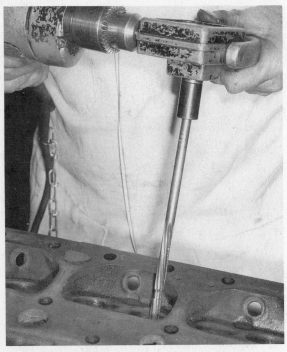

Figure 14-40. The new guide must be reamed to provide the proper clearance and surface finish.

Figure 14-41. Tap the end of the guide, install a bolt, then fit a drift through the guide to contact the bolt.

8. Ream the installed guide to obtain the proper oil clearance, figure 14-40.

CAUTION: Reaming the new guide is a critical operation that requires special consideration. Whether you use an adjustable or a self-aligning reamer, be sure to select the proper size. Do not attempt to remove more than 0.005 inch (0.13 mm) of metal with one pass. Make progressive cuts if necessary. Advance the reamer slowly into the guide and do not stop until the cutting blades have passed all the way through the guide. *Never reverse a reamer.* Reversing dulls the blades and destroys the tool. You can ream cast-iron guides dry, but bronze guides should be lubricated with cutting fluid. Always use the proper speed for the guide material you are working with. Low speed will provide a finer finish and a more accurate clearance tolerance. Note that the tool and the guide will heat up as you use them, especially when working with bronze, so you may find the bore undersize once it cools down. You can minimize heat build up by alternately working with two reamers of the same size. Cool one reamer in a can of oil as you work with the other, switch reamers after you finish each guide.

On some engines, the material around the guide is very thin. If you attempt to remove the guide by pressing, or driving it out through the head you may crack the head. To prevent damage, follow this procedure:

1. Tap threads into the spring side of the valve guide, figure 14-41. The threads should extend approximately one inch (25 mm) into the guide.
2. Screw a bolt into the fresh-cut threads. Then, select a drift that will fit through the guide and contact the end of the bolt.
3. Use a press or steady even hammer blows to drive the guide out by pressing against the bolt. Doing this stretches the guide slightly, making it thinner so it comes out of the casting easier.

This procedure should also be used on valve guides which have a step preventing them from being pressed out through the combustion chamber. If the guide has a snap ring on the spring side it provides the same effect as a step, and you should also use the above procedure. Always check the manufacturer's information for specific requirements. For some of their engines, Toyota recommends you break the spring end of the guide off with a hammer and punch before you remove the insert.

Installing thin-wall valve guide liners

There is no reason to scrap an integral valve guide cylinder head simply because the guides are worn beyond the point of repair by knurling. Many heads can be salvaged by drilling the guide to oversize, then installing a thin-wall liner to bring the internal diameter back to standard. Most liners are made of bronze and have a wall diameter of about 0.010 inch (0.25 mm). If guide wear is within 0.020 inch (0.50 mm), "thin-walling" is a quick and easy repair. Install as follows:

1. Open up the original valve guide bore with a piloting core drill, figure 14-42. The pilot leads the drill along the centerline of the old guide.

Figure 14-42. A piloting drill follows the centerline of the old guide as it cuts the new bore for a thin-wall insert.

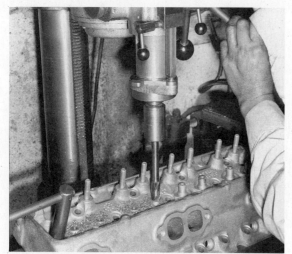

Figure 14-43. Drill out the guide slightly smaller than the outside diameter of the false guide.

2. Drive, or press the liner into the guide using the proper installation tool. Fit the liner to within 1/8 inch (3 mm) of the bottom of the guide.
3. Use the trimmer to cut the top of the insert flush with the head casting.
4. Run the expander broach through the insert to seat it to the head casting, and open up the inside diameter.
5. Measure the inside diameter, ream to size, and the guide is ready for service.

Installing false valve guides
Integral valve guides worn beyond 0.020 inch (0.50 mm) can be repaired by installing a "false guide" to bring the bore back to standard. False guides resemble the guide inserts used with aluminum heads, and can be made of cast-iron or bronze. Bronze is generally preferable because of its superior wall surface and oil retention capacity. A considerable amount of metal must be removed from the cylinder head in order to accommodate the large outside diameter of the guide.

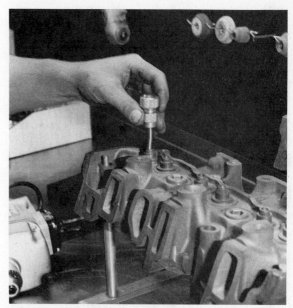

Figure 14-44. Install the coil into the guide by threading it in with the mandrel.

Drill the old guide to the size recommended by the guide manufacturer, figure 14-43. A special fixture is used to align the drill bit and retain the centerline of the original integral guide. Drill the bore slightly smaller than the outside diameter of the sleeve, then open the bore with a reamer to obtain the proper interference fit and provide a smooth surface for good heat transfer.

After preparing the bore, follow the steps outlined under *Replacing Valve Guide Inserts* earlier in this chapter to install the guide and ream it to size. Once installed, the guides are easily replaced during subsequent rebuilds.

NOTE: When purchasing bronze "false guides", inserts, and thin-wall liners always select the proper material. Most are made from phosphor-bronze. However, some are silicon-bronze. Silicon-bronze valve guides are designed for high-performance, heavy-duty, and alternative fuel applications. They are not compatible with stainless-steel valve stems, so you must run valves with hard-chrome stems.

Bronze coil inserts
Another method of repairing worn integral valve guides is to install a bronze coil insert. Bronze coil inserts are similar in design to the helical inserts used to repair damaged threads (see Chapter 10 of this *Shop Manual*). The installed insert reduces the internal diameter of the guide, much the same as after knurling. The spiral finish on the inside diameter, retains oil to lubricate the valve stem. Because of superior oil retention capability, valves can be

Servicing Cylinder Heads and Manifolds

Figure 14-45. Secure the tail of the coil with the special clamp to prevent it from shifting.

Figure 14-46. Drive the swaging tool through the insert to seat the coil to the wall of the guide.

Figure 14-47. Dress the valve guides with a deburring tool before you fit the valves.

fitted to a coil insert with about half the clearance required by solid wall guides.

The main advantage to these inserts is that they do not require any elaborate machinery to install and they give the same service life as a bronze insert guide. They are also easy to replace the next time the head is in for service.

To install a bronze coil insert, make sure the valve guide is absolutely clean and dry, then follow these steps:

1. Run the special piloting tap completely through the guide to cut threads into the wall. *Cut the threads dry*. Do not lubricate the tap.
2. Clear any chips from the guide with compressed air, then thread the insert onto the installation mandrel.
3. Screw the mandrel and insert into the guide until the tool is just slightly recessed from the bottom, figure 14-44.
4. Remove the mandrel. Use diagonal cutters to trim the excess coil from the top of the guide and leave just enough of a tail for the holding clamp to grip.
5. Engage the tail of the coil as you fit the holding clamp, over the guide. Then, tighten the clamp around the guide to prevent the coil from moving, figure 14-45.
6. Drive the swaging tool through the insert to press the coil into the head casting and seat it, figure 14-46.
7. Use a reamer to open up the inside diameter of the insert and provide the proper clearance for the valve stem.

General valve guide service tips

The most important thing in valve guide service is to center the guide so that the valve will be concentric in its seat. After repairing a guide, it is very important to clean it thoroughly before assembling the head. Knurling and reaming a guide leaves burrs and chips in the guide which, if not removed, will cause engine damage. Use a deburring tool to dress irregularities in the guide before you install the valves, figure 14-47.

Typical clearance for an intake guide is 0.001 inch (0.025 mm) and for an exhaust guide 0.015 inch (0.038 mm). Remember, when knurling or installing bronze coil inserts, the valves can operate with about half of the standard clearance.

Many machinists prefer to bring valve guides to final size by honing rather than reaming,

Figure 14-48. Level the cylinder head, use a pilot to align the valve guides with the spindle.

Figure 14-49. Adjust the depth gauge on the machine to the thickness of the new seat.

Figure 14-50. Switch on the spindle motor, then feed the boring tool into the head, cutting to the depth of the new seat.

Replacing valve seats

To replace valve seats of the integral type, use a cylinder head machine to cut a counterbore into the old seat, then a new insert type seat is press-fit into place. You must be familiar with the equipment you are working on. Follow any specific instructions provided by the manufacturer. Replace valve seats following these general steps:

1. Position the cylinder head combustion chamber side up in the machine, then install a pilot into one of the valve guides near the center of the head. Level the cylinder head in the machine to align the pilot with the spindle, figure 14-48.
2. Select, or adjust, a cutting tool to bore a hole 0.005 to 0.008 inch (0.13 to 0.20 mm) smaller than the seat insert to provide an interference fit. Attach the boring head onto the spindle.
3. Place the new seat against the depth gauge on the machine, figure 14-49. Tighten the gauge to the thickness of the new seat. This sets the machine to cut that exact depth so the seat will fit flush.
4. Switch the motor on and feed in the spindle to bore out the old seat and cut a new one for the insert, figure 14-50.
5. Compare the depth of the new seat with the depth of the cut, figure 14-51. Double check this measurement. The new seat must be installed to exactly the same depth as the original.
6. Install the insert with a press, or drive it in with a driver and hammer, figure 14-52. Chilling the seat in a refrigerator or freezer will allow it to go in easier.

because honing allows you to hold tighter tolerances. With a reamer it is difficult to get closer than 0.0005 inch (0.013 mm) to your target diameter, while honing will bring the final bore tolerance to within 0.0002 inch (0.005 mm). Valve guide hones are readily available, and easily used when chucked into a slow-speed drill motor. Work the hone slowly, keeping drill speed to about 350 rpm, and measure frequently to check your progress.

Valve Seat Reconditioning

The valve seat performs two functions. It provides a seal for the valve in the combustion chamber, and it provides a cool surface to carry heat away from the valve. You must replace or resurface the valve seats. Most automotive valves use a 45-degree seating angle. However, some engines have 30-degree intake valve seats. All of our examples are based on a 45-degree seat. When working with a 30-degree seat, your correction angles will be 15-degrees for lowering and 45-degrees for narrowing. All other procedures are identical.

Servicing Cylinder Heads and Manifolds

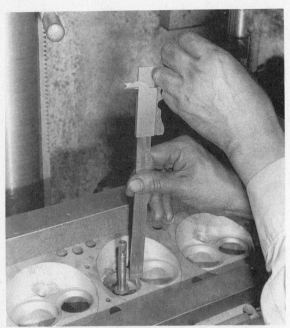

Figure 14-51. Double check your work. Measure to be sure the depth of the cut equals the height of the seat.

Figure 14-53. Once you install the seat, stake it in place to prevent it from working loose.

Figure 14-52. Install the new seat into the cylinder head using an appropriate driver.

Because of the intense pressure on the seat, and the frequent expansion and contraction caused by the wide range of temperatures it undergoes, the seat must be very tight. To ensure that the seat does not come out, stake it after you have replaced it, figure 14-53.

On engines with original equipment insert seats, the old seats must be removed. Some can be driven out with a long punch through the ports. Others can be lifted out with a pry bar. If the seat is very difficult to remove, weld a tiny bead around the inside of the seat. This will cause the seat to shrink and simplify its removal. Always reface valve seats after the valve guides have been reconditioned. This is important because the pilot for the seat grinder or cutter locates in the valve guide bore.

Valve seat grinding
Like valve grinding, seat grinding requires a good surface finish. You can achieve this by dressing the stones frequently with a diamond stone dresser. Valve seat grinding requires three different angles. Grind the 45-degree angle to establish the seat, figure 14-54. Grind the 30-degree angle to lower the seat and position it on the valve face, figure 14-55. Grind the 60-degree angle to narrow the seat and move it up the valve face, figure 14-56. If the seat needs to be narrowed, alternately grind 30-degree and 60-degree angles until the desired width is achieved.

Generally, it is better to leave a wider seat on the exhaust valve because it runs hotter than the intake valve. In theory, the narrower the seat the better the sealing action of the valve. This is true because the spring tension on the valve is applied through less contact area, resulting in more pressure per square inch or millimeter. However, if the seat is too narrow, the valve does not have enough contact area to cool properly. The heat-softened metal deforms due to the increased pressure of the narrow contact.

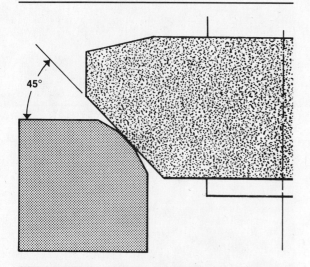

Figure 14-54. Grinding a 45-degree angle establishes the valve seat in the combustion chamber.

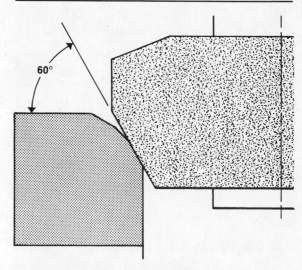

Figure 14-56. Grinding a 60-degree angle removes metal from the bottom to raise and narrow the seat.

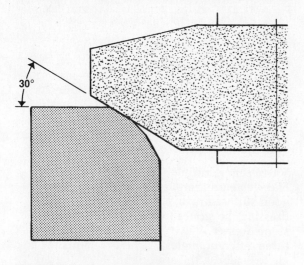

Figure 14-55. Grinding a 30-degree angle removes metal from the top to lower and narrow the seat.

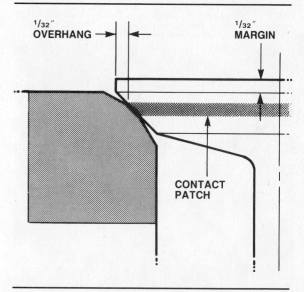

Figure 14-57. The seat must contact evenly around the valve face. For good service life, both margin and overhang should be at least 1/32 inch (0.80 mm).

Keep in mind that grinding will remove metal from the valve seat quite quickly. Usually, just several seconds of grinding is all that is required to recondition a seat. Check your progress frequently when grinding, and never remove more than the minimum amount of metal necessary to attain a good seat.

After you grind the 45-degree angle, remove the pilot, wipe off the valve seat, then install the valve to check the fit. The position and shape of the seat-to-valve contact patch is critical to engine reliability. The seat must contact the valve face evenly along the entire 360-degree circumference of the port. The contact patch should be about 1/16 to 3/32 inch (1.60 to 2.39 mm) wide, with a minimum of 1/32 inch (0.80 mm) overhang between the top of the seat and the edge of the valve face, figure 14-57. You correct the shape and position of the contact patch by "topping" and "throating" the seat using the 60-degree and 30-degree grinding stones respectively. Occasionally, topping and throating makes the seat too narrow. This can be corrected by taking another *light cut* with your 45-degree stone.

Although the following general procedure can be used to grind most valve seats, you must be aware of any specific requirements of the equipment you are using.

Servicing Cylinder Heads and Manifolds

GRINDING VALVE SEATS

1. Insert the valve seat grinder locating pilots into the valve guides. The pilot ensures that the valve seat will be ground concentric to the guide.

2. Select the proper size and grit stone for the seat you will be grinding. Make sure the stone is not chipped, cracked, or broken, then thread it onto a stone holder.

3. Since you will be cutting three different angles with three different stones, It is best to have a separate holder for each stone.

4. Dress each of the stones to the desired angle using the grinder motor and a diamond dressing tool.

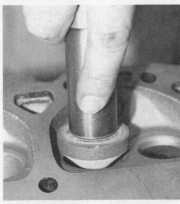

5. Slip the holder onto the pilot to check-fit the stones. They must have a good purchase on the seat and not contact anywhere else on the casting.

6. Remove the holder and slip a "lifting" spring over the pilot, then lubricate the pilot with a drop of motor oil.

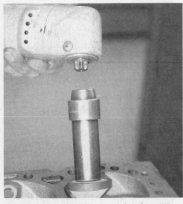

7. Fit the holder with the 45-degree stone onto the pilot and attach the grinder. Hold the driver square to the pilot to prevent pulling the stone into the seat.

8. Turn the motor on and work the stone into the seat with short, momentary contacts. Avoid bearing down. This can increase stone wear and result in a poor finish.

9. Continue to grind until the seat is just clean of blemishes, then make a final light cut to get a good surface finish.

Checking valve seat concentricity and configuration
Special dial indicating fixtures are available for checking valve seat concentricity. Fit the tool pilot into the valve guide, rest the plunger on the valve seat, zero the indicator, then slowly rotate the tool around the seat as you watch the dial. Ideal runout is within 0.001 inch (0.025 mm), although up to 0.002 inch (0.050 mm) is generally acceptable. You can often correct high spots with your 45-degree grinding stone.

A common practice for checking seat configuration is to apply a light coating of Prussian blue to the valve face. Install the valve, hold it to the seat, and rotate the valve several turns. Remove the valve, wipe the face off, reinstall it, and again give it several turns. Now, when you remove the valve you will easily see the seat pattern dyed on the valve face. One drawback to "bluing" the seats is that the dye can clog and dull your grinding stones. Because of this, some machinists prefer to use a felt-tip marking pen on the valve face to show if the seat is the right width and properly located.

Correcting valve seat position and width
High seat contact, too close to the head of the valve, is corrected by "topping" with a 30-degree grinding stone to establish the maximum diameter of the seat. Topping effectively lowers the contact patch on the valve face by removing metal from the combustion chamber side of the seat, figure 14-55. Remember, you must have at least $1/32$ inch (0.80 mm) of valve overhang on the finished seat to provide satisfactory service life. Grind the topping angle to obtain correct overhang following the same procedure you used to cut the 45-degree seating angle. First top the seat to the right height, then "throat" it to bring the contact patch to the proper width.

Throating establishes the minimum seat diameter by removing metal from the bottom edge of the seat to narrow the contact patch, figure 14-56. Using the now familiar grinding procedure, throat with a dressed 60-degree stone, bringing the valve seats to the correct width. Intake valves operate best on a narrow seat, so target the minimum specification, usually about $1/16$ inch (1.60 mm) for seat width. For exhaust valves, the wider the seat the better the seal. Grind exhaust seats as wide as possible without exceeding maximum specification, typically $3/32$ inch (2.39 mm), and still maintaining a margin of $1/32$ inch (0.80 mm) on either side of the valve face, figure 14-57.

CAUTION: The valve seats will disappear quite quickly when you "top" and "throat" them. It is very easy to grind off too much metal and over-correct the seat, then you have to start over. Work the grinder carefully when you make correction cuts. Generally, several quick "bumps" with the stone is all you need to dress the relief angles, shape the seat, and bring it into position.

Inspect the seat face. It must be perfectly smooth without any chatter marks. Double-check seat configuration using Prussian blue or a felt-tip pen as described earlier. When seating is acceptable, wash the valves and cylinder head in clean solvent to remove all trace of grinding dust and dye. Keep the valves in order and store them in a safe place until assembly, then continue with head reconditioning.

Valve seat cutting
An alternative to grinding the valve seats is to use a special cutting tool to shape the seating angles. The tool uses cutting bits, rather than a stone, to remove metal, figure 14-58. The technique works well on bronze seat inserts that are difficult to grind and hard on stone life. Seat cutting can be beneficial with:

- Extremely hard seats that do not grind easily.
- Very soft seats that clog the grinding stone with metal to cause galling.
- Replacement seats that require removal of excessive metal to establish desired angles.

Another advantage of cutters is that they do not require dressing, although a disadvantage is the additional setup time required for different cylinder heads. You must be familiar with the operating procedure for the cutting equipment you are using. There are several designs. Valve seat cutting involves all the same angles as valve seat grinding.

Most seat cutters form one angle at a time and operate similarly to seat grinders. Follow the procedures under *Valve seat grinding*, earlier in this chapter, to establish the seat, then correct the height and width by topping and throating. Seat cutting forms the seat by tearing off bits of metal, so the surface finish is rough, not nearly as smooth as what you get by grinding. Many machinists lightly dress seats with a grinding stone after cutting, this ensures a good surface finish.

A recent development that is gaining in popularity is single point seat cutting. Single point cutting uses three blades on the same head to machine all of the seat angles at once, figure 14-59. This method is extremely precise and quick. Follow the equipment manufacturer's instructions and pay close attention to setting

Servicing Cylinder Heads and Manifolds

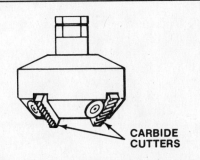

Figure 14-58. This tool uses a cutting bit to remove metal and form the proper seating angles.

up the cutting head. Remember, you only make one pass, so it has to be right.

Valve Lapping

Valve lapping used to be the accepted practice for the general mechanic to restore valve sealing. However, this is no longer true. Lapping uses an abrasive paste to wear a positive seal between the valve seat and the valve face.

Research has shown that, the lapping action wears a groove in the valve. Although the valve and seat may seal perfectly at room temperature, when the engine gets hot and the valve expands, the groove will no longer align with the seat and the valve will not seal well at all. In addition, any of the highly abrasive paste left in the engine can cause instantaneous damage on start-up.

Many machinists stick to the old ways and still lap the valves. However, engine manufacturers do not recommend lapping, and there is absolutely no reason to do so. If all of your machining operations were performed correctly, the valves will seat themselves.

HEAD RESURFACING

You resurface cylinder heads to get a true surface for good gasket sealing. In addition, resurfacing provides "tooth" to hold the gasket, corrects deck misalignment to provide even crush, and cleans up any surface damage. All other cylinder head machining, except OHC cam bearing boring, should be complete before you repair the surface.

The straight edge and feeler gauge reading you took before cylinder head tear-down is a good indicator of what it will take to restore a flat sealing surface. To make sure things did not change as a result of machining and repair operations already performed, take a second measurement now. Use this figure to determine how much machining will be necessary. It is best to straighten aluminum cylinder

Figure 14-59. With single-point cutting, three cutting bits on one tool head machine all of the valve seat angles at the same time.

heads before you resurface them. This keeps metal removal to a minimum and maintains cam timing geometry. It is very important to remove as little metal as possible. Removing too much metal can cause problems, such as:

- The valves may be too close to the top of the piston when fully opened.
- The compression can be raised too high. Compression ratio increases by 0.1 for each 0.010 inch (0.25 mm) of stock removed. Remember, you must also consider any metal that was machined off the deck surface of the block.
- The pushrods may be too long, causing the valves to remain open.
- On overhead cam engines, the cam timing will be retarded as the head is lowered on the block. You can expect timing to retard approximately one degree for each 0.020 inch (0.5 mm) of metal you remove.

Straightening Warped Aluminum Heads

Aluminum cylinder heads have a natural tendency to warp when subjected to extremely high temperatures. Bolting the head to a cast-iron block compounds the problem. The head bolts can only contain metal expansion in the casting up to a certain point. If temperature continues to rise, the head continues to grow away from the block. The cylinder head casting deforms, stressing the metal structure of the head as it pushes its way past the bolt heads. When the head cools it retains the

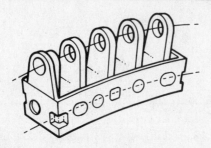

Figure 14-60. Over stressed cylinder heads usually bow up in the middle as the casting tries to push past the head bolts to relieve tension.

stressed shape, usually bowed in the middle with most of the distortion toward the center of the head, figure 14-60. In theory, if you relieve the stress tension, the casting will return to its original shape. In reality, you will never be able to get a head perfectly straight. However, many heads will require only minor machine work to restore the sealing surfaces and bearing bores after straightening.

The best way to relieve stress is by slowly heating and "hot-soaking" the head while applying pressure to remove the bow. A thermal cleaning oven, as detailed in Chapter 9 of this *Shop Manual*, a thick heat-resistant surface plate, and some shim stock are all you need to straighten cylinder heads. Be aware, straightening is time and energy-consuming. Make certain the head is salvageable before you attempt to straighten it. Any valve guide service must be performed now, before the head is heated. You service the valve seats after straightening the head.

Preparing the cylinder head
Begin by accurately measuring the cylinder head thickness with a vernier or dial caliper. Take measurements at various locations on the casting. Keep in mind that after straightening you will be machining the deck surface to true it and provide a good gasket seal. Discard heads that are within 0.005 inch (0.13 mm) of specified thickness. Compare your different measurements. Some cylinder heads are prone to metal migration when they have been overheated. Metal migration will often cause the casting to grow thick in the middle and thin toward the ends. If this is the case, the head may not be salvageable.

The deepest point of the warp will generally be in the center of the casting, so this is where you apply opposing pressure. Coat the bolt heads with anti-seize compound to prevent galling. Loosely attach the cylinder head to the surface plate at the center-most head bolt positions. The plate must be drilled, and possibly

Figure 14-61. Position the shims, apply anti-seize to the bolt heads, install the bolts at the center of the head, and torque to 25 to 30 ft-lb (18 to 22 Nm).

tapped, to accept bolts or studs and nuts. Place shim stock at both ends between the cylinder head and surface plate, then torque the retaining bolts to 25 to 30 ft-lb (18 to 22 Nm). The head is now ready for baking, figure 14-61.

Selecting the correct shim thickness is the key to successful head straightening, but the heating equipment you are using has an effect on results too. If you are unfamiliar with the equipment, talk to your co-workers to find out what shim arrangement has been working well for them. A conservative starting point is with shim stock equal to half the total warp, placed under each end of the head. Shims should fit all the way across the head gasket surface for best results. In general you will find that the thicker the casting, the more shim is required. Unfortunately there are no established rules. You only learn by experience.

Keep a log of all the cylinder heads you straighten. Record how much the head was warped, what shim size you used, the amount of warp after straightening, and how much metal was removed to restore the surface. This information is valuable the next time you work on a similar engine, minor adjustments to shim thickness and bolt torque can get you closer to perfect.

Double-check the cylinder head, make sure all sending units and core plugs have been removed. Then, the head is ready for baking.

Heating and cooling the cylinder head
To relieve stress in the casting the head must be heated to about 500°F (260°C). Maintain the temperature for about five hours, then allow the head to cool slowly. Although a preheated

Servicing Cylinder Heads and Manifolds

thermal cleaning oven works best, the job can be done in any oven that can provide five hours of steady, even heat. If the oven you are using is not thermostat-controlled, you will have to periodically check temperatures with a good surface thermometer or a "heat stick". Keep a close eye on temperatures, because there should never be more than a 10 percent variation. High temperatures can loosen valve seats, and at low temperatures the metal is not pliable enough to relieve tension.

Many shops will fit as many heads as possible into the oven to reduce energy consumption. This practice increases heating time slightly, but is considerably shorter than if you baked each head separately. A practice that works well for cooling is to shut the oven off at the end of the day and leave the heads in the oven overnight.

After the cylinder head has cooled, it must be pressure-tested. See Chapter 10 of this *Shop Manual*. Then, you can measure the remaining surface warp and cam bearing misalignment. If you did the job right, you will take a light cut to true the surface, and the head will be ready to be assembled and put into service. If tolerances are still off by a considerable amount, you can shim and bake the head again. If the head is not straight after two treatments, it is generally more cost effective to simply replace it.

Head Resurfacing

You must have accurate factory specifications for your engine. First, determine how much metal can be safely removed. Keep in mind that the sealing surface on the head has been distorted by disassembly, cleaning, and repairs made to it. Even with a perfectly straight surface, a light cut will bring the surface up to specification. Manufacturers provide a surface finish standard that will establish a tooth to grip and seal the gasket. Always true and prepare the head gasket sealing surface whenever you rebuild an engine.

Heads can be resurfaced either by surface grinding or milling. The equipment and procedures for grinding and milling cylinder heads are similar to those used to surface the block deck. See Chapter 11 of this *Shop Manual*. Because of the similarity, we will not detail them here.

Traditionally, surface grinding yields a much more accurate finish than broach milling because the grinding spindle runs at a much faster speed. The broach may turn slower, but it will remove more metal per pass, so it is substantially faster than grinding. The recent development of milling machines that use CBN (Cubic Boron Nitrate) cutting bits is changing tradition, figure 14-62. CBN milling uses a single point cutting bit rotating at high rpm to remove stock from the workpiece. Set-up and leveling procedures for CBN equipment are generally quick and easy. You can often produce the desired finish and true the surface with one pass.

All resurfacing equipment operates in one of two ways: moving the part over the cutter or stone, or moving the cutter or stone over the part. In both cases the head must be level with the cutter or stone. You must be familiar with the equipment. When resurfacing, keep these points in mind:

- The surface should be parallel to the crankshaft centerline. Acceptable tolerance is usually plus or minus 0.001 inch (0.025 mm). Check the manufacturer's specifications.
- There is usually a specification for minimum head thickness in the shop manual. Make sure you will be within tolerance after machining. If you cut too much, the head is scrap metal.
- You will have to remove more metal than the amount of warpage. If the head is bowed 0.005 inch (0.13 mm), it might take as much as an 0.008 inch (0.20 mm) cut to true the surface.
- Avoid trying to clean up the head with one cut. Multiple passes give you more control and result in a better surface finish. The maximum cut for a grinding stone is 0.005 inch (0.013 mm), while a broach mill can easily remove 0.020 inch (0.50 mm) of stock per pass.
- You must set the machine up to obtain the surface finish specified by the engine manufacturer. Most head surface requirements fall into the 90 to 120 microinch range. However, this is not always the case, so check the service manual.
- Aluminum castings require special considerations. Be sure your machine is properly set up according to the operating instructions.

Overhead Camshaft Bore Alignment

Sometimes simply straightening the cylinder head will bring the camshaft bearing bore alignment back into specification. Check alignment of the saddles using a straight edge and feeler gauge, and record your findings. Then install, tighten, and bring the bearing caps to torque. Measure the bores for taper and out-of-round using a dial bore gauge, inside mi-

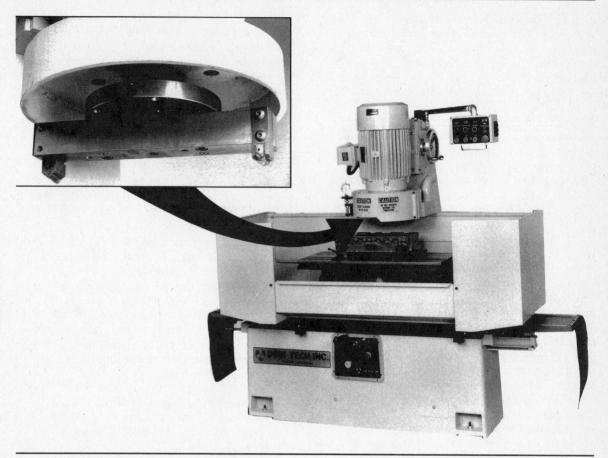

Figure 14-62. This CBN mill uses a high-speed single point cutting bit to quickly and accurately remove stock from the workpiece and produce a good surface finish.

Figure 14-63. Line boring the bearings on an overhead cam cylinder head.

crometer, or telescoping gauge. See Chapter 11 of this *Shop Manual* for details on how to measure bearing bores. Compare your findings with the manufacturer's specifications.

Camshaft bearing bores that are out of specification can be corrected by line boring or align honing. Boring is the more popular method because it is not only faster than honing, but also allows you to remove most of the metal from the cap in order to maintain camshaft position and timing.

You can easily restore the cam bearing bores to standard size if the camshaft is held in the cylinder head by bearing caps. Follow the steps for servicing main bearing bores in Chapter 11 of this *Shop Manual* to:

1. Reduce inside diameter by grinding metal from the parting face of the bearing caps.
2. Install the bearing caps and tighten them to specified torque.
3. Attach the boring equipment according to the manufacturer's instructions. Adjust the tooling to take a minimal cut from the cylinder head side, figure 14-63.
4. After boring, remeasure to make sure everything is up to specification. Thoroughly clean the head, and it is ready for assembly.

There are several methods of restoring bearing alignment when the camshaft is supported directly in the head, without removable caps. The camshaft often rides on bearing inserts with this type of cylinder head design, and the bores can be opened up so that oversize bear-

Servicing Cylinder Heads and Manifolds

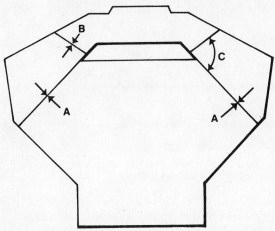

ANGLE	MULTIPLIER
5°	1.1
10°	1.2
15°	1.4
20°	1.7
25°	2.0
30°	3.0
35°	4.0

IF YOU REMOVE .020" (.5MM) OF MATERIAL FROM SURFACE A, AND THE ANGLE OF C IS 10 DEGREES, THEN USE A MULTIPLIER OF 1.2 TO GIVE YOU THE AMOUNT OF MATERIAL TO REMOVE AT B.

EXAMPLE: .020" × 1.2 = .024"
(.5MM × 1.2 = .6MM)

Figure 14-64. A multiplier based on head angle is used to determine the amount of stock you remove from the intake manifold side of the heads.

ing shells can be fitted. This is a quick, easy repair if the engine uses cam bearings.

However, many OHC engines do not use bearings, and the cam rides directly on the cylinder head. This presents problems if the alignment is off. However, there are ways to salvage many of these heads. Replacement camshafts with oversize journals are available for some of the more popular engines. Simply open up the bearing bores to accept the larger journals and provide adequate oil clearance.

Recently, tooling designed to reprofile the bores to accommodate an insert bearing have become available. With this method you open the bearing bores to oversize, fit a special bearing insert, then install a stock camshaft. At the present time, tooling and parts to perform these repairs are limited to a few more popular engine designs.

Some engines, such as the Nissan "L" and "Z" series, have cam towers that bolt onto the cylinder head. Shims can be installed under the towers to compensate for metal removed during head surfacing and keep the valve train and timing components aligned.

Matching Manifold Angles

On V-type engines the intake manifold alignment is affected by cylinder head and block deck resurfacing. The heads fit the block at an angle, so when they move down on the block, they also move closer together. To correct this, you machine the intake manifold side of the heads to maintain the original distance across the valley. The amount of material removed depends on the angle of the intake side of the head to the combustion chamber side. Add how much was removed from the head and the block, then use a multiplier, figure 14-64, to determine how much material to remove from the intake manifold side of the heads.

As explained in Chapter 11 of this *Shop Manual*, on some engines you machine the top part of the block where the ends of the intake manifold seal. If this procedure is overlooked, the intake manifold may not be able to properly align with the heads.

VALVE TESTING AND ASSEMBLY

After all machining operations are complete, the head must be thoroughly cleaned. This is extremely important and cannot be overemphasized. The primary cause of head and valve failure is particle contamination. Clean the head using methods described in Chapter 9 of this *Shop Manual*. After cleaning, scrub the valve guides and oil galleries with a dry brush, figure 14-65, then clear with compressed air.

If any of the galleries seal with core plugs, install them at this time. Be sure the plug is facing in the right direction, and always use the proper tool to drift it into position. Plug fitting instructions can be found in Chapter 16 of this *Shop Manual*.

Some machinists polish the valve stems before they assemble the valve in the head. Polishing can help reduce friction to prevent scuffing and scoring as the parts seat in. Lightly wet-polish the stems using crocus cloth and solvent, then wash off all traces of abrasive dust. Many new valves have part numbers stamped

Figure 14-65. Before you fit the valves, run a dry brush through the guides, then clear them with a blast of compressed air.

Figure 14-66. A vacuum tester will quickly tell you how well the valves are seating.

on the stem. Stamping can raise bits of metal around the edges of the numbers, so be sure to smooth this area down as you polish.

Checking Valve Seat Sealing

Special tools are available that check how well the seat seals using a vacuum, figure 14-66. A rubber-backed plate fits on the head to seal the port, then the port is pumped down to a vacuum. Any leakage around the seated valve will show up as a vacuum loss. Generally there is not a specific vacuum that the port should hold. Look for even figures between the cylinders. The reading for the lowest cylinder should be within 90 percent of the highest.

A quick easy way to check valve seat sealing is with a solvent leakage test. Install a set of spark plugs, then set the cylinder head in stands with the combustion chambers up. Lubricate the valve stems with engine oil and fit all of the valves into the head. Fill the combustion chambers with solvent. Allow the head to sit undisturbed as you watch for any solvent seeping past the valve seats and into the ports.

Installed Height

When you machine valves and seats, the valve sinks deeper into the head. This raises the valve stem, which, in turn, reduces valve spring tension and changes the angle of the rocker arm. Inadequate spring tension can prevent the valve from seating, raise valve operating temperature, and result in poor running engine. You must always measure the installed height of the valve springs. The engine manufacturer provides a specification.

You can restore spring tension and correct installed height by fitting shims under the spring. Shims are available in three sizes: 0.015, 0.030, and 0.060 inch (0.38, 0.76, 1.50 mm). Shims are often serrated on one side to promote heat transfer. These are installed with the serration facing down, toward the cylinder head. Keep in mind that too much shim can lead to spring binding. You cannot compensate for weak springs by adding shims. Check and correct installed valve height as follows:

1. Place the valve in the head and install a spring retainer and a pair of keepers on the valve, without the spring.
2. Measure the distance between the top of the spring seat and the underside of the spring retainer. You can use a telescoping gauge, figure 14-67, caliper, machinist's scale, or special tools.
3. After completing all machining operations, remeasure the installed height and shim the springs the required amount to compensate for the difference.
4. If the valve has moved up more than 0.060 inch (1.5 mm) replace the valve seat to restore the proper installed height.

Cylinder Head Assembly

Inspect all of the parts one last time, and then you are ready to assemble the cylinder head. Pay close attention to the valve keepers. Check fit each one onto the valve stem. If a keeper does not fit securely to the valve, or you notice any scoring, pitting or other signs of wear, replace it. Visually inspect all of the shims, retainers, seats, or other parts that install along with the valves.

All of the parts you assemble, as well as your work area and your hands, must be kept perfectly clean. Remember, particle contamina-

Servicing Cylinder Heads and Manifolds

Figure 14-67. Checking installed valve height, the distance between the top of the spring seat and the bottom of the retainer, with a telescoping gauge.

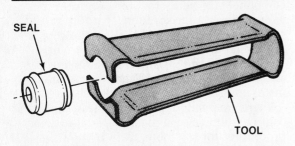

Figure 14-68. Use a seal installation tool to push positive-lock seals onto the valve guide.

Figure 14-69. Install umbrella seals onto the valve stem before you fit the spring.

tion is the number one reason for cylinder head failures.

Installing the valves
Valve installation will be quick and easy if you plan your work and have all of the parts organized. Clear enough space on your workbench to arrange the valves, springs, keepers, shims, seals and other pieces in the order they fit the cylinder head. Be sure the surface of your bench is clean, and have a set of head stands on hand. Fit the valves beginning at one end of the head and work toward the other.

If the engine uses positive-lock type valve stem seals, they must be installed now, before the valve is fitted. Use a seal installation tool, figure 14-68, to press the seal onto the valve guide. Aluminum cylinder heads often have a stamped-steel spring seat that fits between the head casting and the valve spring. The inside diameter of the spring seat may be smaller than the outside diameter of the seal, so you must fit the seat first, then the seal. Install all of the seals, set the cylinder head in stands, then install the valves.

Generously lubricate the valve stem and guide with engine oil. This prevents scuffing the guide when the engine is first started. Then, slip the valve into the guide and hold it against its seat as you fit the shims, springs, and retainer. Install umbrella-type valve guide seals before you fit the spring. Slide the seal all the way down the valve stem until it contacts the guide, figure 14-69. O-ring seals are installed after you compress the valve spring. Use a properly adjusted valve spring compressor to compress the spring and fit the keepers, figure 14-70. Be sure keepers are fully seated into the groove on the valve stem, then slowly release spring tension and remove the valve spring compressor.

Watch out for variable-pitch valve springs. In order to work properly, these springs must be installed right side up. Variable pitch springs are easy to spot because the coils are closer together at one end of the spring than at the other end. The tightly coiled end of a variable-pitch spring always installs toward the cylinder head, figure 14-71.

Figure 14-70. Compress the spring and fit the keepers. Keepers must be fully seated in the valve stem groove with evenly spaced end gaps.

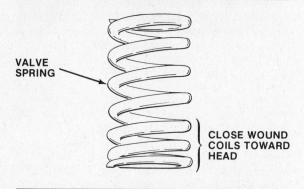

Figure 14-71. The tightly coiled end of a variable pitch spring always installs toward the cylinder head.

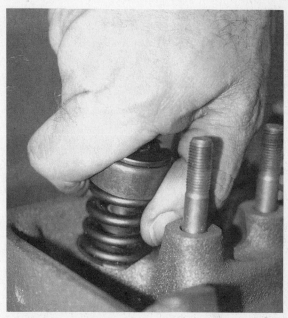

Figure 14-72. You will not be able to rotate the assembled valve spring if it is properly tensioned.

Check keeper seating by giving the valve stem several light raps with a soft faced hammer. Verify assembled spring tension by trying to turn the spring by hand. You should not be able to rotate the spring, figure 14-72. After you install all of the valves, retest seat sealing as previously explained. O-ring type valve guide seals can also be vacuum-tested. Slip a hose from a vacuum pump over the valve stem so it seals to the spring retainer and pump it down. A good seal will hold a vacuum.

Replace any studs that were removed prior to machining. Overhead-valve cylinder heads are now ready for installation on the cylinder block. Wrap the heads in plastic, tag them for easy identification, and store them in a safe place until you are ready to assemble the engine. With some OHC designs, you can fit the camshaft and valve actuation components and adjust valve lash, before you store cylinder heads away for final assembly.

Installing overhead camshafts

Inspect the camshaft and all other actuation components to verify that they conform to the manufacturer's specifications. Inspection and repair procedures for camshafts, followers, lifters, rocker arms, and rocker shafts are in Chapter 15 of this *Shop Manual*. Skip ahead and service these items. Then, return here to install the parts onto the cylinder head.

There are many different overhead camshaft configurations. Some require special tools or procedures for assembly. Keep in mind, as you install the camshaft that some of the valves will be opening and the shaft is under load from valve spring pressure. You must assemble the valve train components according to the factory sequence, adjust, and tighten to specification. Failure to follow proper procedures can result in a bent camshaft.

Verify bearing oil clearance. It must be within tolerance. Generously lubricate the bearings, journals, and other moving parts with engine oil as you assemble the parts. Never tighten down the camshaft with the combustion chamber side of the cylinder head on a flat surface. This can bend the valves as they open. Work with the head supported in stands. Once the cam is installed, lay the head on its side for storage.

Chapter 15

Servicing Camshafts, Lifters, Pushrods, and Rocker Arms

The camshaft, lifters, pushrods and rocker arms are wear items that should be replaced whenever you do a complete rebuild. However, there are times when some, or all, of these parts will be cleaned, measured, inspected, and returned to service. As a general rule, valve actuation components can only be reused when you are working on a well maintained, low-mileage engine that has suffered a failure in an unrelated area. Even in these instances, the parts must be thoroughly inspected before they are reinstalled.

Keep in mind that every time a valve opens, the camshaft, lifter, pushrod and rocker arm are under extreme load. Pressure in excess of 1800 pounds can occur between the camshaft lobe and the lifter, and the camshaft itself must resist the torsional loads from all of the open valves simultaneously. In addition, hydraulic lifter temperature can run as high as 300° F (150° C). The pushrods and rocker arms are under considerable stress as the rotational motion of the camshaft is transferred to the liner action required to open the valves. All of the valve train components must be within specifications or they will quickly wear out.

In this chapter, we explain the procedures you use to evaluate the condition of the camshaft, lifters, pushrods, and rocker arms. We also detail the machining operations for grinding camshafts and refacing rocker arms.

CAMSHAFT SERVICE

The camshaft, a critical part of the engine, is responsible for opening and closing the valves at the proper time. The design of the shaft determines the overall performance of the engine. If a camshaft is not properly configured, the engine will run rough and lack power. Before you install the camshaft, you must be certain that it is designed to meet the needs of the engine. All camshafts, new, used, and reconditioned, must be thoroughly inspected prior to installation.

Inspecting and Measuring the Camshaft

You must inspect the camshaft before installing it. Check the camshaft for:
- Surface finish and condition
- Camshaft straightness
- Journal diameter
- Lobe configuration and runout.

Surface finish and condition

Begin your evaluation with a visual inspection. This is extremely important with a used or re-

Chapter Fifteen

Figure 15-1. Signs of chipping, cracking, or flaking along the edges of a cam lobe indicate poor contact.

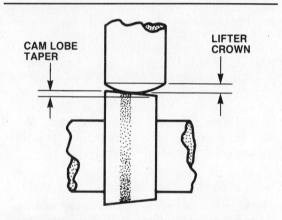

Figure 15-2. A combination of cam lobe taper and lifter face crown creates an off-set contact that spins the lifter in its bore when the engine is running.

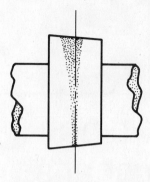

Figure 15-3. Normal camshaft wear is indicated by a pattern slightly off-center on the lobe. The contact patch is widest at the nose and narrowest at the heel.

ground shaft. Look for any signs of chipping, cracking, or flaking along the edges of the cam lobes and the bearing journals, figure 15-1. Also look for any surface irregularities that can lead to accelerated wear and premature failure. If anything at all looks suspicious, replace the camshaft and lifters. You can check a camshaft for cracks using magnetic particle inspection, see Chapter 10 of this *Shop Manual*.

The surface of the cam lobes must be smooth and free of any scoring, galling, or heavy discoloration. The nose of the lobe is where the greatest pressure occurs. Carefully look this area over and run a fingernail across the nose. If you feel any roughness, replace or recondition the camshaft.

Other areas of concern are the drive gears or eccentrics that are part of the shaft. Many engines take the distributor and oil pump drive from the camshaft. This is accomplished by a gear that is often part of the casting. Closely examine the gear for chipping, scoring, or excessive wear on the teeth. If the gear is not up to standard, replace the shaft. The camshaft may also have an eccentric used to drive a mechanical fuel pump. The fuel pump eccentric may be a part of the casting, or a separate piece that bolts to the shaft along with the timing gear. Check to make sure the eccentric is in sound condition. The surface should appear smooth and highly polished. Excessive scoring, galling, or pitting is cause for replacement.

Wear patterns

As explained in the *Classroom Manual*, the surface across the nose of most camshaft lobes is usually not perfectly flat. One exception would be a camshaft that is designed to operate with roller lifters. Cam lobes are tapered by as much as 0.002 inch (0.05 mm). In addition, the lifter face has a convex shape with about 0.002 inch (0.05 mm) height difference between the center and the edges, figure 15-2. The combination of the lobe taper and the spherical lifter face places the point of contact off-center on both pieces. This off-center contact causes the lifter to rotate as it rides up on the cam lobe. Rotating the lifter reduces frictional loss and extends service life. If the lifter face is worn flat or concave, the lifter will ride on the highest point of the cam to create a wear pattern along one edge of the lobe.

A normal camshaft wear pattern will be slightly off-center on the lobe without contacting the edge. The contact patch should be widest at the nose, taper down the ramps, and narrow at the heel, figure 15-3. If the wear pattern extends to the edge of, or completely across the lobe do not install the camshaft. It may perform adequately for a time, but the en-

Servicing Camshafts, Lifters, Pushrods, and Rocker Arms

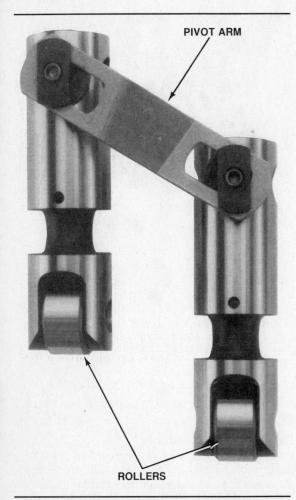

Figure 15-4. Roller lifters contact the camshaft through a small rotating wheel. A pivoting arm links two lifters together to keep the rollers parallel to the cam lobes.

gine will be short-lived because the cam will not be able to rotate the lifters.

Roller lifters, once limited to high performance applications, are now commonplace in production engines. This type of lifter contacts the camshaft through a small rotating wheel that must be kept parallel to the cam lobe at all times, figure 15-4. Since the lifter does not rotate, the nose of the cam lobes are flat, rather than tapered. Roller lifter camshafts have an extremely hard surface finish because the lifter concentrates pressure on a small area of the lobe. Roller camshafts are usually machined from a forged steel billet. A standard camshaft, made of cast-iron, will quickly wear out if installed with roller lifters.

Camshaft straightness

Camshafts, like crankshafts, are prone to bending due to the extreme pressure exerted on them during operation. Before you install a camshaft, you must be certain that any runout

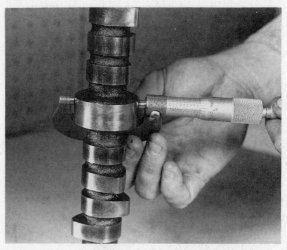

Figure 15-5. Journal diameter can be quickly and accurately measured with an outside micrometer.

is within specifications. To check straightness you will need a dial indicator and V-blocks.

Support the camshaft in the V-blocks by the front and rear bearing journals. Position the dial indicator so the plunger rests on the center bearing journal. Zero the gauge, then slowly rotate the shaft one complete revolution as you watch the dial indicator. The difference between the highest and the lowest reading equals total camshaft runout. Compare your findings to the manufacturer's specifications. As a rule, runout should be less than 0.002 inch (0.05 mm).

A camshaft that is slightly out of limits can usually be straightened. The procedure is detailed later in this chapter. Complete your inspection before you attempt to straighten the shaft. It makes no sense to straighten the shaft now, then find out later that the journals or lobes are worn beyond limits.

Measuring journal diameter

In a running engine, the camshaft journals rotate in the bearing shells supported by a thin film of oil. The amount of clearance between the journal and the bearing is critical. Too little clearance can seize the camshaft, and excessive clearance causes engine oil pressure loss. Camshaft bearings and journals, like those of the crankshaft, have a natural tendency to wear unevenly. Because the camshaft rotates at half crankshaft speed, and does not receive the full force of combustion, camshaft journals do not wear as quickly as crankshaft journals.

Journal diameter can be quickly and accurately measured with an outside micrometer, figure 15-5. Be sure to measure each journal at least twice in different locations around the circumference. Start at one end of the shaft and work your way to the other, record your read-

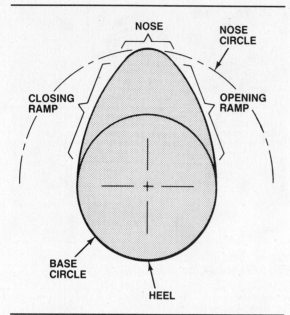

Figure 15-6. You must be familiar with names for various parts of the cam lobe.

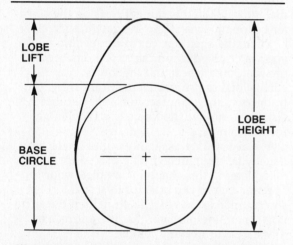

Figure 15-7. Camshaft lift is the difference between the height of the cam lobe and the diameter of the base circle.

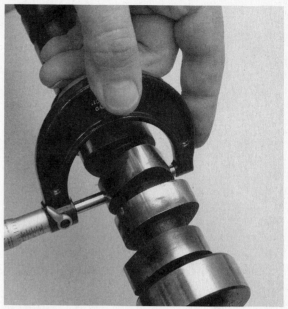

Figure 15-8. To calculate lift with an outside micrometer, first measure across the cam lobe to determine the base circle diameter.

ings as you go. Many camshafts, especially those for overhead valve engines, have different-size journals to allow for easier installation. The largest journal will generally be at the front of the engine.

Compare your measurements to the manufacturer's specifications to determine if the camshaft is within tolerance. If any of the journals are below the limit, replace the camshaft. Undersize and oversize camshaft bearings are available for some engines. If available, these bearings may let you salvage the camshaft or engine block. Camshaft journals can be machined to a smaller diameter and installed with undersize bearings. The engine block bearing bores can be align honed to accept oversize bearings.

Calculating oil clearance
Once you know the exact diameter of the journals, you can use that figure to determine the oil clearance. As explained in Chapter 11 of this *Shop Manual*, oil clearance is the difference between journal outside diameter and bearing inside diameter. Bearing inside diameter must be measured with the bearings installed. If the journals are within standards and you have not yet fitted the bearings, skip this procedure for now. After you install the bearings, be sure to come back to this point. *You must calculate oil clearance* before you fit the camshaft.

Measure bearing inside diameter using a dial bore gauge, inside micrometer, or snap gauge and outside micrometer. See Chapter 11 of this *Shop Manual* for use of these tools. Determine oil clearance by subtracting the journal diameter from the bearing inside diameter.

For example, if the journal measured 1.8673 inch (47.43 mm), and the bearing inside diameter measured 1.8692 inch (47.48 mm), calculate oil clearance as follows:

1.8692 − 1.8673 = 0.0019 inch
47.48 − 47.43 = 0.05 mm

A typical manufacturer's tolerance range for camshaft bearing oil clearance would be 0.001 to 0.004 inch (0.025 to 0.100 mm).

Servicing Camshafts, Lifters, Pushrods, and Rocker Arms

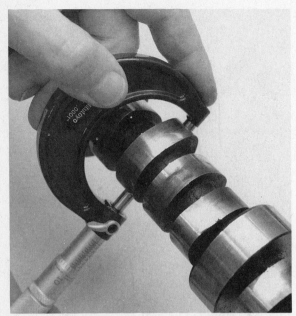

Figure 15-9. Measure the nose to heel distance of the cam lobe, measure at the front and at the rear to determine taper.

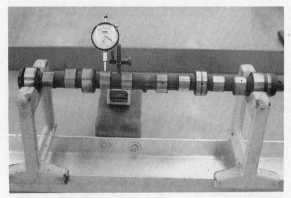

Figure 15-10. With the camshaft in V-blocks, you can measure lift using a dial indicator.

Lobe configuration

Check the cam lobes for correct lift and base circle runout. In order to do this, you must be familiar with the terminology used to describe the various cam lobe features, figure 15-6. Camshaft design and construction is detailed in Chapter 9 of the *Classroom Manual*. Valve lift should always be checked, even if you are installing a new camshaft. You must be certain that the camshaft is compatible with all of the other engine components.

Calculating valve lift

Camshaft lift is the distance that the cam lobes raise the lifters. Lift must be determined before you install the camshaft. Too much lift can cause valve-to-piston interference, while not enough lift will result in poor engine performance. Camshaft lift is the difference between the height of the cam lobe (from nose to heel) and the diameter of the base circle, figure 15-7. On many engines, intake valve lift differs from exhaust valve lift. Camshaft lift can be determined either by measuring with an outside micrometer, or with a dial indicator and V-blocks.

To measure lift with an outside micrometer, first measure across the cam lobe to determine the base circle diameter, figure 15-8. Measure the base circle diameter for all of the lobes. Compare your findings, readings should be identical. This method works for most stock camshafts. However, on many high-performance camshafts lift begins at a point lower than the base circle center line. On these camshafts, it is impossible to measure base circle diameter with a micrometer. Measure lift with a dial indicator as described later.

Once you have established the base circle diameter, use your micrometer to measure the nose-to-heel distance on each of the lobes, figure 15-9. Measure each lobe twice, once at the front and once at the rear. The difference will equal the amount of taper designed into the lobe for lifter rotation. When you calculate lift you use the highest figure because the lifter will ride on the high point of the lobe. Some engines use a different amount of lift for the intake and exhaust valves. Keep this in mind as you compare your findings. All of the intake lobes should be relatively equal in height, exhaust valves should also equal each other. There should never be more than a 0.006 inch (0.15 mm) height difference between lobes.

To calculate cam lift, simply subtract the base circle diameter from the overall height. For example: if the base circle diameter equals 1.230 inch (31.24 mm), and the overall height equals 1.615 inch (41.02 mm), lift would be 0.385 inch (9.53 mm).

You can always measure cam lift with a dial indicator. Measurements can be taken by placing the camshaft in V-blocks, or with the shaft installed in the engine. Set up the dial indicator resting the plunger on the heel of a lobe and zero the gauge, figure 15-10. Watch the indicator needle as you slowly rotate the shaft until the plunger is on the nose. The total dial indicator reading is equal to the cam lobe lift, providing the shaft is straight. A bent camshaft will give incorrect dial indicator readings.

Base circle runout

Checking the base circle runout, or concentricity, is the final measurement you will make on the camshaft. This measurement is taken with a dial indicator while the camshaft is supported in V-blocks. Base circle runout is an important measurement that is often overlooked.

Figure 15-11. Camshafts can be straightened by peening with a blunt chisel and hammer.

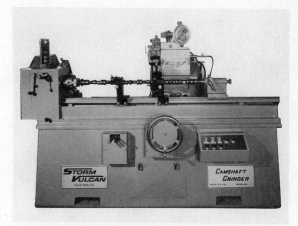

Figure 15-12. A camshaft grinder, an expensive machine tool generally found only in specialty shops, is used to recondition camshafts.

Always check the runout, even when you are installing a new camshaft. Excessive runout can cause low rpm hydraulic pump-up, and also prevent accurate lash adjustment.

Zero the dial indicator with the plunger resting on the heel of a cam lobe, then slowly rotate the shaft to the point where you get your lowest gauge reading. This is where the cam lobe lift ramp begins. Rotate the shaft in the opposite direction, so the plunger crosses the heel, until you reach the start of the closing ramp on the other side. Compare your findings. Total base circle runout should not exceed 0.001 inch (0.025 mm).

Camshaft Straightening

You can often straighten a slightly bent camshaft and return it to service. Straightening procedures are similar to those used for crankshafts, as detailed in Chapter 12 of this *Shop Manual*. Since the camshaft does not absorb the full force of combustion, the amount of bend is usually considerably less than what you will find on a crankshaft. Keep in mind, you are not bending the shaft back into position, you are relieving stress in the metal to allow it to return to its original shape.

You can correct camshaft bend by peening or pressing. Peening is done with the shaft supported between centers, or in V-blocks. Use a dial indicator to find the point of greatest distortion, position the shaft so the high point of the bend faces down. Use a soft tipped blunt chisel and a hammer to strike the shaft in the direction of the bend to relieve stress in the metal, figure 15-11. Recheck shaft straightness frequently, you do not want to overcorrect the bend. The shaft will have a natural tendency to return to its original shape.

A special straightening fixture is needed to press straighten a camshaft. The fixture operates similarly to a crankshaft straightening press. Hydraulic pressure pushes the camshaft back to its original shape. Check your progress frequently to avoid overcorrecting.

Camshaft Grinding

Camshaft grinding is a machining technique that uses a grinding stone to restore the journals and lobes. The grinding process requires a specialized and expensive piece of equipment, and is no longer a common automotive machine shop operation, figure 15-12. Due to the expense, grinding camshafts is generally reserved for high-performance and heavy-duty industrial applications. Inexpensive replacement camshafts are readily available for most automotive engines.

A camshaft with marginal wear can be reconditioned by grinding the lobes undersize. Small equal amounts of metal are removed from each lobe. The finished shaft has the same profile as the original; however, the nose and base circle are slightly smaller, figure

Servicing Camshafts, Lifters, Pushrods, and Rocker Arms

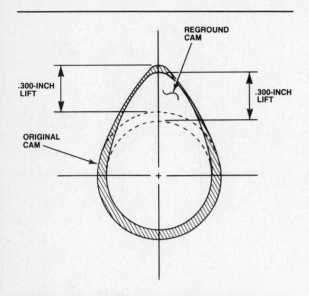

Figure 15-13. A reground camshaft has the same profile as the original, but the nose and base circle are slightly smaller.

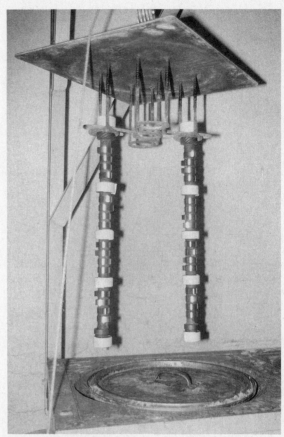

Figure 15-14. Reground camshafts ready to be immersed in a protective surface coating. The process helps the lobes retain lubricant during the first critical minutes after start up.

15-13. On engines with adjustable valve lash, the size difference can be compensated with valve adjustment. If the engine you are working on does not have adjustable valves, you may have to use longer pushrods to get the necessary clearance.

A more accurate method of reconditioning a camshaft is to build up the lobes by welding, then grind them back to standard size. Since the finished shaft is the exact size and profile as the original, there are no lash adjustment complications. As with crankshaft welding, you grind the cam to undersize before welding. This provides a clean, solid surface that ensures good adhesion for your weld. The type of welding equipment you use will vary according to camshaft material and construction. Shops that specialize in camshaft rebuilding generally use a metal spraying process.

Because camshaft grinding is a very exacting process, performed at a limited number of machine shops, we will not detail the operations. To grind a cam, first you must assemble a master shaft that conforms to the final shape of the shaft you are grinding. The workpiece is rotated in the machine, one end supported by the spindle, the other by a tailstock. Once set up, the grinding head follows the profile of the master. The stone moves into and away from the workpiece to precisely finish the lobes as it works its way down the length of the shaft.

When grinding is complete, the bearing journals are taped off for protection and the camshaft is immersed in a solution of molybdenum disulfide or manganese phosphate, figure 15-14. The immersion process applies a surface coating that retains oil to prevent scuffing and promote rapid break-in when the engine is first started. The surface coating has a distinct black color and rough finish. New camshafts also come with a surface coating. Never wash off the coating before installing the cam.

VALVE LIFTER AND CAM FOLLOWER SERVICE

Valve lifters, also known as tappets or cam followers, are cylindrical parts that ride directly on the camshaft to open and close the valves. All valve lifters fall into one of three categories: solid, hydraulic, or roller. Generally, OHV engines use lifters that ride above the cam, DOHC engines have bucket type cam followers that ride below the cams, and SOHC engines use finger-type followers.

Bucket followers, valve lifters, and finger followers all wear to match the specific cam lobe they are riding on and should never be interchanged. If you reinstall the old camshaft, use the old lifters and install them in the same bore they were originally fitted to. If you re-

Figure 15-15. Even after high mileage, this lifter was still rotating in its bore. Note how the circular wear pattern does not extend to the edge of the lifter.

place the camshaft, the lifters must also be replaced. To make sure the new cam and lifters are compatible, always purchase them as a set.

Correcting marginal wear by regrinding the lifter face, once a common practice, is seldom done in the modern repair facility. Current practice is to simply replace the lifters and install a new, or reground camshaft.

Solid Lifters

Solid lifters are actually solid only at the foot where they contact the cam. The pushrod seats inside of the hollow body of the lifter. Solid lifters have not been used in a domestic production engine since the early 1970s. Volkswagen continued using solid lifters in their air-cooled engines until 1980. When you do run across solid lifters, they are easy to deal with. Wash the lifters in clean solvent and inspect them. Examine the camshaft contact face, the body, and the pushrod seat.

The contact face must be smooth and polished, and there cannot be any chips, nicks, or scratches on the surface. Wear patterns should be circular and evenly centered on the lifter, figure 15-15. Patterns that run across the face indicate a lifter sticking in the bore and not rotating. If the wear extends to edge of the face, the lifter face has been worn flat or concave. Remember, the shape of the face must have a high center, or crown, or the lifter will not rotate. Place a straight edge across the face of the lifter. If the straight edge lays flat, and does not rock slightly, replace the lifter. A quick way to check for a crown is to hold two lifters face to face. If you can rock them back and forth, they are good.

The sides of the lifter body should be worn smooth and appear highly polished. Galling or

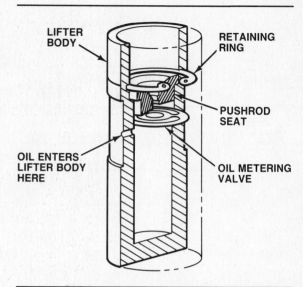

Figure 15-16. Metering-valve lifters are often mistaken for hydraulic lifters. The metering device simply directs pressurized oil up the pushrod.

scoring on the sides of a lifter can indicate a clearance problem, or be the result of abrasive particles in the engine oil. The pushrod seat inside the lifter body bore should also be smooth and polished. The seat should not have any ridges or show signs of pounding from excessive lash clearance.

A variation on the solid design is the metering-valve lifter. These lifters are often mistaken for hydraulic lifters because the inner bore contains an oil reservoir and the pushrod seats on a metering plate, figure 15-16. The oil trapped inside the lifter body does not control valve lash. The metering device simply directs pressurized oil up the hollow pushrod to lubricate the rocker arms.

Hydraulic Lifters

Virtually all OHV automotive engines are now fitted with hydraulic lifters. The hydraulic lifter uses engine oil pressure to take up slack in the valve train so that the valves operate with zero lash, figure 15-17. The principals of hydraulic lifter operation are discussed in Chapter 10 of the *Classroom Manual*. Here we detail only the inspection procedures.

When replacing lifters, always replace them as a set, along with the camshaft. Replacement lifters will not always look identical to the ones you removed from the engine, but this is not a cause for concern. Although appearance will differ between brands, there are four critical dimensions that must be identical to the original. New lifters must have the same outside diameter, operating height, oil feed groove

Servicing Camshafts, Lifters, Pushrods, and Rocker Arms

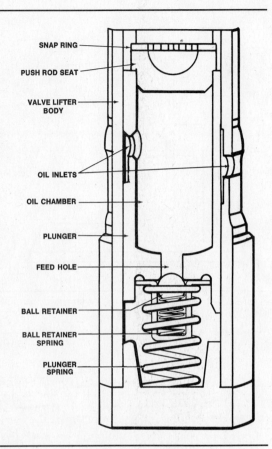

Figure 15-17. The hydraulic lifter consists of an outer body, plunger, and check valve. Engine oil pressure is used to maintain zero valve lash.

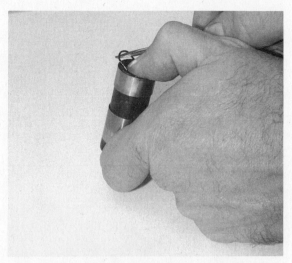

Figure 15-18. Hydraulic lifters can be easily disassembled by removing the snap ring.

width, and oil groove position as the old ones. Keep in mind that oversize lifters may have been installed in production. If this is the case, they must be replaced with oversized units.

If you are refitting the old lifters, you must disassemble, inspect, and leak-down test them before installation. On assembly, the lifters are fitted to their original bore so that they ride on the same cam lobe. Wash the lifters in clean solvent and look over the cam face, body, and pushrod seat as previously detailed for solid lifters. If the lifters pass a visual inspection, your next step is to disassemble and inspect the internal parts. Work with the lifters one at a time. Do not interchange internal pieces, and keep them spotlessly clean.

Disassembling lifters

Hydraulic lifters are held together by a snap ring that fits a groove on the internal bore of the lifter body. The snap ring can be removed with a small screw driver, figure 15-18. Depress the pushrod seat with your thumb to relieve spring pressure on the side of the snap ring. Remove the snap ring and you can lift off the pushrod seat and metering valve disc.

Handle all of the lifter parts with care. Be aware, each part is a precisely machined hydraulic component. They must be kept *perfectly clean*. A tiny piece of dirt can have a severe effect on service life.

The plunger may seem stuck in the bore due to hydraulic lock caused by residual engine oil. Tapping the open end of the lifter on a clean block of wood will generally loosen the seal. Lift out the plunger, then pour the check valve assembly into the palm of your hand. Pay close attention to how the pieces of the check valve are fitted together. There are two basic types, check ball and check disc, but many variations. The ball or disc, springs, and metering valve must be fitted together exactly as they were originally assembled.

Thoroughly wash all the components in clean solvent and look them over for signs of unusual wear, figure 15-19. Replace the lifter if you find any bent, broken, or otherwise damaged parts. If the pieces look good, and are spotlessly clean, you can reassemble the lifter. Slip the pieces of the check valve into the bore, slide the plunger into the body, and rest the metering disc and pushrod seat on top of the plunger. Push down with your thumb, fit the snap ring, and the lifter is assembled. Now you can set the lifter aside and go on to the next one. Remember, lifters must be kept in their original order. After you service all of the lifters, you can check how well they will perform in service by doing a leak-down test.

Leak-down testing

Leak-down testing is a way to check the ability of a lifter to hold hydraulic pressure and maintain zero valve lash. The test is performed with a leak-down tester using a special fluid. Be

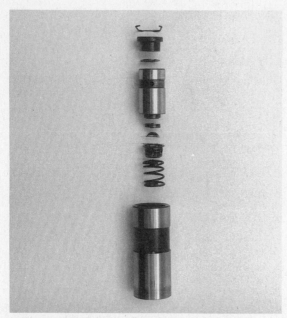

Figure 15-19. All the internal lifter parts must be cleaned, inspected, and reassembled in the exact same order as the original.

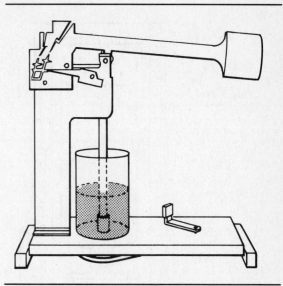

Figure 15-21. Fit the ram into the push rod bore of the lifter. Be sure everything is centered.

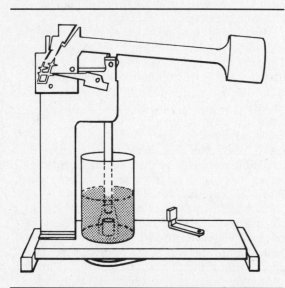

Figure 15-20. Place the lifter to be tested into the center of the cup. The lifter must be fully submerged in the special test fluid.

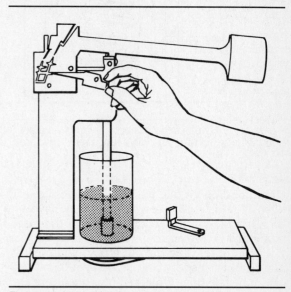

Figure 15-22. Adjust the ram length to align the pointer with the set mark. Tighten the jam nut to maintain the setting.

aware, the tester is calibrated to the special fluid. Substituting any other liquid will give inaccurate test results. Lifter performance is rated by the number of seconds it takes the tester to overcome the hydraulic pressure of the lifter. Perform the test by following these steps:

1. Fill the cup of the tester with enough of the special fluid to cover the top of the lifter. Place the lifter to be tested into the center of the cup. It must be fully submerged, figure 15-20.

2. Fit the ram into the pushrod bore of the lifter, figure 15-21. Some testers use a small steel ball between the end of the ram and the pushrod seat. Be sure everything is centered on the tester.
3. Adjust the ram length so that the pointer aligns with the set mark on the lever, figure 15-22. Tighten the jam nut to maintain the setting.
4. You must purge the air from the hydraulic chamber of the lifter to get an accurate reading. This is done by slowly pumping the weight arm of the tester to operate the

Servicing Camshafts, Lifters, Pushrods, and Rocker Arms

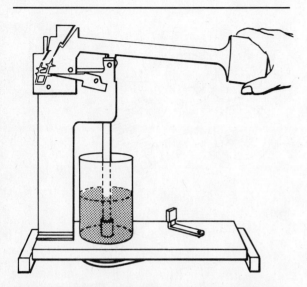

Figure 15-23. Pump the arm of the tester to operate the lifter through its full range of travel and purge the air from the hydraulic chamber.

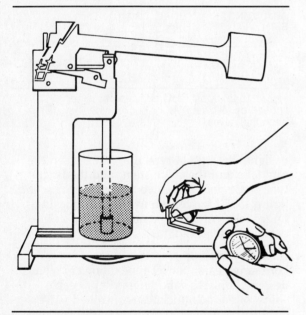

Figure 15-24. Time how long it takes the weighted arm to fully compress the lifter. Turn the hand crank to rotate the fluid cup every two seconds.

lifter through its full range of travel, figure 15-23. Continue pumping until there is no trace of air bubbles escaping from the lifter. You will feel resistance build as the air is expelled from the lifter.

5. After purging the air, raise the weight arm so that the lifter can expand to full height. Now, lower the ram onto the pushrod seat and time how long it takes the weighted arm to compress the lifter, figure 15-24.

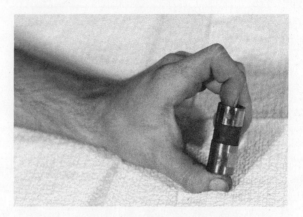

Figure 15-25. Depress the plunger with your finger and quickly release it. If the lifter is good the plunger will snap back into position.

Turn the hand crank to rotate the fluid cup one complete revolution every two seconds.

Manufacturers provide leak-down rates for their lifters. Normal leak-down range is from 20 to 90 seconds. If a lifter tests marginal, retest it. Trapped air or dirt could be the problem. Never fail a lifter after one leak-down check. When the lifters pass, keep them in order and store them in a protected place to keep them clean until you are ready to assemble the engine.

Kickback test

If you do not have a leak-down tester available, you can perform a kickback test. Although not an accurate way to evaluate lifters, the test lets you know if any of the parts are binding. Simply hold the lifter upright and press down on the pushrod seat with your finger, figure 15-25. Quickly release the plunger, it should snap back into a fully extended position immediately. If not, disassemble, clean, reassemble, and repeat the test.

Roller Lifters

A small rotating wheel maintains lifter to camshaft contact on a roller lifter, figure 15-5. The rolling action greatly reduces frictional losses for improved efficiency and increased performance. Roller lifters have been preferred by race and high performance engine builders for many years because they offer more positive valve actuation and permit more radical camshaft timing and profile. Until recently, these hot-rod lifters were only available on the aftermarket as a solid design. Many current production engines are assembled at the factory

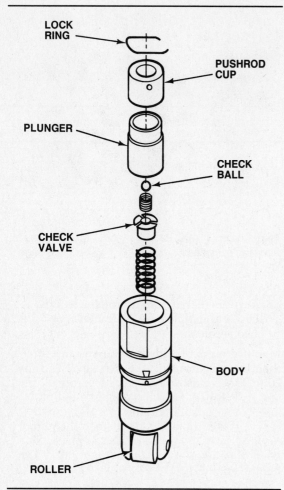

Figure 15-26. Many late-model engines use hydraulic roller lifters to reduce frictional losses and improve fuel economy and performance.

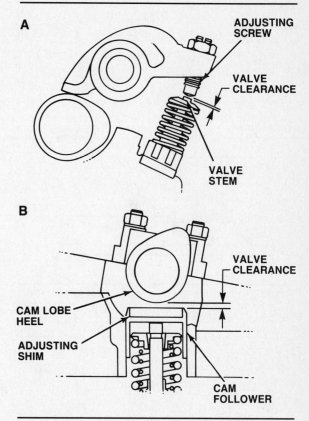

Figure 15-27. Most OHC engines use either finger-type followers (A), or bucket-type lifters (B) to open the valves.

with hydraulic roller lifters, figure 15-26. Hydraulic roller lifters provide a maintenance-free way of reducing parasitic drag to improve fuel economy, as well as performance.

The roller of the lifter must be parallel with the cam lobe at all times. A roller lifter allowed to turn in the bore will quickly destroy itself and the camshaft. Linking two lifters together with a pivoting arm is a popular way of preventing rotation.

Some designs allow you to replace a defective roller head, but many do not. Hydraulic units can be taken apart, cleaned, inspected, and reassembled as described for standard hydraulic lifters. Replace defective lifters as a set, and install a new compatible camshaft.

Over Head Cam Lifters

The cam followers of an OHC engine perform a function similar to that of the lifters of an OHV engine. There are two basic cam follower designs: the bucket-type that fit between the cam lobe and the valve stem, and the finger-type that opens the valve through lever action, figure 15-27. Valve lash with either bucket or finger followers may be mechanically or hydraulically adjusted.

Bucket-type lifters

For many years, bucket-type followers were designed only as solid lifters. Replaceable shims of various thickness controlled valve lash. As explained in Chapter 7 of this *Shop Manual*, special tools are used to remove and replace the shims to attain the proper valve clearance. Hydraulic bucket cam followers, figure 15-28, were first installed in production beginning in the mid 1980s.

Mechanical bucket lifters

When the shim rides on top of the follower, the shim is the only actuation component that comes into contact with the camshaft. Because of this feature, you do not have to replace the lifters when installing a new camshaft. The cam lobe contact surface of the shim is perfectly flat. The nose of the lobe is tapered and slightly off-set from the followers centerline to provide the necessary rotation.

Servicing Camshafts, Lifters, Pushrods, and Rocker Arms

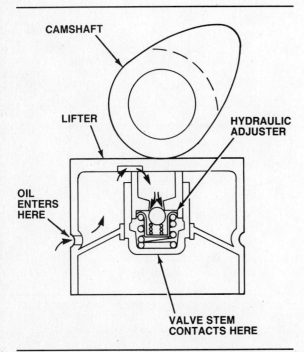

Figure 15-28. Many late-model bucket type followers feature hydraulic lash adjustment.

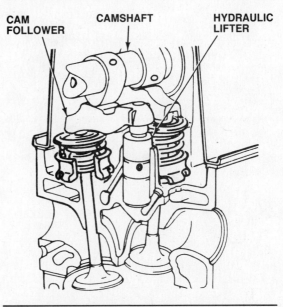

Figure 15-29. Valve lash with this OHC design is maintained by a hydraulically adjusted remote pivot.

Wash the cam followers in clean solvent for inspection. Be sure to keep them in order so they can be returned to the same location in the cylinder head. Inspect the followers for signs of scoring and galling, and replace any defective units.

Hydraulic bucket lifters

Hydraulic bucket followers are not a serviceable item. Defective followers must be replaced. Because the camshaft acts directly on the follower, you must replace the camshaft and followers as a set. When reinstalling used followers, they must be returned to their original position in the engine.

Proper handling is important. Wipe the followers down with a clean rag and look them over for signs of seizing, galling, scoring, or other damage. When you remove the followers from the engine, store them upside down in a container of clean engine oil. This prevents the hydraulic chamber from bleeding-down, and helps to eliminate excessive valve clearance when you first start the engine.

Finger-type cam followers

Finger-type cam followers are similar in design to the rocker arms of an OHV engine. We detail servicing the fingers later in this chapter under Rocker Arm and Shaft Service. At this point we will discuss the methods used to take up slack in the valve train. Valve train slack can be compensated for either mechanically or hydraulically. Chapter 7 of this *Shop Manual*, explains in detail how you set the valve lash for mechanically adjustable followers.

Hydraulic adjusters fall into two categories: the remote pivoting type, figure 15-29, and the rocker arm mounted type, figure 15-30. Of the two, the remote pivot is the more common.

Remote-pivot hydraulic adjusters

Remote-pivot adjusters fit into a bore in the cylinder head, with a socket on one end of the finger fitting onto the top of the adjuster. The opposite end of the finger rests on the end of the valve stem. The camshaft rides on top of the finger. Hydraulic pressure pushes the adjuster up to eliminate slack and keep the lever in constant contact with the camshaft. Because the fingers will wear to match the cam lobes, they must be replaced whenever you install a new camshaft.

Service procedures for the hydraulic unit are similar to that previously detailed for hydraulic lifters. With most designs, you can disassemble, clean, inspect, and leak-down test the adjusters, figure 15-31. Follow the steps we outlined previously in this chapter.

Rocker arm mounted hydraulic adjusters

Rocker arm mounted hydraulic valve adjusters are a relatively new trend in engine design. The tiny hydraulic unit fits into a recess in the rocker arm in place of an adjusting screw, figure 15-30. These are not a serviceable item, so you must replace defective units. When returning units to service, they must be matched to the valve they originally operated with.

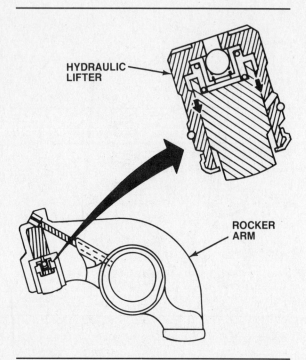

Figure 15-30. Recent technology led to development of the rocker arm mounted hydraulic adjuster.

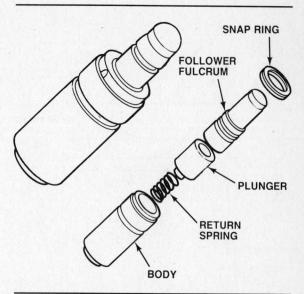

Figure 15-31. You can disassemble, service, and test remote pivot type OHC hydraulic adjusters. The procedure is similar to that used for OHV lifters.

PUSHROD SERVICE

There are two basic pushrod designs: hollow and solid. Solid pushrods are not actually solid, but are a length of steel tubing with press-fit end caps. The term solid applies to the fact that the caps on either end of the tube are a solid piece. Hollow pushrods have an oil passage drilled through the end caps. The drilled passage directs engine oil from the lifter, through the pushrod, and up to the rocker arm assembly.

The biggest variable in pushrod design can be found in the end cap shape, figure 15-32. The lower end of the pushrod is a convex ball that seats into the lifter. The top end of the pushrod may be either ball shaped or socket shaped. A ball-shaped end is used when the pushrod seats into the rocker arm. Socket ends are used when the pushrod seats to a screw-type lash adjuster. Some older engines used an adjustable pushrod with a threaded insert at the end. Turning the insert alters the length of the rod, and a jam nut locks it in place.

Inspecting the Pushrods

Wash the pushrods in solvent to prepare them for inspection. Remember to keep them in order so you can return them to their original position in the engine. Cleaning the inside of hollow pushrods is important. Flush the internal passage with solvent, run a thin wire through the pushrod to dislodge any trapped particles, figure 15-33, then flush again with solvent. The *pushrods must be perfectly clean*. In a running engine, any debris left inside the pushrod will work its way free to restrict or block the oil flow to the top end. Once the pushrods are clean, look them over for signs of wear and damage.

A good place to start your inspection is at the ends. The ball shaped ends must be smooth and well rounded. Replace the pushrod if you find nicks or grooves, a rough surface finish, any deformation, or signs of excessive wear. With hollow pushrods, carefully inspect the oil drill holes. The hole must be perfectly round with a well defined edge. If the oil hole is oval shaped, the pushrod is worn out, and you must replace it. Carefully inspect the pushrod seat of the lifter, and the rocker arm, when you find end damage.

Look the sides of the pushrod over for signs of rubbing and scuffing. To prevent sideways movement of the pushrod, some engines have a guide slot cast into the cylinder head, while others use a guide plate that bolts on. Replace the pushrod if any metal has been rubbed away, and check the guide for wear and alignment. You must use hardened pushrods if a guide plate is installed. The steel plate will quickly cut through the sides of a non-hardened pushrod.

Pushrods are the weak link in the valve train. Valve train problems, such as a sticking valve, improper valve adjustment, incorrect

Servicing Camshafts, Lifters, Pushrods, and Rocker Arms

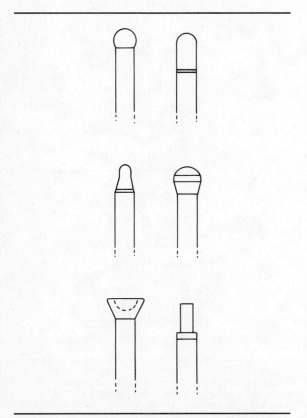

Figure 15-32. The shape of the pushrod end caps varies for the specific engine application.

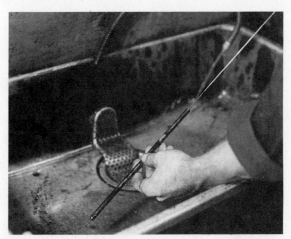

Figure 15-33. Clean hollow pushrods by running a length of wire through them and flushing with solvent.

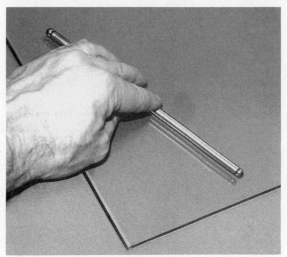

Figure 15-34. Check for straightness by rolling a pushrod across a surface plate, thick plate of glass, or any other perfectly smooth surface.

valve timing, or improper assembly, can cause a pushrod to bend. Check all of the pushrods for straightness before you install them in the engine. A quick and easy way to check straightness is to roll each pushrod across a surface plate, thick plate of glass, or other perfectly smooth surface, figure 15-34. A straight pushrod will roll smoothly across the surface, while a bent rod will tend to hop as you roll it. You can also check pushrod runout with a special fixture or a dial indicator and V-blocks. When using one of these methods, total indicated pushrod runout should be less than 0.003 inch (0.08 mm).

ROCKER ARM AND SHAFT SERVICE

The rocker arms perform two important functions: they are the mechanical link that opens the valves, and they increase the effect of camshaft lift for wider valve opening. Rocker arm construction and function is discussed in detail in Chapter 10 of the *Classroom Manual*. As with other engine components, a wide variety of rocker arm designs are in use. They can be classified into two categories: stud mounted and shaft mounted. Many of the service considerations are similar for both types.

Stud-Mounted Rocker Arms

Most stud-mounted rocker arms are made of stamped steel. The contact areas for the pushrod and valve stem are hardened for improved service life. The lightweight stamped arms are mounted to the cylinder head on a stud, or with a bolt and pivot on a fulcrum. Design particulars of various rocker arm assemblies are detailed in Chapter 10 of the *Classroom Manual*. Regardless of design, you must thoroughly clean and inspect all of the rocker arm components before you assemble the engine.

Disassembling and inspecting

Rocker arms may be fastened to the cylinder head with either a lock nut or a bolt. Both types require a fulcrum seat between the fas-

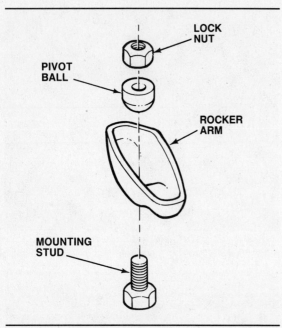

Figure 15-35. This rocker arm pivots on a ball-shaped fulcrum, the lock-nut attaches the assembly to a stud mounted in the cylinder head.

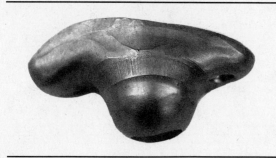

Figure 15-36. The valve stem contact pad, pushrod seat, and pivot area, are the prime wear zones of a rocker arm.

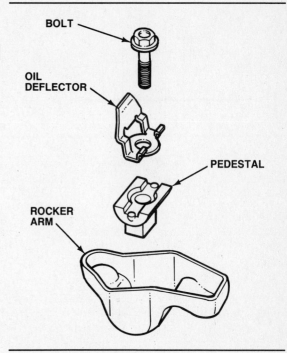

Figure 15-37. This Ford rocker arm pivots on a fulcrum that bolts to a pedestal cast into the cylinder head. The deflector directs oil flow from the hollow pushrod to the rocker.

tener and the rocker to allow the arm to pivot. The fulcrum allows the arm to move smoothly as it follows the profile of the cam lobe. Loosen the nut, or bolt, and remove all the pieces of the assembly. Keep the parts in order, wash them in solvent, and inspect them.

Rocker arms that are held in place with a lock nut are usually a three-piece assembly: lock nut, pivot ball, and rocker arm, figure 15-35. Some pivots are lubricated through an oil passage drilled into the mounting stud. Severe scuffing on the pivot ball and rocker arm indicates a lack of lubrication. The oil passage may be restricted and the stud should be replaced. Rocker arm stud replacement is detailed in chapter 14 of this *Shop Manual*. Use new lock nuts on new studs. With old studs, if the threads are in good shape, you can reuse the lock nuts, but most engine builders automatically replace them rather than take the chance of having one work loose when the engine is running.

The face of the pivot ball, as well as the contact area of the rocker arm, must be smooth and shiny. Replace both the ball and rocker arm if you find any scoring, galling, or pitting. Some pivots are relieved to direct oil flow. Replace the pivot if wear is to the bottom of any machined oil relief.

Closely inspect the rocker arm itself. Pay close attention to the valve stem contact area, the pushrod seat, and the pivot area, figure 15-36. These are the three high-stress zones on any rocker arm. Valve stem wear should be evenly centered in the arm. If wear is off-center or shows grooves, pits, scores, or other irregularities, replace the arm. The pushrod seat should be round and worn to a smooth shiny finish. Look for a ridge built up around the circumference, galling, pitting, scoring, or indications of hammering. Inspect the sides of the rocker around the stud opening for stress cracks and uneven wear. Avoid a potential comeback by replacing a rocker arm if anything at all looks questionable.

Bolt-on rocker arms generally pivot on a fulcrum that attaches to a pedestal cast into the cylinder head, figure 15-37. Pedestal-mounted

Servicing Camshafts, Lifters, Pushrods, and Rocker Arms

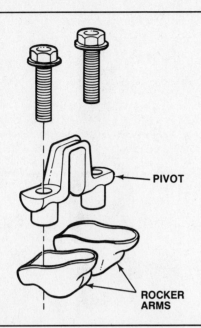

Figure 15-38. This Oldsmobile design maintains rocker arm alignment by linking two adjacent arms with a one-piece pivot.

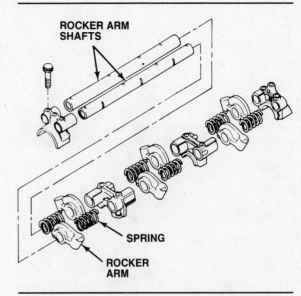

Figure 15-39. Shaft mounted rocker arms are held in position by an assortment of springs, spacers, supports, and washers.

rockers are generally lubricated through a hollow pushrod, and often have an oil deflector that fits between the bolt head and the fulcrum. Some designs use a retainer to link the pivots of two adjacent valves together to keep the rocker arms in alignment, figure 15-38. Look over the contact surfaces of both the fulcrum and the rocker arm. Any sign of scoring, galling, or excessive wear is an indication that you should replace the part. Replace the fulcrum if wear reaches to the bottom of any oil reliefs machined into the contact surface. Replace any oil deflectors that are broken, bent, cracked, or otherwise damaged. Inspect the rocker arms as previously described.

Stamped steel rocker arm assemblies should be replaced if you are doing a complete rebuild. Avoid jeopardizing the quality of your work by reusing these comparatively inexpensive items. However, if you are overhauling and plan to reuse the rocker arms, *keep them in order*. They must be assembled to their original position on the engine.

Shaft-Mounted Rocker Arms

Most Chrysler and many import engines, both OHV and OHC, use shaft-mounted rocker arms. Although some shaft-mounted rockers are made from stamped steel, the majority are either cast iron or cast aluminum. The rocker arm has a hole bored through the side to allow it to pivot on the shaft. An assortment of springs, spacers, and washers fit between the rockers to hold them in position, figure 15-39. Disassemble and carefully inspect the rocker arms, shafts, and attachment hardware.

Disassembling and inspecting the rocker shaft
The rocker shaft assembly on most engines can be unbolted and lifted off the engine as a unit. You must handle the assemblies with care, as the pressure of the valve springs is pushing against the rockers. Careless handling can result in a bent shaft. Always loosen and tighten the retaining bolts in small increments, and in sequence, to equalize valve spring pressure on the shaft. Follow the manufacturer's recommended sequence to remove the bolts.

Prepare for disassembly by clearing a space on your work bench, because you will need enough room to spread the parts out and keep them in order. Remember, if you are reusing the parts, they must be assembled to operate the same valve, and fitted to the same pushrod. Wash the assembly in clean solvent to remove any heavy deposits and make disassembly and inspection easier. Flush oil passages with a good stream of solvent. Solvent should flow freely around all of the arms. If flow is restricted, there is an obstruction and you will have to take a closer look to determine the cause after tear down.

Rocker arm assemblies are often held together by a cotter pin, roll pin, or circlip, although some designs are simply held together by the bolts attaching the assembly to the cylinder head. Remove the cotter pin, roll pin, or any other fastener that holds the shaft in place. Be

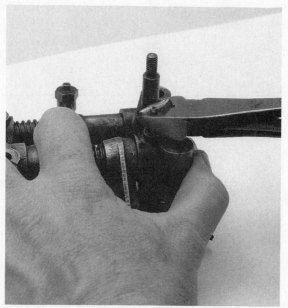

Figure 15-40. Begin your disassembly be removing the cotter pin, roll pin, or other fastener that holds the assembly together.

Figure 15-41. Inspect the rocker shaft, looking for signs of scoring, galling, and excessive wear.

Figure 15-42. The worn surface on this cast-iron rocker arm can be resurfaced with a fixture on a valve grinding machine.

sure to have a firm grip on the end tower as you remove the fastener, figure 15-40. If not held in check, the spacer springs could send parts flying across the shop. In theory all of the parts should slip easily off of the shaft. In reality, the towers will tend to stick due to an accumulation of sludge and varnish. Slight twisting as you work the tower off is usually enough to break the seal, and a light rap with a soft hammer will free up even the most difficult assemblies. As you slip the parts off the shaft, lay them out in order on the work bench for inspection.

Inspecting the rocker shaft

Wipe the shaft down with a clean rag and inspect for wear. The surface of the shaft should be smooth and free of any galling, scoring, or signs of excessive wear, figure 15-41. A good shaft will have a highly polished appearance without any built-up ridges where the rockers ride. Most shaft failures are a result of inadequate lubrication, or simply wear from high mileage. Due to valve spring and pushrod pressure, normal wear patterns will be greatest on the underside of the shaft facing toward the cylinder head. Shaft wear can be accurately measured with an outside micrometer. However, most experienced engine builders rely on a visual inspection. Replace the shaft if you find any noticeable wear. Avoid compromising your work by using questionable parts.

Wash the shaft with clean solvent. You can soak the shaft in carburetor cleaner if the engine has a lot of sludge. Some rocker arms are lubricated by an internal passage in the shaft, make sure oil flow is not restricted. After washing, use compressed air to dry the shaft and blow out any trapped debris. If the shaft is good, set it aside and continue by inspecting the rocker arms. If you are replacing the shaft, it is good practice to also replace the rocker arms.

Rocker arm inspection

Most of the inspection techniques previously detailed for stud-mounted rocker arms can also be applied to shaft mounted arms. The

Servicing Camshafts, Lifters, Pushrods, and Rocker Arms

Figure 15-43. The notch machined on the end of this shaft must face down toward the cylinder head, to line up the oil passages.

Figure 15-44. Rocker shafts being assembled in a production shop using all new parts.

main difference is that you must closely scrutinize the shaft bore. The majority of rocker arms ride directly on the shaft. However, some are fitted with a soft metal bushing made of bronze or brass. Although the bushing can be replaced, it is often more cost effective to simply replace the entire arm. An advantage to cast rocker arms is that you can recondition a worn valve contact face by grinding, figure 15-42. This refacing procedure is detailed later in this chapter.

Look the bore over for any signs of unusual wear. Scoring, pitting, or galling is cause for replacement. Wear patterns should be even across the bore. You must also check the arm-to-shaft fit of each rocker. Manufactures provide an oil clearance specification. As a general rule, there should be no more than 0.005 inch (0.13 mm) of clearance. A tight fit can result in seizure, and a loose fit will cause accelerated wear on both the arm and shaft.

Spring and spacer inspection

Springs and spacers are used to keep the rocker arms properly aligned with the pushrods and valve stems. These parts can usually be refitted because they are not highly stressed and seldom wear out. However, they must be inspected. Look the springs over for any obvious damage. Keep an eye out for any deformation or signs of bending, binding, or cracking. Spacers should slip readily off, and spin freely on the shaft. Look for indications of binding, galling or scoring on the internal bore, and signs of undercutting on the end faces. Clean and inspect the parts, and replace any defective pieces. Cotter pins, roll pins, or other small fasteners that hold the assembly together should always be replaced.

Rocker shaft assembly

At this point, all of the rocker shaft components should be clean, inspected, and ready for reassembly. If the contact faces of the rocker arms are to be resurfaced, skip ahead to the next section, machine the rocker arms, then return here for final assembly.

Look for any direction indicators, such as a notch on the shaft, figure 15-43. The shaft must be positioned correctly or the oil passages will not line up. An easy method of assembly is to: lightly lubricate the shaft, stand it on end, then stack the rocker arms, springs, spacers, and support towers in place, figure 15-44. Check your work. Make sure everything is properly aligned and in the correct order. Secure the assembly with a new fastener as required.

NOTE: When working with rocker arms, pay close attention to their original position. Often the rockers all look the same, but close inspection will reveal a slight difference. Some engines have a different end-to-end offset for the intake and exhaust rocker arms. If assembled incorrectly, the geometry will be off, and the engine will not run properly. Always *double check your work.*

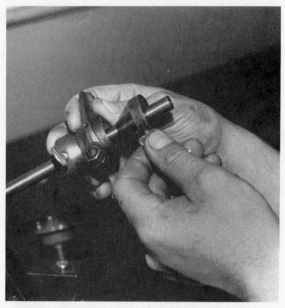

Figure 15-45. Position the rocker arm on the pivot between the two attachment collars. Tighten the jam nuts on the collars to hold the assembly in position.

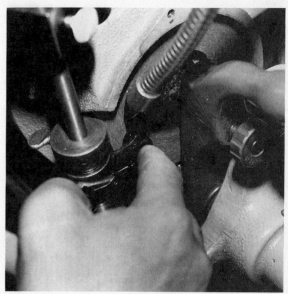

Figure 15-47. Work the arm into the stone by pivoting it back and forth. Continue until the surface is ground smooth.

Figure 15-46. Swing the pivot arm so the rocker contacts the stone, and adjust the coolant pipe to get a good flow of cutting fluid on the workpiece.

Refacing Cast Rocker Arms

Cast rocker arms that have marginal scuffing or wear on the valve stem contact pad can be reconditioned by grinding. You perform the operation using a valve grinding machine. Never attempt to resurface stamped steel arms, as the grinding stone will quickly remove the hardened surface. You may get a surface that looks good, but it will quickly wear out. Also, avoid trying to resurface rockers by hand or with a bench grinder. You will never be able to get the proper face radius or surface finish.

To recondition the rockers you should have them clean and in their proper order. Then, follow these steps:

1. Place a rocker arm, along with the two attachment collars, onto the pivot arm of the valve grinding machine, figure 15-45.
2. Center the rocker arm and collars on the pivot for good stone contact, then tighten the jam nuts to hold the collars in place.
3. Swing the pivot arm into position so the arm contacts the grinding stone. Make sure there is a good supply of cutting fluid directed to the workpiece, then turn the machine on, figure 15-46.
4. Pivot the rocker arm back and forth making light contact with the stone. Continue until the surface is smooth, figure 15-47.
5. Repeat the procedure for the remaining rocker arms. After machining, thoroughly wash the arms in solvent to remove any metal chips and residual grinding dust. Then assemble the arms on the shaft.

PART FOUR

Engine Assembly, Reinstallation, and Break-In

Chapter Sixteen
Engine Assembly

Chapter Seventeen
Engine Installation, Break-in, and Delivery

Chapter 16
Engine Assembly

GENERAL ASSEMBLY PRACTICES

At this point, all of the internal parts of the engine, except the oil pump, have been cleaned and inspected, repaired, reconditioned, or replaced. You are now ready to assemble the engine. Cleanliness is the most important aspect of engine assembly. Work in an area that is free of any air-born dirt or dust. Keep in mind that even tiny particles of dirt that find their way into the engine are abrasive and cause accelerated wear. Many shops have an assembly room that is separated from the service bays. High-performance engine builders often have a special "dust proof" assembly room. The room has a sealed door and an air filtration system that traps even microscopic dust particles. Take your work seriously and make it a priority to keep it clean.

Before you begin, review your notebook. Make sure you have corrected any unusual conditions that you recorded during disassembly. Also check over your parts. Assembly goes much smoother and quicker if you have everything you need, including special tools, on hand. You must have the manufacturer's assembly and torque specifications for the particular engine you are working on. Every bolt and nut on the engine requires a specific torque, and engine failure can result from fasteners not properly tightened. In addition, there may be some unique assembly procedures to follow, or special tools you need.

In this chapter we explain the steps to build a complete engine, or long block, as well as oil pump inspection procedures. Although we recommend always replacing the oil pump, we detail inspection procedures to determine if a used pump can be returned to service.

OIL PUMP SERVICE

The oil pump is the heart of the engine. It creates the oil flow necessary to lubricate all of the moving parts. Because it is such a critical component, many technicians prefer to replace the pump and pick-up screen with a new unit whenever an engine is torn down. However, on occasion you may have to disassemble, inspect, and recondition the oil pump and return it to service.

As explained in Chapter 6 of the *Classroom Manual*, there are two basic oil pump designs, the rotor type and the gear type. Gear type pumps generally bolt to the engine block or to a main bearing cap. The pump is driven by a shaft that is geared to the camshaft or to a layshaft. Some rotor-type pumps are similar in design and take their power off the camshaft.

Engine Assembly

Figure 16-1. Replace the oil pump shaft if the drive end has rounded edges, or if it does not fit snugly into the pump.

Figure 16-2. Many oil pumps use a roll pin to hold the pressure relief valve in the housing. Remove the pin.

Others mount to the timing cover and are driven directly by the crankshaft.

If the pump is shaft-driven, begin your inspection by examining the drive shaft. Look the shaft over for signs of twisting, bending, or other distortion. Replace the shaft if anything appears questionable. The end of the shaft must fit snugly into the pump with little play, figure 16-1. If the end of the shaft is worn or rounded off, install a new one.

All oil pumps use a pick-up tube and screen to filter large particles out of the intake oil. Any debris that can pass through the screen is small enough to go through the pump without locking up the gears or rotors. The pick-up may be bolted, threaded, or pressed onto the pump body or engine block. Remove the pick-up assembly, wash it in solvent, and inspect the screen. If the screen shows any sign of damage, replace it. It is a good habit to simply replace the pick-up as a precautionary measure, even if you are reinstalling the pump.

The pressure relief valve is another component that is common to all oil pumps. A relief valve is usually a spring-loaded plunger that bleeds off excess oil to maintain optimum pressure. The valve generally fits a machined bore in the pump housing, and is held in place by a cotter pin, roll pin, or threaded plug. Remove the fastener that holds the relief valve on the pump, figure 16-2. Some pumps have a small core plug installed behind the roll pin. To remove the plug, center-drill it and use a slide hammer to pull it out.

Remove the plunger and spring from the bore. Pay close attention to how the pieces are assembled. Installing them incorrectly can result in no oil pressure. Wash the check valve parts in clean solvent and inspect them. If you find any signs of scoring or excessive wear on the plunger, replace the oil pump. Measure the spring length and tension, then compare your findings to the specifications.

Servicing Rotor-Type Pumps

As stated earlier, rotor pumps can be mounted to the front of the engine on the timing cover, or bolted to the bottom of the block inside the oil pan. All rotor pumps work on the same principal. An inner rotor is positioned off-center in the pump body and driven by an external power source. The outer rotor fits around the inner rotor and is driven by the inner rotor. The clearance between the two rotors is constantly changing. Oil is carried from large to small clearance areas between the rotors, then forced out of the pump and into the galleries under pressure.

Most pumps can be readily disassembled for inspection. If the pump body is in good condition, you can install a rebuild kit and return the unit to service. A kit generally includes new rotors, relief valve spring, seals and gaskets.

Camshaft driven pumps

Begin by removing the bolts holding the end housing or cover plate to the pump body. Look over the end faces of the two rotors for indexing marks. If there are no index marks, you have to provide them. The rotors must be reassembled to their original positions. Now remove the rotors from the body and wash all the parts in clean solvent.

Look over all the contact surfaces of the rotors, body, and end housing for signs of wear. Replace the pump if you find any deep scoring, chips, cracks, or other deformation. If the parts are in good condition, reinstall the rotors so you can check clearances.

Your first measurement is to check the end clearance. This measurement is taken with a feeler gauge and a straightedge. Place the

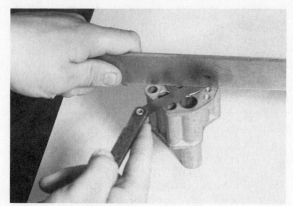

Figure 16-3. Measure the oil pump end clearance with a straightedge and a feeler gauge.

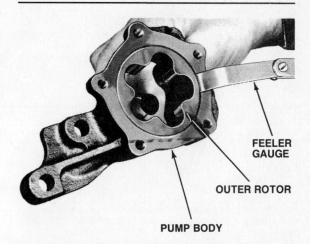

Figure 16-5. Rotor-to-housing clearance is measured by inserting a feeler gauge between the outer rotor and the pump body.

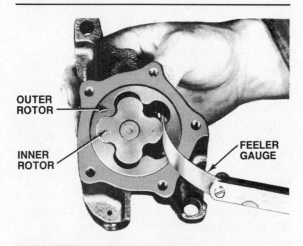

Figure 16-4. With two lobes facing each other, measure the rotor-to-rotor clearance of the pump with a feeler gauge.

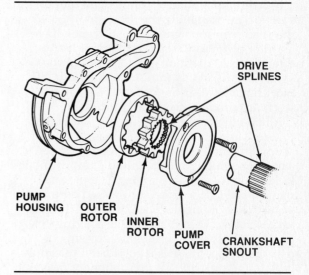

Figure 16-6. This pump assembles into the front timing cover and is driven directly by the crankshaft.

pump on the workbench with the rotors facing up and lay your straightedge across the pump. Clearance is equal to the largest feeler gauge you can fit between the rotor faces and the straightedge, figure 16-3. Most manufacturers provide an end clearance specification in their service manuals. In general, end clearance should be less than 0.003 inch (0.08 mm). If greater, replace the pump.

Next, measure rotor-to-rotor and rotor-to-housing clearance. Both measurements are taken with a feeler gauge. Position the rotors so that a lobe on the inner rotor faces a lobe on the outer rotor, then measure the clearance between the lobes with your feeler gauge, figure 16-4. Compare your findings with the manufacturers specifications. As a rule, rotor-to rotor clearance should be less than 0.010 inch (0.25 mm). Now use your feeler gauge to measure the clearance between the outer rotor and the pump body, figure 16-5, and compare your reading to the specifications. If the clearance exceeds 0.012 inch (0.30 mm), replace the pump or install a rebuild kit.

The end plate of the pump housing will show scoring and wear from contacting the rotors. You can resurface the end plate if scoring is less than 0.003 inch (0.08 mm) deep. Resurface the plate by lapping on a flat surface using wet or dry sandpaper lubricated with engine oil. After you true the surface, thoroughly wash the end plate with clean solvent. Be sure to remove all traces of grit and abrasive, because any debris left in the pump will cause engine damage.

Once again, clean all of the parts before you reassemble the pump. Begin your assembly by installing the outer rotor into the pump body.

Engine Assembly

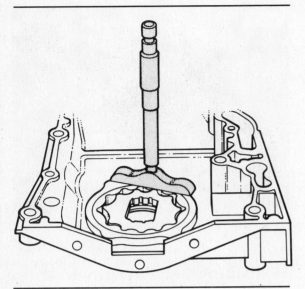

Figure 16-7. Rotor drop is measured with the rotors installed in the pump body using a depth micrometer.

Now install the inner rotor, making sure the index marks are aligned. Bolt the end housing onto the pump body using a new gasket and tighten to specified torque.

CAUTION: The gasket that seals the end housing to the pump body is manufactured to a specific thickness to provide the proper end clearance. *Do not substitute* any other gasket. Too much end clearance results in an oil pressure loss, while too little clearance can cause binding and premature failure. Install the gasket without any sealer because using sealer can alter clearance. In addition, any excess sealer can be drawn into the pump, restrict oil flow, and lead to engine damage.

After fitting the end plate, you can reassemble the pressure relief valve. Be sure you have the plunger properly positioned and pointing in the right direction. The spring should be replaced with a new one. Rebuild kits generally come with a new spring. If the valve is held in place with a cotter pin or roll pin assemble the valve using a new fastener.

The pump is now ready to be installed. Place the pump in a plastic bag and store it in a safe place to keep it clean until you are ready for final assembly.

Crankshaft driven pumps

Crankshaft driven rotor pumps mount to the engine timing cover. The crankshaft snout fits through a splined opening machined into the center of the inner rotor and the assembly is held in place by a cover plate that bolts to the timing cover, figure 16-6. The pressure relief valve is often not part of the pump itself, but is remotely mounted in the timing cover or oil filter flange. Do not overlook the relief valve simply because it is not part of the pump. Regardless of where it mounts to the engine, the relief valve must be disassembled, cleaned, and inspected.

The service procedures for crankshaft-driven pumps are similar to those outlined for camshaft-driven pumps. Begin by removing the cover plate bolts. Lift off the plate, then remove the inner and outer rotors. Wash the parts in solvent and inspect them for wear. If you find any sign of excessive wear, replace the pump. Once the parts are clean, measure end clearance, rotor-to-rotor clearance, and rotor-to-housing clearance.

It is often impossible to measure end clearance using a straightedge because the pump assembly is recessed into the front cover. Instead, you can check end clearance, also known as rotor drop, with a depth micrometer. Install the rotors into the body and position the micrometer so that it rests on the cover mounting flange and straddles the opening, figure 16-7. Rotate the micrometer spindle to record the distance from the rotor face to the cover plate mounting surface and compare your findings to the manufacturer's specifications. Acceptable tolerance is generally in the 0.001 to 0.004 inch (0.025 to 0.10 mm) range.

Rotor-to-rotor, or rotor tip clearance, and rotor-to-body, or rotor side clearance is measured with a feeler gauge. The procedure for measuring is the same as described for camshaft driven pumps. However, tolerance is normally tighter. You must check the specifications for the pump you are working on. You can expect tip clearance readings in the 0.004 to 0.009 inch (0.110 to 0.240 mm) range. Side clearance should be slightly more, with an upper limit of about 0.015 inch (0.38 mm).

Assembly procedures vary between manufacturers. Always check the service manual for your engine and follow any special requirements. General Motors recommends you pack the rotors with petroleum jelly when installing, but many other manufacturers do not. Align the index marks and fit the two rotors into the body. Make sure the mating surfaces are perfectly clean, install the pump cover, and draw the fasteners up to torque. Store the pump in a clean, safe place until you are ready for final assembly.

Servicing Gear-Type Pumps

Gear-type pumps work in much the same way as rotor pumps, except oil flow is provided by the meshing of two gears instead of rotors.

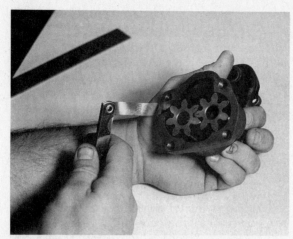

Figure 16-8. For gear type pumps, measure gear tooth to housing clearance with a feeler gauge.

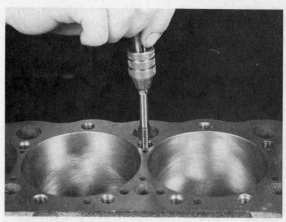

Figure 16-10. Run a bottoming tap through all threaded holes to clean and true the threads.

Figure 16-9. Use a polishing stone to clean up any burrs, nicks, and indentations.

One of the gears is driven and the other idles. Disassembly, cleaning, and inspection procedures are so similar to rotor pumps that we will only point out the differences here. Make sure there are clear index marks on the gears before you remove them from the pump body.

There are only two measurements to take on a gear pump: end clearance and gear-to-housing clearance. Gear-to-housing clearance equals the largest feeler gauge you can fit between a gear tooth and the housing, figure 16-8. Measure both gears at several locations, and rotate the gears to get readings on different teeth. In general, clearance should not exceed 0.005 inch (0.13 mm). Check the manufacturer's specifications for your particular engine.

Once you are sure all the parts are up to standards, reassemble and store the pump as previously described. At this point, all of the internal components of the engine have been looked after. Now you are ready to assemble the engine.

ASSEMBLY PREPARATION

The most important aspects of engine assembly are rechecking measurements and keeping internal parts clean. Studies indicate that over 42 percent of all bearing failures result from dirt or other foreign particles in the engine. Meticulously clean assembly practices produce long-lasting high-mileage engines, satisfied customers, and a good reputation for your skills as an engine builder.

After completing all the machining operations, prepare the block for assembly following these steps:

1. Inspect all gasket surfaces for burrs, nicks, and indentations. Scrape off any remaining gasket material or sealer, and dress any irregularities in the metal with a polishing stone, figure 16-9.
2. Use a bottoming tap to clean the threads of all bolt holes that take a torqued bolt, figure 16-10. You must also clean the threads for the oil gallery plugs, as well as any threaded blind holes on the block.
3. Slightly chamfer the edges of the cylinder head bolt holes, figure 16-11. This prevents metal from pulling up to interfere with gasket sealing as the bolts are brought up to torque.
4. If the top edge of the cylinder bores have not been chamfered, do it now. Be careful not to damage the cylinder walls. Chamfering not only makes ring installation easier, it also helps eliminate hot spots in the combustion chamber.
5. Look over the cam bearing bores for any sharp edges. A rough spot can easily snag and gall a bearing during installation. Dress any sharp edges and high spots with a fine file, polishing stone, or other deburring tool.

Engine Assembly

Figure 16-11. The head bolt holes in the block must be chamfered to prevent head gasket sealing problems.

Figure 16-12. Use a rifle brush to give the oil galleries a final cleaning.

6. Use the same technique to deburr and dress the lifter bores. Clean up any irregularity that might prevent lifter rotation.
7. Inspect the core plug bores. The bore surface must be smooth and clean. Use emery paper to lightly polish the contact area and knock down any burrs or high spots.
8. Carefully look the block over to make sure all required machining has been completed. If so, the block is now ready for a final cleaning.

Cleaning Engine Components

Do not neglect to give the block a final cleaning simply because it looks clean. During the various machining operations, highly abrasive particles wore off the cutting tools and stones. This debris, combined with small metal chips machined from the block, can find its way into the oil galleries and other internal passages of the engine. The cutting fluids used during machining can turn this abrasive dirt into a paste that adheres to the metal of the block. If not removed now, the contaminants will be circulated through the engine by oil and coolant. The result is rapidly accelerated wear and premature engine failure.

For optimum results, clean the block by soaking in a hot tank, or cycling it through a spray booth. Follow up with steam cleaning or a high pressure wash. Be sure to flush out all oil galleries, coolant passages, and threaded holes. Run a clean brush through the galleries and lifter bores, figure 16-12, to make certain all of the debris has been removed.

The freshly honed cylinder bores require special consideration. Wash the cylinders with hot soapy water. Any trace of dirt left behind can cause hot spots and prevent ring seating. Laundry detergent or dishwashing soap works well in the cylinder bores. Always use a good quality rag, sponge, or soft brush that will not leave lint behind. Contaminants in the engine turn the soapy water gray. Continue cleaning with fresh solution until the soap suds no longer change color. Use fresh clean water to rinse off all the detergent.

Once the bores are clean, dry the block using compressed air. Immediately spray all the machined surfaces with a light lubricant to prevent corrosion. If you intend to paint the block, do it now. Take care not to get any overspray on the cylinder walls, in the bearing journals, or on sealing surfaces.

Now that the block is perfectly clean, you must keep it that way. Cover the block with a plastic bag to prevent contamination whenever you are not working on it.

The most convenient way to assemble an engine is to use an engine stand. If you have a stand available, firmly bolt the block to it at this point. Once the block is on the stand, slip the plastic bag over it for protection. If you are going to assemble the engine on a workbench, make sure the surface is clean. A work bench with a wooden top works best. The soft wood will not scratch or damage sealing surfaces on the block. Be sure the wood is seasoned and hard enough so that it cannot chip, gouge, or splinter when you reposition the engine during assembly.

ASSEMBLING THE BLOCK AND BOTTOM END

Begin by building up the bottom end. The assembled bottom end, block, crankshaft, camshaft, and pistons, is often referred to as a

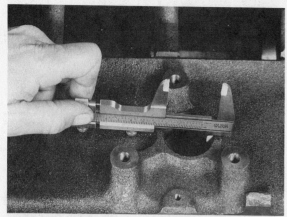

Figure 16-13. Measure the core plug bore and the core plug with a vernier caliper to verify fit.

Figure 16-14. Apply sealer to the threads, then install the core plug tightly into the block.

Figure 16-15. You must use the correct driver to install press-fit core plugs into the block. Never substitute any other tool.

"short block". The order of assembly will be as follows:

- Oil and gallery plugs
- Camshaft bearings and camshaft
- Main bearings and crankshaft
- Piston and connecting rod assemblies

The two most important aspects of assembly are to *keep your work clean* and *pay attention to detail*. By now you have invested too much time and effort to allow a careless mistake or oversight to destroy your engine. Remember, dirt is your number one enemy and must be eliminated. Avoid letting an anxious customer pressure you into working at a faster pace than you are comfortable with. Keep in mind that until you deliver the engine, you own it. Treat it as your own, work at a comfortable pace that allows you to be thorough. Never forget that you are the responsible party. Establishing good work habits now will help you develop a reputation as a quality engine builder.

Installing Core and Oil Gallery Plugs

Sealing off the oil galleries and installing the core plugs in the water jackets are your first steps. Remember, these parts will be under pressure when the engine is running. The plugs must seal tightly and completely, to prevent leakage and failure. Check the parts before you install them.

Measure threaded gallery plugs with a thread pitch gauge and check the fit by running them in by hand. Most plugs have a taper pipe thread for positive sealing. Expect to feel resistance as you screw them in. You will not be able to fully seat them by hand. You are simply checking thread fit.

Core plugs and the plug bores on the block can be quickly and accurately measured with a vernier caliper, figure 16-13. Watch for oversize plugs installed at the factory. Oversize plugs are generally stamped with an "OS" on the cup face, and must be replaced with identical units. Core plugs are available in steel or brass. Brass is preferable because of its superior corrosion resistance.

Installing oil gallery plugs

Oil gallery plugs can be either the threaded type that screws in, or the disc type that is driven in. Both types are easy to install.

Threaded plugs

Most threaded oil gallery plugs have an internal hex-socket head and can be installed with an Allen wrench. The threads must be sealed to prevent the possibility of leakage. Apply a liquid pipe-thread sealer or teflon tape to the plug. Keep sealant off the inner end of the plug where it could dissolve in the oil and circulate through the engine. Start the plug by

Engine Assembly

Figure 16-16. Press fit plugs must be installed into the block with a slight recess. Stake the casting at several locations to lock the plug in.

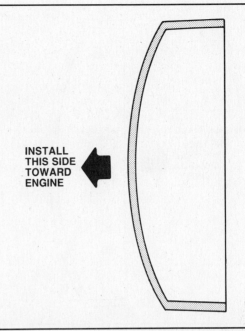

Figure 16-18. Cup plugs have a deep tapered flange. The flange tapers out to lead the plug into the bore.

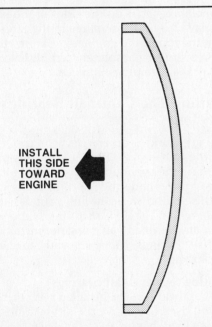

Figure 16-17. Dish, or disc, plugs flatten out to form a positive seal as you install them.

hand, then run it in with a wrench to the bottom of the threads, figure 16-14. Install the plugs one at a time, and double check to make sure none have been overlooked.

Disc-type plugs

Disc-type plugs must be installed with a special driver that fits the exact size of the plug you are installing. Do not try to substitute with any other tool the plug must be properly seated or it can leak or blow out.

Fit the plug onto the driver, figure 16-15, coat the sides with an oil resistant sealer, and position the tool and plug squarely in the bore. Strike the end of the driver with a hammer to seat the plug into the engine block. A properly seated plug is recessed slightly in the bore, below the machined surface on the block, figure 16-16. Once the plug is seated, stake the metal at several locations around the plug with a sharp punch. This expands the metal over the edge of the plug to prevent high oil pressure from forcing it out of the block.

Repeat the process to install the rest of the plugs. Make sure all of the oil galleries have been capped off.

Installing core plugs

Core plugs are driven into the engine block with a special tool to seal off the water jackets. The back side of the camshaft bore is also sealed by a core plug. At this point you install only the cooling system plugs. The camshaft plug is not installed until after you have fitted the cam bearings and camshaft. Core plugs are available in three designs, and each type requires special handling.

1. Dish-, or disc-, type plugs, figure 16-17, are placed into the block with the convex side facing *out*. Rest the end of the installation tool on the crown of the plug, strike the opposite end of the tool with a hammer to seat the plug. Driving the plug in flattens out the crown and pushes the sides of the plug out to provide a good, tight seal.
2. Cup-type plugs, figure 16-18, are installed with the convex side facing *into* the block,

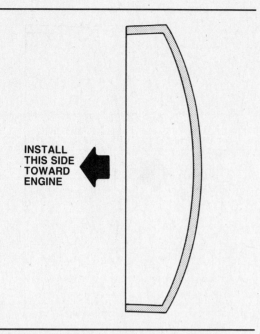

Figure 16-19. Expansion plugs generally have a shorter flange than cup plugs. The flange tapers into the bore and expands as the plug is driven in.

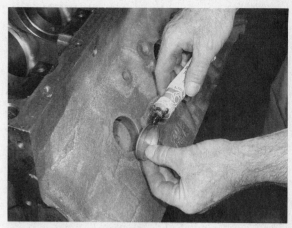

Figure 16-20. Coat the outside of the plug flange with a good water-resistant sealer.

and the flange facing out. The sides of the flange have a slight taper, the widest diameter on the outer edge. The taper helps lead the plug into the bore straight. If the block opening is chamfered, you must seat the cup below the chamfer for effective sealing.

3. Expansion-type plugs, figure 16-19, look similar to cup plugs. However, the flange is generally much shallower and tapers in the opposite direction. The widest point of the plug is at the base, and the sides of the flange taper in. This type plug is installed with the lip of the flange facing *into* the block. As the plug is driven in, pressure on the face causes the flange to expand, straightening out the taper, to provide a positive seat. As with cup plugs, the outer edge must be positioned below any chamfer on the bore.

Prepare the core plug for installation by coating the sides with a good water-resistant sealer, figure 16-20. Core plugs are driven into the block with a special tool, figure 16-21. Each plug design requires a different type of driver. Always use the proper tool. Installing a core plug without the correct tool can result in a poor fit that leads to leakage or failure.

Double-check to make sure that all the core plugs, except the one for the camshaft, have been installed before you move on. Many manufacturers use core plugs to seal openings on the back of the engine, behind the flywheel, and in the cylinder heads, in addition to the obvious ones on the sides of the block. You may also find core plugs in the lifter valley on the top of the engine. Look the entire engine over to be absolutely certain that you have all the core plugs in place.

Installing The Camshaft Bearings and Camshaft

If you are working on an overhead cam engine, the camshaft is either installed as part of rebuilding the cylinder head, or you fit it after you bolt the cylinder head on. In either case, skip this section and continue your assembly with *Installing The Crankshaft*. If you are working on an overhead valve engine, install the camshaft bearings, camshaft, and camshaft core plug now.

Installing the camshaft bearings

Camshaft bearings are installed with the same tool you used to remove the old bearings. As explained in Chapter 10 of this *Shop Manual*, there are two styles of cam bearing tools available, the forcing-screw type and the adjustable-driver type. With either tool you must work carefully. The soft bearings can be easily damaged if not seated properly.

Begin by examining the bearings. Be absolutely certain that they are the correct parts for your engine. Be aware that the camshaft bearings are usually not all the same diameter. They must be installed in their proper position in the block. Take a final measurement of the bores using a dial bore gauge, inside micrometer, or telescoping gauge. Once you are sure you have the correct parts, lay the bearings out in order on the workbench. Standard procedure is to install the rear most bearing first, then work your way toward the front of the engine. For most engines, you start with the

Engine Assembly

Figure 16-21. Always use the proper tool to drive the core plugs into the block.

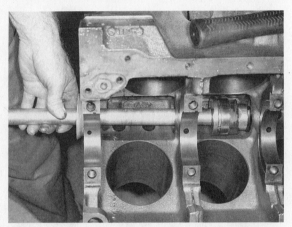

Figure 16-22. Slip the bearing driver through the bores and fit the bearing onto the mandrel. Make sure the oil passage in the block lines up with the drilled hole of the bearing.

Figure 16-23. Carefully guide the camshaft into the block. Avoid any cam lobe-to-bearing contact.

smallest diameter bearing and finish with the largest.

Camshaft bearings are assembled into the block dry. Do not apply any type of lubricant. Wipe the bearing bores down with a clean rag, being careful not to leave any lint on the seating area. Install the bearings by fitting the tool through the bores. Slip a bearing onto the tool, and *align the oil hole in the bearing with the oil passage in the block*, figure 16-22. Once the oil holes align, drive the bearing into the bore either by striking the tool with a hammer, or tightening the forcing screw. Most tools have a positive stop to prevent overcentering the bearing. Do not use excessive force as you can easily damage the soft bearing shells. Check the manufacturer's specifications for any specific fitting requirements of the engine you are working on.

Remove the installation tool and check to make sure the block and bearing oil passages are aligned. You can quickly check alignment with a small piece of wire. You should be able to easily pass the end of the wire through the drill hole in the bearing and into the oil gallery of the block. Install the remaining bearings, and you are ready to fit the camshaft.

Installing the camshaft

As mentioned in Chapter 15 of this *Shop Manual*, new and reconditioned camshafts are given a special surface treatment to reduce wear during the critical break-in period. The process applies a coat of molybdenum disulfide or manganese phosphate to the shaft. The coating gives the shaft surface an irregular black appearance. *Do not remove* the coating from the cam lobes. Immediate wear and premature failure will result. However, you must clean off any coating material that has found its way onto the bearing journals. Wipe the journals down with a clean solvent-soaked rag, then follow up with a clean dry rag to remove any residue.

Lubricate the bearings and the journals with a light film of engine oil before you fit the camshaft. Installing the camshaft can be a bit awkward. You must pass the shaft through the bearings. The sharp edge of a cam lobe can easily gouge or scratch the soft bearing surface if you allow them to contact. Work the camshaft in slowly and carefully, feed the shaft in with one hand as you guide it with the other, figure 16-23.

NOTE: A good way to get a little extra grip on a camshaft during installation is to screw two long bolts into the threaded holes for the drive gear on the front of the shaft. Use the bolts as a handle to install the shaft. This provides you

Figure 16-24. Install the retaining plate that holds the camshaft to the block.

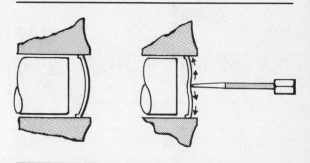

Figure 16-26. After installing a disc-type camshaft plug, strike the center of the plug with a punch to increase side tension for a positive seal.

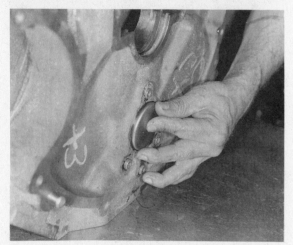

Figure 16-25. Run a bead of sealer along the edge of the camshaft core plug, then position it in the block. Be sure to use the proper driver to seat it in the block.

Figure 16-27. Use a rifle brush to give the crankshaft oil galleries a final cleaning.

with a firm hand hold and gives better leverage to steady the shaft as you guide it in.

Once the shaft is in place, spin it on the bearings. The shaft should turn freely without any resistance. If you feel any binding, or tight spots, remove the shaft and inspect the bearings. Recheck the bearing and journal dimensions, calculate oil clearance, and make sure you are within specifications. A minor high spot on an individual bearing can often be corrected by taking a light cut with a bearing scraper. If all the bearings are tight, you have most likely installed the wrong parts. Remove the bearings and replace them with the proper ones. If the bearings "feel good", and the cam rotates freely, secure the camshaft to the block by installing the retaining plate, figure 16-24.

New and reconditioned camshafts generally come with a tube of special high-pressure lubricant. The lubricant is provided to reduce lobe to lifter friction, promote rapid break-in, and prevent wear. If not provided, or when reinstalling the old cam, camshaft lubricants are readily available on the aftermarket and *must be used*. Apply the lubricant to the lobes after the camshaft is in place in the engine block. Liberally coat the lobes as you rotate the shaft, but avoid getting lubricant on the bearings and journals.

Installing the cam core plug

Finish off your camshaft installation by fitting the core plug into the back of the block. Occasionally your engine stand can prevent you from getting a good straight line with the plug driver. If this is the case, remove the engine from the stand to install the plug. *Do not attempt to install the plug with your tool at an angle.*

Engine Assembly

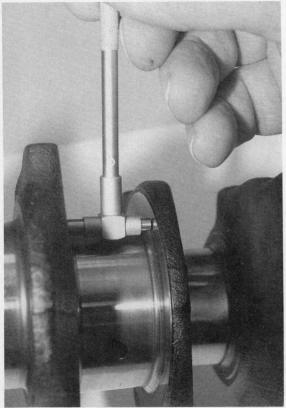

Figure 16-28. Measure the distance across the thrust journal, then subtract the thrust bearing total width to calculate clearance.

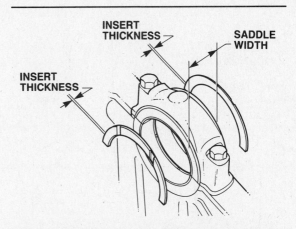

Figure 16-29. When separate thrust bearings are installed, measure the insert thickness and saddle width to determine thrust bearing clearance.

Apply sealer to the sides of the plug and place it in the block, figure 16-25. Most camshafts use a disc-type plug so they install with the convex side out. Drive the plug into the block by striking the installation tool with a hammer. Once in place, strike the center of the plug a sharp blow using a dull punch and a hammer, figure 16-26. This dimples the plug at the center and increases tension at the edges for good sealing. This last step is important. Since the core plug receives full oil pressure, the additional tension at the edges helps to keep it in place. Rotate the camshaft, it must turn freely. Make sure you have not driven the plug in so far that it contacts the shaft and impedes movement.

Installing The Crankshaft

You must give the crankshaft and the main bearing bores a thorough final inspection before you install the shaft. Begin by running a rifle brush through the oil passages of the shaft to clear any trapped debris, figure 16-27. Oil passage openings must be chamfered and perfectly smooth. Make sure any sludge traps are in place and properly staked. Look over the journals for proper surface finish and check the fillets for proper radius. Measure the main and rod journals with an outside micrometer. Compare your readings to those you recorded earlier in your notebook, and note any differences. Carefully wipe down the journals and the bearing bores to clean off all trace of dust and dirt.

NOTE: Many engine rebuilding manuals tell you to wipe the internal parts down with a "lint-free rag". In our estimation, such a rag does not exist. Any woven material, no matter how high quality, can leave behind traces of lint and fuzz. These small traces of cloth can circulate in the oil, work their way through the engine, and cause damage. We recommend that you use a chamois to clean critical internal engine parts. Since a chamois is made of hide, not a woven fabric, it will not leave lint, even if it snags on a sharp edge. In addition, buffing with the soft leather gives bearing journals a desireable highly-polished surface.

Torque the main bearing caps, without bearings, onto the engine block. Measure the bearing bore diameter and compare your findings to those previously recorded in your notebook. Calculate the main bearing oil clearance as detailed in Chapter 11 of this *Shop Manual*. Be certain that all of the bearing clearances are within the manufacturer's specified tolerance range. If so, remove the bearing caps and continue. If not, take the necessary steps to bring the clearance to standard before proceeding.

If you are installing a reground crankshaft, measure the distance from the thrust bearing surface, across the journal, to the opposite cheek, figure 16-28. Also measure the assembled width of the thrust bearing. Subtract the second measurement from the first to calculate

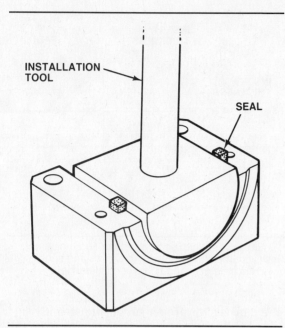

Figure 16-30. Use a special tool or other round object to roll the seal to the bottom of the groove.

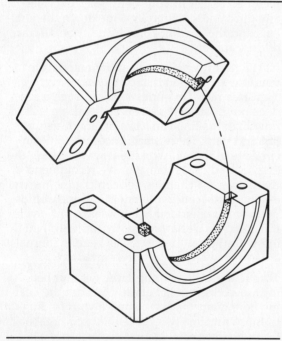

Figure 16-31. Many engine builders stagger the parting lines of a split seal. The ends must be trimmed to an exact length.

thrust bearing clearance. Your findings must be within the specified tolerance. For engines with separate thrust inserts, figure 16-29, take a micrometer reading across the saddle from seating groove to seating groove, then measure the thickness of both thrust inserts. Add the three figures together to determine assembled bearing width.

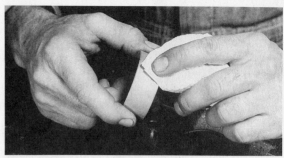

Figure 16-32. Wipe all trace of dirt, dust, oil, or any other debris from the back side of the insert.

Installing a split-type rear main seal

All engines require a seal at the back end of the crankshaft. The seal, often referred to as the flywheel seal or rear main seal, prevents pressurized oil from escaping from the rear main bearing journal. Modern automobile engines use one of two seal designs: the two-piece split type or a one-piece lip seal. Install a split-type seal now. Whether rope, braided fabric, or rubber, install the seal before you fit the bearings. You install one-piece lip seals after the crankshaft is in the block.

Installing a two-piece seal is a simple procedure. However, if not done properly, you end up with a leaky engine. To install a split seal, regardless of the material it is made of, follow these steps:

1. Dip the seal in oil. This is important, especially with the fabric type. The oil not only makes installation easier, it also expands the seal to give a good, tight fit.
2. Lay the seal into the machined groove on the block, then push it in slightly by hand.
3. Use a round object, such as a large socket or piece of pipe, to roll the seal into place. Work from one end of the seal to the other. Make sure the seal seats to the bottom of the groove, figure 16-30.
4. Trim the excess ends of the seal flush to the block using a sharp knife or razor blade. If the seal has a wire core, trim it down with a pair of cutters.
5. Clean off any oil that squeezes onto the bearing bore. Make sure to get all of the oil. You do not want to leave any trace of oil behind the bearing shell.
6. Repeat the procedure to install the other half of the seal into the bearing cap.

NOTE: Many high performance engine builders stagger the ends of split-type seals so the seal ends do not meet at the bearing bore parting line, figure 16-31. Fit the seal into the groove, start the end below the parting line, roll the seal in, then trim the other end so an equal amount protrudes past the parting line.

Engine Assembly

Figure 16-33. Lightly squeeze the ends of the shell together so you do not scrape the back of the bearing on the saddle as you install it.

Install the other seal half, precisely fit and trim the ends. *The seam must be perfect*. Any gap causes an oil leak. If the seal is too long, you cannot properly torque the bearing cap.

Installing the main bearings

Give your hands a good scrubbing before picking up the main bearings. Both sides of a bearing shell are delicate and easily damaged by dirt. Microscopic particles on the front become an abrasive that moves at crankshaft speed once the engine is running. Anything left on the back side can interrupt heat transfer and create a high spot that reduces oil clearance. When working with bearings, you, your tools, and your work area *must be meticulously clean*.

Arrange the bearing inserts, along with the bearing caps, in the order they fit the block. Wipe off the bearing contact surface on the block saddles and the bearing caps with your chamois. Be sure to get all traces of dirt, dust, and oil. Fit all of the bearing inserts to the block first, then to the caps. The block side insert must have a drilled hole that aligns with the oil gallery opening in the saddle. Some manufacturers machine a relief into the upper shell that directs oil flow to the parting line at either side of the bearing. Be sure these bearing halves are installed into the engine block, not into the cap. Install main bearings one at a time in the following steps:

1. Wipe the back side of the bearing shell clean, figure 16-32. Be sure to remove all traces of dirt.

2. Hold the insert between your thumb and forefinger, figure 16-33. Place the shell up to the saddle, make sure any lock tang and groove are in alignment. *Gently squeeze* in on the bearing ends as you slip the insert straight down into the saddle.
3. Push in on the ends to make sure the shell is all the way seated in the saddle.
4. After you install all the bearings in the block, fit the inserts to the caps in the same manner.
5. If the engine has separate thrust bearing inserts, fit them last. Place the inserts into the grooves. A couple of dabs of assembly lube on the outside edge of the insert helps hold it in place. Avoid getting any lubricant behind the thrust bearings.
6. Wipe off the working surface of the bearings, then lubricate them with a thin film of engine oil. If you intend to measure oil clearance with plastigauge®, as detailed later in this chapter, do not oil the cap side of the bearings. Now you are ready to install the crankshaft.

Installing the crankshaft

Your hands should be clean and dry before you pick up the crankshaft. Be careful. Although the crankshaft is heavy, it can easily be damaged, or cause damage, should it slip out of your hands. A good way to lift a crank is by grasping the snout firmly with one hand, rest the rear flange in the palm of your other hand, and insert a finger into the pilot bore to prevent the shaft from slipping. Another trick is to thread some of the flywheel bolts in place on the flange. Then you can use the bolt heads as a handle to get a good grip and keep the shaft under control.

Remember, the crankshaft journals must be perfectly clean. Wipe them down once again before you place the crank into the block. Be sure to remove all trace of dirt, dust, or other debris. Journals with a proper finish have a highly polished, mirror-like appearance. Any imperfection can distress the bearings. Once the journals are spotless, check to be sure the bearing inserts are fully seated, oil passages are aligned, and there is a film of engine oil covering the entire contact surface. Now you are ready to fit the crankshaft.

Keep the crankshaft parallel to the bearing bores as you place it in the engine. Work carefully. Lower the crank straight down into position slowly, keeping it square to the saddles at all times, figure 16-34. Keep an eye on the thrust bearing because the crank can easily gouge the bearing and destroy it. On engines that have separate thrust bearing inserts, be

Figure 16-34. Gently lower the crankshaft into the bearings, be careful not to scrape or dislodge the thrust bearing.

Figure 16-36. Register the cap to the saddle by lightly tapping the sides with a soft hammer. This allows the cap to find its natural location.

Figure 16-35. Install the bolts and run them up as tight as possible by hand.

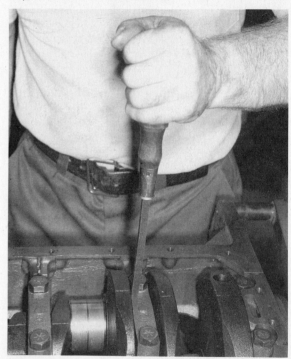

Figure 16-37. Align the thrust bearings by using a large screwdriver to shift the crankshaft back and forth in the block.

careful not to dislodge them from their seat as you lower the crank into position. The rear flange on most shafts does not provide much of a hand hold, so be careful not to pinch your fingers between the crank and the bearing.

The installed crankshaft must fit squarely, and be solidly supported by the main bearings. The shaft must rotate freely and easily without resistance. If not, there is a problem. Binding may be caused by a bent shaft, misaligned bearing bores, and wrong or improperly installed bearing shells. Correct any problems now. Do not install the bearing caps if the crankshaft does not spin smoothly when resting in the block saddles.

Installing the main bearing caps

Install the main bearing caps starting at the rear of the engine (flywheel end) and working your way towards the front. Look the caps over before you begin. Be sure the bearings are installed properly, the caps are in order and matched to their correct saddle, and that any directional indicators point the right way. If you measured the journals and bearings, and calculated oil clearance, and found them all to be within specification, lubricate the entire surface of all the bearings with engine oil. If you will be verifying oil clearance with plastigauge®, *do not lubricate the bearings*. The bearing surface and the journal must be perfectly dry to get an accurate plastigauge® reading.

Engine Assembly

Figure 16-38. Bring the bolts to final torque in stages. Alternate from side to side to get good, even seating.

Look up the main bearing tightening torque and sequence in the manufacturer's specifications. Also review the service literature for any specific requirements. Some manufacturers recommend you lightly oil or apply a sealing compound to the threads of the retaining bolts. Others require you to install the bolts dry. Follow the instructions for the particular engine you are working on. Install the main bearing caps following these steps:

1. Fit the bearing caps over the journals and push down to mate them to the saddles. If you staggered the rear main seal parting line, add a drop of gasket sealer to the seal ends. Be sure to fit both ends of the seal precisely into the grooves.
2. Lubricate the bolt threads as required, then install the bolts and draw them up hand tight, figure 16-35.
3. Rotate the crankshaft as you lightly tap the sides of each cap with a soft-face hammer, figure 16-36. This is an important step, as it registers the cap to its natural position and squares it in the saddle.
4. After you register the cap, tighten the bolts and draw them up as far as you can using *hand pressure only*.
5. Place a large screwdriver between a cap and crankshaft cheek and pry the crankshaft back and forth to properly align the thrust bearing, figure 16-37. *Do not pry on the bearing cap that holds the thrust bearing.*
6. Rotate the crankshaft at least one complete revolution to check for free movement. It should spin easily without binding.
7. Tighten the bolts in increments, alternating from side to side to bring them to final torque, figure 16-38. Tighten the bearing caps one at a time, and rotate the shaft after tightening each bearing. If you feel any binding, you know which bearing is causing the problem. If the manufacturer provides a tightening sequence, follow it. If not, start with the center main bearing, then work your way alternately toward either end of the engine.

Measuring oil clearance with plastigauge®
Plastigauge® is a string-like piece of plastic manufactured to a precise diameter. A strip of the plastic is placed between the bearing and the journal and the cap is tightened to final torque, then carefully removed. Since the diameter of the gauging material is exact, tightening the cap crushes the string a specific amount. The amount of crush can be measured to determine oil clearance.

This method of determining clearance is not nearly as accurate as measuring with precision tools. However, on final assembly it is a quick and easy way to double-check the measurements you made earlier. Plastigauge® can be used to check main bearings, rod bearings, and some cam bearings. In fact, most any split-type bearing clearance can be checked with plastigauge®.

There are two important features to keep in mind when using plastigauge®. First, the plastic material is not compatible with engine oil. Both bearing and journal *must be perfectly dry*. Secondly, *never turn the crankshaft* with plastigauge® installed. This can smear the plastic and possibly damage the bearing.

We detail how to check main bearing clearance using plastigauge®, the procedure for checking connecting rod, or other split-type, bearings is similar.

Checking crankshaft end play

Once all the main bearings are torqued and the crankshaft spins freely, you can measure the crankshaft end play. End play, or end float, is how far the shaft can move lengthwise in the block. Measure end play using a dial indicator as follows:

1. Attach the dial indicator to the front of the engine so that the plunger can rest on the end of the crankshaft snout, figure 16-39.
2. Move the crankshaft as far back in the block as you can by prying with a screw-

CHECKING MAIN BEARING CLEARANCE WITH PLASTIGAUGE®

1. Plastigauge® is available in several sizes. Select the correct gauging material based on the specified oil clearance tolerance of your engine.

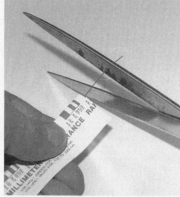

2. Slip the string out of the wrapper and cut off a piece just long enough to fit all the way across the journal.

3. Make sure the journal is clean and dry. Then position the length of plastic across the journal.

4. Wipe any oil off the bearing surface. It must be perfectly dry. Then install the cap.

5. Alternately tighten the bolts in increments as you bring them up to the specified final torque.

6. Loosen the bolts, then lift the cap straight off. Be careful not to jiggle the cap, as you could smear the plastic and spoil the reading.

7. The crushed string should stay on the journal. Compare the width of the crushed string to the scale printed on the product wrapper.

8. Measure the crush at both ends of the string. Any difference between readings indicates bore or journal taper.

9. Thoroughly clean *all* of the plastigauge® off of the journal and bearing. Then oil the bearing and install the cap.

Engine Assembly

Figure 16-39. Attach a dial indicator to read off the end of the crankshaft snout to check end play.

Figure 16-40. Use a large screwdriver to move the crankshaft as far back as it will go in the block. Zero the indicator, then pry forward and read end play.

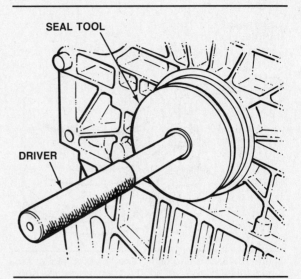

Figure 16-41. Always use the proper driver to install a main seal. Never pound directly on the seal.

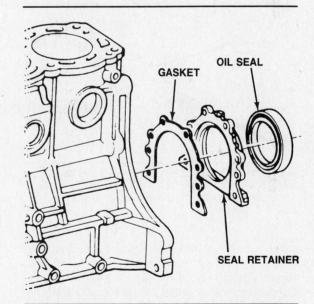

Figure 16-42. The rear main seal for this engine mounts in a retainer plate. The retainer is then bolted to the engine block.

driver between a bearing web and crank throw, figure 16-40.
3. Zero the dial indicator.
4. Now, move the crankshaft as far forward in the block as you can by prying in the opposite direction.
5. Read total end play on the dial indicator. End play must be within the manufacturer's specified tolerance.

Installing a one-piece main seal
A one-piece lip-type rear main seal can generally be fitted to the flywheel end of the engine now. Occasionally, the position of the engine in the stand does not allow enough room for you to install the seal. If the engine stand prevents clear access to the seal, skip this procedure for now. Return and install the seal once you have the engine off the stand. The seal must go in straight or it will leak. Do not attempt to install the seal if space is limited.

Some engines have a groove machined directly into the engine block for the seal to seat into. Half of the groove is cut into the main bearing cap, the other half in the block casting. This type of seal is quickly and easily installed with a seal driver. Inspect the seating groove. It must be clean and free of burrs, scratches, or rough spots as a unit. Carefully clean up any irregularities with a polishing stone. Avoid getting any filings or dust into the engine. Check the parting lines. They must fit together perfectly without any high spots or overhangs. Lubricate the seal lip with fresh engine oil, slip it over the crankshaft flange, and position it straight over the machined groove. Fit the seal driver over the seal, then seat it by striking the driver with a hammer, figure 16-41. *Never hammer directly on the seal*, this can distort it and result in an oil leak.

Many import engines have the rear main seal mounted in a retaining plate. The plate is then bolted to the engine block, figure 16-42. Inspect the seal seat of the retainer plate, and the mating surfaces of both the retainer and the engine block. Make sure there are no traces of gasket or sealer on the mating surfaces. Use a seal driver to install the seal into the retainer. It must fit perfectly square. Al-

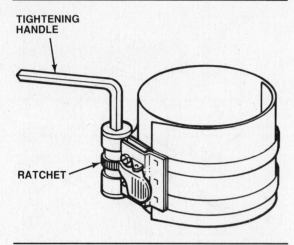

Figure 16-43. This ring compressor uses a ratchet to contract the spring steel band and compress the rings into their grooves. Since the tool diameter expands, it can be used with a variety of piston sizes.

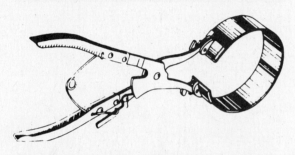

Figure 16-44. This plier-like tool is used to close the metal band around the piston to collapse the rings. An assortment of bands are available to service different size pistons.

Figure 16-45. When threaded onto the rod bolts, these guides not only help align the rod, they also protect the threads, hold the bearing shell in place, and the soft ends will not damage the journal.

ways use the proper tool. Do not hammer directly on the seal. A thin gasket is used to seal the retainer to the block. Apply a film of non-hardening gasket sealer to the block, allow the sealer to set up, then fit the gasket onto the engine. Lightly oil the lip of the seal, then install the retainer and seal as a unit. Carefully guide the seal over the crankshaft flange, line up the bolt holes, install the bolts, and tighten to specified torque.

Installing Pistons, Rings, and Rods

Procedures for assembling the pistons, connecting rods, and rings, as well as checking ring gap and bearing clearance, were detailed in Chapter 12 of this *Shop Manual*. At this point, the piston/rod assemblies should be ready to install. Play it safe and recheck the rod bearing clearance. You can measure clearance with precision instruments now, or you can use plastigauge® once the pistons are installed. Remember, if you will be using plastigauge®, do not oil the bearings for assembly.

To install the pistons, you need:
- A good quality ring compressor
- Protective caps for the rod bolts
- A hammer with a soft handle, preferably wood or polyurethane
- A socket that fits squarely on the rod nuts
- A torque wrench.

Piston ring compressors squeeze the rings into the grooves on the piston so the assembly slips into the bore easily without damaging the rings or ring lands. Ring compressors are available in several different designs. The ratchet-type compressor, figure 16-43, is a thin metal band that tightens to squeeze the rings into their grooves. A plier-like handle is used to close a metal band over the rings with a clamp-type compressor, figure 16-44. Some compressors are a simple funnel-shaped device with tapered walls. The bottom diameter is the same as the cylinder bore. The rings are compressed simply by sliding the piston through the tool. The only disadvantage is that the tool works with only one particular size piston, and you need an assortment of tools unless all of the engines you work on are identical.

You can purchase inexpensive plastic covers that slip over the rod bolts to protect the

Engine Assembly

Figure 16-46. This socket has been relieved to prevent interference when torquing connecting rod caps. If the socket does not fit squarely on the nut, it will effect torque reading and can lead to failure.

crankshaft journal and the bolt threads from most bearing manufacturers. You can also fit short lengths of fuel hose onto the bolts for protection. Special threaded guides are also available to protect the journal and threads, figure 16-45. These long guides thread onto the bolts. The wide diameter holds the bearing shell in place and the ends are soft so they cannot mar the crankshaft. Never attempt to install pistons without capping the bolt ends.

Most tool manufacturers sell special piston installation hammers, or piston poppers. These are convenience items that are nice to have if you work in a production shop and install pistons on a daily basis. For occasional use, a conventional hammer handle will work just as well for driving pistons into the bore.

Connecting rod bolt torque is critical to engine reliability. The bolts take the full force of the power and inertia transfer as the piston changes direction. *Rod bolts that are not properly torqued will fail*. Make sure the socket you use seats square and fully on the nut, figure 16-46, because any interference gives erroneous torque wrench readings.

Installing piston assemblies

Once you have all the tools you need on hand, set the pistons out in the order they fit into the engine. Double-check the pistons for proper assembly. Make sure the pistons and rods will all face the proper direction when they are installed. As detailed in Chapter 12 of this *Shop Manual*, there are directional indicators on both the connecting rod and the piston. Be certain that everything will face the right way on final assembly. The pistons should also be clearly marked as to what bore they fit. Remember, each bore was sized to fit a particular piston, and the rings were end-gapped to fit a specific cylinder bore.

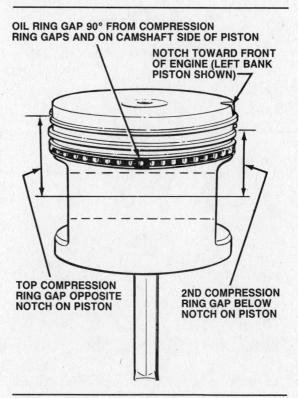

Figure 16-47. Be sure the ring end gaps are in position before you install the ring compressor.

The connecting rod bearing oil clearance must be double-checked. Although the best way to do this is by measuring with precision tools as detailed in Chapter 12 of this *Shop Manual*, it can also be accomplished using plastigauge®, as described earlier in this Chapter. When using plastigauge®, install the bearings dry and do not apply thread sealer to the bolts. Check using the old rod nuts. After verifying clearance, lubricate the bearing, apply thread locking compound, install the new rod nuts, then tighten them to final torque.

Follow these steps to install the pistons:

1. Separate the rod caps from the connecting rods. Be sure to keep them in order. Carefully wipe off all trace of dust and dirt from the bearing seating surfaces and install the bearing shells. Fit the bearing inserts by gently squeezing the ends together, making sure the tang on the shell aligns with the notch on the rod or cap, and that the bearing is fully seated.
2. Slip the protective covers over the connecting rod bolts.
3. Lightly lubricate the cylinder wall, piston rings, ring lands, skirts, and the bearings with engine oil. Rotate the piston rings so the end gaps are properly positioned, figure 16-47.

Figure 16-48. Carefully guide the rod into the cylinder so the bottom of the ring compressor rests on the deck surface.

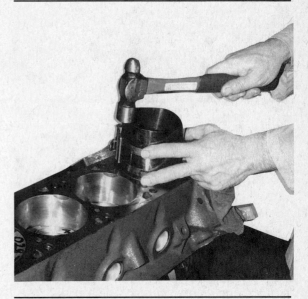

Figure 16-49. With a ratchet-type compressor, lightly tap the top edge of the tool with a hammer to true it on the deck surface.

Figure 16-50. Push the piston into the bore using the hammer handle or piston installing tool.

4. Rotate the crankshaft so that the rod journal of the piston to be installed is at bottom dead center.
5. Place the piston into the ring compressor and tighten the tool to squeeze the rings into the ring grooves. Work slowly and carefully to avoid cocking the rings and damaging the ring lands on the piston.
6. Fit the end of the rod into the cylinder, figure 16-48, be sure the piston faces the right direction, and guide it down until the lower edge of the ring compressor contacts the deck surface.
7. Check the compressor to make sure it is tight and the rings cannot snag on the block as they pass into the bore. When using a ratchet-type compressor, lightly tap the top edge with a hammer to set the tool squarely on the deck surface, figure 16-49.
8. Slip the piston into the bore by applying pressure to the crown with your hammer handle, figure 16-50. As the piston moves into position, reach up inside the bottom of the block to guide the connecting rod around the crankshaft journal. Be sure the bearing insert is still in position, then push the piston all the way in until the rod is seated on the crankshaft.
9. Remove the protective bolt caps, then slip the rod cap onto the connecting rod. Check to make sure the bearing is fully seated in the cap, and that the rod and cap are properly aligned so the number stamps are both on the same side.

Engine Assembly

Figure 16-51. Tighten the nuts in small increments, alternating from side to side as you bring them up to their final torque.

Figure 16-52. After installing all the pistons, check rod side clearance with a feeler gauge.

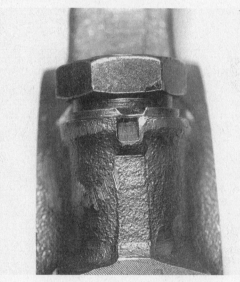

Figure 16-53. Watch for special features. This Volkswagen rod nut has a flange that is peened into a groove on the cap to lock it in position.

10. Apply a small drop of thread locking compound to each of the rod bolts. Although not recommended by all manufacturers, this is good insurance to prevent the nuts from working loose. Install *new rod nuts* and run them up by hand.
11. Bring the nuts up to final torque in small increments, alternating from side to side, figure 16-51. Be sure the socket seats squarely on the nut.
12. Rotate the crankshaft at least one complete revolution to make sure there is no binding. The piston rings create some resistance, but the engine should turn smoothly. Position the crankshaft so the journal of the next piston to be installed is at bottom dead center. Repeat the process until all of the pistons are installed.

Next, check the connecting rod side clearance. Side clearance is the distance between the crankshaft cheek and the side of the connecting rod, or the space between the two connecting rods that share a journal on a V-type engine. Manufacturers provide side clearance requirements in their specification data. The largest feeler gauge that fits into the gap between the rod and crankshaft, or two adjacent rods, equals the side clearance, figure 16-52.

If the side clearance is too tight, you can correct it by removing the piston assembly and precisely grinding the required amount of metal from the sides of the rod. If there is too much side clearance, replace the connecting rod. Failure to do so can result in an oil pressure loss.

After checking and correcting side clearance, retorque the rod nuts. If the rod nuts have a locking flange and a groove is machined into the rod, figure 16-53, lock the nut in place by peening the flange with a small punch.

Measuring deck clearance

Once the pistons are in place you can check the deck clearance. Deck clearance is the distance a piston crown, at top dead center, is below, or above, the deck surface of the block. Most automotive engines are manufactured with a considerable amount of deck tolerance. If the surface cuts you made to the block deck and cylinder heads are within specification this step can often be overlooked. However, for diesel engines that operate at high compres-

Figure 16-54. Deck height is measured with a dial indicator and a special straddle bracket.

sion, or for gasoline engines that have been modified for performance, checking the deck clearance is critical. Improper deck clearance can create poor combustion characteristics that result in a lack of performance. In addition, too little clearance can cause valve-to-piston interference that will immediately destroy the engine on start up.

Deck clearance, or deck height, is measured with a dial indicator and a special straddle bracket. The bracket spans the cylinder bore and allows the indicator plunger to rest on the piston crown, figure 16-54. Slowly rotate the crankshaft as you watch the gauge, to locate the piston at exact top dead center. The sweep of the indicator needle momentarily dwells as the piston changes direction at TDC.

Once the piston is up, adjust the straddle bracket so the plunger rests on the highest point of the piston crown. Then zero the dial indicator. Move the straddle so the plunger comes to rest on the block deck surface along side the cylinder. Take an indicator reading. The figure displayed is the deck clearance. Deck height may be positive or negative depending on engine design.

Compare your findings to the manufacturer's specifications. Occasionally, too little clearance can be corrected by installing a thicker head gasket or a head gasket shim. You may have to replace the cylinder head or engine block, alter the stroke by installing a different crankshaft, install shorter connecting rods, or replacing the pistons with ones that have higher pin bores.

Installing Valve Timing Components

Now that the pistons are in place, you can install the cam timing gear on an OHV engine. You must measure camshaft end play and runout, install the rest of the timing components, then measure gear backlash. These operations were discussed in detail in Chapter 7 of this *Shop Manual*, and will not be repeated here. Follow the appropriate steps in Chapter 7 to assemble the timing mechanism, then return here for final bottom end assembly.

FINAL BOTTOM END ASSEMBLY

From this point on, every part you bolt up to the engine block requires some sort of gasket or seal. Properly sealing the engine is important. Gaskets and seals hold back pressurized oil, coolant, and combustion. Handle seals and gaskets carefully and always check the fit. Incorrect installation can result in failure.

Fel-Pro®, an automotive gasket manufacturer, provides the following general rules for installing gaskets of any type:

- Read the gasket label and instruction forms before beginning the job.
- Clean the sealing surfaces.
- Check the sealing surfaces for damage, warpage, or distortion.
- Never reuse a gasket.
- Check the new gasket for proper fit.
- Clean and prepare all threads on assembly bolts.
- Follow the proper torquing sequence and torque to specifications. *Do not overtorque*.

Even a good gasket cannot seal if installed incorrectly. Clean surfaces in good repair and proper torque are the main requirements.

You must be aware that the electronic controls on some engines can be effected by certain gasket sealing compounds. Sealants cure in a running engine and as they do, spent chemicals are emitted. These contaminants can circulate in the engine and cause faulty signals from various sensors. Oxygen sensors are especially vulnerable to certain RTV silicone sealers. Always follow the engine manufacturer's recommendation for use of gasket sealers. Use the correct type and only where specified.

At some point of engine assembly, many shops cycle the partially assembled engine on a test station. The test station may be a simple device that checks oil flow and prelubricates the engine. More elaborate test equipment can gauge bearing clearance and alignment, check compression pressures and piston clearances, and measure the torque required to rotate the engine, figure 16-55. These machines provide

Engine Assembly

Figure 16-55. This Sim-Test engine test bench prelubricates the engine and checks bearing and piston clearances, connecting rod alignment, and frictional resistance before the engine leaves the shop.

an excellent way to isolate a problem before the engine leaves the shop. As with most shop equipment, there are a variety of test stations available and procedures vary. In general, the engine is tested with all the moving parts assembled, but without the oil pan, timing cover, valve covers, and manifolds installed. Filtered oil is circulated through the engine while the crankshaft is rotated. The flow and drain back rate of the oil is used to evaluate engine condition. You must follow the manufacturer's instructions for the particular piece of equipment you are using.

Installing The Oil Pump

If the oil pump bolts to the bottom of the block, install it now. Pumps that are driven by the crankshaft generally install with the front cover. However, you may have to bolt the pickup tube to the block. Some oil pickup tubes and screens must be attached before you install the pump. If so, assemble the pickup.

Check the fit of the pump drive shaft. Whether it uses a hexagon or a dog, the shaft must fit snugly. If the fit is sloppy, replace the drive shaft. Some engine designs require you to fit the drive shaft before the pump is installed. Others allow you to fit it from above after the pump is in place. Install the parts in proper order according to your engine.

You must *prime the oil pump* before you install it. If not primed, the pump will not circulate oil immediately on start up, or it may not be able to pick up oil at all. The result can be failure or serious reduction in engine life. Submerge the pump in a container of clean engine oil and spin the rotors or gears by hand until the pump discharges a good stream of oil. Unless specifically instructed to do so by the engine manufacturer, *do not pack a pump with grease*, assembly lube, or other heavy lubricant. Prime the pump with engine oil only.

Most oil pumps bolt on with a thin paper gasket between the machined surfaces of the pump flange and the block. *Do not use sealer* of any type on the oil pump gasket. Sealant can be drawn into the oil, restrict flow, and cause damage. Simply soak the gasket in oil. This softens the gasket, allowing the material to crush and form a tight seal as the bolts are brought to torque.

Check-fit the pump before you install the gasket. Make sure everything lines up perfectly. The machined surfaces of the pump and block must butt together without a gap. Some Chrysler, and other manufacturer's engines have pumps with a mounting neck that fits a bushed hole in the engine block. Tightening the bolts with the neck not seated can damage the pump housing. Remember, a mistake here may prove to be time consuming and costly.

Once you are certain the pump fits, install it along with the oil-soaked gasket. Make sure the drive shaft or any other attachment is in place if it must be fitted before the pump. Install the bolts, run them in as far as possible by hand, then alternately bring them to specified torque, figure 16-56. Double-check the bolt torque, then install or adjust the pickup.

Installing the pickup screen

The pickup screen assembly may be threaded or press-fit into the pump housing, or it may bolt on. All designs require careful handling. Any leak here can be a disaster. Improperly press-fitting a screen can distort the pump housing and lead to check valve failure.

Press-fit screens are installed with a special tool, figure 16-57, that fits around the intake tube. The collar of the tool equalizes pressure on the flange to drive the tube straight into the

Figure 16-56. Hand-tighten the oil pump bolts, then bring them evenly up to final torque.

Figure 16-58. Measure the distance from the bottom of the pickup screen to the oil pan rails.

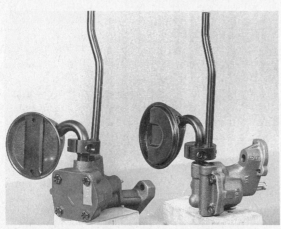

Figure 16-57. These special tools install press-fit oil pickup screens without distorting the pump body.

housing bore. Follow any specific instructions for the tool you are using. *Never try to drive in a pickup without the correct tool.*

Pick-up tubes that thread into the pump housing can be installed with a pipe wrench. Seal the threads with pipe sealer or teflon tape. Avoid oil contamination by keeping sealer off the leading threads of the tube. Screw the tube into the housing. Most have a taper-thread so resistance builds as it goes in. Run the tube in tightly, then continue tightening to position the screen parallel to the pan rails.

Bolt-on pickups are the easiest to deal with. They may use either a gasket or an O-ring to seal to the pump. *Do not use gasket sealer*. Soak gaskets in oil before installation and lightly lubricate O-rings to be sure they seat properly. Install the bolts, then alternately bring them up to specified torque.

Checking screen to pan clearance

Upon assembly, the pickup screen must be positioned parallel to the bottom of the oil pan, with about 1/4 to 3/8 inch (6 to 10 mm) clearance between the screen and the pan. You can quickly measure clearance by taking two readings with a ruler. Measure from the screen opening to the block pan rails, figure 16-58, and then from the floor of the sump to the flange on the oil pan, figure 16-59. The difference between the two measurements is the clearance. Clearance must be within specification. Too little clearance causes oil starvation, while too much causes the oil to aerate. Both result in bearing damage. Correct pickup screen position, then install the oil pan.

Another way to check pan-to-screen clearance is with a dowel pin or gauge block. The thickness of the pin or block is equal to the desired clearance. Lay the gauge across the pickup screen and carefully fit the oil pan. You can feel the pan just contacting the gauge if the clearance is right.

NOTE: Engine vibrations can shake a press-fit oil pickup loose from its mounting, disrupting oil flow to the bearings, and leading to imme-

Engine Assembly

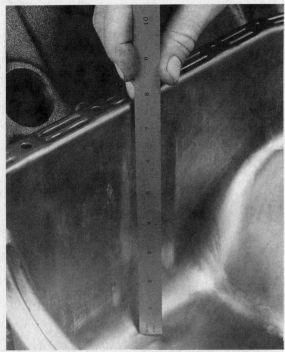

Figure 16-59. Measure from the oil pan gasket flange to the floor of the sump. Subtract this reading from the previous one to determine clearance.

Figure 16-60. Install the water pump now if the pump bolts are also used to attach the timing cover to the engine block.

diate engine failure. Vibration damage is more likely to occur in a high-performance, heavy-duty, or other heavily stressed engine. To prevent loosing the oil pickup, many engine builders braze or weld the pickup tube to the housing. The pickup *must be aligned* properly, and the clearance must be checked before you weld. Also, remove the pressure relief valve. Heat transfer from your torch may damage the valve if you leave it in place.

Installing the oil pan and timing cover
The bottom end of the engine can now be finished off by installing the timing cover and oil pan to seal the crankcase. The order of assembly varies by engine design. Cast timing covers are often installed before the oil pan. The installation procedures for both oil pans and timing covers are covered in detail in Chapter 7 of this *Shop Manual*, and are only highlighted here. Double-check all your work before you seal the bottom end.

Timing covers
Make sure the timing gears, chains, guides, and tensioners are properly assembled. All timing marks must line up properly, and the piston of number one cylinder must be at TDC. The sealing surface of the cover must be perfectly clean and flat. Remember to *lubricate the chain and gears* before you fit the cover and to install a new crankshaft seal.

On some engines that use a cast front cover, additional machining may be required. If you have machined the block deck or surfaced the cylinder head, you may have to remove an equal amount of metal from the top of the timing cover. Failure to do so can result in leaks.

With stamped timing covers, make sure you get a good fit by flattening the area around the bolt holes. Pulled bolt holes are straightened by resting the cover on a flat surface and striking with a ballpeen hammer.

Whether cast or stamped, all traces of old gasket and sealer must be scrapped from both the cover and the engine block. Cover gaskets should be sealed with a good non-hardening liquid sealing compound.

Check fit the bolts because many covers attach with several different length bolts, and they must be installed in their proper place. On some engines, the water pump shares some of the timing cover mounting bolts. With this design, install the water pump now. Doing so equalizes pressure between the cover and block, and promotes effective sealing, figure 16-60. Always install a new water pump; do not compromise your work by using questionable parts. Draw the bolts up in stages, alternating from side to side and torque them to the manufacturer's specification.

Finish by installing the harmonic balancer or front pulley. Be sure to fully seat the woodruff key, or any other alignment device, into the crankshaft. Slip the balancer or pulley onto the crankshaft snout, taking care not to dislodge the key. Apply thread locking compound to the bolt threads and install the bolt. Tighten the bolt to specified torque. The crankshaft must be held to prevent it from turning as you tighten the bolt. A variety of special tools are available to lock up the crank. If there is a timing pointer that bolts to the front cover, install it now. Be sure the pointer aligns with the

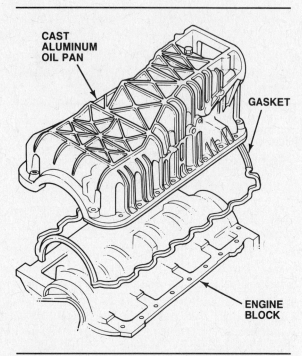

Figure 16-61. Many late model engines use a cast aluminum oil pan that is an integral stressed member of the engine.

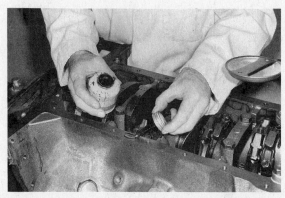

Figure 16-62. Spread a thin film of non-hardening gasket sealer on the oil pan rails of the engine block.

Figure 16-63. Allow the sealer to set up, then position the gasket on the block as you carefully line up all of the bolt holes.

TDC mark on the balancer, and that the number one piston is at exact top dead center.

Oil pans

For many years, most oil pans were a simple stamped steel part that was easy to fit and presented little challenge. This has been changed by recent design technology. Many late-model engines use a cast aluminum oil pan, or sump, that is an integral stressed member of the engine. This type of pan is much more than a reservoir to hold engine oil. Aluminum pans can be designed to transfer heat from the block, increase block rigidity, and absorb torsional twist, figure 16-61. Some engines have a two-piece assembly where an aluminum sump attaches to the engine block and a small stamped steel oil pan bolts to the sump. Properly fitting an aluminum oil pan is a critical operation. Some require a special sealing compound, so check the service literature and do not make substitutions.

Oil pans, regardless of design, require a perfectly clean and flat surface in order to seal. Scrape any bits of old gasket and sealer from the pan rails on the block and the flange of the pan. Straighten any dimpled holes on a stamped pan using a ballpeen hammer. The threads of the bolt holes should have been previously cleaned. If not, do it now. Be careful to keep any debris out of the engine. Check-fit the oil pan gasket before you install it.

Gasket designs vary. Some engines use a one-piece cork or rubber gasket, while others use a multi-piece assembly with cork gaskets along the sides and rubber sealing strips at the ends. If the mating surfaces have been properly prepared, most pan gaskets can be installed without sealer. Should you decide to use sealer, use only a non-hardening type, and use it sparingly. Spread the sealant on the block rails, and allow it to set up several minutes and become tacky, figure 16-62. Now lay the gasket into position as you carefully align the bolt holes, figure 16-63. With multi-piece gaskets, place a small dab of RTV silicone at the seams on each corner, figure 16-64.

Once the gasket is in place, carefully fit the pan. Start all of the bolts, then snug them up in a criss-cross pattern to eliminate distortion. Check the specifications and *tighten the bolts to torque*. Overtightening the oil pan bolts can cause the gasket to split and result in a leak. Most oil pan gasket failures are the result of bolts being tightened too much.

Be aware that aluminum pans may be fitted without a gasket, but a special sealing compound is used to join the parts. *Use only the*

Engine Assembly

Figure 16-64. Apply a small dab of RTV silicone to the seams of the end seals and gaskets.

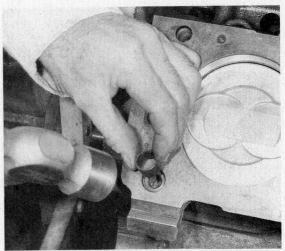

Figure 16-65. Fit the cylinder head alignment dowels into the engine block. Set them in place with a light hammer tap.

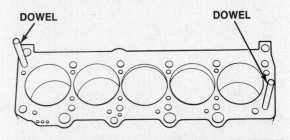

Figure 16-66. These alignment dowels are temporarily fitted in the head bolt holes to align the gasket and cylinder head during installation.

correct sealer. The aluminum sump is a fully stressed engine component and must be properly bonded to the engine block. Use of the wrong sealant can promote leaks, disrupt heat transfer, and alter the torsional rigidity.

ASSEMBLING THE TOP END

Now that the crankcase is sealed, turn the engine over to assemble the top end. Begin by rotating the crankshaft through several complete cycles. Make sure the engine spins freely. You can expect to feel some resistance due to friction, but any binding indicates a problem. A helpful hint is to add about a quart of oil to the crankcase to prevent the oil pump from drawing air and loosing its prime.

Inspect the gasket sealing surfaces on the block deck and the cylinder head. They must be perfectly flat and free of any gasket material, dirt, debris, or other irregularity. The head bolts must fit easily through the cylinder head. Overheating and warping, especially on aluminum heads, can distort the bolt holes. If the bolts hang up in the head, they cannot be properly torqued, and the block-to-head seal is jeopardized. The threads on the bolts and in the block must be clean and in sound condition. In addition, the bolt holes in the block must be properly chamfered. If not, the upper threads can pull as you tighten the bolt and interfere with sealing. Remember, torque-to-yield bolts may need to be replaced with new ones. Once everything checks out, cover the cylinders with a rag to keep any debris out, then clear the bolt holes with a blast of compressed air. Wipe the sealing surfaces off with a clean dry rag.

For most overhead-cam engines, the camshaft can be installed in the cylinder head and valve lash adjusted before the head is bolted to the block. Installing overhead camshafts is covered in detail in Chapter 14 of this *Shop Manual*. Before you install the head, rotate the camshaft to align timing marks and set the number one piston at top dead center. The number one cylinder should be ready to fire, the piston up at full compression, and both valves closed. Keep in mind that some European engines use marks timed to the number four cylinder. Check the specifications and have everything properly aligned before you install the cylinder head.

Installing The Cylinder Heads

Most engine manufacturers provide locating dowels to hold the gasket in place and align

Figure 16-67. Fit the head gasket onto the block. Be sure it is facing the right direction and all the coolant and oil passages align.

Figure 16-69. Install all of the head bolts. Run them in hand tight.

Figure 16-68. Lower the cylinder head straight down onto the block deck. The dowels will fit into the head to lock it into position.

Figure 16-70. Bring the head bolts to final torque by tightening them in stages following the manufacturer's specified sequence.

the bolt holes as you install the cylinder head, figure 16-65. Fit the dowels into the engine block before you install the gasket. The dowels may fit into an undercut on a head bolt hole, have a special dedicated drill hole, or install into the oil gallery that leads to the top end. Be sure the dowels are in their proper place, then firmly seat them with light hammer taps.

Some engines do not use locating dowels. Instead, the gasket is held in place during assembly with a special alignment tool. The tool is simply a threaded rod, in which the threads match those of the head bolts and are long enough to pass completely through the cylinder head, figure 16-66. Position the head gasket on the deck, then screw in the alignment tools, one at each end of the block. Sliding the cylinder head down over the rods mates it squarely to the head gasket and block deck. Install the head bolts hand tight, remove the alignment tools, then install the bolts.

Once the locating dowels are in place, you are ready to install the cylinder head. All engine designs vary, so you must have the correct technical data for the one you are working on. You can install most cylinder heads following these steps:

1. Look the head gasket over and check the fit. Most head gaskets only fit in one way. If upside down or reversed they may bolt up but can restrict coolant or oil flow. Many head gaskets are stamped "top" or "front" to make proper installation easy. Check the gasket surface over for any damage from improper handling. Install head gaskets dry. Fit the head gasket onto the block deck and make sure all the passages line up, figure 16-67.
2. Get a good grip on either end of the cylinder head. Heads can be heavy and awkward, so you must have a firm hand-hold.

Engine Assembly

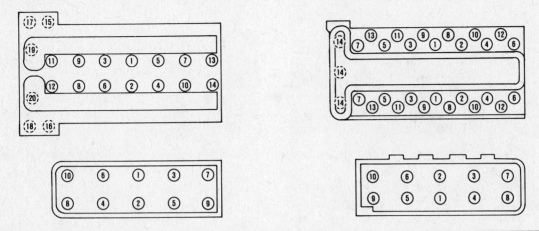

Figure 16-71. These are some typical cylinder head bolt torquing sequences. You must follow the correct procedure for the engine you are working on.

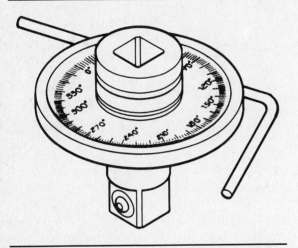

Figure 16-72. Torque angle can be measured using a special adaptor. Tighten the bolts the specified number of degrees.

Following proper cylinder head bolt tightening procedures is critical to engine integrity. Figure 16-71 shows some typical torquing sequences, which are very important. They ensure even tension over the entire head and minimize the possibility of gasket failure. Most head bolts are brought from snug-fit to final torque in three stages. Torque-to-yield bolts often tighten to a specific torque, then tighten an additional amount. The additional tightening is measured in degrees. A special angle gauge, figure 16-72, is used to precisely measure how far you turn the bolt.

Your next assembly steps vary depending whether the engine is an OHC or OHV design. With an overhead valve engine, you have to install the valve train. An overhead cam engine should already have the valve gear in place, so you can skip ahead to *Installing Valve Timing Components*.

Installing The Valve Train

The valve train consists of the lifters, pushrods, and rocker arm assemblies. Also included would be pushrod guides, oil baffles, or any other related parts. Specific fitting procedures apply to individual engines. The following steps provide a general guideline:

1. Take a last close look at the lifter bores, make sure they are free of burrs and spotlessly clean, figure 16-73.
2. Lightly oil the lifter bores, or dip the lifters in clean oil as you install them.
3. Spin the lifters as you slowly insert them straight into their bores, figure 16-74. Remember, they must rotate freely, or the lifters and camshaft quickly wear out once the engine is running.

Lower the head straight down into position on the block, figure 16-68, and fit it over the dowels to lock it in place.

3. Lightly oil the threads of the head bolts. Be aware, any oil that runs off the end of the bolt and into a blind hole can create a hydraulic lock which can keep the bolt from reaching torque. Be stingy with the oil. Some engines require dry bolts. Check the specifications.
4. Install all of the head bolts, figure 16-69. Run them in as tight as you can by hand, then snug them with a short wrench in torque sequence.
5. Bring the head bolts to final torque in stages, following the manufacturer's recommended sequence, figure 16-70. After all the bolts are tight, double-check torque following the same sequence.

Figure 16-73. Make sure the bores are clean and free of burrs before you install the lifters.

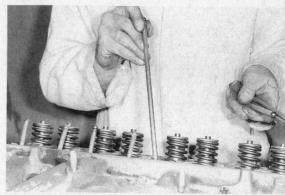

Figure 16-75. Install the pushrods so they rest squarely in the lifter seat.

Figure 16-74. Lubricate the lifters, then slip them into the engine block. Be sure they spin freely in the bore.

Figure 16-76. Oil baffles and push rod guides are installed before the rocker assemblies.

Figure 16-77. Assemble the rocker arms onto the cylinder head. Be sure the pushrod ends are seated.

4. Insert the pushrods, making sure the ends rest squarely in the pushrod seat of the lifters, figure 16-75.
5. If the engine uses an oil baffle or pushrod guides, figure 16-76, install them now.
6. Install the rocker arms, figure 16-77, or rocker shaft assembly, figure 16-78. Make sure the pushrods seat at both ends.
7. Adjust the valve lash as detailed in Chapter 7 of this *Shop Manual*. Keep an eye on the pushrods as you rotate the crankshaft making sure they do not contact the cylinder head or guide plate as they move.
8. Now is a good time to check oil flow through the galleries. On many engines you can operate the oil pump by turning the drive shaft with a speed-handle or a low-speed drill motor. Make sure there is enough oil in the sump to avoid drawing air. After the air is purged, oil should flow freely to lubricate all of the rocker arms. If not, there is a restriction, and you must recheck your work.

Engine Assembly

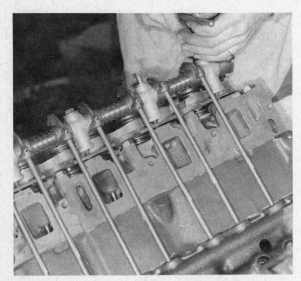

Figure 16-78. Rocker shafts are installed as an assembly. Be sure the pushrods are seated, then draw the bolts up slowly and in increments to avoid distorting the shaft.

9. Your next step is to fit the valve cover gasket. If you are using gasket sealer, apply it to the cover side only, not the head side.
10. Fit the valve cover and gasket to the cylinder head, figure 16-79. Install the bolts by hand and bring them to specified torque. Do not overtighten the bolts, this can distort or crack the cover and cause leaks.

Installing Valve Timing Components

If you are working on an overhead-valve engine, the following procedures do not apply. Skip this section and continue with the final assembly procedures in Chapter 17 of this *Shop Manual*. For an overhead cam engine you have to install the timing belt or chain, gears or sprockets, tensioners, guide rails, front cover and seal, as well as the crankshaft pulley and valve cover. All of these operations were covered in detail in Chapter 7 of this *Shop Manual*, and are not be repeated here. Go back to the appropriate section for specific instructions, then return here to continue assembly.

When installing overhead cam timing components, keep these tips in mind:

- Make sure all the marks, including those on idler or balance shafts, line up before fitting the chain or belt.
- Always install a new timing belt.
- Make sure guide rails are correctly positioned and cannot snag or catch the belt or chain.

Figure 16-79. Install the valve cover and gasket, then tighten the bolts to torque. Do not overtighten.

- If a chain tensioner is used, install a new one. Be sure it is properly adjusted.
- Double-check chain or belt tension.
- Secure any lock tabs on the gear or sprocket retaining nuts.
- Check operation by sight and feel as you rotate the engine through at least one complete cycle.
- Generously lubricate timing chains before you fit the cover.

Your engine is now nearing completion. In fact, engines are often delivered at this point, leaving it up to the customer to fit the remaining parts and install the engine in the vehicle. Production engine rebuilders often ship their engines in this manner. The unit is sold as a "long block". It includes all of the moving parts, but none of the attachments. This allows the builder to reduce his inventory and cover a variety of model applications with a single engine. Many car models use the same base engine with different attachments and the long block can be custom-fitted to the specific requirements of the vehicle.

Chapter

17

Engine Installation, Break-in, and Delivery

At long last, the engine is near completion and the end of the job is in sight. Now is not the time to drop your guard and let any oversights ruin the fruits of your labor. The customer is anxious, wants the car back, and is probably phoning every half hour to check your progress. Do not let the customer's anxiety push you into rushing the final stages of engine assembly and installation.

Keep in mind that the customer will soon forget any additional wait to get the car back, if the job is perfect. On the other hand, customers can become annoyed, upset, angry, and even hostile when they have to return the car for unscheduled repairs and adjustments. Remember, a satisfied customer may tell somebody about the fine job you did when rebuilding his engine, while dissatisfied customers tend to tell everybody they know about the poor quality of your work. Doing the job right and paying attention to detail will result in a good-running, long-lasting engine and a happy customer.

In this chapter, we detail the steps to assemble a complete engine from a long block, install the engine in the vehicle, run the engine to break it in, and deliver the car to the customer. Even if you are simply delivering a long block, you must be familiar with these procedures. Installation errors can cause engine damage. It is your responsibility to protect your engine, so make sure the installation is done properly.

FINAL ASSEMBLY PROCEDURES

Before you begin, review your tear-down notes for any special features or situations you found during disassembly. You will want to install as many parts as possible to the engine now, while it is out of the car and easy to work around.

Remember that, you must tighten every part on the engine to specific torque. *Proper tightening is extremely important.* For instance, overtightening bellhousing bolts can pull the rear cylinder bore out of round by as much as 0.006 inch (0.15 mm). Exceeding torque on other external components can have a similar effect on the internal parts of the engine. Do the job right and leave the impact wrench on the bench. Take the time to use your torque wrench on all of the attachment bolts.

Installing Manifolds

The manifolds, both intake and exhaust, seal to the cylinder head with a gasket. In addition, the intake manifold on a V-type engine seals to the engine block to cap off the lifter valley.

Engine Installation, Break-in, and Delivery

Figure 17-1. Some intake manifolds use small seals for the coolant passages that install with the gasket.

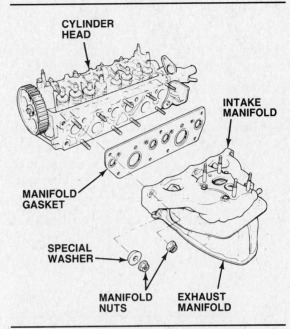

Figure 17-2. Many in-line engines use a common gasket for both the intake and exhaust manifolds.

Sealing surfaces must be clean, flat, and in good condition. Inspect all of the surfaces before you install the manifolds. Be sure that the bottom of the manifold, underneath the heat shield, and all passageways are free of carbon and any other deposits. Check-fit the gaskets to make sure they will seal properly and all of the openings line up.

Manifold designs vary from engine to engine, as well as from year to year on the same engine. Although there are many different manifolds in the field, they all attach to the engine in a similar manner. Here we will detail basic installation steps that can be applied to most manifolds. We also highlight precautions you can take to ensure successful sealing.

Installing intake manifolds

The intake manifold gasket not only forms an air-tight seal around the ports, but also forms a water-tight seal for the coolant passages. Check-fit the gasket before you install it. All port, coolant passage, and bolt openings must be perfectly aligned. Some engines use small O-ring type seals for the coolant passages in addition to the manifold gasket, figure 17-1. Make sure they are in place. Failure to do so will result in a leak.

Intake manifold gaskets normally install dry, without any sealer. Although not always required, a thin bead of RTV silicone can be used around coolant passages for extra protection. Be careful not to use too much. Excess silicone can restrict coolant flow and prevent the manifold from properly seating.

The easiest intake manifolds to deal with are those for an in-line engine with a crossflow cylinder head. Simply bolt the manifold and gasket to the side of the cylinder head. Be sure to have any brackets that are held in place by the manifold bolts properly positioned. Draw the bolts up in increments, and follow the proper sequence as you bring them to specified torque.

Intake manifolds on engines that have the intake and exhaust ports on the same side of the cylinder head are installed in a similar manner. These designs often use a single common gasket for the intake and exhaust, so you install both manifolds at the same time, figure 17-2. When the two manifolds share a common stud, a large thick washer is used to hold the assembly in position. Be sure to use the correct washers and install them in the proper position. Failure to do so will create an uneven clamping force that prevents sealing.

Installing the intake manifold on a V-type engine requires special consideration. The manifold must seal to both cylinder heads, as well as to the engine block. Gasket designs vary, but all are multiple-piece assemblies. Some engines use two composition gaskets, one for each head. Others use a single valley pan, made of thin stamped steel, to seal both heads and the lifter valley. All use small rubber or cork strips to seal the ends of the lifter valley. In our example we detail the installation of a valley pan type intake gasket. You can fit composition gaskets in a similar manner. To install the intake manifold gasket, follow these steps:

1. Check all of the sealing surfaces on the manifold, cylinder heads, and engine

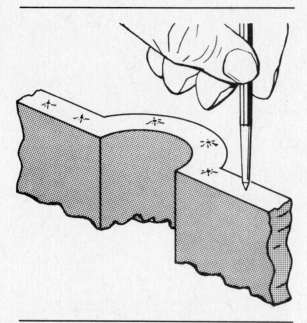

Figure 17-3. Staking the engine block before fitting the end seals raises small bits of metal to hold the seal in position when you install and tighten the manifold.

Figure 17-4. Run a bead of RTV silicone along the ends of the seals where they fit to the cylinder heads.

Figure 17-5. Slightly squeeze in on the sides of the valley pan as you lower it straight down into position on the engine.

17-3. Staking raises small amounts of metal that will catch the seal to prevent it from slipping out as you install the manifold and tighten bolts.

3. Once the seals are in place, run a thin bead of RTV silicone across the ends of the seals where they join the cylinder heads, figure 17-4. Also run a thin bead of RTV silicone around the coolant passages on the cylinder heads. Remember not to use too much silicone. It can prevent seating, restrict flow, and interfere with oxygen sensor function.
4. Fit the valley pan onto the engine. You may have to squeeze the sides in as you place the pan onto the block. Avoid shifting the pan from side-to-side as you line up the ports and bolt holes, figure 17-5. A shift can smear the sealant and dislodge the end seals. Some engines have dowel pins in the cylinder heads to help hold the pan in place.
5. Carefully lift the intake manifold and place it onto the engine, figure 17-6. *You must lower the manifold straight down.* Any side-to-side movement can displace the end seals. Not only is a cast-iron manifold heavy and awkward but the entire bottom butts to the engine so you cannot get a firm handhold from underneath. Get a good grip on the top of the manifold so you can gently set it in place and avoid dropping it onto the engine.
6. Install all of the manifold bolts hand-tight. There are probably several different length bolts. Make sure they are in the proper position. Check your disassembly notes and be sure you install any brackets that are held in place by the manifold bolts.
7. Check the manufacturer's specifications for tightening sequence and torque. Tighten the bolts in sequence and bring them to final torque in stages, figure 17-7.

block. Make sure they are clean and free from any old gasket material and sealer. If you resurfaced the block deck or cylinder heads, you must also machine the manifold for a proper fit. These procedures are detailed in Chapter 14 of this *Shop Manual*.
2. Fit the end seals into position on the engine block. Rubber seals often have small tabs on the bottom that fit into holes on the block to hold them in position. To prevent cork gaskets from shifting when you install the manifold, they can be glued to the block with non-hardening gasket sealer. A technique you can use to help hold the seals in place is to stake the metal underneath them with a small punch, figure

Engine Installation, Break-in, and Delivery

Figure 17-6. Carefully lower the manifold onto the engine. Avoid dislodging the gaskets and smearing the sealant.

Figure 17-7. Manifold bolts are often different lengths. Make sure you install them in their proper position, then tighten to torque.

Wipe off any sealant that squeezed out from the end seals. Check the seals to make sure they did not shift during assembly. It is much easier to catch a potential leak now than to have to deal with it on a running engine. Double-check the bolt torque, then install the exhaust manifold.

Installing exhaust manifolds

Installing exhaust manifolds is easy, but make sure you check the sealing surfaces, gasket fit, and hardware condition.

Look over the mating surface on the cylinder head for any problems that were not corrected during reconditioning. Threaded holes must be chamfered and there cannot be any pulled or damaged threads. Check the straightness of the manifold flanges. A quick easy way to do this is to run a file straight across each flange. If flat, the file will scuff the entire surface area. You can take the surface down with the file if the flange is slightly warped.

Exhaust manifolds seal to the head with either a single gasket, an individual gasket for each port, or a combination intake/exhaust gasket, figure 17-8. Some gaskets are stainless steel on one side and a composition material

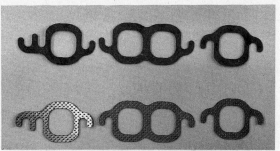

Figure 17-8. Some exhaust manifold gaskets are stainless steel on one side, and a composition material on the other. Install the steel surfacing towards the manifold.

on the other. These are installed with the steel side toward the manifold. The manifold will expand and contract at a faster rate than the cylinder head, the stainless steel allows free movement without gasket damage as this occurs. Always check-fit the gaskets, many can easily bolt onto the head either upside down or backward, but will not seal. Make sure you face the gasket in the right direction.

Replace all exhaust manifold mounting hardware, studs, nuts, and bolts with new pieces. The extreme temperatures these fasteners are exposed to result in corrosion and a loss of tensile strength. Over time the metal becomes brittle and will easily break. Always use brass or copper-plated nuts. These will not bind on the threads as the metals expand and contract.

Once everything passes inspection, install the exhaust manifold and gasket. Start all of the nuts, or bolts, and run them up hand-tight. Be sure any brackets or shrouds that attach with the manifold are in place. Then bring the fasteners up to specified torque in stages.

Installing Flywheels and Flexplates

If the engine is on a workbench, or your engine stand allows clear access to the back of the block, your next step will be to install the flywheel or flexplate. Many engines have a spacer plate that fits between the engine block and the transmission bellhousing. These plates are held in place by the engine-to-transmission bolts and must be fitted before you bolt up the flywheel. In these instances, you cannot possibly install the plate when the engine is in a stand. If this is the case, or access is limited, skip this section for now and continue your assembly. Then, return here and install the flywheel or flexplate once you remove the engine from the stand.

When disassembling the engine, you should have stamped indexing marks on the flywheel and crankshaft. It is very important that these are aligned during assembly. Failure to do so

Figure 17-9. The crankshaft must be held in place as you bring the flywheel/flexplate bolts up to torque.

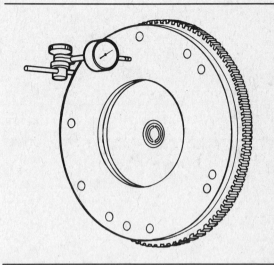

Figure 17-10. Using a dial indicator to check flywheel runout ensures that the flywheel is true and seated.

can upset engine balance and result in a vibration when the engine is running. To eliminate this possibility, the bolt holes on many flexplates, flywheels, and crankshaft flanges are unevenly spaced so that the parts will only fit together one way.

The crankshaft must be held to prevent it from turning as you tighten the attachment bolts. Always use the proper tool to hold the shaft in place. Should the crankshaft shift, your torque readings will be inaccurate. A loose flywheel on a running engine can cause severe damage.

Look the parts over one last time before you bolt them on. Check the teeth on the ring gear. If any are missing, chipped, or otherwise damaged, replace the ring gear as detailed in Chapter 12 of this *Shop Manual*. Inspect flexplates for cracks, fractures, or other signs of stress on the surface, especially around the mounting flange. Make sure flexplate balance weights are in place and tightly attached. Inspect the clutch contact surface on a flywheel, and correct any irregularities as described in Chapter 12 of this *Shop Manual*. Look over the seal area on the back of the flange. High mileage can cause a lip seal to wear a groove into the flange, which can prevent sealing and lead to an oil leak.

Once the parts pass inspection, fit the flexplate or flywheel onto the crankshaft flange to line up the bolt holes. *Make sure your index marks line up*. Lightly coat the bolt threads with a thread locking compound, then run them in by hand. Tighten the bolts in a star pattern, bring them to final torque in stages, figure 17-9.

Checking runout

Check flywheel or flexplate runout with a dial indicator to make sure it is not warped and is properly seated. Attach the indicator stand to the engine block and position the plunger to rest on the clutch surface of the flywheel or the ring gear face on the flexplate, figure 17-10. Pry the flywheel away from the block with a screwdriver to shift the crankshaft and remove end play, then zero the indicator dial. Rotate the crankshaft one complete revolution as you watch the indicator. The highest reading is total runout. Compare your findings to the manufacturer's specifications.

Be aware that, the crankshaft will have a tendency to float back and forth as you turn it. This will give erroneous runout readings. To prevent the shaft from moving, keep your screwdriver in place and apply *light* pressure as you *slowly* rotate the crankshaft.

If runout is beyond specifications, remove the flexplate or flywheel. Inspect the mating surfaces, as burrs, nicks, or debris can prevent seating and alter runout readings. If the mating surfaces are clean and true, the flywheel or flexplate is warped and must be replaced.

Installing pilot bearings and bushings

If the engine will be fitted to a manual transmission, your next step is to install the clutch pilot bearing bushing. Before you do, inspect the crankshaft bore, then check-fit the bearing or bushing to the transmission. The bore must

Engine Installation, Break-in, and Delivery

Figure 17-11. Hold the new pilot bushing so that one end is sealed off by a finger.

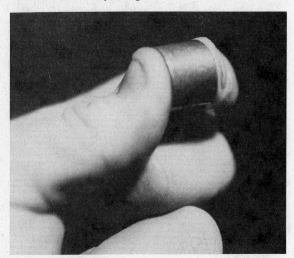

Figure 17-12. Fill the bushing with oil, seal the open end with your thumb, and gently squeeze to force oil into the porous surface.

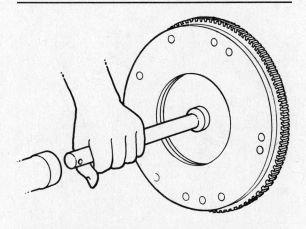

Figure 17-13. Use a driver to seat the bushing square into the flywheel bore.

be clean and free of any burrs, chips, or gouges. If the engine uses a bearing held by a snap ring, the ring groove must also be free of all debris and deposits. Check the internal bore on the bearing or bushing by sliding it onto the transmission input shaft.

All pilots, whether bearing or bushing, are installed in a similar manner using a special driver. To install a soft metal bushing follow these steps:

1. Place the bushing on your finger to seal off the hole at one end, figure 17-11.
2. Fill the inside of the bushing with engine oil, then cover the open end with your thumb, figure 17-12.
3. Press your finger and thumb together tightly and hold for several minutes. This creates pressure that forces oil into the porous surface of the bushing.
4. Fit the bushing onto the driver, then position the tool so the bushing leads straight into the bore, figure 17-13.
5. Seat the bushing into the crankshaft. Tool design varies, so you may have to strike the end of the tool with a hammer, or turn it in with a forcing screw. With all tools, you must drive the bushing straight into the crankshaft.

Installing clutch assemblies

A good practice is to always replace the clutch assembly, including the throw out bearing, with a new one. A clutch is a wear item and if the engine ran long enough to require rebuilding, the clutch is generally worn out as well.

Before you install the clutch, take a close look at the input shaft seal on the front of the transmission. If you notice any signs of leakage, or seal deterioration, replace the seal. Fitting a new seal now is much easier than it will be once the engine is installed.

A clutch alignment tool is used to center the clutch disc to the pilot bearing during assembly. If the disc is off-center, the engine and transmission will not fit together. To install the clutch assembly follow these steps:

1. Look the flywheel over and make sure the bolt holes are clean and threads are in good shape. Replace any alignment dowels that were removed to machine the flywheel. Clean the clutch attachment bolts and apply a small drop of thread locking compound to the threads.
2. Slip the splined hole of the clutch disc onto the alignment tool, figure 17-14. Then insert the tool into the pilot bushing so the disc rests against the flywheel. Be sure the disc is facing in the right direction.

Figure 17-14. Fit the clutch disc onto a clutch alignment shaft to center it on the flywheel.

Figure 17-15. Once the clutch parts are in position, install the bolts and bring them to torque in an alternating pattern.

Figure 17-16. This fuel pump insulating block also serves as a guide to keep the pushrod in alignment.

3. Fit the pressure plate over the disc and onto the flywheel. Push the pressure plate onto the alignment dowels to line up the bolt holes, then install the bolts hand tight.
4. Make sure the disc is centered. Then, tighten the bolts in increments. Use a star pattern as you bring the bolts up to final torque, figure 17-15.
5. Install a new throwout bearing onto the transmission clutch fork.

Installing Accessories and External Parts

Now you can bolt on any accessories, brackets, or external parts that attach to the engine. Review your disassembly notes and only install items that will not interfere with engine installation. What parts can be attached now, and what must be attached after the engine is in the chassis will vary for each job. Some of the items you may be able to install at this point are the:

- Fuel pump
- Engine mounts and brackets
- Sending units
- Electronic sensors

Fuel pump installation

Most mechanical fuel pumps attach to the engine block with two bolts or studs. The pump is activated either by a pivoting arm attached to the pump housing, or a separate push rod that is inserted into the block before the pump is attached. Fuel pumps often require a phenolic insulating block between the pump and the engine, figure 17-16. The insulating block may also be used to guide the push rod and keep it properly aligned.

Thin paper gaskets are used to seal the fuel pump to the engine. When an insulating block is used there will be two gaskets, one for each side of the insulator. Install the gaskets using non-hardening gasket sealer and be sure all the sealing surfaces are perfectly clean.

Coat the end of the pivot arm or push rod with assembly lube. Then fit the pump to the engine, install the bolts or nuts, and tighten them evenly to specified torque, figure 17-17. Cap off all of the fuel ports on the pump to prevent any contaminants from falling in during engine installation.

Engine mounts and brackets

Inspect the engine mounts for cracks, breaks, or deteriorating rubber. If any of these conditions exist, replace the mounts. Bolt the engine

Engine Installation, Break-in, and Delivery

Figure 17-17. Install the gasket and fuel pump. Apply firm hand pressure to compress the plunger, then install and tighten the fasteners.

Figure 17-19. A combination wrench can be used to install readily accessible sending units.

Figure 17-18. Bolt the engine mount brackets to the sides of the block.

Figure 17-20. Special sockets are available for installing and removing sending units.

mount brackets into position on the sides of the engine block, figure 17-18. For two piece assemblies, bolt the other half of the mounts to the chassis.

Some or all of the mounting brackets for the alternator, power steering pump, air pump, and air conditioning compressor can be installed now. Remember, you want to attach as many parts as possible to the engine while it is in the stand, but avoid fitting any parts that might catch or snag on the body work when you install the engine.

Sending units

All engines use at least two sending units, one for coolant temperature and one for oil pressure. These are critical items that warn the driver of engine malfunctions. It is good practice to replace the oil and coolant senders with new ones. Additional sending units may be used to monitor internal engine conditions and provide input signals for electronic fuel injection and emission control systems. All of the sending units can be installed at this time.

Most sending units have a tapered pipe thread, and are easily installed into the block, cylinder head, or intake manifold with a wrench, figure 17-19. Do not overtighten because sending units are fragile and can easily distort. Special sockets are available to hold a sending unit without damaging it, figure 17-20.

Electronic sensors

Late-model engines all use electronic sensors to transmit engine operating condition to the electronic control unit, figure 17-21. These sensors are delicate and must be handled with care. Electronic controls vary between manufacturers, from engine to engine, and from

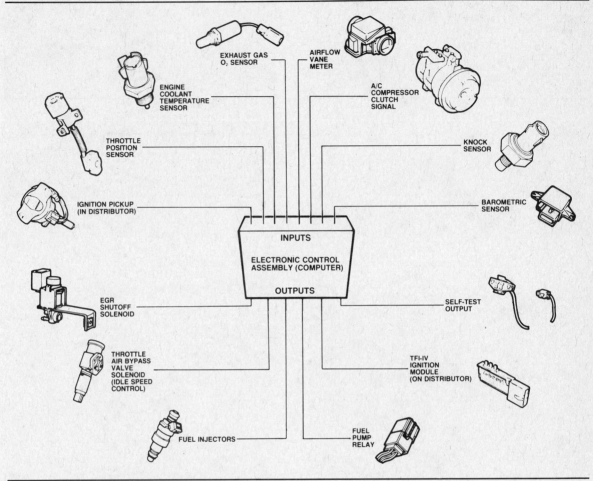

Figure 17-21. Modern computer controlled engines use a variety of sensors, actuators, and electronic controls to keep the engine running at peak efficiency.

year to year. They all have one thing in common: If they are not installed correctly, faulty signals can be generated and the engine will not run properly.

You must follow the engine manufacturer's service information when installing electronic components. Some will need to be tightened to a specific torque, while others will have to be adjusted with a precise amount of clearance. Simply follow the procedures in the service manual and always handle electronic sensors with care.

ENGINE PREOILING

You must preoil the engine before you start it. Preoiling pressurizes the entire lubrication system to fill the oil galleries and flood the main, rod, and cam bearings with engine oil. This is an extremely important procedure. If not performed, the dry bearings will be severely scuffed on start-up causing immediate and irreversible damage.

If the engine was previously checked out on a test station, it has already been preoiled. However, during final assembly much of the oil will have drained off of the bearings. It is good practice to preoil the engine again so you are sure the moving parts will be well lubricated for start up. If you are unable to install the rear main seal because the engine is in a stand, you will have to preoil the engine after you remove it from the stand and have the seal and flywheel or flexplate in place.

There are two methods to preoil the engine: manually using common hand tools, and automatically using a special pressurized oiler. Both techniques will be detailed here.

Manual Preoiling

Make sure all external oil passages are sealed off. The oil filter and oil sending unit must be in place, and there must be enough oil in the crankcase (about three quarts or liters) to pre-

Engine Installation, Break-in, and Delivery

Figure 17-22. The oil pump can be operated with a special shaft driven by a low speed drill motor to prelubricate the engine.

Figure 17-24. Mark the firing position of the number one cylinder on the distributor housing, then install the distributor so the rotor arm points to your index mark.

Figure 17-23. Rotate the engine in 90 degree increments as you preoil. Continue the process for two complete crankshaft revolutions.

vent the pump from drawing air. To preoil the engine, follow these steps:

1. Insert an oil pump priming tool into the distributor drive hole. A variety of special tools are available to service different oil pump designs.
2. Operate the pump by turning the shaft with a speed handle, or use a low-speed drill motor, figure 17-22. As the system pressurizes, you will begin to feel resistance in the pump.
3. Continue to operate the pump for about 20 to 30 seconds after you first feel resistance. This allows oil to fill the galleries and reach all of the moving parts.
4. Rotate the crankshaft 90 degrees, figure 17-23, then repeat steps 2 and 3.
5. Continue the process until you have turned the crankshaft through two complete revolutions. This ensures that all of the internal parts are oiled, and the hydraulic lifters are fully charged.

Installing the distributor

Once the engine is preoiled, you can remove the oil pump drive tool, install the distributor, and static-time the engine.

Rotate the crankshaft so that the number one cylinder is at top dead center on the compression stroke, with both valves closed. Place the cap on the distributor and mark the position of the number one tower on the side of the housing, then remove the cap. Replace any O-rings, seals, or gaskets that seal the distributor body to the engine block. Lightly lubricate O-rings with engine oil to allow easy installation. Turn the distributor shaft so the rotor arm points to your index mark, then fit the distributor into the engine, figure 17-24.

Most distributors are gear driven. Because the gear teeth are cut at an angle, the rotor will rotate slightly as the gears are engaged. The amount of rotation will vary according to the number of gear teeth. You can expect the rotor to twist about 30 degrees, usually clockwise, as you install the distributor. To counteract this rotation, position the rotor so it points about 30 degrees before your index mark. As you fit the distributor, the shaft will turn so the rotor and index mark line up once the housing seats. If not, note how far off the mark you are, remove the distributor, position the rotor accordingly, then refit the distributor.

Once the distributor is in position, install the holddown bracket and run the nut or bolt in as tight as you can by hand. The distributor must be free to move so you can adjust the timing once you start the engine.

Automatic Preoiling

You must have a specialized piece of equipment to automatically preoil the engine, figure 17-25. The preoiler uses compressed air to force engine oil through the galleries and lubricate the moving parts. Most units feed oil into the engine through the oil sending unit port. Prepare the engine by installing the oil filter and making sure all external oil openings, except the sending unit port, are sealed off. Since you will not be operating the oil pump, the distributor can be installed.

There are a variety of preoilers available. Although most operate in a similar fashion, you must follow the manufacturer's instructions for the particular unit you are using. The following steps are provided as a general guideline when preoiling:

1. Fill the oil chamber of the preoiler with the correct quantity, grade, and viscosity engine oil for your engine.
2. Charge the pressure chamber of the unit with compressed air, figure 17-26.
3. Connect the oil feed line of the preoiler to the sending unit port on the engine block.
4. Adjust the pressure regulator on the unit so that feed pressure will not exceed 40 psi (275 KPa). Excessive pressure can damage the oil seals in the engine.
5. Open the feed valve to allow pressurized oil to flow into the engine.
6. As oil flows into the engine, slowly turn the crankshaft two complete revolutions, pausing for about 20 to 30 seconds at each 90 degree interval.
7. After two complete crankshaft revolutions, close the feed valve, disconnect the oil hose, and install the sending unit.

ENGINE INSTALLATION

Your engine is now ready to be installed into the vehicle. Take a few minutes to look things over. Make sure that all necessary parts are in place on the engine, and all fasteners are tightened to specified torque. Check the timing marks on the engine, make sure they are well defined and clearly visible. Any timing tags or pointers that attach to the engine must be accurately positioned.

NOTE: A quick and easy way to highlight timing marks cut into the front pulley or flywheel

Figure 17-25. This automatic preoiler uses air pressure to force engine oil through the galleries.

is to paint them. Brush the graduated area with white, yellow, or other bright color paint, then quickly wipe the surface off with a clean rag. The rag will rub the paint off of the surface, but leave it in the machined grooves, figure 17-27. This procedure makes the timing marks stand out once the engine is installed, even under poor lighting conditions.

After passing final inspection, your engine is ready to be installed in the vehicle. Set the engine aside as you clean and inspect the engine compartment to prepare the chassis.

Preparing the Chassis

If you steam-cleaned or pressure-washed the engine before removing it, as described in Chapter 8 of this *Shop Manual*, the engine compartment will be fairly clean. Now is the time for a final cleaning. You want your new engine to look good when the customer raises the hood. Appearance is very important to your customers. Having the entire engine compartment looking like it just came off of the showroom will instill confidence. They may understand very little of what goes on inside the engine, but they know what looks good. Paying attention to this small detail is a good indicator of the fine quality craftsmanship you used when assembling the engine.

If you have steam cleaning or pressure washing facilities on the premises, give the engine compartment a thorough cleaning. An alternative is to hand wash with solvent and a brush. Be sure to clean the underside of the air conditioning compressor, power steering

Engine Installation, Break-in, and Delivery

Figure 17-26. Charge the preoiler with compressed air, and set the regulator to limit feed pressure to 40 psi (275 KPa).

Figure 17-27. White, or brightly colored paint will highlight timing marks and make them clearly visible after you install the engine.

pump, or other components that are still attached to the vehicle. While you are at it, wash down the fan shroud, air cleaner housing, battery tray and holddown, and any other loose pieces that assemble to the engine. Repaint the parts as necessary.

Once the engine compartment is clean, look over the attachment points for the engine, radiator, and other assemblies you will be bolting on. Repair any damaged threads following the procedures in Chapter 10 of this *Shop Manual*. Be sure there is enough room for the engine to slide in easily. The air conditioning compressor, power steering pump, exhaust pipe, and all wiring harnesses must be securely tied out of the way.

Move the vehicle into the shop and set it up on a hoist or jack stands. Support the transmission with a floor jack, and remove any ties you installed to hold the transmission in place while the engine was out. If the car has an automatic transmission, fit the torque converter onto the transmission. Make sure the converter seats all the way onto the front pump, then secure it in place.

Lifting the Engine

You learned about attaching slings and operating lifting equipment when you removed the engine. The same principles apply to installation. For a refresher, return to Chapter 8 of this *Shop Manual*. Never attempt to install an engine by yourself. Two people are required: one to operate the hoist and the other to keep the engine under control and guide it in.

Attach a sling to the engine, then connect the hoist to the sling slightly forward of the mid-point of the engine. Raise the hoist to take the slack out of the sling and bear the weight of the engine. You want the engine to hang with the transmission end angled down slightly for easy installation. If the engine is in a stand, slip the head from the base of the stand. If the angle is good, unbolt the engine stand head from the engine. Install the rear main seal and flywheel or flexplate, then preoil the engine as previously described.

If the engine and transaxle install as a unit, lower the hoist so the engine rests on the floor, be careful not to damage the oil pan. Then attach the transaxle and reposition the lifting sling to balance the additional weight.

Move the hoist to position the engine directly in front of the vehicle. When using a cherry picker, remember you want to keep the engine close to the ground as you wheel the hoist around the shop.

Installing the engine

Raise the hoist so the engine is lifted just enough to clear the front bodywork. Then move the hoist so the engine can come straight down into the compartment without hitting the transmission or body. Have your partner *slowly lower the hoist* as you make fine position adjustments with the engine to line everything up, figure 17-28. Stop lowering when the en-

Figure 17-28. Have an assistant slowly lower the hoist as you guide the engine into position.

Figure 17-29. The "L" stamped on this engine mount indicates that it installs on the left side of the engine. Always check engine mounts for directional indicators.

gine is just slightly above the chassis mounts and roughly in line with the bellhousing.

Adjust the transmission by raising or lowering the floor jack so the angle matches that of the engine. The input shaft, or torque converter hub, must be perfectly aligned with the crankshaft pilot. Push back on the engine to mate it up to the transmission, you may have to twist the engine slightly to line up the bolt holes and splines. Most engines use alignment dowels between the engine and transmission. Once the dowels engage, they hold the engine in place while you install the bolts. If no dowels are fitted, you may have to hold the engine while your partner installs the bolts.

NOTE: You cannot install an engine to a manual transmission unless the input shaft splines are perfectly aligned to the clutch disc splines. The fewer splines, the fewer positions the two assemblies can fit together. Plan ahead to make the job easier. Have the transmission in gear and have a socket and ratchet on hand to fit the front pulley nut. *Very slowly* rotate the crankshaft as you push back on the engine, stop turning when you feel the splines engage. Check to see the bolt holes and dowels align, then push the engine in to mate it to the transmission. You may have to rock or jiggle the engine a bit to work it down the splines before it will butt up to the transmission.

Install the engine-to-transmission retaining bolts hand tight, then remove the floor jack from underneath the transmission. The hoist will support the engine, holding it in position just above the engine mounts.

Securing the engine in the chassis

Before you lower the engine, check the mounts to be sure they are correctly installed. The left and right side mounts often look the same, however, the internal damping characteristics are different. Because a running engine has a tendency to twist in the chassis, engine mounts are often constructed to counteract these twisting forces. An engine mount installed on the wrong side is destined for early failure. Look for an "L" or "R", an arrow, or other directional indicator stamped on the side of the engine mounts, figure 17-29.

Have your assistant slowly lower the hoist as you guide the engine into position on the mounts. When the mounts line up, install the retaining bolts, or nuts, and draw them up hand tight. At this point, the engine will be supported by the chassis. Lower the hoist far enough to disconnect it from the sling, then move it out of your way. Remove the lifting sling from the engine and set it aside.

Draw the transmission-to-engine bolts up and tighten them to their specified torque. Proper tightening is critical, as overtightening can pull the rear cylinder out-of-round, and leaving the bolts loose can cause unwanted movement that will crack the bellhousing. Remember to install any brackets or guides that are held in place by the bellhousing bolts. Many engines have the engine-to-body ground strap attached at the rear of the engine. Never overlook this important connection because it can create many assorted electrical problems if not attached.

Engine Installation, Break-in, and Delivery

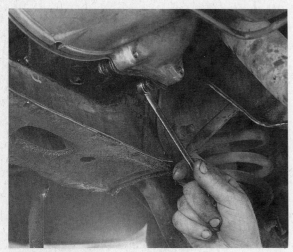

Figure 17-30. Fit the starter. Be sure to install any required shims, then run the bolts in tight.

Next, tighten the engine mount bolts to their specified torque. Then raise the vehicle and continue your installation from underneath.

Installing the starter
Install the starter motor before you attach any other parts that might limit access. Slip the starter into position, check to make sure any adjustment shims or spacers are in place, then install the bolts hand tight. Check that the starter is properly seated, then tighten the bolts to specified torque, figure 17-30.

Connecting the exhaust
Fit a new gasket, without sealer, to the exhaust flange. Install the exhaust pipe onto the flange, fit new brass or copper-plated nuts, and tighten to specification. Finish securing the exhaust by installing or tightening any brackets, hangers, or connections that were removed or loosened to remove the engine.

Look over the undercarriage to make sure all the engine and transmission mounts are in place, the fasteners are tightened to specification, and the assembly is properly aligned.

Next, reinstall the clutch slave cylinder or linkage if it is accessible from underneath. For vehicles with an automatic transmission, install the bolts that hold the torque converter to the flexplate and tighten them to specification, figure 17-31. Refit any flywheel or flexplate dust shields at this time. Then, lower the vehicle and continue your assembly from above.

Connecting the Engine

Now that the engine is in the chassis, you will be attaching all of the parts necessary to restore it to running condition. There is no specific order in which parts must be assembled.

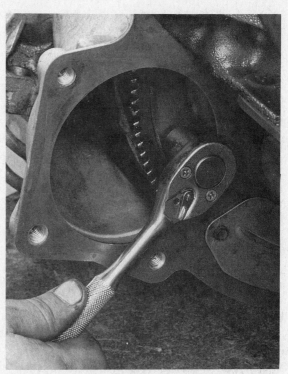

Figure 17-31. Bolt the torque converter to the flexplate if the engine is mated to an automatic transmission.

The engine and chassis configuration may dictate that certain items are assembled before others, but for most installations, assembly order is a matter of personal preference. Generally, you can simply reverse the order in which the parts were removed.

We prefer to install the cooling system components before others for two reasons. First, the cooling system is sealed. Once installed and filled, you can pressurize the system so leaks will become visible while you continue attaching other components. Second, you must eliminate trapped air pockets that can cause overheating once the engine is running. The longer the filled system sits, the more time the air has to work its way to the top so there is less bleeding required when you start the engine.

Final installation procedures will differ for each engine and chassis you work on. The following steps are provided as a general guideline only. Always consult the factory service manuals for specific information that applies to the job at hand.

Connecting the cooling system
A rebuilt, or overhauled, engine will operate at higher temperatures than the old worn out engine it replaces. Tighter tolerances create more friction, the increased friction generates more heat that must be dissipated. If the cooling system components were marginal to begin

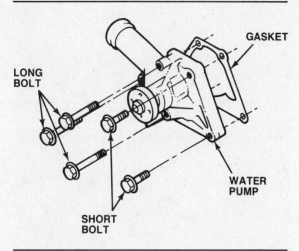

Figure 17-32. Water pumps often attach to the engine with different length bolts. Make sure you install them in the correct position.

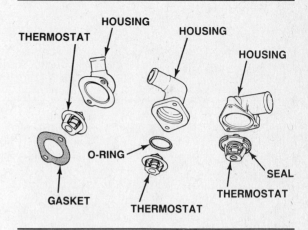

Figure 17-33. The thermostat may be sealed by a gasket or an O-ring. Some use a rubber seal that fits around the outer edge of the thermostat flange.

with, they cannot possibly cope with the additional heat generated by your new engine. Make sure the cooling system is capable of handling the higher temperatures.

The best way to ensure adequate cooling is to replace the water pump, thermostat, and all of the hoses with new ones. Also send the radiator out for service. Have it boiled out, rodded, or recored to make sure it is up to the task. Of course, all of these item will add to the total cost of the job.

Most customers are hesitant to spend money unless they absolutely have to. On the other hand, you cannot afford to compromise your work simply by trying to save the customer a few dollars. Be up front with your customers and include all anticipated costs in your original estimate. Explain in terms that they can understand what the added costs are for, and why they are necessary.

Installing the water pump
Check fit a new water pump, as well as the gasket, before you install it. Look over the hardware. Water pump bolts corrode easily and often need to be replaced. Many water pumps use several different length bolts that must be assembled in their original locations, figure 17-32. The pump bolts may also be used to hold accessory brackets or harness clamps, so be sure to install any of these items now. Removing a bolt later can break the seal and cause a leak.

Make sure the mating surfaces on the block and pump are perfectly clean. Use water-resistant, non-hardening sealer on the gasket, then fit the pump. Install the bolts, along with any attachments. Tighten bolts in increments and follow an alternating pattern as you bring them to final torque. Once the bolts are tight, wipe off any gasket sealer that seeped out.

Installing the thermostat
Always replace the thermostat with a new one. Most thermostats fit into a machined bore in the engine block. A housing bolts over the thermostat to seal the cooling system and direct flow to the radiator. The thermostat housing may be sealed to the block with a gasket, an O-ring, or a rubber seal that fits around the thermostat, figure 17-33. Install gaskets using a water-resistant non-hardening sealer, but install O-rings and seals dry.

Thermostats must be installed with the temperature sensing side facing into the engine. Watch out for thermostats that are slightly offset from center or have a small by-pass valve or vent. These units must be correctly positioned in the engine or they will not function properly and restrict flow.

Like water pump bolts, thermostat housing fasteners are prone to corrosion. Replace any damaged fasteners, and tighten them to their specified torque.

Installing the fan, shroud, and radiator
As explained in Chapter 7 of this *Shop Manual*, radiators, fans, and fan shrouds come in a wide variety of designs and sizes. Assembly order and procedures will vary. Check your tear-down notes for any specific requirements, then install the parts, figure 17-34. Remember to reattach the condenser to the radiator if the car has air conditioning.

Make sure that any rubber mounts for the radiator and condenser are in good condition. Replace any mounts that show signs of cracking or deterioration. If sealing washers are used for the radiator drain plug or thermal fan switch, replace them with new ones.

Engine Installation, Break-in, and Delivery 405

Figure 17-34. Install the radiator, fan shroud and fan. Make sure any rubber mounts are in good condition.

Figure 17-35. Fill the cooling system, then check for leaks with a pressure tester.

Once the radiator is in place, attach the transmission cooler lines to the radiator, if so equipped. Make sure the cooler lines are properly routed. Any mounting brackets that hold the lines must be in position and tight.

Installing hoses
Install the new radiator hoses. Replace the heater, water pump by-pass, and any other coolant hoses with new ones. Install the hoses with new clamps and tighten the clamps firmly, but not too tight. Double-check your work. Be sure all the coolant passages are sealed off before you fill the system.

Filling and bleeding the cooling system
Slowly fill the cooling system with a mixture of coolant and water. Be sure to use a coolant that meets the specific requirements of the engine manufacturer. Blend coolant with distilled water to provide adequate protection for the local climate. Using distilled water will help prevent mineral deposit build-up in the water jackets of the engine and the radiator.

Many late-model engines require you to follow a particular step-by-step procedure to purge the air from the cooling system. Some systems can only be bled with the engine running. Follow the manufacturer's instructions to top-off and bleed the system.

Connect a cooling system pressure tester to the radiator filler neck. Pump the tester to charge the system to normal operating pressure, then check for leaks, figure 17-35. Leave the tester in place, with the system charged, as you continue assembly. This will help force any trapped air to the surface. You can expect pressure to drop off gradually over a period of time. However, a rapid drop indicates a leak you will have to locate and correct.

Installing belt driven accessories
All automotive engines have at least one belt-driven accessory, such as the alternator, that takes its power off the front crankshaft pulley. In addition, the water pump is generally powered by an accessory drive belt. Most engines also use belt drives for the air conditioning compressor, power steering pump, and air pump. Early designs use individual belts for each driven item, figure 17-36, while late model engines often use a single elaborately routed serpentine belt to power all of the accessories, figure 17-37. At this point, you can install the mounting and adjusting brackets, fit the alternator, compressor, and pumps, then install and adjust a *new* set of belts.

Begin your assembly by inspecting the brackets. Replace any brackets that show signs of bending, cracking, or other distortion. Brackets may attach to the engine block, front cover, cylinder head, or any combination of the three. The brackets will only fit the engine correctly in one way, although it is often possible to attach them incorrectly in several ways. If installed incorrectly the belts will not align properly, and belt or bearing failure can result. Mounting bolts often fit through small cylindrical spacers that fit behind the brackets. Several different size spacers may be used to position the bracket and keep all of the pulleys parallel. Check-fit the brackets and spacers. Once you are certain everything will line up correctly, install the bracket mounting bolts and tighten them to the specified torque.

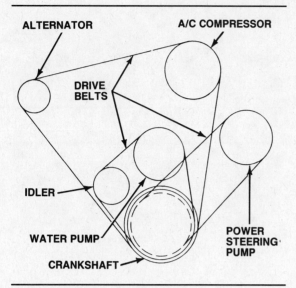

Figure 17-36. Many older engines use individual drive belts to power each accessory.

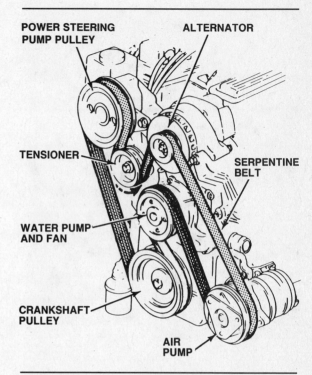

Figure 17-37. Most modern engines use a single serpentine belt to operate all of the crankshaft-driven accessories.

Once the mounting brackets are in place, install any adjustment brackets and idler pulleys. Leave the attachment bolt on the adjustment brackets slightly loose at this time, and tighten it after you fit the accessory and are sure it has full travel. Idler pulleys are often used on the belt that drives the air conditioning compressor, or other belts with a long run, to adjust the belt tension. Inspect the idler to make sure it is in good shape. Spin the bearing, feel for roughness and listen for abnormal noise. If the bearing shows wear, replace it. Some units allow you to remove a snap-ring, press the old bearing out, and install a new one. Other idlers are a sealed unit, so if the bearing is bad you replace the entire assembly. Loosely bolt the idler into position.

Fit the accessories to their brackets one at a time. Install all the retaining bolts, nuts, and spacers. Tighten the bolts snugly, but not too tight. Pivot the accessory in the bracket to make sure it can travel through its full adjustment range. Then tighten the adjustment bracket mounting bolt to its specified torque.

When all the accessories are in position, install the drive belts one at a time, beginning with the innermost belt. Adjust belt tension according to the manufacturer's service literature, then firmly tighten the adjustment fasteners, figure 17-38. Engines that use a single serpentine belt will often have an automatic tensioning device. Check the tensioner for free movement on its adjustment travel, and make sure the bearing is in good shape. Fit the belt, then install the tensioner.

Any hydraulic lines or pressure hoses removed from the accessories can be reinstalled at this time. Replace any O-rings or seals with new ones. Remember to top off the fluid level in the power steering reservoir. If the air conditioning was disconnected, you can hook up your manifold gauges and draw the system down with an evacuation pump while you continue with assembly.

Connecting the fuel system

Fuel may be supplied to the engine either by a carburetor, throttle-body injection, or port injection. Although new cars are no longer equipped with carburetors, you can expect to see carburetted engines in for rebuilding for many years to come. Fuel system designs cover a broad spectrum, and we cannot possibly detail the installation steps for every engine. Here we will simply highlight general procedures that apply to all installations.

Installing carburetors

For best results, rebuild the carburetor. As fuel circulates through the system it carries contaminants with it. When the car is driven, these contaminants remain suspended in the fuel where they cause little harm. However, when fuel is no longer circulating, they settle, dry, and harden to form varnish deposits. Varnish deposits in the carburetor not only restrict internal passages, but can also hold the needle and seat valve open, allowing raw fuel to flood

Engine Installation, Break-in, and Delivery

Figure 17-38. New, as well as old, belts must by adjusted to the correct tension. Manufacturers provide belt tension specifications.

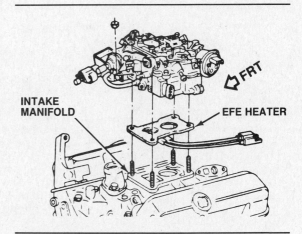

Figure 17-39. The insulator block fits between the carburetor base and the intake manifold.

the engine when you try to start it. The only effective way to remove varnish is to disassemble the carburetor and soak the parts in carburetor cleaner. Include the cost of a carburetor overhaul in your original estimate, as this will eliminate the need to make embarrassing phone calls to tell your customers that they must part with more money.

Most carburetors fit over four studs on the intake manifold, and are held in place with nuts and lock washers. A phenolic insulating block fits between the carburetor base and the manifold flange. The insulator block provides a barrier to restrict engine heat from vaporizing fuel in the float bowl, figure 17-39. The insulator block on many late model engines also contains a heating grid for the Early Fuel Evaporation (EFE) system. Inspect the insulator. It must have clean mating surfaces and be free of any cracks or chips. Check the EFE grid also. If it is burned through or broken, replace it.

Install the insulator, then fit the carburetor. Some engines use two paper gaskets, one on either side of the insulating block. Others have neoprene seals built into the insulator and are installed without gaskets. With either design, do not use gasket sealer, assemble the carburetor and insulator to the manifold dry. Fit all the pieces, then tighten the nuts to specification in an alternating pattern.

Once the carburetor is in place, connect the throttle, choke, and kickdown linkages, as well as the fuel feed and return lines. Replace any vacuum hoses that connect to the carburetor, insulator block, and intake manifold with new ones. Refer to your tear down notes to double check vacuum hose routing, or use the underhood decal if one is attached.

Installing throttle bodies

Throttle body injection feeds fuel into the engine from a central location on the intake manifold, figure 17-40. The throttle body supports the fuel injector (most V-type engines use two injectors mounted in one throttle body) and regulates air flow with butterfly valves. The throttle body assembly resembles a carburetor without a float bowl. The assembly mounts to the manifold with an insulator block. Installing a throttle body is similar to fitting a carburetor, so refer to the previous section.

Installing injection manifolds

Port fuel injection uses a separate injector to service each individual cylinder. The injectors are linked together by a manifold, also known as a fuel rail, and fit into bores machined on the intake manifold near where the runners mate to the cylinder head ports. Air flow is regulated by butterfly valves installed upstream in the manifold. Fuel injectors are delicate instruments, always handle with care.

There is a wide variety of fuel injection systems in the field. You must be familiar with the setup for the particular engine you are working on. Most port injectors can be attached to the fuel rail, then assembled to the engine as a unit. Neoprene seals or O-rings fit over the ends of the injectors to seal them to the manifold and keep unmetered air out of the intake charge, figure 17-41. Always replace the seals with new ones whenever you remove the injectors. *Lightly lubricating* the outside of

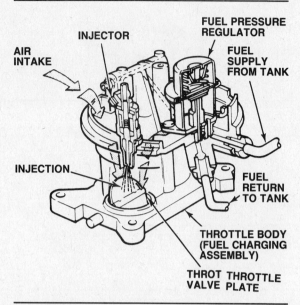

Figure 17-40. Throttle body, or single point injection sprays high pressure fuel through a throttle body in place of the carburetor.

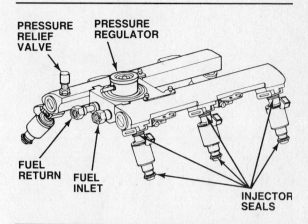

Figure 17-42. Double check all connections on the fuel rail. They must be secure, but not overtightened.

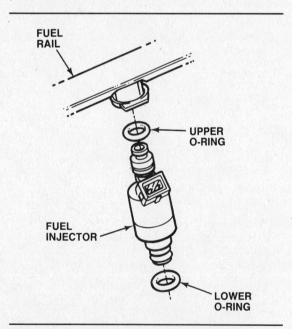

Figure 17-41. Replace the seals whenever you remove injectors. A leak here can be disastrous.

the seal can make installation easier and prevent seal damage. The bores must be perfectly clean for the seals to seat properly.

Small bolts or cap screws generally secure the fuel rail to the manifold and hold the injectors in position. Proper tightening torque on these fasteners is critical. Overtightening can bend and stress the fuel rail. If the bolts are loose, the assembly can vibrate. In a running engine, either condition can crack the fuel rail and cause a leak.

WARNING: A high-pressure fuel leak is a disaster. When installing fuel injection components, *You must follow* the specific requirements provided by the manufacturer of the engine you are working on, and all fasteners must be torqued to specification.

Attach the fuel lines to the rail. Make sure all hoses, pipes, lines, and fittings are in perfect condition, figure 17-42. If required, install a new hose between the pressure regulator and vacuum source.

Intake air management and throttle control system designs are common, and installation procedures are often quite different. Consider what other components remain to be installed, and how they connect to the engine. Attach them now before you cover them up with the manifolding. Fit and secure the air intake according to the manufacturer's procedure and specifications.

Connecting the electrical system

Pay close attention to the electrical connections. Make sure they are clean and tight. With multi-plug connectors, be sure all of the pins make good contact, and the two plug halves effectively seal and lock into place, figure 17-43. Replace the harness if any of the plugs are burned or melted. Work your away around the engine compartment in an orderly sequence to make sure you get all of the connections. An easy way to avoid confusion is to leave any identification tags you installed during disassembly in place until all of the connections are made. Double check to make sure everything is hooked up correctly, then remove your locating tags. Always charge the battery before you install it. Make sure all oth-

Engine Installation, Break-in, and Delivery 409

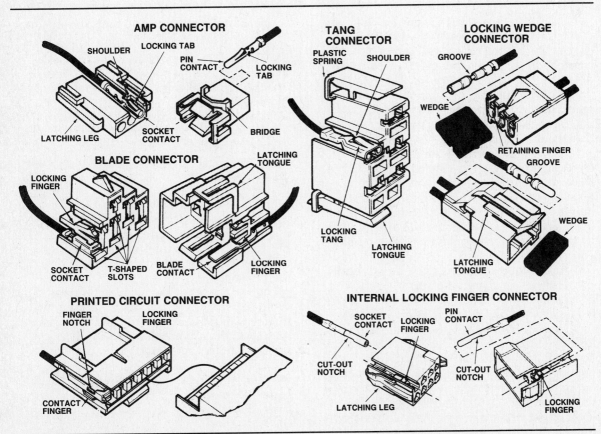

Figure 17-43. Engines use a variety of electrical connectors. Make sure all of the contacts are in good condition. You must seat the plug halves together so that all locking and latching devices snap into place.

er electrical connections have been made, then connect the battery.

Remember, disconnecting the battery on a computer-controlled engine erases the long-term memory in the electronic control unit. This memory stores the small adjustments to the fuel and ignition systems that are unique to each individual car. The computer "learns" these settings during everyday operation, and stores them for use whenever the car is running. When memory is lost, the car may run poorly until the computer relearns the settings that make it run best.

Normally, just driving the car a few miles will give the computer all the information it needs to reestablish memory. Some manufacturers describe specific procedures that you can perform while driving the vehicle to reset the computer quickly. Typically, cycling the engine and test driving the car will establish parameters for the engine to operate efficiently.

Final Checks and Adjustments

Check and top off the engine oil and coolant. Attach a remote starter and perform a dry compression test. The compression test lets you know the valves are seating correctly, and there are no major problems with valve timing or combustion sealing. Compression between cylinders should be even, but expect readings to be low because the piston rings have not yet seated. In addition, cranking the engine circulates oil to prime the moving parts and helps charge the fuel system. Install and torque a new set of spark plugs. Fit the distributor cap and attach the spark plug wires, then the engine is ready to fire.

Take the time to carefully go over the entire installation making sure everything is in place, properly adjusted, and tightened. Correct any potential problems now, before any damage is done. Then you are ready to start your engine.

STARTING THE ENGINE

The first few minutes after initial start-up is the most critical time in an engine's life. Remember, camshaft-to-lifter seating begins immediately. If the engine is not started and run-in properly, there will be irreversible damage. Be prepared. Have everything ready, and all the tools you will need at hand.

If everything is perfect, the engine will start on the first crank as soon as the starter is engaged. Most of the time it will take a bit of cranking to get a good enough fuel charge to fire. You may have to make fine timing adjustments by rotating the distributor before the engine will start. Engage the starter in short bursts, as excessive cranking can overheat and damage the starter.

The moment the engine starts, open the throttle and bring engine speed up to 1,400 to 2,000 rpm. Block the throttle open, as you must maintain and hold this speed for 15 to 20 minutes. Remember these are general guidelines. Follow any specific instructions provided by the manufacturer of the camshaft and lifters you installed.

With the engine running, look for any leaks, loose fittings, or other problems. Reinstall the hood and any other parts that remain to be fitted. Allow the camshaft and lifters sufficient time to seat, then release the throttle and return the engine to idle speed. Check and adjust ignition timing, idle speed, and fuel mixture. Service the air conditioning system if required. Shut the engine off, allow fluids to settle, then check and top-off all fluid levels — including the transmission. Make sure everything is in perfect order before you close the hood and test-drive the vehicle.

CYCLING THE ENGINE

During the initial test drive it is important that you cycle the engine to help seat the moving parts. Cycling is merely a prescribed driving method that systematically varies the internal loads on the engine. Cycling can be done on public roads, preferably in light traffic. Plan your test drive time. You do not want to start out in the height of rush hour. Cycle the engine by the following steps:

1. Engage third gear with a manual transmission, or select second gear if the vehicle has an automatic transmission. Accelerate from 20 to 50 miles per hour at full throttle.
2. When a speed of 50 miles per hour is attained, release the throttle allowing the engine to coast. Let engine braking slow the vehicle down to about 20 miles per hour. Do not use the foot brake and leave the transmission in the preselected gear.
3. After vehicle speed drops to 20 miles per hour, accelerate to 50 miles per hour at full throttle, then release the accelerator.
4. Repeat the process several times. Keep an eye on the rear view mirror as you accelerate. The amount of oil smoke from the tail pipe will diminish as the rings seat. Continue cycling the engine until you no longer see heavy oil smoke.

Cycling breaks the engine in by varying the load on the internal parts. As the vehicle accelerates, the piston rings are forced into the cylinder walls to make them seat rapidly. Coasting with the throttle plates closed reduces ring loading and produces high manifold vacuum. The high vacuum draws engine oil up toward the combustion chambers past the relaxed rings. The engine oil acts as a coolant to lower piston ring and ring land temperatures. It also washes away any residual metal particles or other debris present in the cylinders. Avoid any sustained engine loading and high speeds. Never run the engine at higher than the speed necessary to cycle it and seat the piston rings.

The surface of the new engine bearings is very soft, so it quickly wears to conform with the crankshaft journal profile and compensate for any slight imperfections. In addition, tiny foreign particles can embed themselves into the bearings without damaging the bearing or the journal. The new bearings will gradually season themselves and harden as the engine runs. The seasoning process takes place during the first several hundred miles of vehicle operation. For this reason, it is important the vehicle be operated at moderate speeds, below 50 miles per hour, for several hundred miles to season the bearings.

After cycling the engine, and there is no visible oil smoke, drive the vehicle in normal fashion to complete your test drive. Monitor the gauges as you go, and watch for tell-tale signs of any abnormal conditions. Take your time with the test drive. Once around the block is not enough to thoroughly evaluate engine condition. Try to include a variety of left and right hand turns, as well as uphill and downhill grades. Make note of any unusual noises or vibrations that must be corrected before you deliver the vehicle to the customer.

Return the vehicle to the shop and adjust the timing, idle speed, and fuel mixture. Shut the engine off, then check and adjust belt tension, electrical connections, and fluid levels. Once again look over the installation. Check hose and cable routings, tighten all hose clamps, secure any loose wiring or hoses with tie wraps, check fastener torque, and carefully look for any signs of leakage. Review the manufacturer's service information for any engine specific requirements. Some engines require you to retorque the cylinder head and adjust the valves after the initial run-in. Follow any requirements particular to your engine.

Engine Installation, Break-in, and Delivery

Make sure you have not left any tools, rags, or other items under the hood, and there are no greasy hand prints or smears on the body work or interior. A simple gesture that will please the customer is to wash the car and vacuum the interior. You must deliver a product the customer will be proud to drive.

RELEASING TO THE CUSTOMER

Make sure the vehicle is in perfect working order before you telephone the customer to pick up the car. Repair any minor problems, such as a burned out light bulb or an empty windshield washer bottle. Even though these may not be related to the work you did on the engine, it is a good way to promote customer relations. Correcting these problems now, will prevent the customer having to return later to have the repairs made.

Have all of the paper work completed and on hand when you make the phone call. Speak in a friendly and courteous manner. You must be prepared to answer any question that arises. Have the totaled work order, itemized parts list, your warranty, and any parts warranties in front of you while you are talking. Be sure to clearly note on the work order any parts or services you recommended that the customer refused to purchase.

Educating the Owner

When you talk to your customers, keep in mind that most of them know very little of what goes on inside an engine. Using a lot of technical terms will confuse, and not impress the average customer. Speak to the customer as an individual on his own level, without being condescending. Before driving off, the customer must be made aware of the following:

1. You have machined, fitted, and assembled the engine components with utmost care. All of the manufacturer's procedures have been followed, everything is within tolerance standards, and the engine should provide many miles of trouble free driving.
2. Once the vehicle leaves the shop, you can no longer monitor operating conditions and fluid levels. These are now the customer's responsibility.
3. No matter how exacting you were when assembling, installing, and cycling the engine, it will take several hundred miles of operation before all the parts can season and seat themselves. The engine must be operated at moderate speeds and loads for at least the first two hundred miles.
4. A new engine will burn oil as the parts break in. It is the owners responsibility to frequently check the crankcase oil and coolant levels. Make sure they know the proper way to do this, or have them come by the shop so you can do it for them.
5. The new engine must be serviced after the first 500 miles of operation. The engine oil and filter must be changed to remove debris and contaminants that are a normal result of breaking in. Additional service requirements, such as valve adjustment, will vary from engine to engine.
6. Carefully explain your warranty policy. Be specific as to what is covered and what is not. Be sure the customers are aware of their responsibilities. Failure to maintain vital fluid levels, improper break-in, misuse, or abuse can void warranties.
7. Once every detail has been explained, and understood, have your customer sign the work order. Then, once again stress the importance of the first service as you hand over the keys.

Follow Up Service

If you did the job right, and your customer drove off with a smile, the new engine will perform trouble-free for a long time. However, the job is not yet complete.

Several days after the car has been delivered, *telephone the owner* to make sure everything is functioning properly. After-the-fact concern and reassurance from you can establish a bond of trust between you and your customer. Be sure to answer any questions, ask how many miles the car has been driven, and schedule a time for the first service.

When the vehicle does come back for the initial service, take your time. Be sure to perform all of the operations required by the manufacturer. Thoroughly check the engine over.

Most shops perform the first service at no charge to the customer, and include the cost into the price of the rebuild. This is a good practice, as the customer recently spent a lot of money, and will be hesitant to part with any more. If you did the job right and treated the customer fairly, you will be servicing the engine on a regular basis until it is ready for another rebuild.

Index

Air Conditioning System
disconnecting, 146
reconnecting and servicing, 410
recycling machines, 147

Align Honing
main bearing bores, 230-233
machine, 45, 230, 246
precautions, 231
technique, 231-233

Alignment Dowels
cylinder head, 140, 385-386
engine to transmission, 402

Aluminum Cylinder Bores
reconditioning, 251-252

Arbor Press, 54, 113, 320

Automatic Transmission
cooler fittings, 110, 148, 405

Balance Pad
connecting rod, 300
piston, 299-300

Balancing Machine, 297-298

Balancing (see: Engine Balancing)

Ball Hone (see: Glazebreaker)

Ball-Anvil Micrometer, 219

Barrel Taper, 257-258

Bead Blasting (see: Parts Cleaning)

Bearing (see also: specific applications)
insert thickness, 217-219
measuring inside diameter, 217-219

Bearing Housing Bore
diameter, 217
measuring, 219
line boring, 234

Bearing Journal (see: Crankshaft Journal)

Bearing Leakage Tester, 92

Bearing Wear
connecting rod, 272-273
main bearing, 222
patterns, 171-173

Bench Grinder, 29

Blind Bearing Puller, 255-256

Block Deck (see: Deck Surface)

Boring Bar (see: Cylinder Boring Machines)

Bolts (see: Fasteners)

Bottom-End Assembly, 363-385

Broken Bolt Removal, 187

Bronze Coil Valve Guide Inserts, 322-323

Cam Follower (see: Valve Lifter)

Camshaft
advance, 114
base circle runout, 341-342
calculating lift, 341
core plug removal, 175
drive, 108-126
duration, 128
grinding, 342-343
high-pressure lubricant, 368
in-car repair, 108-137
inspection, 337-342
installation, OHC, 336
installation, OHV, 367-368
journal diameter, 339-340
lobe taper, 338
nose taper, 254
removal, OHC, 307-308
removal, OHV, 165-166
specifications, 112
straightening, 342
surface coating, 343, 367
thrust plate removal, 165
wear patterns, 338-339

Camshaft Bearing
installation, 366-367
oil clearance, 340
removal, 175-176
removal tools, 40-41
undersize and oversize, 340

Camshaft Bearing Bore
deburring, 362
OHC inspection, 314-315
reconditioning, 331-333

Camshaft Core Plug
deburring plug bores, 362
installing, 368-369

Camshaft Drive
backlash, 112
belts, 122-126
endplay, 112-113, 125
gears, 110-114
OHC chains, 117-122
OHV chains, 114-119
runout, 111-112

Camshaft Grinder, 342

Camshaft Straightening Fixture, 342

Carburetor Installation, 406-407

Casting Marks and Numbers, 78-79

Cleaning (see: Parts Cleaning)

Clutch Alignment Tool, 395-396

Clutch Linkage Installation, 403

Clutch Assembly
removal, 164
installation, 395

Clutch Slave Cylinder, 146, 151, 403

Coil Expanders (see: Piston Rings)

Combustion Chamber Deposits, 103, 163

Compression
gauge, 98
pressure, 98
specifications, 98
testing, 98-100, 409
ratio, 235

Compression Rings (see: Piston Rings)

Computer Balancing Machine, 303-304

Connecting Rod
assembling to piston, 286-288
bearing wear analysis, 272-273
bend and twist, 275-276
bolt protectors, 376-377
bore resizing, 279-283

Index

directional indicators, 286-287, 377
fasteners, 273-274
grinding machine, 230-231
honing machine, 281
inspection, 272-278
installing rod caps, 377-378
knock, 106
machining for balance, 300
matched set, 267-268
measuring, 274-276
offset, 286
oil clearance, 285, 377
preliminary inspection, 169
reconditioning, 278-283
side clearance, checking, 379
storing, 267-268
torquing, 378-379
wear patterns, 272-274

Connecting Rod Bearings
in-car replacement, 141-143
inspection, 170-174
locating tangs, 143

Connecting Rod Cap
removal, 143, 169
numbering, 167

Connecting Rod Service
big-end bore reconditioning, 279
big-end bore honing, 280-282
grinding parting faces, 279
rod bolts, removing and installing, 278-280
pin bushing replacement, 283-284
small-end bore reconditioning, 282
straightening, 278-279

Connecting Rod Service Tools
alignment fixture, 275
aligning tool, 52
assembly vise, 280
dial bore gauge, 276
grinder, 53, 230-231, 279
heaters, 54, 288
honing machine, 52, 280

Cooling System
connecting components, 403-404
filling and bleeding, 405
pressure tester, 405

Core Plug
design, 365-366
driver, 366
installation, 366
removal, 175

Crack Detection Techniques, 203-208

Crack Repair
pinning, 209
stop-drilling, 209
welding, 208

Crankshaft
balancing, 48-49
bend, 259
cleaning oil passages, 369
crack detection, 256
disassembly, 255-256
endplay, 106, 111
fillet radius, 264
final inspection, 369-370
fitting to block, 371-375
grinding, 261-265
in-car repair, 143
inspection, 113, 115, 143
journal inspection, 257-259
journal diameter, 218-219
keyway repair, 256
measuring end play, 373-374
measuring journals, 369-370
measuring thrust surface, 369-370
polishing, 265-266
runout specifications, 112
service, 255-266
sludge trap, 170, 255, 369
standard undersize, 257-258
straightness, measuring, 259
straightening, 259-261
storage, 170
thrust bearing surface, 258-259, 265
tolerances, 257-259
wear patterns, 257-259
welding, 261

Crankshaft Grinding
grinding machine, 48, 261
indexing the crankshaft, 262-263
main journal set up, 264
procedure, 265
rod journal set up, 263
steady rest, 261
stone preparation, 264-265
thrust surface grinding, 265

Crankshaft Journal
undersize, 79
inspection, 143
out-of-round, 143
taper, 143

Crankshaft Kit, 84

Crankshaft Polishing
hand lapping, 265
machine, 50, 265
micro polishing, 265-266
surface finish, 266

Crankshaft Pulley Removal, 111

Crankshaft Seal
installing split-type, 370-371
installing lip-type, 375-376
removal, 167, 170

Crankshaft Straightening
peening, 260
pressing, 260
technique, 259-261

Crankshaft Welder, 261

Cross Hatch, 240

Cycling the Engine, 410-411

Cylinder
inspection, 227-229
indexing fixture, 242
out-of-round, 228-229, 242
preassembly cleaning, 363
reconditioning, 241-254
sleeves, 252-254
taper, 228-229
wear patterns, 228-229

Cylinder Block (*see:* Engine Block)

Cylinder Boring
calculating overbore, 241
chamfering bores, 362
boring machines, 241
boring to oversize, 241
overbore standards, 228
specifications, 242

Cylinder Bore Service Tools
dial bore gauge, 42
glaze breaker, 44
ridge reamer, 42, 168
torque plate, 46

Cylinder Deck (*see:* Deck Surface)

Cylinder Deglazing, 240

Cylinder Gauge (*see:* Dial Bore Gauge)

Cylinder Head
assembling, 333-336
chamfering bolt holes, 362
cleaning, 308-309
gasket, 386-387
gasket surface inspection, 310
installing, 385-387
locating dowels, 385-386
loosening sequence, 162
preliminary inspection, 163
pressure testing, 206-209, 211
removal, in-car, 137-141
removal, on bench, 162-163
reconditioning, 305-329
resurfacing, 329-331
resurfacing tips, 331
straightening, 329-331
torque sequence, 387
valve inspection, 311-314
valve seat inspection, 314

Cylinder Head Gasket, 234, 386-387

Cylinder Head Service Tools, 38-40

Cylinder Head Straightening
preparation, 330
procedure, 329-330
theory, 329-330

Cylinder Head Surface, 235

Cylinder Honing
align honing, 229-230
automatic honing, 246-247
hand honing, 251
machine, 246-248
procedure, 248-251
technique, 243-246

Cylinder Leakage Tester, 100-102

Cylinder Sleeves
design and construction, 252-253
installing, 253-254

Deck Clearance, 163, 379-380

Deck Plate (see: Torque Plate)

Deck Resurfacing
effect on valve timing, 236
effect on compression, 235-236
reconditioning, 234-240
surfacing considerations, 234-235
surface finish, 226, 235

Deck Surface
checking flatness, 226-227
measuring for straightness, 226-227
inspecting, 226-227
limits of distortion, 227

Deglazing Tool, 240

Depth Micrometer, 361

Diagnostic Tests
compression, 98-100
cylinder leakage, 100-110
interpreting test results, 103
power balance, 102-103
vacuum, 93-98

Dial Bore Gauge, 42, 223, 313, 315, 331, 340

Dial Bore Gauge, Valve Guide, 313

Dial Bore Gauge, Using
adjusting the setting fixture, 223-224
adjusting the gauge, 224-225
connecting rod, 276
description, 223
measuring a bore with, 225-226
setting fixture, 223

Dial Caliper, 63-64, 330

Dial Indicator, 47, 69-70, 111, 259, 314, 339

Dial Indicator Straddle Bracket, 229

Diamond Stone Dresser, 264

Die Grinder, 188

Die Stock, 197

Distributor Drive, 118

Distributor Installation, 399-400

Drive Belts, Installing, 406

Drive Shaft, Oil Pump, 118

Dry Sleeve (see: Cylinder Sleeve)

Double Plunge Grinding, 265

Dye Penetrant Testing, 205-206, 220

Dynamic Balance, 302-303

Easy-Outs (see: Screw Extractors)

Early Fuel Evaporation, 407

Electrical Discharge Machining (EDM), 192

Electrical System Connections, 408

Electronic Programed Memory, 409

Electronic Sensors Installing, 397

End-to-End Taper, 257

Engine Accessories
inspecting and installing, 397, 405-406

Engine
assembly, 358-389
assembly preparation, 362-363
balancing, 297-304
casting marks, 78
casting numbers, 79
cleaning, 145-148
diagnosis, 93-108
disassembly, 160-176
final assembly, 390-389
final cleaning, 363
identification, 75-80
in-car service, 141-144
installation, 396-409
noises, 103-107
overhaul versus rebuild, 82
oversize/undersize codes, 79
prelubricating, 380-381, 398-400
production number, 79
removal, 7, 145-158
rotation direction, 110
specifications, 80
testing, 86-107

Engine Balancing
adding weight, 304
additional rotating parts, 304
calculating bob weight, 300-302
common crank pin short cut, 302
equalizing connecting rod weight, 300
equalizing piston weight, 299-300
factor, 301-302
precautions, 297-298
procedure, 303-304
reciprocating weight, 299-300
records, 299
removing weight, 303-304
rotating weight, 302-304

Engine Block
preliminary inspection, 176
bottom end assembly, 363-385
bottom end disassembly, 163-176
top end assembly, 385-389
top end disassembly, 160-162

Engine Compartment
cleaning, 400-401
inspection, 401
preparing for installation, 400

Engine Installation
attaching external parts, 396-397
belt-driven accessories, 405-406
final checks and adjustments, 410
Installing and connecting, 401-409

Engine Lifting Procedure, 157-158

Engine Mounts, 397, 402

Engine Kits, 84

Engine Oil (see: Oil)

Engine Service Tools
analyzer, 102
bearing rollout pin, 143
engine dolly, 155
hoist, 7, 155-158
flywheel lock, 164
lifting slings, 156-157
prelubricator, 92
stand, 108, 110, 158, 363
test station, 380-381
transaxle holding fixture, 155

Engine Vacuum (see: Manifold Vacuum)

Epoxies, 212

Exhaust Manifold
gaskets, 161
installing, 403
removing, 161

Index

Expander Broach
valve guide, 322
wrist pin bushing, 283-284

Externally Balanced Engine, 303

Eye Protection, 2-3

Fan and Fan Shroud
installation, 404-405
removal, 110, 148

Fasteners
Allen-head, 21
thread identification, 193-194
torque-to-yield, 73, 137
torx-head, 20

Feeler Gauge, 61-62, 112, 130, 331

Firing Intervals, 134

Flexplate
checking runout, 394
inspection, 164, 292
installation, 393-394
removal, 164

Flywheel
checking runout, 394
grinder, 293
inspection, 164, 292
installation, 393-395
machining, 293-296
removal, 164
stepped flywheels, 295

Freon (*see also:* Air Conditioning)

Front Pulley, Installing, 383

Fuel Injection
installing manifolds, 407-408
installing throttle bodies, 408

Fuel Pressure, Relieving, 147

Fuel Pump, Installation, 396

Gasket
installation tips, 380
oil pan, 384-385
sealing compounds, 380
surfaces, 121
valve cover gasket, 388-389

Gear Drive (*see:* Camshaft Drive)

Glaze Breaker, 44, 240, 254

Green Casting, 83, 220

Hand Tools, 13-27

Harmonic Balancer
inspection, 164
installing, 383
puller, 47, 111, 164
removing, 111, 164
repair sleeve, 164

Hazardous Waste Disposal, 184-185

Head Gasket Installation, 386

Helical Insert (*see:* Thread Repair Procedures)

Helicoil (*see:* Thread Repair Procedures)

Honing Equipment
bearing bore, 45
connecting rod, 52
cylinder, 44-45, 46
valve guide, 36

Honing Hints, 250-251

Honing Plate (*see:* Torque Plate)

Hourglass Taper, 257

Hydraulic Lifter (*see:* Valve Lifter)

Hydraulic Press, 55, 287

Hypereutectic Casting (*see:* Aluminum Cylinder Bores)

Ignition Problems, 103

Ignition Timing, 97, 236

Initial Start-Up, 409-410

Initial Test Drive, 410

Inside Micrometer, 64, 143, 219, 222, 228, 315, 332, 340

Intake Manifold
alignment, 236-237
in-car removal, 138
inspection, 161
installation, 391-393
machining, 333
removing, 160

Interference Engine, 110, 235-236

Internally Balanced Engine, 303

International Standards Organization (ISO), 193

Jack Stands, 148

Jet Cleaning Booths (*see:* Parts Cleaning)

Keyways, 113

Leak-Down Tester
cylinder leakage tester, 100
hydraulic lifter, 345

Left-Hand Threads, 186

Lifter (*see:* Valve Lifter)

Lifter Bore Reconditioning, 254

Linerless Engines (*see:* Aluminum Cylinder Bores)

Long Bock, 82, 389

Magnaflux (*see:* Magnetic Particle Inspection)

Magnetic Particle Inspection, 203-205, 220, 338

Magnetic Field, 204

Main Bearing
align honing, 230-233
bore misalignment, 220
bore stretch, 221
bore tolerances, 216-218, 229-230
calculating oil clearance, 369-370
crush relief, 221
in-car replacement, 143
inspection, 170, 369
installation, 369-371
noise, 107
parting line pinch, 221
saddle inspection, 219
seating, 410
wear patterns, 222, 231

Main Bearing Cap
grinding, 230-231
installing, 372-373
numbering, 167
registering, 373
removal, 143, 170
tightening, 373

Manifold
general removal, 160
general installation, 390-393

Manifold Vacuum
diagnosing engine problems, 94-98
locating leaks, 95
measuring, 93-94

Measurement Tools, 60-71

Mechanical Lifter (*see:* Valve Lifter)

Metal-Coating, 211

Metallic Plastics, 212

Metric System, 57-60

Metric threads, 193

Micrometer (*see also:* specific type)
caring for, 67-68
design, 64-65
using, 65-67

Milling Machine, 235-236

Multi-Plug Connectors, 408-409

Numbered Punches, 167

Nut Splitter, 188

Nuts (*see:* Fasteners)

Oil
  checking flow, 380-381, 388
  consumption, 86-90, 240
  control rings, 88
  deflector clips, 136
  deposits, 86
  dilution, 89
  leaks, 90
  pressure relief valve, 89, 359-361
  pressure test, 91-92
  service life, 86

Oil Clearance
  calculating, 217-219
  measuring with plastigauge, 373-374

Oil Control Rings (see also: Piston Rings)
  design, 289-291
  installing three piece assemblies, 290
  installing two piece assemblies, 291
  positioning end gaps, 290
  removing, 268

Oil Gallery Plugs
  removal, 174
  installation, 364-365

Oil Pan
  gasket, 121
  in-car removal, 141-142
  installation, 384-385
  removal, 166

Oil Pickup
  checking screen clearance, 382-383
  driver, 381-382
  installing, 381-382

Oil Pump
  assembling, 360-361
  clearance specifications, 359-361
  crankshaft driven, 361
  drive shaft inspection, 359
  drive shaft installation, 381
  end plate resurfacing, 360
  gear type, 361-362
  in-car service, 144
  index marks, 359
  installation, 381-383
  pressure relief valve, 359, 361
  priming, 144, 381
  rebuild, kit, 359
  removal, 118, 167
  rotor type, 359-361
  servicing, 358-362

Oil Wedge, 217

Outside Micrometer, 219, 223, 257, 271, 312, 340

Parts Cleaning
  airless shot blasting, 183
  bead blasting, 181-182
  chemical dip tanks, 181
  hot tanks, 179
  parts washer, 4, 178
  pressure washer, 145
  shaker units, 183
  steam cleaning, 145, 159, 177, 306
  thermal ovens, 183-184, 208, 331
  tumblers, 183

Parts Retrieval Tools, 27

PCV Valve and System Testing, 89, 94-95

Penetrating Oil, 160, 187

Pilot Bearing and Bushing
  bore inspection, 255-256
  removal, 255-256
  installation, 394-395

Piston
  cleaning, 269-270
  clearance problems, 235
  directional indicator, 286, 377
  disassembly, 266-268
  failure analysis, 268-269
  inspection, 270-272
  installation, 377-379
  nominal size, 245
  knurling, 272
  machining for balance, 299-300
  measuring diameter, 271
  measuring side clearance, 271-272
  oil bleed passages, 269-70
  offset, 286
  pin boss, 270
  preliminary inspection, 169
  reconditioning, 268-272
  removal, 169
  rings, 87-89, 268, 270-271, 285-286, 289-292
  ring land, 87, 268-272
  ring groove, 268-272
  stress fractures, 270
  wear patterns, 268-270

Piston Pin (see: Wrist Pin)

Piston Ring
  cast iron, 240
  compression ring design, 88, 289
  compression ring installing, 291-292
  diagnosis, 87-89
  hard chrome, 240
  installation tips, 89
  molybdenum, 240
  oil control ring design, 290-291
  oil control ring installing, 290-291
  checking end gap, 285-286
  checking side gap, 289-290
  removing from piston, 268
  sealing, 87, 240
  seating, 410
  specifications, 271

Piston Ring Groove
  caution, 268
  cleaning, 269-70
  shimming, 272
  specifications, 271

Piston Service Tools
  knurling machine, 51, 272
  flycutter, 40
  pin drift, 50
  ring compressor, 55, 376, 378
  ring end gap grinder, 52
  ring expander, 50, 291
  ring gap cutter, 285
  ring groove cleaner, 51, 269
  ring groove lathe, 272
  ring groove wear gauge, 271
  wrist pin press, 267

Piston Skirt
  measuring, 271
  knurling, 272

Plastigauge®, 143, 373-374

Pliers, 21

Plunge Grinding, 265

Pneumatic Tools, 29-30

Power Balance Test, 102-103

Power Steering Pumps
  connecting, 406
  disconnecting, 149

Pressure Testing, 206-209, 220

Power Tools, 28

Preoiling (see: Engine Prelubricating)

Propane Torch, 267

Punches, 25

Pushrod
  adjustment, 135
  cleaning, 350
  inspection, 162, 350-351
  installing, 388
  oversize/undersize, 236
  removal, 113, 162

Radiator
  installation, 404-405
  servicing, 404
  removal, 110, 148

Radius Ride, 273

Reciprocating Mass Percentage (see: Balance Factor)

Remote Starter, 99

Index

Replacement Engines, 82

Ring Gear
inspection, 293
replacement, 292-293

Ring Land (*see:* Piston Ring Land)

Rocker Arm
adjustable, 135
designs, 353
disassembly and inspection, 162, 351-354
fulcrum and fulcrum seat, 352-353
oil deflector, 353
installing, 388
locknuts, 111, 352
noise, 105
pivot ball, 352
removal, 306
refacing cast rocker arms, 356
service, 351-356
stud removal, 307

Rocker Shaft
assembly, 355
disassembly and inspection, 353-354
direction indicators, 355
installing, 388
spring and spacer inspection, 355
removing, 111, 353

Rocking Couple, 303

Rod (*see:* Connecting Rod)

Roller Lifter (*see:* Hydraulic Valve Lifter)

Rotary Grinder, 30

Safety
chemical, 4
electrical, 4-5
fire, 3-4
hand tool, 5
machine tool, 11
personal, 2-3
vehicle and underhood, 4

Screw Extractors, 189-190

Screwdrivers, 19

Seal Installation Tools, 26, 116, 122, 375

Seasoned Casting, 83

Seat Insertion Tool, 37

Serpentine Belt (*see:* Drive Belt)

Sending Unit, Installation, 397

Short Block, 82, 363

Single Point Cutting (*see:* Valve Seat Cutting)

Sludge Trap Plug, 170, 255

Small-Hole Gauge, 71, 313

Snap Gauge (*see:* Telescoping Gauge)

Socket Wrenches, 17-19

Solid Lifter (*see:* Valve Lifter)

Spark Plug Deposits, 87

Specifications, Locating, 80-81

Splayed Crankpin, 262-263

Split-Ball Gauge (*see:* Small-Hole Gauge)

Spray Booths (*see:* Parts Cleaning)

Springs (*see:* Valve Springs)

Spring Tension Gauge, 313

Sprockets, 110, 115, 122-126

Starter Motor, Installation, 403

Static Balance, 302-303

Steam Cleaner (*see:* Parts Cleaning)

Steel Rule, 60

Straightedge, 60, 222, 331

Strobe Balancing Machine, 304

Stud Pullers and Removers, 187-188

Surface Grinding
considerations, 235-237
machines, 236-237
procedure, 237-239

Surface Roughness Indicator, 235

Sweep Grinding, 265

Tachometer, 97

Tap (*see:* Thread Repair Tools)

Tap Extractor, 192

Tapered Reamer (*see:* Valve Guide Service Tools)

Taping Threads, 194-197

Tappet (*see:* Valve Lifter)

Teflon Tape, 382

Telescoping Gauge, 71, 216, 222, 276, 315, 332, 334

Thermal Cleaning Ovens (*see:* Parts Cleaning)

Thermostat, Installation, 404

Thread Locking Compound, 379, 383

Thread Repair Procedures
helical insert, 199-200
Helicoil, 199
key-locking inserts, 202
thread chasing, 196-198
threaded inserts, 201-202
thread pitch gauge, 193
thread restorers, 198
self-tapping inserts, 201
solid bushing inserts, 201

Thread Repair Tools
bottoming taps, 194
plug taps, 194
taper taps, 194
thread chasers, 198
thread files, 198
thread pitch gauge, 193
thread restorers, 198
threading die, 197

Throating (*see:* Valve Seat Grinding)

Throttle Body Fuel Injection, 407

Throttle Linkage, Connecting, 407

Thrust Bearing
aligning, 373
calculating clearance, 369-370
in-car replacement, 144
installing inserts, 371
measuring, 369-370

Thrust Forces, 228, 270

Thrust Surfaces, 114

Thrustplate, 113

Timing Belt
adjusting tension, 126
advantages, 108
construction, 122-123
cover removal, 123
guides, 125
inspecting, 123
replacing, 123-126
tensioner, 125

Timing Chain (*see also:* Camshaft Drive)
inspection, 114, 118-120
lubrication, 108
service, 114-122
specifications, 115
tensioner, 114, 119
preload, 121

Timing Chain Tunnel, 121

Timing Components
installing, 380
OHC installation tips, 389

Timing Cover (*see also:* Camshaft Drive)
bolt torque, 116
construction, 116-117

gasket, 121
inspection, 116
installing, 383-384
machining cast covers, 383
oil seal, 121
sealing, 383
straightening bolt holes, 383

Timing Gear (*see also:* Camshaft Drive)
construction, 110
locating dowels, 116
service, 111-113

Timing Marks, 108, 109, 113, 115-119, 124-125, 236

Timing Tag, 111, 116

Tipping (*see:* Valve Grinding)

Tolerance Values, 216-218

Tool Holder, 243

Tool Sharpener, 243

Top-End Repair, 126

Topping (*see:* Valve Seat Grinding)

Torque Angle Gauge, 387

Torque-to-Yield (*see:* Fasteners)

Torque Plate, 46, 245

Total Indicated Runout (TIR), 259

Toxic Waste Disposal, 184-185

Transaxle Removal, 152-155

Undersize/Oversize Codes, 79

V-Blocks, 47, 259, 339

Vacuum Modulator, 94

Valve Clearance, Adjusting
engine cold, 129, 388
engine off, 132
engine running, 131
non-adjustable rockers, 134
hydraulic lifters, 133
running mates method, 134
shims, 132
Stud-mounted adjustable rocker arms, 135

Valve Cover
design, 126
gasket, 141
inspection, 127
installation, 388-389
replacing, 126-127
tightening torque, 127

Valves, Engine
cleaning, 308
clearances, 128
compression testing, 98-102
deposits, 90
face runout, 318
grinding, 315-318
inspection, 311-312
installing, 335-336
interference designs, 108-110
lapping, 329
margin, 312, 317-318
oversize stems, 80
removal, 306-308
sodium-filled, 315-316
vacuum testing, 96

Valve Grinding
interference angle, 316
stem grinding, 318
valve facing, 316-318
valve grinding machine, 312, 315

Valve Guide
clearance, 313
insert, 35, 319-322
inspection, 313-314
oil consumption, 89
reaming, 36, 321
repair, 35, 319-324
service tips, 323-324

Valve Guide Bore Gauge, 34

Valve Guide Driver, 320

Valve Guide Reamer, 319, 321, 322, 323, 324

Valve Guide Repair
installing false guides, 322
installing thin wall liners, 321-322
knurling, 319
replacing valve guide inserts, 319-320

Valve Lifter
bleed down, 136
bore deburring, 362
bore inspection, 387
bucket followers, 348-349
disassembly, 345
finger followers, 349
function, 128-129
hydraulic, 128-129, 133-136, 344-348
in-car service, 129-136
inspection, 343-344
installing, 387-388
kickback test, 347
leak-down testing, 345-347
mechanical, 128-129
metering-valve type, 344-345
oversize, 79
remote-pivot, 349
removal, 113, 165
removal tool, 38, 165
roller type, 347-348
rocker arm hydraulic adjusters, 349-350
rotation, 387
solid, 343-344

Valve Seals
function, 90
in-car replacement, 136-137
installation, 335
removal, 307

Valve Seat
checking concentricity, 328
checking sealing, 334
configuration, 325-326
correcting position, 328
cutting, 328-329
designs, 314
grinding, 325-328
inspection, 314
reconditioning, 324-329
replacing, 324-325
valve overhang, 328

Valve Seat Grinding
precautions, 328
lowering, 328
narrowing, 328
procedure, 327-328
seating angles, 325

Valve Service Tools
guide insert, 35
knurling tool, 35
lash adjustment, 130-133
spring compressor, 31, 136
spring tension tester, 31
seat insertion tool, 37
spring tension tester, 31
valve grinding machine, 33
valve seat cutter, 33
valve seat grinder, 32
valve guide bore gauge, 34
valve spring compressor, 31, 136

Valve Spring
assembling, 335-336
free length, 313
in-car repair, 136
inspection, 312-313
installed height, 334
load, 111
removal, 306-307
replacing, 136
shims, 334
tension, 313
vacuum testing, 96
variable-pitch, 335-336

Valve Spring Compressor, 31, 136, 306, 308, 335

Valve Spring Tension Tester, 31, 313

Valve Stem
grinding, 318
measuring, 312
oversize, 316
polishing, 333-334

Valve Timing, 114

Index

Vernier Caliper, 63, 312, 330

Vernier Scale, Reading, 63-64, 66

VIN (*see:* Engine Identification)

Water Pump
    installation, 404
    OHC removal, 118
    OHV removal, 110

Welding Equipment, 10, 209, 258, 261

Wet Sleeve (*see:* Cylinder Liners)

Woodruff Keys, 113, 256, 383

Wrenches, 14-16

Wrist Pin, Full-Floating
    bushing installation, 283-284
    installation, 287
    clearance, 266, 285
    removing, 266-267
    snap-ring, 267

Wrist Pin Press, 54-55, 287

Wrist Pin, Press-Fit
    clearance, 267, 283, 285
    heat fitting, 288
    installing, 286-288
    oversize, 275, 282
    press fitting, 287-288
    removing, 267-268
    selecting press attachments, 267, 288